VOLUME 1

SUSTAINABLE
MICRO IRRIGATION
Principles and Practices

VOLUME 1

SUSTAINABLE MICRO IRRIGATION

Principles and Practices

Edited by
Megh R. Goyal, PhD, PE

Apple Academic Press

TORONTO   NEW JERSEY

Apple Academic Press Inc.	Apple Academic Press Inc.
3333 Mistwell Crescent	9 Spinnaker Way
Oakville, ON L6L 0A2	Waretown, NJ 08758
Canada	USA

©2015 by Apple Academic Press, Inc.

First issued in paperback 2021

Exclusive worldwide distribution by CRC Press, a member of Taylor & Francis Group
No claim to original U.S. Government works

ISBN 13: 978-1-77463-337-3 (pbk)
ISBN 13: 978-1-77188-016-9 (hbk)

Library of Congress Control Number: 2014940501

Library and Archives Canada Cataloguing in Publication

Sustainable micro irrigation: principles and practices/edited by Megh R. Goyal, PhD, PE.

(Research advances in sustainable micro irrigation; volume 1)

Includes bibliographical references and index.
ISBN 978-1-77188-016-9 (bound)
1. Microirrigation. I. Goyal, Megh Raj, author, editor
II. Series: Research advances in sustainable micro irrigation; v. 1

S619.T74S98 2014	631.5'87	C2014-903394-X

Apple Academic Press also publishes its books in a variety of electronic formats. Some content that appears in print may not be available in electronic format. For information about Apple Academic Press products, visit our website at **www.appleacademicpress.com** and the CRC Press website at **www.crcpress.com**

CONTENTS

List of Contributors .... *vii*

List of Abbreviations .... *ix*

List of Symbols.. *xi*

Preface... *xiii*

Hymn: "Micro Irrigation" .... *xvii*

Foreword by Gajendra Singh .... *xix*

Foreword by Miguel Muñoz Muñoz.. *xxiii*

Foreword by Marvin J Jensen .... *xxvii*

Book Series: Research Advances in Sustainable Micro Irrigation .... *xxix*

About the Senior Editor-in-Chief... *xxxi*

Warning/Disclaimer... *xxxiii*

PART I: STATUS AND POTENTIAL OF MICRO IRRIGATION

1. **Pressurized Irrigation Technology: Global Scenario** 1
   S. A. Kulkarni, F. B. Reinders, and F. Ligetvari

2. **Micro Irrigation Technology: Global Scenario**... 21
   Felix B. Reinders

3. **Affordable Micro Irrigation for Small Farms** 31
   Jack Keller and Andrew A. Keller

4. **Micro Irrigation Potential in the Developing Countries**............................. 49
   Tushaar Shah and Jack Keller

5. **Low Cost Micro Irrigation in Kenya** 77
   Jack Keller

6. **Micro Irrigation Potential in India** 87
   R. K. Sivanappan

7. **Advanced Technologies for Sugarcane Cultivation Under Micro Irrigation: Tamil Nadu** 121
   R. K. Sivanappan

8. **Design and Management of Irrigation Systems: Chile** 147
   Eduardo A. Holzapfel, Alejandro Pannunzio, Ignacio Lorite, Aureo S. Silva de Oliveira, and István Farkas

9. **Subsurface Micro Irrigation: Review** 161
    Leonor Rodriguez-Sinobas and Maria Gil-Rodriguez

PART II: PRINCIPLES OF MICRO IRRIGATION

10. **Modernization of Floating Pumping Station for Irrigation** 185
    Rares Halbac-Cotoara-Zamfir and Alina Gabor

11. **Principles of Micro Irrigation** 195
    Megh R. Goyal

12. **Chemigation** 235
    Megh R. Goyal

13. **Fertigation: Venturi Injection Method** 263
    Mukesh Kumar, T. B. S. Rajput, and Neelam Patel

14. **Chemigation Uniformity in Micro Irrigation System** 273
    Larry Schwankl

15. **Micro Irrigation: Filtration systems** 285
    Megh R. Goyal

16. **Micro Irrigation Design Using Hydrocalc Software** 301
    Rares Halbac-Cotoara-Zamfir and Emanuel Hategan

    Glossary of Technical Terms *315*
    Appendices *447*
    Index *461*

LIST OF CONTRIBUTORS

István Farkas
Gödöllő University, Department of Physics and Process Control, 2103 Pater Karoly utca 1 Gödöllő, Hungary, Email: Farkas.Istvan@gek.szie.hu.

Alina Gabor
Politehnica University of Timişoara, 2nd Victoriei Square, 300006, Timişoara.

María Gil-Rodríguez
Agricultural Engineering School, Technical University of Madrid, Ciudad Universitaria, 28040 Madrid (Spain), Web: http://hideriego.upm.es.

Megh R. Goyal
Retired Professor in Agricultural & Biological Engineering, University of Puerto Rico – Mayaguez Campus, PO Box 86, Rincon – PR – 00677; and Senior Acquisitions Editor and Senior Editor-in-Chief, Apple Academic Press Inc., New Jersey. Email: goyalmegh@gmail.com.

Rares Halbac-Cotoara-Zamfir
Assist. Prof., Politehnica" University of Timişoara, 2nd Victoriei Square, 300006, Timişoara, Email: raresh_81@yahoo.com

Emanuel Hategan
Politehnica" University of Timişoara, 2nd Victoriei Square, 300006, Timişoara.

Eduardo A. Holzapfel
Universidad de Concepción, Facultad de Ingeniería Agrícola, Departamento de Recursos Hídricos, Av. Vicente Méndez 595, Chillán, Chile, Email: eholzapf@udec.cl.

Marvin E. Jensen
Retired Research Leader USDA-ARS, 1207 Springwood Drive, Fort Collins, CO, 80525, Email: mjensen419@aol.com.

Andrew A. Keller
Vice President of Keller-Bliesner Engineering, LLC, 78 East Center, Logan, UT 84321, Tel.: 435-753-5651 Fax: 435-753-6139, Email: akeller@kelbli.com

Jack Keller
Emeritus Professor of Irrigation Engineering, Utah State University; and Chief Executive Officer of Keller-Bliesner Engineering, 78 East Center, Logan, UT 84321, Tel.: 435-753-5651 Fax: 435-753-6139, Email: jkeller@kelbli.com

Suresh A. Kulkarni
Executive Secretary – ICID, 48 Nyaya Marg, Chanakyapuri, New Delhi 110021, India. Tel.: +91 11 26115679, +91 11 26116837, Fax: +91 11 2611 5962; Email: icid@icid.org Website: http://www.icid.org

Mukesh Kumar
School of Agriculture, School of Agriculture, Indira Gandhi National Open University, Block G2, Academic Complex, Maiden Garhi, New Delhi-110068, India, Email: mukeshop@gmail.com

Ferenc Ligetvari
Vice President Hon., ICID, Chairman, Hungarian National Committee on ICID (HUCID Gödöllo, P.O. Box 303, H-2103; Hungary, E-mail: ligetvari.ferenc@mkk.szie.hu

Ignacio Lorite
Consejería de Agricultura y Pesca IFAPA, Centro Alameda del Obispo, Junta de Andalucía, Córdoba, Spain. Email: ignacioj.lorite@juntadeandalucia.es.

Miguel A. Muñoz-Muñoz
Ex-President of University of Puerto Rico, University of Puerto Rico, Mayaguez Campus, College of Agriculture Sciences, Call Box 9000, Mayagüez, PR. 00681-9000. Tel. 787-265-3871, Email: miguel.munoz3@upr.edu

Alejandro Pannunzio
Universidad de Buenos Aires, Departamento de Agronomía, Av. Chorroarin 280, Ciudad Autónoma de Buenos Aires, Buenos Aires, Argentina, Email: alejandropannunzio@gmail.com.

Neelam Patel
Senior Scientist, Water Technology Centre (WTC), Indian Agricultural Research Institute (IARI), New Delhi, Delhi 110012, India, Email: neelam@iari.res.in

T.B.S. Rajput
Principal Scientist, Water Technology Centre (WTC), Indian Agricultural Research Institute (IARI), New Delhi, Delhi 110012, India, Email: tbsraj@iari.res.in

Felix B. Reinders
Vice President – ICID, Chairman, Agricultural Research Council, Institute for Agricultural Engineering, Private Bag X519, SILVERTON 0127 South Africa, Email: reindersf@arc.agric.za

Leonor Rodríguez-Sinobas
Professor of Irrigation and Drainage Engineering, Agricultural Engineering School, Technical University of Madrid, Ciudad Universitaria, 28040 Madrid (Spain), Tel.: +34 913365675, Fax: +34913365845, Email: leonor.rodriguez.sinobas@upm.es.

Lawrence J. Schwankl
CE Irrigation Specialist, UC Kearney Agricultural Research & Extension Center, 9240 S. Riverbend Ave., Parlier, CA 93648, Tel.: (559) 646-6569, Email: ljschwankl@ucanr.edu

Tushaar Shah
Senior Fellow of the Colombo-based International Water Management Institute, C/o Indian Natural Resources Economics and Management (INREM) Foundation, Near Smruti Apartment, Behind IRMA Mangalpura, Anand 388001, Gujarat, India, Tel/Fax: +91 2692 263816/817. Email: t.shah@cgiar.org

Aureo S. Silva de Oliveira
Universidade Federal do Recôncavo da Bahia (UFRB), Núcleo de Engenharia de Água e Solo (NEAS), 44380-000 Cruz das Almas, Bahia, Brasil, Email: aureo@ufrb.edu.br.

Gajendra Singh
Former Vice President, Asian Institute of Technology, Thailand. C-86, Millennium Apartments, Plot E-10A, Sector-61, Noida, U.P. – 201301, India, Mobile: (011)-(91)-997-108-7591, Email: prof.gsingh@gmail.com.

R. K. Sivanappan
Former Dean and Professor, Tamil Nadu Agricultural University at Coimbatore, Tamil Nadu, India; and Consultant, 14, Bharathi Park, 4th Cross Road, Coimbatore 641 043, India. Email: sivanappan@hotmail.com

LIST OF ABBREVIATIONS

°C	degree Celsius
AAP	Apple Academic Press Inc.
AC-FT	acre foot
ASABE	American Society of Agricultural and Biological Engineers
B-C	Blaney-Criddle
BCR	benefit-cost ratio
cfs	cubic feet per second
cfsm	water depth, cubic feet per second per square mile
CGDD	cumulative growing degree days
CIAE	Central Institute of Agricultural Engineering Bhopal—India
cm	centimeter(s)
CU	coefficient of uniformity
CWSI	crop water stress index
DAP	days after planting
DAT	days after transplanting
DOY	day of the year
DU	distribution uniformity
EPAN	pan evaporation
ET	evapotranspiration
ET_c	crop evapotranspiration
FAO	Food and Agricultural Organization, Rome
FC	field capacity
GOI	Government of India
gph	gallons per hour
gpm	gallons per minute
ha	hectare(s)
HRG	Hargreaves
IARI	Indian Agricultural Research Institute
ICAR	Indian Council of Agriculture Research
IDE	International Development Enterprises
ISAE	Indian Society of Agricultural Engineers
k_c	crop coefficient
K_p	Pan coefficient
KSA	Kingdom of Saudi Arabia
LAI	leaf area index
LEPA	low energy pressure irrigation system
lps	liters per second
MAD	maximum allowable depletion
Mha	million hectare(s)

MI	micro irrigation
MSL	mean sea level
NABARD	National Bank for Agriculture and Rural Development India
NCPA	National Committee on the use of Plastics in Agriculture in India
NGO	Non Government Organization
PE	polyethylene
PET	potential evapotranspiration
pH	acidity/alkalinity measurement scale
PM	Penman-Monteith
ppb	one part per billion
ppm	one part per million
psi	pounds per square inch
PVC	poly vinyl chloride
PWP	permanent wilting point
RA	extraterrestrial radiation
RH	relative humidity
RMAX	maximum relative humidity
RMIN	minimum relative humidity
RMSE	root mean squared error
RS	solar radiation
SAR	Sodium absorption rate
SCS-BC	SCS Blaney-Criddle
SDI	subsurface drip irrigation
SIMINET	Smallholder Irrigation Market Initiative
SWB	soil water balance
TE	transpiration efficiency
TEW	total evaporable water
TMAX	maximum temperature
TMIN	minimum temperature
TR	temperature range
TUE	transpiration use efficiency
USDA	US Department of Agriculture
USDA-SCS	US Department of Agriculture-Soil Conservation Service
VPD	vapor pressure deficit
VWC	volumetric water content
WATBAL	water balance
WISP	wind speed
WS	Weather Stations
WSEE	weighed standard error of estimate
WUE	water use efficiency
µg/g	micrograms per gram
µg/L	micrograms per liter

LIST OF SYMBOLS

A	area in square meters
C	concentration of the chemical in the solution
Cl_2	chlorine gas
CVm	manufacturing variation
CVq	discharge variability
CvU	coefficient of variation uniformity
D	lateral internal diameter
D_c	cylindrical section
D_i	superior exit
EU'	emission uniformity
F	reduction factor
Fe+++	insoluble iron compounds
FV, %	flow rate variation
g	rate of injection
H+	hydrogen ions
H_2SO_4	concentrated sulfuric acid
Ha	pressure head
hf	lateral friction
hf_{i-i+1}	head loss between two successive emitters
HL	lateral inlet pressure head
$HOCl^-$	solutions of hypochlorite
h_s	positive pressure
ID	internal diameter
IR	rate of chlorine injection
k	emitter characteristics
K_c	crop coefficient
K_p	pan factor
K_s	hydraulic conductivity
LL	lateral length
lph	liters per hour
m	meters
M	weighted average irrigation rate
MI	drip
N	number of emitters
PE	pan evaporation
pH	acidity
Q	flow rate of the system
q	emitter flow rate
Q_0	lateral inlet flow

q_a	average rate
$Q_{i\text{-}i+1}$	flow conveying
QL	total lateral flow rate
QS	system discharge
r_0	radius
r_o	cavity radius
S	surface
sd	estimated standard deviation
Se	emitter spacing
T	effective irrigation time
t_2	chemigation time
t_f	daily operating time
t_r	irrigation duration
u	normal random
v	hand calculator
Mha-m	million hectare-meter
WR	water requirement
α	fitting parameter
σ_q	mean absolute flow

PREFACE

Due to increased agricultural production, irrigated land has increased in the arid and subhumid zones around the world. Agriculture has started to compete for the water use with industries, municipalities and other sectors. This increasing demand along with increments in water and energy costs have made it necessary to develop new technologies for the adequate management of water. The intelligent use of water for crops requires understanding of evapotranspiration processes and use of efficient irrigation methods.

The "http://newindianexpress.com/cities/bangalore/Micro-irrigation-to-be-promoted/2013/08/17/" weblink published an article on the importance of micro irrigation in India. Every day, similar news appear all around the world indicating that government agencies at central/state/local level, research and educational institutions, industry, sellers and others are aware of the urgent need to adopt micro irrigation technology that can have an irrigation efficiency up to 90% compared to 30–40% for the conventional irrigation systems. I share here with the readers the news on 17 August of 2013 by *Indian Express Newspaper*,

> *"In its efforts to increase the irrigated area by efficiently distributing the available water in the Cauvery basin, The Cauvery Neeravari Nigama Limited (CNNL) is planning to undertake pilot projects on micro irrigation at four places. The CNNL Managing Director Kapil Mohan said, 'the Cauvery water disputes tribunal has permitted the state to irrigate up to 18.85 lakh acres of land in the Cauvery basin. Therefore, we have to judiciously use the available water to increase the irrigated area. In the conventional irrigation method, a lot of water is required to irrigate even a small piece of land. Therefore, we are planning to undertake pilot projects to introduce micro irrigation in four or five places in the Cauvery basin'. Kapil further said that unless the farmers are willing to embrace micro irrigation, it would be difficult for the project to succeed. Therefore, the CNNL is holding discussions with the farmers in different villages of the basin to select the villages in which the project would be undertaken. The CNNL is also in the process of finalizing the technology that should be adopted while undertaking the pilot project. 'If everything goes as planned we should implement the pilot project within this financial year. If the project yields the desired result, we will think of extending it to the other areas in the basin', Kapil added. According to the official sources, water would be supplied through micro sprinklers instead of canals in the micro irrigation system. Therefore, one can irrigate more than two acres of land through the system with the water that is used to irrigate one acre of land in the conventional canal irrigation system."*

Evapotranspiration (ET) is a combination of two processes: evaporation and transpiration. Evaporation is a physical process that involves conversion of liquid water into water vapor and then into the atmosphere. Evaporation of water into the atmosphere occurs on the surface of rivers, lakes, soils and vegetation. Transpiration is a physical process that involves flow of liquid water from the soil (root zone) through the trunk, branches and surface of leaves through the stomates. An energy gradient is created during the evaporation of water, which causes the water movement into and out of the plant stomates. In the majority of green plants, stomates remain open during the day and stay closed during the night. If the soil is too dry, the stomates will remain closed during the day in order to slow down the transpiration.

Evaporation, transpiration and ET processes are important for estimating crop water requirements and for irrigation scheduling. To determine crop water requirements, it is necessary to estimate ET by onsite measurements or by using meteorological data. Onsite measurements are very costly and are mostly employed to calibrate ET methods using climatological data. There are a number of proposed mathematical equations that require meteorological data and are used to estimate the ET for periods of one day or more. Potential ET is the ET from a well-watered crop, which completely covers the surface. Meteorological processes determine the ET of a crop. Closing of stomates and reduction in transpiration are usually important only under drought or under stress conditions of a plant. The ET depends on four factors: (1) climate, (2) vegetation, (3) water availability in the soil and (4) behavior of stomates. Vegetation affects the ET in various ways. It affects the ability of the soil surface to reflect light. The vegetation affects the amount of energy absorbed by the soil surface. Soil properties, including soil moisture, also affect the amount of energy that flows through the soil. The height and density of vegetation influence efficiency of the turbulent heat interchange and the water vapor of the foliage.

Micro irrigation, also known as trickle irrigation or drip irrigation or localized irrigation or high frequency or pressurized irrigation, is an irrigation method that saves water and fertilizer by allowing water to drip slowly to the roots of plants, either onto the soil surface or directly onto the root zone, through a network of valves, pipes, tubing, and emitters. It is done through narrow tubes that deliver water directly to the base of the plant. It is a system of crop irrigation involving the controlled delivery of water directly to individual plants and can be installed on the soil surface or subsurface. Micro irrigation systems are often used in farms and large gardens, but are equally effective in the home garden or even for houseplants or lawns. They are easily customizable and can be set up even by inexperienced gardeners. Putting a drip system into the garden is a great do-it-yourself project that will ultimately save the time and help the plants grow. It is equally used in landscaping and in green cities.

The mission of this compendium is to serve as a textbook or a reference manual for graduate and undergraduate students of agricultural, biological and civil engineering; horticulture, soil science, crop science and agronomy. I hope that it will be a valuable reference for professionals who work with micro irrigation and water management; for professional training institutes, technical agricultural centers, irrigation centers, agricultural extension services, and other agencies that work with micro irrigation programs.

After my first textbook on *Drip/Trickle or Micro Irrigation Management*, published by Apple Academic Press Inc., and response from international readers, I was motivated to bring out for the world community this series on *Research Advances in Sustainable Micro Irrigation*. This book series will complement other books on micro irrigation that are currently available in the market, and my intention is not to replace any one of these. This book series is unique because it is a complete and simple, a one stop manual, with worldwide applicability to irrigation management in agriculture. Its coverage of the field of micro irrigation includes historical review; current status and potential; basic principles and applications; research results for vegetable/row/ tree crops; research studies from Chile, Colombia, Egypt, India, Mexico, Puerto Rico, Saudi Arabia, Spain, and U.S.A.; research results on simulation of micro irrigation and wetting patterns; development of software for micro irrigation design; micro irrigation for small farms and marginal farmers; studies related to agronomical crops in arid, humid, semiarid, and tropical climates; and methods and techniques that can be easily applied to other locations (not included in this book).

This book offers basic principles, knowledge and techniques of micro irrigation management that are necessary to understand before designing/developing and evaluating an agricultural irrigation management system. This book is a must for those interested in irrigation planning and management, namely, researchers, scientists, educators and students.

Volume 1 in this book series is titled *Sustainable Micro Irrigation: Principles and Practices*, and includes 16 chapters consisting of: Pressurized Irrigation Technology: Global Scenario *by S. A. Kulkarni, F. B. Reinders, and F. Ligetvari*; Micro Irrigation Technology: Global Scenario *by Felix B. Reinders;* Affordable Micro Irrigation for Small Farms *by Jack Keller and Andrew A. Keller*; Micro Irrigation Potential in the Developing Countries *by Tushaar Shah and Jack Keller*; Low Cost Micro Irrigation in Kenya *by Jack* Keller; Micro Irrigation Potential in India *by R. K. Sivanappan*; Advanced Technologies for Sugarcane Cultivation under Micro Irrigation: Tamil Nadu *by R. K. Sivanappan*; Design and Management of Irrigation Systems: Chile *by Eduardo A. Holzapfel, Alejandro Pannunzio, Ignacio Lorite, Aureo S. Silva de Oliveira, and István Farkas*; Subsurface Micro Irrigation: Review *by Leonor Rodriguez-Sinobas and Maria Gil-Rodriguez*; Modernization of Floating Pumping Station for Irrigation *by Rares Halbac-Cotoara-Zamfir and Alina Gabor;* Principles of Micro Irrigation *by Megh R. Goyal*; Chemigation *by Megh R. Goyal*; Fertigation: Venturi Injection Method *by Mukesh Kumar, T. B. S. Rajput and Neelam Patel;* Chemigation Uniformity in Micro Irrigation System *by Larry Schwankl*; Micro Irrigation: Filtration Systems *by Megh R. Goyal*; Micro Irrigation Design using Hydrocalc Software *by Rares Halbac-Cotoara-Zamfir and Emanuel Hategan*; Glossary of Technical Terms in Micro Irrigation; and Appendices.

Volume 2 in this book series is titled *Research Advances and Applications in Subsurface Micro Irrigation and Surface Micro Irrigation*, and includes 16 chapters. Volume 3 in this book series is titled, *Sustainable Micro Irrigation Management for Trees and Vines*. We expect to publish ten volumes in the book series.

The contribution by all cooperating authors to this book series has been most valuable in the compilation of this compendium. Their names are mentioned in each chapter. This book would not have been written without the valuable cooperation of these investigators, many of who are renowned scientists who have worked in the field of evapotranspiration throughout their professional careers.

I will like to thank editorial staff, Sandy Jones Sickels, Vice President, and Ashish Kumar, Publisher and President, at Apple Academic Press, Inc., (http://appleacademicpress.com) for making every effort to publish the book when the diminishing water resources is a major issue worldwide. Special thanks are due to the AAP Production Staff for typesetting the entire manuscript and for the quality production of this book. We request that reader to offer us your constructive suggestions that may help to improve the next edition.

I express my deep admiration to my family for understanding and collaboration during the preparation of this book series. With my whole heart and best affection, I dedicate this book series to my wife, Subhadra Devi Goyal, who has supported me during the last 44 years. We both have been trickling on to add our drop to the ocean of service to the world of humanity. Without her patience and dedication, I would not have been a teacher with vocation and zeal for service to others. I present here the Hymn on Micro Irrigation by my students. As an educator, there is a piece of advice to one and all in the world: *"Permit that our almighty God, our Creator and excellent Teacher, irrigate the life with His Grace of rain trickle by trickle, because our life must continue trickling on . . ."*

—Megh R. Goyal, PhD, PE, Senior Editor-in-Chief
May 14, 2014

HYMN: "MICRO IRRIGATION"

CHORUS
Tomorrow in the morning
Irrigate your plants and your flowers
And surely with our God
We are going to do the best.

Chant. 1
The rain is falling from the sky,
Drop by drop…
Irrigating a lot of knowledge
To all the people.

CHORUS
Tomorrow in the morning
Irrigate your plants and your flowers
And surely with our God
We are going to do the best.

Chant. 2
With drip irrigation
We a gonna save a lot
Leaving us without any worries.

CHORUS
Tomorrow in the morning
Irrigate your plants and your flowers
And surely with our God
We are going to do the best.

Chant. 3
If your crops are drying out
You're in a big dilemma
And surely with drip irrigation
Your problems will be solved.

CHORUS
Tomorrow in the morning
Irrigate your plants and your flowers
And surely with our God
We are going to do the best.

Chant. 4
If you want to have a best crop,
And save lot of money
Irrigate drop by drop.

CHORUS
Tomorrow in the morning
Irrigate your plants and your flowers
And surely with our God
We are going to do the best.

Chant. 5
Now I'm saying good bye
With the taste of fluid drops
And that is why I repeat
That Goyal, drip man is with you …

CHORUS
Tomorrow in the morning
Irrigate your plants and your flowers
And surely with our God
We are going to do the best.

Ruddy A. Mateo and Edagardo Perez
April 4, 2007
(Dedicated to Megh R. Goyal, Our Guru)

FOREWORD

Since 1978, I have been a research assistant at Agricultural Experiment Substation – Juana Diaz, a soil scientist, Chairman of Department of Agronomy and Soils in the College of Agricultural Sciences at the University of Puerto Rico – Mayaguez Campus; and President of University of Puerto Rico (February 2011 to June 2013). I was also an Under-Secretary (1993–1997) and Secretary of the Puerto Rico Agriculture Department (1997–2000). I am privileged to write a foreword for Goyal's book series that is titled "*Research Advances Micro Irrigation.*"

I have known Dr. Megh R. Goyal since 1st of October of 1979 when he came from Columbus – Ohio (later I went to study at the OSU to complete my MSc and PhD in Soil Fertility during 1981–1988) to Puerto Rico with his wife and three children. According to his oral story, he had job offers from Texas A&M, Kenya, Nigeria, University of Guelph, and my university. He accepted the lowest paid job in Puerto Rico. When I asked why he did so, his straight-forward reply was "challenges in drip irrigation offered by this job". With no knowledge of Spanish, Megh survived. He also started learning the Spanish language and tasting Puerto Rican food (of course no meat, as he with his family is vegetarian till today).

Within four months of his arrival in Puerto Rico, the first drip irrigation system in our university for research on water requirements of vegetable crops was in action. Soon, he formed the State Drip Irrigation Committee consisting of experts from universities, suppliers, and farmers. He published his first 22-page Spanish publication titled, "Tensiometers: Use, Service and Maintenance for Drip Irrigation." Soon, he would have graduate students for their MSc research from our College of Agricultural Sciences. I saw him working in the field and laboratory hand in hand with his students. These students would later collaborate with Megh to produce a Spanish book on *Drip Irrigation Management* in 1990. I have personally read this book and have found that it can be easily adopted by different groups of readers with a high school diploma or a PhD degree: farmers, technicians, agronomists, drip irrigation suppliers and designers, extension workers, scientists. Great contribution for Spanish speaking users!

Megh is a fluent writer. His research studies and results started giving fruits with at least one peer-reviewed publication on drip irrigation per month. Soon, our university researchers will have available basic information on drip irrigation in vegetable and tree crops so that they could design their field experiments. Megh produced research publications not only on different aspects of drip irrigation, but also on crop evapotranspiration estimations, crop coefficients, agroclimatic data, crop water requirements, etc. I had a chance to review his two latest books by Apple Academic Press Inc.: *Management of Drip/Trickle or Micro Irrigation*, (2013) and *Evapotranspiration: Research Advances and Applications for Water Management*, (2014) and wrote a foreword for both books. I am impressed with the professional organization of the contents in each book that indicates his relationship with the world educational community. Now he has written the first book of the series on Research Advances in Sustainable Micro Irrigation. My appreciation to Megh for his good work and contribution to micro irrigation; and for this he will always be remembered among the educational fraternity today, tomorrow and forever.

Conventional irrigation systems, such as gravity irrigation and flooding, tend to waste water as large quantities are supplied to the field in one go, most of which just flows over the crop and runs away without being taken up by the plants. Micro irrigation keeps the water demand to a minimum. It has been driven by commercial farmers in arid regions of the United States of America and Israel in farming areas where water is scarce. Micro irrigation, also known as trickle irrigation or drip irrigation or pressurized irrigation, saves water and fertilizer by allowing water to drip slowly to the roots of plants, either onto the soil surface or directly near the root zone, through a network of valves, pipes, tubing, and emitters. It is done through narrow tubes that deliver water directly to the base of the plant. Primitive drip irrigation has been used since ancient times.

Ancient Methods: Centuries ago in the Middle East, farmers developed an efficient way of irrigating trees in desert soil with a minimum of water. If poured directly on the ground, much water flows away from the plant and seeps beyond the reach of the roots. To control the flow of water, farmers buried special unglazed pots near the trees and periodically filled them with water. The water seeped through the clay walls slowly, creating a pocket of wet soil around the tree. Trees grew as well with the pot method as in orchards watered by trench irrigation.

Clay Pipes: Predecessors of today's drip irrigation systems included experiments with unglazed clay pipe systems in Afghanistan in the late 1800s. Research conducted by E. B. House at Colorado State University in 1913 showed that slow irrigation could target the root zone of plants. In Germany during the 1920s, researchers devised controlled irrigation systems based on perforated pipe. None of these systems proved as efficient as modern drip irrigation technology.

Plastics: In the 1950s, plastics molding techniques and cheap polyethylene tubing made micro irrigation systems possible for the first time. Though researchers in both England and France experimented with controlled irrigation, the greatest advancements came from the work of a retired British Water Agency employee—Symcha Blass. In Israel, Blass found inspiration in a dripping faucet near a thriving tree and applied his knowledge of microtubing to an improved drip method. The Blass system overcame clogging of low volume water emitters by adding wider and longer passageways or labyrinths to the tubing. Patented in 1959 in partnership with Kibbutz Harzerim in Israel, the Blass emitter became the first efficient drip irrigation method.

Expansion: By the late 1960s, many farmers in the Americas and Australia shifted to this new drip irrigation technology. Typical water consumption decreased from 30% to 50%. In the 1980s, drip irrigation saw use in commercial landscaping applications. Because the drip irrigation emitter technology focused water below ground in the root zone, these systems saved labor costs by reducing weed growth. With drip systems yards and gardens flourished without sprinklers or manual watering.

Plantations: One of drip irrigation's bigger success stories involves the sugar plantations of Hawaii. Sugar cane fields require irrigation for two years before harvesting, and in Hawaii, the hillside fields make ditch irrigation impractical. Producers abandoned sprinkler systems in favor of the low-flow drip irrigation methods—a conversion which took 16 years. In 1986, 11 sugar plantations in Hawaii had completely shifted to drip irrigation. One plantation spanned 37,000 acres of drip-irrigated sugar cane fields. The total cost of the conversion reached $30 million. Typically, these commercial irrigation systems consist of a surface or buried pipe distribution network using emitters supplying water directly to the soil at regular intervals along the pipework. They can be permanent or portable.

Many parts of the world are now using micro irrigation technology. The systems used by large commercial companies are generally quite complex with an emphasis on reducing the amount of labor involved. Small-scale farmers in developing countries have been reluctant to take up micro irrigation methods due to the initial high investment required for the equipment.

Today, drip irrigation is used by farms, commercial greenhouses, and residential gardeners. Drip irrigation is adopted extensively in areas of acute water scarcity and especially for crops such as coconuts, containerized landscape trees, grapes, bananas, berries, eggplant, citrus, strawberries, sugarcane, cotton, maize, and tomatoes.

Properly designed, installed, and managed, drip irrigation may help achieve water conservation by reducing evaporation and deep drainage when compared to other types of irrigation such as flood or overhead sprinklers since water can be more precisely applied to the plant roots. Some advantages of drip irrigation system are: Fertilizer and nutrient loss is minimized due to localized application and reduced leaching; water application efficiency is high; field leveling is not necessary; fields with irregular shapes are easily accommodated; recycled nonpotable water can be safely used; moisture within the root zone can be maintained at field capacity; soil type plays less important role in frequency of irrigation; soil erosion is minimized; weed growth is minimized; water distribution is highly uniform, controlled by output of each nozzle; labor cost is less than other irrigation methods; variation in supply can be regulated by regulating the valves and drippers; fertigation can easily be included with minimal waste of fertilizers; foliage remains dry, reducing the risk of disease; usually operated at lower pressure than other types of pressurized irrigation, reducing energy costs.

The disadvantages of drip irrigation are: Initial cost can be more than overhead systems; the sun can affect the tubes used for drip irrigation, shortening their usable life; if the water is not properly filtered and the equipment not properly maintained, it can result in clogging; drip irrigation might be unsatisfactory if herbicides or top dressed fertilizers need sprinkler irrigation for activation; drip tape causes extra cleanup costs after harvest; and users need to plan for drip tape winding, disposal, recycling or reuse; waste of water, time and harvest, if not installed properly; highly technical; in lighter soils subsurface drip may be unable to wet the soil surface for germination; requires careful consideration of the installation depth; and the PVC pipes often suffer from rodent damage, requiring replacement of the entire tube and increasing expenses.

Modern drip irrigation has arguably become the world's most valued innovation in agriculture. Drip irrigation may also use devices called microspray heads, which spray water in a small area, instead of emitters. These are generally used on tree and vine crops with wider root zones. Subsurface drip irrigation (SDI) uses permanently or temporarily buried dripper-line or drip tape located at or below the plant roots. It is becoming popular for row crop irrigation, especially in areas where water supplies are limited or recycled water is used for irrigation. Careful study of all the relevant factors like land topography, soil, water, crop and agro-climatic conditions are needed to determine the most suitable drip irrigation system and components to be used in a specific installation.

The main purpose of drip irrigation is to reduce the water consumption by reducing the leaching factor. However, when the available water is of high salinity or alkalinity, the field soil becomes gradually unsuitable for cultivation due to high salinity or poor infiltration of the soil. Thus drip irrigation converts fields in to fallow lands when natural leaching by rainwater is not adequate in semiarid and arid regions. Most drip systems are designed for high efficiency, meaning little or no leaching fraction. Without sufficient leaching, salts applied with the irrigation water may build up in the root zone, usually at the edge of the wetting pattern. On the other hand, drip irrigation avoids the high capillary potential of traditional surface-applied irrigation, which can draw salt deposits up from deposits below.

This new book series brings academia, researchers, suppliers and industry partners together to present micro irrigation technology to partially solve water scarcity problems in agriculture sector. The compendium includes key aspects of micro irrigation principles and applications. I find it user-friendly and easy-to-read and recommend being to be on shelf of each library. My hats high to Apple Academic Press Inc. and Dr. Megh R. Goyal, my long-time colleague.

Miguel A Muñoz-Muñoz, PhD
Ex-President of University of Puerto Rico, USA
Professor and Soil Scientist
University of Puerto Rico – Mayaguez Campus
Call Box 9000
Mayaguez, P.R., 00681-9000, USA
Email: miguel.munoz3@upr.edu

February 14, 2014

FOREWORD

With only a small portion of cultivated area under irrigation and with the scope to the additional area, which can be brought under irrigation, it is clear that the most critical input for agriculture today is water. It is important that all available supplies of water should be used intelligently to the best possible advantage. Recent research around the world has shown that the yields per unit quantity of water can be increased if the fields are properly leveled, the water requirements of the crops as well as the characteristics of the soil are known, and the correct methods of irrigation are followed. Significant gains can also be made if the cropping patterns are changed so as to minimize storage during the hot summer months when evaporation losses are high, if seepage losses during conveyance are reduced, and if water is applied at critical times when it is most useful for plant growth.

Irrigation is mentioned in the Holy Bible and in the old documents of Syria, Persia, India, China, Java, and Italy. The importance of irrigation in our times has been defined appropriately by N.D. Gulati: "In many countries irrigation is an old art, as much as the civilization, but for humanity it is a science, the one to survive." The need for additional food for the world's population has spurred rapid development of irrigated land throughout the world. Vitally important in arid regions, irrigation is also an important improvement in many circumstances in humid regions. Unfortunately, often less than half the water applied is used by the crop—irrigation water may be lost through runoff, which may also cause damaging soil erosion, deep percolation beyond that required for leaching to maintain a favorable salt balance. New irrigation systems, design and selection techniques are continually being developed and examined in an effort to obtain high practically attainable efficiency of water application.

The main objective of irrigation is to provide plants with sufficient water to prevent stress that may reduce the yield. The frequency and quantity of water depends upon local climatic conditions, crop and stage of growth, and soil-moisture-plant characteristics. Need for irrigation can be determined in several ways that do not require knowledge of evapotranspiration [ET] rates. One way is to observe crop indicators such as change of color or leaf angle, but this information may appear too late to avoid reduction in the crop yield or quality. Other similar methods of scheduling include determination of the plant water stress, soil moisture status, or soil water potential. Methods of estimating crop water requirements using ET and combined with soil characteristics have the advantage of not only being useful in determining when to irrigate, but also enables us to know the quantity of water needed. ET estimates have not been made for the developing countries though basic information on weather data is available. This has contributed to one of the existing problems that the vegetable crops are over irrigated and tree crops are under irrigated.

Water supply in the world is dwindling because of luxury use of sources; competition for domestic, municipal, and industrial demands; declining water quality; and losses through seepage, runoff, and evaporation. Water rather than land is one of the limiting factors in our goal for self-sufficiency in agriculture. Intelligent use of water will avoid problem of sea water seeping into aquifers. Introduction of new irrigation methods has encouraged marginal farmers to adopt these methods without taking into consideration economic benefits of conventional, overhead, and drip irrigation systems. What is important is "net in the pocket" under limited available resources. Irrigation of crops in tropics requires appropriately tailored working principles for the effective use of all resources peculiar to the local conditions. Irrigation methods include border-, furrow-, subsurface-, sprinkler-, sprinkler, micro, and drip/trickle, and xylem irrigation.

Drip irrigation is an application of water in combination with fertilizers within the vicinity of plant root in predetermined quantities at a specified time interval. The application of water is by means of drippers, which are located at desired spacing on a lateral line. The emitted water moves due to an unsaturated soil. Thus, favorable conditions of soil moisture in the root zone are maintained. This causes an optimum development of the crop. Drip/micro or trickle irrigation is convenient for vineyards, tree orchards, and row crops. The principal limitation is the high initial cost of the system that can be very high for crops with very narrow planting distances. Forage crops may not be irrigated economically with drip irrigation. Drip irrigation is adaptable for almost all soils. In very fine textured soils, the intensity of water application can cause problems of aeration. In heavy soils, the lateral movement of the water is limited, thus more emitters per plant are needed to wet the desired area. With adequate design, use of pressure compensating drippers and pressure regulating valves, drip irrigation can be adapted to almost any topography. In some areas, drip irrigation is used successfully on steep slopes. In subsurface drip irrigation, laterals with drippers are buried at about 45 cm depth, with an objective to avoid the costs of transportation, installation, and dismantling of the system at the end of a crop. When it is located permanently, it does not harm the crop and solve the problem of installation and annual or periodic movement of the laterals. A carefully installed system can last for about 10 years.

The publication of this book series and this volume is an indication that things are beginning to change, that we are beginning to realize the importance of water conservation to minimize the hunger. It is hoped that the publisher will produce similar materials in other languages.

In providing this resource in micro irrigation, Megh Raj Goyal, as well as the Apple Academic Press, is rendering an important service to the farmers, and above all to the poor marginal farmers. Dr. Goyal, Father of Irrigation Engineering in Puerto Rico, has done an unselfish job in the presentation of this compendium that is simple and thorough. I know Megh Raj since 1973 when we were working together at Haryana Agricultural University on an ICAR research project in "Cotton Mechanization in India."

Gajendra Singh, PhD., prof.gsingh@gmail.com,
Tel.: +91 99 7108 7591
Adjunct Professor, Indian Agricultural Research Institute, New Delhi
Ex-President (2010-2012), Indian Society of Agricultural Engineers,
Former Vice Chancellor, Doon University, Dehradun, India.
Former Deputy Director General (Engineering),
Indian Council of Agricultural Research (ICAR), New Delhi
Former Vice-President/Dean/Professor and Chairman,
Asian Institute of Technology, Thailand.

New Delhi
February 14, 2014

FOREWORD

The micro irrigation system, more commonly known as the drip irrigation system, was one of the greatest advancements in irrigation system technology developed over the past half century. The system delivers water directly to individual vines or to plant rows as needed for transpiration. The system tubing may be attached to vines, placed on or buried below the soil surface.

This book, written by experience system designers/scientists, describes various systems that are being used around the world, the principles of micro irrigation, chemigation, filtration systems, water movement in soils, soil-wetting patterns, crop water requirements and crop coefficients for a number of crops. It also includes chapters on hydraulic design, emitter discharge and variability, and pumping station. Irrigation engineers will find this book to be a valuable reference.

Marvin E. Jensen, PhD, PE
Retired Research Program Leader at USDA-ARS; and
Irrigation Consultant
1207 Spring Wood Drive, Fort Collins, Colorado 80525, USA.
E-mail: mjensen419@aol.com

February 14, 2014

BOOK SERIES: RESEARCH ADVANCES IN SUSTAINABLE MICRO IRRIGATION

Volume 1: Sustainable Micro Irrigation: Principles and Practices
Senior Editor-in-Chief: Megh R. Goyal, PhD, PE

Volume 2: Sustainable Practices in Surface and Subsurface Micro Irrigation
Senior Editor-in-Chief: Megh R. Goyal, PhD, PE

Volume 3: Sustainable Micro Irrigation Management for Trees and Vines
Senior Editor-in-Chief: Megh R. Goyal, PhD, PE

Volume 4: Management, Performance, and Applications of Micro Irrigation
Senior Editor-in-Chief: Megh R. Goyal, PhD, PE

ABOUT THE SENIOR EDITOR-IN-CHIEF

 Megh R. Goyal received his BSc degree in Engineering in 1971 from Punjab Agricultural University, Ludhiana, India; his MSc degree in 1977 and PhD degree in 1979 from the Ohio State University, Columbus; his Master of Divinity degree in 2001 from Puerto Rico Evangelical Seminary, Hato Rey, Puerto Rico, USA. He spent a one-year sabbatical leave in 2002–2003 at Biomedical Engineering Department, Florida International University, Miami, USA.

Since 1971, he has worked as Soil Conservation Inspector; Research Assistant at Haryana Agricultural University and the Ohio State University; and Research Agricultural Engineer at Agricultural Experiment Station of UPRM. At present, he is a Retired Professor in Agricultural and Biomedical Engineering in the College of Engineering at University of Puerto Rico – Mayaguez Campus; and Senior Acquisitions Editor and Senior Technical Editor-in-Chief in Agriculture and Biomedical Engineering for Apple Academic Press, Inc.

He was the first agricultural engineer to receive the professional license in Agricultural Engineering in 1986 from the College of Engineers and Surveyors of Puerto Rico. On September 16, 2005, he was proclaimed as "Father of Irrigation Engineering in Puerto Rico for the twentieth century" by the ASABE, Puerto Rico Section, for his pioneer work on micro irrigation, evapotranspiration, agroclimatology, and soil and water engineering. During his professional career of 45 years, he has received awards such as: Scientist of the Year, Blue Ribbon Extension Award, Research Paper Award, Nolan Mitchell Young Extension Worker Award, Agricultural Engineer of the Year, Citations by Mayors of Juana Diaz and Ponce, Membership Grand Prize for ASAE Campaign, Felix Castro Rodriguez Academic Excellence, Rashtrya Ratan Award and Bharat Excellence Award and Gold Medal, Domingo Marrero Navarro Prize, Adopted son of Moca, Irrigation Protagonist of UPRM, Man of Drip Irrigation by Mayor of Municipalities of Mayaguez/Caguas/Ponce and Senate/Secretary of Agriculture of ELA, Puerto Rico.

He has authored more than 200 journal articles and textbooks including: *Elements of Agroclimatology* (Spanish) by UNISARC, Colombia; two *Bibliographies on Drip Irrigation.* Apple Academic Press Inc. (AAP) has published his books, namely, *Biofluid Dynamics of Human Body*, *Management of Drip/Trickle or Micro Irrigation*, *Evapotranspiration: Principles and Applications for Water Management*, and *Biomechanics of Artificial Organs and Prostheses.*" With this volume, AAP will publish 10-volume set on *Research Advances in Sustainable Micro Irrigation* by Readers may contact him at: goyalmegh@gmail.com.

WARNING/DISCLAIMER

The goal of this compendium is to guide the world community on how to manage the Sustainable Surface and Subsurface Drip/Trickle or Micro Irrigation system efficiently for economical crop production. The reader must be aware that the dedication, commitment, honesty, and sincerity are most important factors in a dynamic manner for a complete success. It is not a one-time reading of this compendium. Read and follow every time, it is needed. To err is human. However, we must do our best. Always, there is a space for learning new experiences.

The editor, the contributing authors, the publisher and the printer have made every effort to make this book as complete and as accurate as possible. However, there still may be grammatical errors or mistakes in the content or typography. Therefore, the contents in this book should be considered as a general guide and not a complete solution to address any specific situation in irrigation. For example, one size of irrigation pump does not fit all sizes of agricultural land and to all crops.

The editor, the contributing authors, the publisher and the printer shall have neither liability nor responsibility to any person, any organization or entity with respect to any loss or damage caused, or alleged to have caused, directly or indirectly, by information or advice contained in this book. Therefore, the purchaser/reader must assume full responsibility for the use of the book or the information therein.

The mentioning of commercial brands and trade names are only for technical purposes. It does not mean that a particular product is endorsed over to another product or equipment not mentioned. Author, cooperating authors, educational institutions, and the publisher Apple Academic Press Inc. do not have any preference for a particular product.

All weblinks that are mentioned in this book were active on October 31, 2013. The editors, the contributing authors, the publisher and the printing company shall have neither liability nor responsibility, if any of the weblinks is inactive at the time of reading of this book.

PART I
STATUS AND POTENTIAL OF MICRO IRRIGATION

PRESSURIZED IRRIGATION TECHNOLOGY: GLOBAL SCENARIO*

S. A. KULKARNI, F. B. REINDERS, and F. LIGETVARI

CONTENTS

1.1 Introduction 2
1.2 Pressurized Irrigated Area in the World...2
1.3 Regional Spread of Sprinkler and Micro Irrigated Area...................................3
1.4 Present Global Coverage of Sprinkler and Micro Irrigation 7
1.5 Top Twenty Countries Having Sprinkler and Micro Irrigated Areas................8
1.6 Development of Sprinkler and Micro Irrigation in India and USA..................11
1.7 Non-Conventional Use of Sprinkler and Micro Irrigation 12
1.8 Future Developments...14
1.10 Conclusions 14
1.11 Summary..15
Keywords...16
References..16
Appendix I 18
Appendix II..19

*Modified from, "*S.A. Kulkarni, F.B. Reinders, and F. Ligetvari. Global Scenario of Sprinkler and Micro Irrigated Areas. International Commission on Irrigation and Draiage, www.icid.org. Keynote address at the opening of the 7th International Micro-Irrigation Congress in Kuala Lumpur, Malaysia, 13–15 September 2006.*"

1.1 INTRODUCTION

Historically, there are four basic irrigation systems of on-farm water application: (1) surface/gravity, (2) subsurface, (3) sprinkler, and (4) micro irrigation. The sprinkler and micro irrigation methods together are generally categorized as 'Pressurized systems.' Sprinkler method includes hand move, hose pull, side roll/wheel move, stationary big guns, and solid state/permanent types, continuous move types such as center pivots, linear move, and travelers. Water application through drip (also called trickle), micro sprinklers (spinners and rotators), microjets (static and vibrating), microsprayers, bubblers, drip tapes (both surface and subsurface) referred to as micro irrigation. There is no definite distinction between conventional sprinklers and emitters used in micro irrigation, emitters having flow rates up to 200 L per hour (L/h) can be regarded as micro emitters [15]. Generally, drippers have flow rates from 2 to 8 L/h. Micro-sprayers and static micro jets are non-rotating type with flow rates ranging from 20 to 150 L/h. Whereas, microsprinklers are rotating type with flow rates ranging from 100 to 300 L/h.

Although sprinkler irrigation is known to the world since more than 80 years its wide scale use for field crops commenced in 1950s due to availability of better sprinklers, aluminum   pipes and more efficient pumps [18]. Today, a variety of sprinkler systems comprising periodically move systems, fix system and continuous move system are in use worldwide. By the end of 1990s, the total area under sprinkler irrigation in the world was about 21.6 million hectares (Mha) [12]. Further data of the worldwide coverage of sprinkler irrigation is not available.

The large-scale use of drip irrigation system started in 1970s in Australia, Israel, Mexico, New Zealand, South Africa, and USA to irrigate vegetables, orchards and its coverage was reported as 56,000 ha. The micro irrigated area grew slowly but steadily and it was 0.41 Mha in 1981, 1.1 Mha in 1986, 1.77 Mha in 1991 [13], and about 3.0 Mha in 2000 [24].

In this chapter, efforts have been made to compile the latest available data about area coverage of both sprinkler micro irrigation. The data is compiled mainly from National Committees of ICID, FAO-AQUASTAT, websites and other published documents.

1.2 PRESSURIZED IRRIGATED AREA IN THE WORLD

In the last worldwide survey of micro irrigation (1991) carried out by International Commission on Irrigation and Drainage (ICID), data of 35 countries was reported. During the last 15 years the numbers of countries using micro irrigation have more than doubled. In this chapter, data for sprinkler and micro irrigated areas in 77 countries, arranged by region and sub region, is presented in Table 1. The column 9 in Table 1 shows the year of reporting the data for each country along with the source of information. The data pertains to different years from 1989 to 2005 and thus may not represent the current status in case of some countries. Data for 2011–2013 is shown in Appendix I at the end of this chapter.

1.3 REGIONAL SPREAD OF SPRINKLER AND MICRO IRRIGATED AREA

Use of pressurized irrigation systems vary considerably from region to region and country   to country. Table 2 shows the region-wise area coverage by pressurized systems. Proportion of micro and sprinkler irrigated areas in each of the five regions is shown in Fig. 1. Major aspects of sprinkler and micro irrigation development in the five world regions are briefly discussed as follows:

1.3.1 AFRICA

In African continent area covered under full/partial control irrigation technique is 12.48   Mha, which is 6% of the total cultivated area of the continent [8, 9]. Of which almost 50% is concentrated in Northern Africa and about 75% are in five countries viz.,   South Africa, Egypt, Madagascar, Morocco, and Sudan. The FAO has categorized the 53   African countries into seven regions. However, for the purpose of this chapter, these are grouped   under five subregions (Table 1).

Most of the data used in this chapter is drawn from the latest FAO—AQUASTAT Survey-2005 [8, 9]. As can be seen from the Table 2 that, in Africa, pressurized irrigation systems are used on 2.3 million hectares or 22% of the total irrigated area reported in this chapter and 18% of the total area equipped for irrigation of the continent. Within the pressurized systems, sprinkler irrigation covers 1.86 million ha (18%) and micro irrigation 0.43 million ha (4%). Both sprinkler and micro irrigation are concentrated mainly in the Northern and Southern regions.

1.3.2 AMERICAS

American region as a whole has the largest area (14.68 million ha) under sprinkler and micro irrigation among five regions of the world. More than 85% of the area coverage is in three Northern countries viz. Canada, Mexico, and USA. The sprinkler irrigation is the dominant method of irrigation and is practiced on 88% of the total reported irrigated area (37.2 million ha), while micro irrigation is used on 12% of the irrigated area. The USA not only has the   world's largest area under pressurized systems but also has varieties of them.

In Latin America, the pace of adoption of new irrigation technologies is rather slow but steady. In arid and semiarid zones of countries like Argentina, Chile, Peru, Ecuador the drip system has become popular among fruit, vegetable and flower growers. In countries like Mexico and Peru governments are extending liberal technical, financial and training support to farmers in adoption of modern irrigation techniques. Peru is the world's biggest paprika exporter, the 2nd biggest asparagus exporter, and the 3rd biggest onion provider. Argentina is among the top 10 world's biggest apple exporters. Colombia is the second largest exporter of flowers in the world and also one of the biggest banana exporters. Ecuador is the world's largest banana exporter and second largest flower producer in South America. Chile is a leading wine exporting country. In all these countries, an increase of sprinkler and drip irrigation is likely to happen in the years to come [New AG International, Ref. 26].

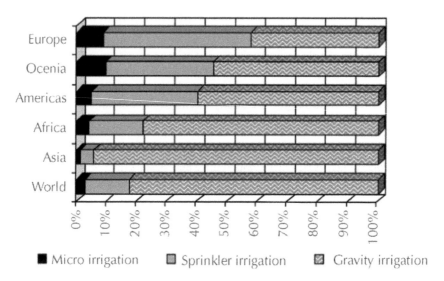

FIGURE 1   Proportion of the micro and sprinkler irrigated areas in the five regions.

1.3.3   ASIA

Although the Asian region shares 70% of the total world irrigated area, its share in the world's sprinkler and micro irrigated area is about 30%. At present, the proportion of sprinkler and micro irrigated area to the irrigated area is rather low. It is only 6% of the reported irrigated area in this chapter (148.6 million ha). China, India, and Pakistan together constitute almost half of the world's irrigated land. About 90% of the world's rice grown area is in Asia (118 million hectares), of which about 70 million ha (60%) are estimated as irrigated. The dominance of rice area could be one of the reasons for the low share of the pressurized irrigation in the region.

Both China and India have emerged as the major users of pressurized irrigation technologies and occupy 3rd and 4th position, respectively, in the world as regards the area coverage (Table 1). In China, water withdrawals for irrigation were 310 km³ in the year 2003. Water conservation and increased water use efficiency have become the driving force of China's irrigation development. About one-third of the 55.9 million hectares of irrigated area was equipped with various water-saving irrigation techniques. As a result average volume of water application declined from 9,405 m3/ha in 1980 to 5,539 m³/ha in 2003 [2]. Due to increasing water scarcity in many regions of China and India, it is expected that during forthcoming years the major expansion of pressurized systems will occur in these counties. In other major irrigated Asian countries viz. Bangladesh, Indonesia, Myanmar, Pakistan, Vietnam data on area coverage by pressurized systems is not available. In Baluchistan Province of Pakistan, drip irrigation is practiced by a few orchard growers as the groundwater is being mined.

TABLE 1　Country-wise sprinkler and micro irrigated area.

Region	Sr. No.	Country	Irriga-ted area (Mha)	Sprinkler irrigated area (ha)	Micro irrigated area (ha)	Total sprinkler and micro irrigated area (ha)	% of irrigated area, as per column	Reporting year and the source of information
(1)	(2)	(3)	(4)	(5)	(6)	(7)	(8)	(9)
AFRICA	1	Algeria	0.56	40,000	NA	40,000	7.1	2001(FAO)[1]
Northern	2	Egypt	3.4	450,000	104,000	554,000	16.3	2000(ICID)[3]
	3	Libya	0.47	470,000	NA	470,000	100.0	1990(FAO)[2]
	4	Morocco	1.65	189,750	8,250	198,000	12.0	2003(ICID)[3]
	5	Tunisia	0.39	90,000	62,000	152,000	39.0	2000(FAO)[1]
	Sub-total		6.08	1,239,750	174,250	1,414,000	23.3	
	6	Benin	0.01	4,570	1,360	5,930	59.3	2002(FAO)[1]
	7	Burkina Faso	0.03	3,900	NA	3,900	13.0	2001(FAO)[1]
Western	8	Chad	0.03	3,754	NA	3,754	12.5	2002(FAO)[1]
	9	Cote d'Ivoire	0.07	36,000	NA	36,000	51.4	1994(FAO)[1]
	10	Ghana	0.03	6,300	NA	6,300	21.0	2000(FAO)[1]
	Sub-total		0.17	54,524	1,360	55,884	32.9	
Central	11	Cameroon	0.02	5,430	NA	5,430	27.0	2000(FAO)[1]
	12	Ethiopia	0.29	6,355	NA	6,355	2.2	2001(FAO)[1]
	Sub-total		0.31	11,785	0	11,785	3.8	
	13	Kenya	0.1	61,986	2,000	63,986	64.0	2003(FAO)[1]
Eastern	14	Madagascar	1.09	2,400	NA	2,400	0.2	2000(FAO)[1]
	15	Mauritius	0.02	17,028	1,822	18,850	94.0	2002(FAO)[1]
	Sub-total		1.83	104,984	3,822	108,806	5.9	
	16	Malawi	0.06	43,193	5,450	48,643	81.0	2000(ICID)[3]
	17	Namibia	0.01	3,276	1,347	4,623	46.2	2002(FAO)[1]
Southern	18	South Africa	1.49	255,000	220,000	475,000	31.9	2000(ICID)[3]
	19	Swaziland	0.05	20,905	3,051	23,956	47.9	2000(FAO)[1]
	20	Zambia	0.16	17,570	5,628	23,198	14.5	2002(FAO)[1]
	21	Zimbabwe	0.15	112,783	13,881	126,664	84.4	2002(FAO)[1]
	Sub-total		1.92	452,727	249,357	702,084	36.6	
AFRICA TOTAL			10.31	1,863,770	428,789	2,292,559	22.2	
AMERICAS	22	Canada	0.87	683,029	6,034	689,063	79.2	2006(ICID)[3]
Northern	23	Mexico	6.2	400,000	200,000	600,000	9.7	1999(ICID)[3]
	24	United States	21.3	10,906,006	1,209,757	12,115,763	56.9	2003(USDA)[4]
	Sub-total		27.50	11,306,006	1,409,757	12,715,763	46.2	
	25	El Salvador	0.05	4,949	0	4,949	9.8	1997(FAO)[5]
Central	26	Costa Rica	0.11	3,900	13,700	17,600	16.0	1997(FAO)[5]
	27	Cuba	0.87	402,663	19,492	422,155	48.5	1997(FAO)[5]
	Sub-total		1.03	411,512	33,192	444,704	43.1	
	28	Argentina	1.44	65,207		65,207	4.5	1988(FAO)[5]
	29	Brazil	3.44	1,373,000	338,000	1,711,000	49.7	2003(ICID)[3]
	30	Chile	1.9	30,523	62,153	92,676	4.9	1996(FAO)[5]
Southern	31	Colombia	0.9	37,271	6,036	43,307	4.8	1992(FAO)[5]
	32	Panama	0.03	8,322	572	8,894	29.6	1997(FAO)[5]
	33	Peru	1.19	11,980	7,700	19,680	1.7	1998(FAO)[5]
	34	Venezuela	0.57	72,800	20,800	93,600	16.4	1989(FAO)[5]
	Sub-total		9.47	1,599,103	435,261	2,034,364	21.5	
AMERICAS TOTAL			38.00	13,316,621	1,878,210	15,194,831	40.0	
	35	Saudi Arabia	1.17	716,000	198,000	914,000	78.1	2004(ICID)[3]
ASIA	36	Yemen	0.66	25,000	NA	25,000	3.8	2005 (NWRA)[6]
	37	Iran	8.1	290,000	160,000	450,000	5.6	2003(ICID)[3]
	38	Israel	0.23	60,000	170,000	230,000	100.0	2000(ICID)[3]
	39	Jordan	0.07	4,900	47,600	52,500	75.0	1995(ICID)[7]
Near East	40	Lebanon	0.1	21,000	13,000	34,000	34.0	1993(FAO)[5]
	41	Syria	1.28	93,000	62,000	155,000	12.1	2000(ICID)[3]
	42	Turkey	5.2	145,340	43,000	188,340	3.6	2003(ICID)[3]
	43	Azerbaijan	1.45	149,000	2,618	151,618	10.5	1995(FAO)[8]
	Sub-total		18.26	1,504,240	696,218	2,200,458	12.1	
	44	Kazakhstan	2.31	549,600	NA	549,600	23.8	1993(FAO)[8]
Central	45	Kyrgyzstan	1.05	141,000	NA	141,000	13.4	1990(FAO)[2]
	46	Uzbekistan	4.31	NA	4,510	4,510	0.1	1994(FAO)[8]
	Sub-total		7.67	690,600	4,510	695,110	9.1	
	47	China	55.9	2,634,000	371,000	3,005,000	5.4	2005(ICID)[3]
	48	Chinese Taipei	0.46	12,300	13,400	25,700	5.6	2001(ICID)[3]
Southern and Eastern	49	India	57	1,634,997	589,251	2,224,248	3.9	2005(INCPAH)[9]
	50	Japan	2.57	243,000	55,000	298,000	11.6	1999(FAO)[10]
	51	Malaysia	0.36	NA	10,000	10,000	2.8	2005(ICID)[3]
	52	Mongolia	0.08	43,900	NA	43,900	54.9	1999(FAO)[10]
	53	Philippines	1.41	7,175	6,635	13,810	1.0	2003(ICID)[3]
	54	Thailand	4.9	N.A.	41,150	41,150	0.8	1991 (ICID)[3]
	Sub-total		122.68	4,575,372	1,086,436	5,661,808	4.8	
ASIA TOTAL			148.61	6,770,212	1,787,164	8,557,376	5.8	
EUROPE	55	Finland	0.09	85,000	700	85,700	99.7	2000(ICID)[3]
	56	Austria	0.08	76,000	3,000	79,000	98.8	1994(ICID)[3]
	57	France	1.58	1,379,800	103,300	1,483,100	94.2	2000(ICID)[3]
Northern and Western	58	Germany	0.54	525,000	5,000	530,000	100.0	2005(ICID)[3]
	59	The Netherlands	0.36	NA	3,000	3,000	0.5	1991 (ICID)[11]
	60	United Kingdom	0.15	140,000	10,000	150,000	100.0	2001(ICID)[3]

TABLE 1   *(Continued)*

Region	Sr. No.	Country	Irrigated area (Mha)	Sprinkler irrigated area (ha)	Micro irrigated area (ha)	Total sprinkler and micro irrigated area (ha)	% of irrigated area, as per column	Reporting year and the source of information
		Sub-total	3.00	2,205,800	125,000	2,330,800	77.7	
(1)	(2)	(3)	(4)	(5)	(6)	(7)	(8)	(9)
Central	61	Bulgaria	0.04	34,574	333	34,907	87.3	2001(ICID)
	62	Czech Rep.	0.16	153,000	1000	154,000	99.4	2000(ICID)
	63	Hungary	0.22	197,000	6,100	203,100	92.3	2002(ICID)
	64	Poland	0.08	4,974	5,000	9,974	12.5	2003(ICID)
	65	Romania	3.18	500,000	NA	500,000	15.7	2000 (ICID)
	66	Slovak Rep.	0.31	310,000	2,650	312,650	99.9	2000(ICID)
		Sub-total	3.99	1,199,548	15,083	1,214,631	30.3	
Eastern	67	Lithuania	0.05	44,518	NA	44,518	98.9	2004(ICID)
	68	Russia	4.45	3,560,000	200,000	3,760,000	84.5	2005(ICID)
	69	Ukraine	0.73	700,000	NA	700,000	95.9	2005 (ICID)
		Sub-total	5.23	4,304,518	200,000	4,504,518	86.1	
Medi-terranean	70	Cyprus	0.04	2,000	35,600	37,600	94.0	1995(ICID)
	71	Greece	1.43	239,012	128,583	367,595	25.7	2004(MOA)
	72	Italy	2.54	1,051,201	366,038	1,417,239	55.9	2000(ISTAT)
	73	Macedonia	0.16	100,000	500	100,500	61.4	2000(ICID)
	74	Portugal	0.63	40,000	25,000	65,000	10.3	1999(ICID)
	75	Spain	3.58	905,377	914,112	1,819,489	50.8	1999(ICID)
		Sub-total	8.38	2,337,590	1,469,833	3,807,423	45.5	
		EUROPE TOTAL	20.69	10,132,456	1,810,616	11,943,072	57.7	
OCEANIA & PACIFIC	76	Australia	2.0	660,000	180,000	840,000	42.0	2005 (DAFF)
	77	New Zealand	0.47	188,000	47,000	235,000	50.0	2002(MOA)
		OCEANIA & PACIFIC TOTAL	2.47	848,000	227,000	1,075,000	43.5	
		ALL REGION TOTAL	220.08	32,931,059	6,131,779	39,062,838	17.7	

TABLE 2   Region-wise sprinkler and microirrigated area (in descending order).

Region	Irrigated area as per FAO(2003)	Irrigated area reported in this paper	Sprinkler irrigated area	Micro irrigated area	Total sprinkler and micro Irrigated area	Proportion of Sprinkler and micro irrigated area w.r.t. column 3, %
1	2	3	4	5	6	7
Americas	41.91 (35)*	38.00 (13)*	13.32	1.88	15.20	40
Europe	25.22 (35)	20.69 (21)	10.13	1.81	11.94	58
Asia	193.87 (46)	148.61 (20)	6.78	1.79	8.56	6
Africa	12.48 (53)	10.31 (21)	1.86	0.43	2.29	22
Oceania	2.55 (5)	2.47 (2)	0.85	0.23	1.08	44
World Total	276.03 (174)	220. 08 (77)	32.93	6.13	39.06	18

Note: Area in million hectares.
* Numbers in parenthesis indicate the number of countries.

In Central Asian Countries, sprinkler irrigation was widely used during the former Soviet Union period. However, subsequently due to technological and financial constraints, deterioration of irrigation infrastructure has taken place [25]. It is therefore likely that the area under pressurized systems reported in this chapter might have been declined further. In Azerbaijan, Iran, Saudi Arabia, Syria and Turkey sprinkler irrigation is by far the most predominant, while in Israel and Jordan micro irrigation is the most widely used technique, being practiced on over 70% of their irrigated land.

1.3.4   EUROPE

Amongst all the five regions, Europe has the highest proportion of its irrigated area under   pressurized systems. Of the total 25 million ha of irrigated land about 47% (12 million ha) is covered by sprinkler and micro irrigation. Of the 21 countries reported in this chapter, 13 countries have more than 75% of their irrigated land under pressurized systems.

At the beginning of the 1990s, in the Central and Eastern European Countries (CEEC) the   period of transition from a central planning economy to the market economy was commenced. Due to privatization of agricultural land and reduction of government support to irrigation   sector, deterioration of irrigation and drainage infrastructure took place. As a result there was a drastic reduction in irrigated areas in most of the CEEC. For example, in Bulgaria, the irrigated area prior to 1991 was 1.25 million ha, which declined to about 40,000 at present. Similarly in Poland, the present irrigated area is 83,000 while it was more than 3.2 million ha prior to 1991. Prior to 1990 in CEEC, large-scale sprinkler irrigated farms under cooperative and state sectors were in vogue. However, with the establishment of smaller farming units and diversification of crops, the earlier large-sized mechanized sprinkler systems were replaced by appropriate small-scale systems such as hand move and hose-reel systems [3]. The area under pressurized irrigation in the CEEC has therefore dropped substantially as compared to pre-1990s.

1.3.5   OCEANIA

Australia and New Zealand are the major irrigated countries in Oceania and Pacific regions. In New Zealand almost all irrigation is in private hands. About 80% of the sprinkler and micro irrigated area of the region is located in Australia alone.

1.4   PRESENT GLOBAL COVERAGE OF SPRINKLER AND MICRO IRRIGATION

Owing to year-to-year fluctuations in the actual irrigated areas and also nonavailability of information in some countries, it is difficult to have an accurate worldwide statistics of irrigated area and so also of pressurized irrigation. The FAO Production Yearbook [6, 7] has reported the total irrigated area of 174 countries across the world as 276 million ha. In this chapter   data pertaining to sprinkler and micro irrigation in 77 countries covering 220 million ha is presented.

Worldwide experience during the last 50 years has shown that sprinkler and micro irrigation can be used for all types of crops excepting rice-paddy. Although there are some experimental evidences of use of sprinkler irrigation for rice production in Pakistan [17], Turkey, USA and Zimbabwe [1], its application is yet to spread on farmers' fields.   Thus while reporting the proportion of pressurized irrigation to the total irrigated area of a country/region or at global scale, the corresponding irrigated rice area may be excluded and then the percentage is computed. This is further explained as follows:

Rice is grown in 113 countries around the world with a total harvested area of about 150 million hectares. About 40% of the total rice area is classified as rainfed [8, 9]. Thus it can be estimated that about 90 million hectares is an irrigated rice area worldwide. Of the total irrigated area of 276 million ha, about 186 million ha can be considered as nonrice irrigated area, of which presently pressurized systems are practiced on about 39 million ha. Consequently, the proportion of area coverage by pressurized systems with respect to the total nonrice irrigated area of the world works out to be 21% (Fig. 2). This proportion becomes 30% if it is computed with respect to the nonrice irrigated area (132 million ha) worked out on the basis of total irrigated area of 77 countries (220 million ha) reported in this chapter.

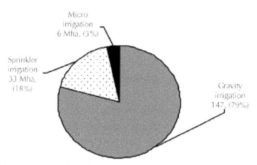

FIGURE 2   Worldwide coverage of sprinkler and micro irrigation (in nonrice irrigated area).

1.5   TOP TWENTY COUNTRIES HAVING SPRINKLER AND MICRO IRRIGATED AREAS

The area coverage of pressurized systems varies widely between countries. Tables 3 and 4 show the top 20 countries having sprinkler and micro irrigation, respectively.

TABLE 3   The top 20 countries with sprinkler irrigated area.

Sl. No	Country	Sprinkler irrigated area (ha)	% of irrigated area
1.	United States	10,906,006	51
2.	Russia	3,560,000	80
3.	China	2,634,000	5
4.	India	1,634,997	3
5.	France	1,379,800	87
6.	Brazil	1,373,000	40
7.	Italy	1,051,201	41
8.	Spain	905,377	25
9.	Saudi Arabia	716,000	61
10.	Ukraine	700,000	96
11.	Canada	683,029	79
12.	Australia	660,000	33
13.	Kazakhstan	549,600	24
14.	Germany	525,000	99
15.	Romania	500,000	16
16.	Libya	470,000	100
17.	Egypt	450,000	13
18.	Cuba	402,663	49
19.	Mexico	400,000	10
20.	Slovak Rep.	310,000	99
	Total	29,810,673	17

TABLE 4 The top 20 countries with micro irrigated area.

Sl. No.	Country	Micro irrigated area (ha)	% of irrigated area
1.	United States	1,209,757	6
2.	Spain	914,112	26
3.	India	589,251	1
4.	China	371,000	1
5.	Italy	366,038	14
6.	Brazil	338,000	10
7.	South Africa	220,000	15
8.	Russia.	200,000	4
9.	Mexico	200,000	3
10.	Saudi Arabia	198,000	17
11.	Australia	180,000	9
12.	Israel	170,000	74
13.	Iran	160,000	2
14.	Greece	128,583	9
15.	Egypt	104,000	3
16.	France	103,300	7
17.	Chile	62,153	3
18.	Syria	62,000	5
19.	Tunisia	62,000	16
20.	Japan	55,000	2
	Total	5,693,194	3

Top 20 countries with major area irrigated by pressurized irrigation are shown in Table 5. The top 10 and top 20 countries respectively share about 75% and 90% of the total area   irrigated by the pressurized systems worldwide.

Today there are eight countries having more than a million hectare under pressurized irrigation systems viz. USA, Russia, China, India, Spain, Brazil, France, and Italy. Countries having   more than 75% of their irrigated land under pressurized irrigation systems are shown in Table 6. However, these 20 countries together share about 25% of the total sprinkler and micro   irrigated area of the world. In five countries viz. Austria, Israel, Libya, Slovak Republic and United Kingdom, irrigation is accomplished entirely through pressurized systems. In nine countries viz. Czech Republic, Cyprus, Finland, France, Germany, Hungary, Lithuania Mauritius, and Ukraine pressurized irrigation covers more than 90% of their total irrigated land.

TABLE 5 The top 20 countries having pressurized irrigation systems.

Sl. No.	Country	Irrigated area (Mha)	Total sprinkler and micro irrigated area (ha)	% of irri-gated area
1.	United States	21.3	12,115,763	57
2.	Russia	4.45	3,760,000	85
3.	China	55.9	3,005,000	5
4.	India	57.0	2,224,248	4
5.	Spain	3.58	1,819,489	51
6.	Brazil	3.44	1,711,000	50
7.	France	1.58	1,483,100	94
8.	Italy	2.54	1,417,239	56
9.	Saudi Arabia	1.17	914,000	78
10.	Australia	2.0	840,000	42
11.	Ukraine	0.73	700,000	96
12.	Canada	0.87	689,063	79
13.	Mexico	6.2	600,000	10
14.	Egypt	3.4	554,000	16
15.	Kazakhstan	2.31	549,600	24
16.	Germany	0.54	543,000	100
17.	Romania	3.18	500,000	15.7
18.	South Africa	1.49	475,000	32
19.	Libya	0.47	470,000	100
20	Iran	8.1	450,000	6
	Total	180.25	34,820,502	19

TABLE 6 Countries having more than 75% of their irrigated land under pressurized irrigation systems.

Sl. No.	Country	Irri-gated area (Mha)	Total sprinkler and micro irrigated area (ha)	% of irri-gated area
1.	Slovak Rep.	0.31	312,650	100.0
2.	Libya	0.47	470,000	100.0
3.	Israel	0.23	230,000	100.0
4.	United Kingdom	0.15	150,000	100.0
5.	Austria	0.08	79,000	98.8
6.	Germany	0.54	525,000	98.1
7.	Czech Re.	0.16	154,000	96.3
8.	Ukraine	0.73	700,000	95.9
9.	Finland	0.09	85,700	95.2
10.	Mauritius	0.02	18,850	94.3
11.	Cyprus	0.04	37,600	94.0
12.	France	1.58	1,483,100	93.9
13.	Hungary	0.22	203,100	92.3
14.	Lithuania	0.05	44,518	89.0
15.	Bulgaria	0.04	34,907	87.3
16.	Russia	4.45	3,760,000	84.5
17.	Zimbabwe	0.15	126,664	84.4
18.	Malawi	0.06	48,643	81.1
19.	Canada	0.87	689,063	79.2
20.	Saudi Arabia	1.17	914,000	78.1
	Total	11.41	10,066,795	88.5

1.6 DEVELOPMENT OF SPRINKLER AND MICRO IRRIGATION IN INDIA AND USA

1.6.1 INDIA

India has world's largest irrigated area and presently 57 million hectares (90 million ha harvested) are irrigated. About two-third of the area is irrigated by groundwater and one- third from surface water resources. The ultimate irrigation (harvested area) potential is estimated as 139 million ha without the 'River linking project' and 174 million ha upon its implementation. Present irrigation withdrawals are 534 km³ and are estimated to be 611 km³ in 2025 and 807 km³ by the year 2050 (Govt. of India, 10).

In India, the sprinkler-irrigated area was 0.23 Mha in 1985, which grew to 0.67 Mha by 1998 [10]. As per the 2005 survey carried out by the National Committee on Plasticulture Application in Horticulture (NCPAH), the sprinkler area has dramatically increased to 1.63 million ha [20]. In 1985 only 1000 hectares were under drip/micro irrigation, which rose to 35,000 hectares in 1990, 152,930 hectares in 1995, 350,850 hectares in 2000, and about 500,000 hectares by 2005. With a few exceptions, almost all sprinkler and micro irrigation systems are served by groundwater structures (open dug wells and tube wells). Although the micro irrigated area has grown 15 times during the last 15 years, the rate of adoption is far from its potential. The Ministry of Agriculture has estimated an ultimate potential area (harvested) for micro and sprinkler irrigation as 27 million ha and 42.5 million ha, respectively [10]. The Ministry has proposed to bring 17 million ha under pressurized irrigation in the country, comprising 12 million ha under micro irrigation and 5 million ha under sprinkler irrigation by the end of 11th Five Year Plan period (2007–2012). This is expected to result in an annual water savings of 58.6 billion cubic meters.

1.6.2 UNITED STATES OF AMERICA

The USA has the third largest irrigated area in the world but ranks first in area coverage under pressurized systems. In 1992, 23.5 million ha land was irrigated, of which 54.5% was irrigated by gravity systems, 42.5% by sprinklers and 2.9% by micro irrigation [13]. The Farm and Ranch Irrigation Survey is conducted every five years by the National Agricultural Statistics Service (NASS) of the Department of Agriculture. As per the 2003survey, 21.3 million hectares were irrigated in the USA, of which 12.1 million hectares (57%) were covered by sprinkler and micro irrigation [26]. Although overall irrigated land area has not changed substantially during the past two decades, the area under surface irrigation methods has declined from the 62% in 1979 to 43% in 2003. The micro irrigated has doubled during the last decade—from 0.56 million hectares in 1990 to 1.2 million hectares in 2003.

In USA, wide range of sprinkler systems, ranging from hand-moved to large-scale self-propelled center pivot systems, are used by farmers. The self-propelled center pivot, which gained popularity in the 1960s, has given a fillip to rapid increase in the area under sprinkler irrigation in the country. Today, about 80% of all sprinkler irrigated lands or about 50% of the total irrigated area in the USA uses center pivots.

There has been rising trend towards use of low-pressure center pivots known as "Low Energy Precision Application (LEPA)" systems. The LEPA system has drop tubes fixed on the lateral that extend towards soil surface where a low-pressure bubbler is attached in place of a sprinkler. Water is applied directly to the furrow and not over the crop canopy. Self-propelled sprinkler systems like center pivot and linear move are particularly amenable to high level of automation and effective precision irrigated crop management. Table 7 shows area irrigated by different types of sprinkler, micro and other irrigation methods in USA in 1998, and 2003.

TABLE 7   Area irrigated by different irrigation methods in USA (1998 and 2003).

Irrigation Method	Area irrigated , hectares	
	1998	2003
Sprinklers	10,066,859	10,906,006
Center pivot -	7,568,947	8,620,686
• Low pressure	3,761,952	3,925,883
• Medium pressure	3,003,809	3,909,860
• High pressure	803,186	784,943
Linear move towers	115,286	139,337
Solid set and permanent	495,013	476,904
Side roll	823,411	739,231
Big gun or traveler	310,038	256,351
Hand move	754,164	673,498
Drip/ trickle/ low-flow	914,646	1,209,757
Sub-irrigation	222,532	113,167
Gravity flow	11,041,870	9,361,996
Total	21,800,843	21,590,926

1.7   NON-CONVENTIONAL USE OF SPRINKLER AND MICRO IRRIGATION

Besides conventional use of sprinkler and micro irrigation for field crops and horticultural crops, there are three major nonconventional applications such as: (1) Low cost systems for smallholders, (2) Greenhouse irrigation, and (3) Landscape and park irrigation. Status of these uses is briefly given as follows:

1.7.1   LOW COST MICRO AND SPRINKLER IRRIGATION SYSTEMS

Generally, conventional "state-of-the-art" micro irrigation system is viewed as the technology for large commercial farms engaged in high value agriculture and not appropriate and affordable for small and marginal farm families in many developing countries [18]. The prospectus for the expansion of this technology has brightened with the development of a range of low cost/affordable drip systems fitting to different income level farmers and farm sizes [19, 23].

The so called 'Affordable Micro Irrigation Technologies' (AMITs) are low cost and low pressure systems and have same technical advantages as conventional micro irrigation system but the technology is packaged and marketed as kits suitable for small fields (25 m^2 to 4000 m^2). The AMITs were developed by Chapin Watermatics in the USA and Africa, Underhill International, USA, IDE International in the USA and India [14], Netafim and Ein Tal of Israel. The AMITs include bucket-kits, drum-kits, and customized kits. Irrigation is accomplished through short length laterals fitted with micro tubes, microsprinklers [14]. Sometimes a single lateral is used to irrigate several rows of plants by shifting it manually. An Israeli firm NETAFIM has developed gravity pressurized 'Family Drip Systems' which are marketed in China, India and Africa [15]. In West Africa, a 'Spray-head Irrigation' technology consisting of a small petrol or pedal pump with a conventional lay-flat hose of about 40 mm diameter and a hand-held spraying-head is popular among farmers [13]. The AMIT has the specific advantage of affordability, easy to understand, rapid pay back, divisibility and expandability.

Farms under two hectares represent 98% of the farms in China, 96% in Bangladesh, 95% in Vietnam, 90% in Egypt, 88% in Indonesia, 87% in Ethiopia, 80% in India, 75% in Tanzania, and 74% in Nigeria (22). The AMITs are targeted towards small and marginal farmers in countries of Bangladesh, China Cambodia, Ethiopia, India, Kenya, Myanmar, Mexico, Nepal, Nicaragua, Niger, Sri Lanka, Tanzania, Vietnam, Zambia, and Zimbabwe.

As per the International Development Enterprise, India (IDEI, 14), worldwide over 250,000 smallholders have adopted the low cost drip irrigation system covering about 50,000 hectares (Personal communication). There are about 150 million rural poor families with smallholdings in India. During the last decade, over 100,000 smallholder farmers have adopted the low- cost drip systems. The IDEI envisages reaching 5 million farmers by 2020 in India. The AMITs have unlocked the benefits of pressurized irrigation systems to millions of resource-poor farmers having access to a limited water supply and small land.

1.7.2 GREENHOUSE IRRIGATION

Greenhouses (plastic houses, glasshouses) and high and low tunnels are increasingly used not only in production of vegetable and ornamental crops, but also in variety of plant seedlings in many countries world over. Greenhouses are now better understood as a system of 'Controlled Environment Agriculture' (CEA) with precise control over air and root temperature, water, humidity, plant nutrition, carbon dioxide, and light. In greenhouses water, fertilizers, plant nutrition, insecticides and fumigants are usually applied through drip system.

Countries having major area under greenhouses are USA, China, Spain, Japan, Italy, Greece, Portugal, Hungary, Morocco, Turkey, Algeria, Colombia, and Korea. The largest contiguous area under greenhouses in the world (popularly known as 'Plastic Sea'), with an estimated coverage of 141,700 hectares, is located in Almeria and Murcia regions of the Spain [11]. At the end of the nineties, greenhouses covered about 0.75 million hectares around the world [26]. The growth in area is likely to continue in the foreseeable future.

1.7.3   LANDSCAPE AND PARK IRRIGATION

The micro irrigation has become popular for irrigating lawns gardens/recreational parks in around urban areas, golf courses, commercial, industrial, institutional, and public areas in many countries. Much of the hardware used for landscape irrigation is the same as that used for agricultural irrigation systems. However, many special fittings, piping system, valves, timers and controller that are automatically or remotely operated are added to the irrigation system.

1.8   FUTURE DEVELOPMENTS

The world's population is projected to increase from the present 6.45 billion to an estimated 8 billion by 2030, which will put a tremendous pressure on already stressed water resources. Out of the total world's irrigated area, the 81 developed countries account for about quarter area (69 million ha) and the 93 developing countries share the three-fourth area (202 million ha). The focus of future irrigation development is expected to take place in developing countries due to high demographic growth. The developing countries are expected to expand their irrigated area by 20%, that is, from 202 million ha in 1997/99 to 242 million ha by 2030. Corresponding irrigation water withdrawal however, is expected to grow by about 14% from the current $2,128 \text{ km}^3$ to $2,420 \text{ km}^3$ by 2030 [7]. In future, due to increasing demand from the higher valued domestic and industrial sector, and also from environmental purposes, there will be gradual decline in the amount of overall water used for irrigation. The pressurized systems therefore will play a critical role in meeting the expected irrigated area expansion with diminishing water availability. The irrigation expansion is projected to be the highest in South Asia, East Asia, Near East and North Africa. Most of the increase in the South Asia is expected to occur in India and China. In Sub-Saharan Africa (SSA) and Latin America, irrigated areas are not projected to expand quickly, despite their potential.

In Central and East European Countries (CEEC) the process of water sector restructuring and transfer is in progress and it is hoped that the pace of use of sprinkler and micro irrigation will pick up in near future.

Non availability of local technicians, skilled laborers, equipment manufacturers, suppliers and trained extension personnel; lack of training to farmers are some of the stumbling blocks in rapid expansion of the pressurized irrigation systems in many developing countries. Quality control and standardization will help improve quality of products. Both, the government support in providing institutional and financial assistance and active involvement of private organizations (manufacturers, system suppliers, and designers) are crucial for the further growth of the pressurized irrigation in developing countries.

1.10   CONCLUSIONS

Worldwide, sprinkler and micro irrigation for field crops is practiced in 77 countries on over 39 million ha, comprising 33 million ha under sprinkler and 6 million ha under micro irrigation. Among pressurized systems, sprinkler irrigation has much greater coverage compared to micro irrigation in almost all countries with the exception of Chinese Taipei, Cyprus, Israel, Jordan, and Spain, where the micro irrigated area has out-weighed the sprinkler irrigation. Region-wise spread of sprinkler-irrigated area is: Americas (13.3 Mha), Europe (10.1 Mha), Asia (6.8 Mha), Africa (1.9 Mha), and Oceania (0.9 Mha). The sprinkler system will continue to be popular for irrigating cereals, oilseed crops and fodder crops.

The decade 1990–2000, witnessed a quantum leap in expansion of micro irrigation technology, both in developed and developing countries. In case of micro irrigation, the highest coverage is in Americas (1.9 Mha) followed by Europe and Asia (1.8 Mha each), Africa (0.4 Mha), and Oceania (0.2 Mha). The area under micro irrigation increased almost six fold during last 20 years—from1.1 million ha in 1986 to 6.1 million ha at present. Micro irrigation, owing to its many unique agronomic, water and energy conservation benefits, and low labor requirement, its use is expected to enhance further in the near future.

Low cost and simple pressurized irrigation systems are becoming popular among the small and poor farmers in developing countries. These systems not only help improve crop productivity but also provide other secondary benefits such as income generation, employment opportunities, and food security to large number of smallholders.

There is a need to carry out a comprehensive worldwide survey of sprinkler and micro irrigation to understand the present field level scenario, issues and challenges. Joint efforts of International organizations like ICID, FAO, and the World Bank would be useful in providing much needed vision and support for the sustainable expansion of the sprinkler and micro irrigation technologies worldwide.

1.11 SUMMARY

During the last three decades, sprinkler and micro irrigation systems owing to their capability to apply water efficiently, low labor requirement, and increase in quantity and quality of crop yield/produce have made a breakthrough in many countries around the globe. Uses of sprinkler irrigation for field crops commenced in 1950s and today, varieties of sprinkler systems ranging from simple hand move to large self-propelled systems are in use worldwide. Drip irrigation was first used about 40 years ago but its wide-scale adoption commenced in 1970s when it was used on 56,000 ha.

In this chapter, efforts have been made to compile the worldwide area coverage of both sprinkler and micro irrigation using ICID, FAO databases and other sources. It is seen that at present, the global coverage of sprinkler and micro irrigated areas is 33 million hectares (Mha) and 6 Mha, respectively. Region-wise spread of sprinkler-irrigated area is: Americas (13.3 Mha), Europe (10.1 Mha), Asia (6.8 Mha), Africa (1.9 Mha), and Oceania (0.9 Mha). The top ten sprinkler irrigated countries are: USA, Russia, China, India, France, Brazil, Italy, Spain, Saudi Arabia and Ukraine. These countries together constitute 75% of total sprinkler irrigated area. In case of micro irrigation, the highest coverage is in Americas (1.9 Mha) followed by Europe and Asia (1.8 Mha each), Africa (0.4 Mha), and Oceania (0.2 Mha). The top 10 countries in micro irrigated areas are: USA, Spain, India, China, Italy, Brazil, South Africa, Russia, Mexico, and Saudi Arabia. These countries share 77% of the total micro irrigated area of the world. In five countries viz. Austria, Israel, Libya, Slovak Republic and United Kingdom, irrigation is accomplished solely through pressurized systems.

Besides the micro Irrigation systems that are of the state-of-the-art, low cost drip irrigation systems which are also called as "Affordable Micro irrigation Technologies" (AMITs) are being increasingly used in some developing countries. These systems have same technical advantages as conventional micro irrigation systems and are available as packaged kits suitable for small fields (0.02 to 0.4 ha). Low cost drip systems are used on some 50,000 ha by over 250,000 small holders mainly in developing countries. Micro irrigation is increasingly used in greenhouses and irrigation of lawns, parks and golf courses. The Greenhouses cover about 0.75 Mha around the world and the growth in acreage is likely to continue in the coming years.

This chapter provides information on country and region-wise coverage of sprinkler and micro irrigated areas, a brief about the AMITs and micro irrigation in greenhouses. Case studies are briefly described highlighting growth and trends of sprinkler and micro irrigation in USA and India.

KEYWORDS

- **Affordable Micro Irrigation Technologies**
- **Chinese National Committee on Irrigation and Drainage**
- **FAO**
- **global scenario**
- **gravity irrigation**
- **green house**
- **India**
- **International Commission on Irrigation and Drainage**
- **irrigation**
- **irrigation survey**
- **lawn irrigation**
- **micro irrigation**
- **pressurized irrigation**
- **Soviet Union**
- **Sprinkler irrigation**
- **surface irrigation**
- **USA**
- **water saving**

REFERENCES

1. Cakir, R, Surek H, Aydin H and H. Karaata (1998), Sprinkler Irrigation—A Water Saving Approach in Rice, Proceedings of the 1st Inter-regional Conference on Environment and water, Lisbon, Portugal

2. Chinese National Committee on Irrigation and drainage (CNCID): Irrigation and drainage in China, China Water Power Press, Beijing, 2005

3. Dirksen, W, and W. Huppert (Editors): Irrigation sector Reform in Central and Eastern European Countries, European Regional Working Group of ICID (EWRG), GTZ GmbH, Germany, 2005

4. FAO, 1997a, Irrigation in the Near East Region in Figures, FAO Water Report 9 FAO.

5. FAO. 1997b, Irrigation in the Countries of the Former Soviet Union in Figures, FAO, Water Report 15.

6. FAO, 2003a, Production Yearbook, Vol.27, FAO Statistics Series No.177 FAO.

7. FAO, 2003b, World Agriculture: Towards 2015/2030, An FAO Perspective.

8. FAO, 2005a, Irrigation in Africa in Figures, AQUASTAT Survey 2005, FAO Water Report 29.

9. FAO, 2005b, International Rice Commission Newsletter, 2005, Vol.54

10. Government of India, 2004, Salient Findings and Recommendations of Task Force on Micro Irrigation, Ministry of Agriculture, Department of Agriculture and Cooperation, New Delhi, http://agricoop.nic.in>.

11. Hecht, Albert, 2006, A View from Above: How Spanish Growers Manage to Stay Afloat in the Plastic Sea, International Water & Irrigation Magazine, Vol. 26. No.2

12. Indian National Committee on Irrigation and Drainage, 1998, Sprinkler Irrigation in India, INCID, Ministry of Water Resources, India.

13. International Commission on Irrigation and Drainage, 1993, Proceedings of the Workshop on Micro Irrigation Worldwide, 15th Congress on Irrigation and Drainage, The Hague, The Netherlands.

14. International Development Enterprise, India (IDEI), New Delhi: <http://www.ide-india.org>

15. Israel National Committee on Irrigation and Drainage, 2006, Guidelines for Planning and Design of Micro irrigation in Arid and Semi-Arid Areas (Draft), ISCID, Israel.

16. Jensen, M.H., 2000, Plasticulture in the Global Community—View of the Past and Future, Proceedings of the 15th International Congress for Plastic in Agriculture, 2000, Arizona, USA;<www.plasticulture.org>.

17. Kahlown, M.A and Abdul Raoof, 2005, Growing Rice in the Indus Basin with Sprinkler Irrigation, Proceedings of the 19th ICID Congress, Q.52, R 4.04, Beijing, China.

18. Keller, Jack, 2002, Evolution of Drip/ Micro Irrigation: Traditional and Non-Traditional Uses, Keynote address at the International Meeting on Advances in Drip/ Micro Irrigation, December 2002, Spain

19. Keller, Jack and Andrew Keller, 2005, Mini–irrigation Technologies for Smallholders, Proceeding of the World Water and Environmental Resources Congress, 15–19 May 2005, Alaska, USA.

20. Kulkarni, S.A., 2005, Looking Beyond Eight Sprinklers, Proceedings of the National conference on Micro Irrigation, June 2005, G.B. Pant University of Agriculture and Technology, Pantnagar, Uttaranchal, India

21. Ministry of Water Resources, Government of India, 1999, Report of the National Commission for Integrated Water Resources Development Plan (NCIWRDP), Volume-1

22. Polak, Paul, 2006, From Groundwater to Wealth for One-Acre Farmers, Proceedings of the International Symposium on Groundwater Sustainability (ISGWAS), Spain, 24–27 January 2006.

23. Postel, Sandra, Paul Polak, Fernando Gonzales, and Jack Keller, 2001, Drip Irrigation for Small Farmers—A New Initiative to Alleviate Hunger and Poverty, Water International, IWRA, Vol.26, No.1

24. Reinders, F.B., 2000, Micro Irrigation: A World Overview, Proceedings of the 6th International Micro Irrigation Congress, South Africa, October 2000.

25. The Netherlands Water Partnership, 2003, Smart Water Solutions, The Practica Foundation The World Bank, 2006, Reengaging in Agricultural Water management—Challenges and Options.

26. United States Department of Agriculture (USDA): http//:www.nass.usda.gov/census Worldwide Trends in Greenhouse Technology, NEW AG INTERNATIONAL Magazine, France, June 2003.

APPENDIX I

Sprinkler and Micro Irrigated Area (Arranged in descending order of the total sprinkler and micro irrigated area).

S. No.	Country	Total irrigated area	Sprinkler irrigation	Micro Irrigation	Total sprinkler and micro irrigation	Share in total irrigated area (%)	Year of reporting
		(Mha)	Hectares				
DEVELOPED COUNTRIES							
1	USA	24.7	12,348,178	1,639,676	13,987,854	56.5	2009
2	Spain	3.47	782,508	1,658,317	2,440,825	70.3	2011
3	Italy	2.67	981,163	570,568	1,551,731	58.1	2010
4	France	2.9	1,379,800	103,300	1,483,100	51.1	2011
5	Australia	2.38	690,200	214,200	904,400	38.0	2005
6	Canada	1.053	683,029	6,034	689,063	65.4	2004
7	Germany	0.54	525,000	5,000	530,000	98.1	2005
8	Japan	2.47	430,000	60,000	490,000	19.8	2012
9	Romania	1.5	448,000	4,000	452,000	30.1	2008
10	Slovak Republic	0.313	310,000	2,650	312,650	99.9	2000
11	Hungary	0.22	185,000	7,000	192,000	87.3	2008
12	UK	0.11	105,000	6,000	111,000	100.0	2005
13	Finland	0.07	60,000	10,000	70,000	100.0	2010
14	Portugal	0.63	40,000	25,000	65,000	10.3	1999
15	Bulgaria	0.588	21,000	3,000	24,000	4.1	2008
16	Czech Rep.	0.153	11,000	5,000	16,000	10.5	2007
17	Poland	0.1	5,000	8,000	13,000	13.0	2008
18	Slovenia	0.0073	8,072	733	8,805	100.0	2009
19	Lithuania	0.0044	4,463	—	4,463	100.0	2010
20	Estonia	0.002	100	500	600	30.0	2013
	Sub-total	43.9	19,017,513	4,328,978	23,346,491	53.16	
EMERGING/DEVELOPING COUNTRIES							
1	India	60.9	3,044,940	1,897,280	4,942,220	8.1	2010
2	China	59.3	2,926,710	1,669,270	4,595,980	7.8	2009
3	Uzbekistan	4.186	4,300,000	2,000	4,302,000	100.0	2013
4	Brazil	4.45	2,313,008	327,866	2,640,874	59.3	2006
5	Russia	4.5	2,500,000	47,000	2,547,000	56.6	2012

6	Ukraine	2.2	2,450,000	52,000	2,502,000	100.0	2013
7	Kazakhstan	1.2	1,400,000	17,000	1,417,000	118.1	2013
8	Iran	8.7	744,000	547,000	1,291,000	14.8	2013
9	South Africa	1.67	920,059	365,342	1,285,401	77.0	2007
10	Turkey	5.73	680,000	340,000	1,020,000	17.8	2012
11	Saudi Arabia	1.62	716,000	198,000	914,000	56.4	2004
12	Azerbaijan	1.433	610,000	100	610,100	42.6	2013
13	Mexico	6.2	400,000	200,000	600,000	9.7	1999
14	Korea Rep.	1.010	200,000	400,000	600,000	59.4	2009
15	Egypt	3.42	450,000	104,000	554,000	16.2	2000
16	Israel	0.231	60,000	170,000	230,000	99.6	2000
17	Morocco	1.65	189,750	8,250	198,000	12.0	2003
18	Moldova Rep.	0.228	145,000	15,000	160,000	70.2	2012
19	Syria	1.28	93,000	62,000	155,000	12.1	2000
20	Chile	1.09	16,000	23,000	39,000	3.6	2006
21	Chinese Tai-pei	0.38	18,850	8,750	27,600	7.3	2009
22	Philippines	1.52	7,175	6,635	13,810	0.9	2004
23	Malaysia	0.38	2,000	5,000	7,000	1.8	2009
24	Macedonia	0.055	5,000	1,000	6,000	10.9	2008
	Sub-total	**173.3**	**24,191,492**	**6,466,493**	**30,657,985**	17.69	
LEAST DEVELOPED COUNTRIES							
1	Malawi	0.055	43,193	5,450	48,643	88.4	2000
2	Nepal	1.18	—	—	5,000	0.4	2012
	Sub-total	**1.235**	**43,193**	**5,450**	**53,643**	4.34	

APPENDIX II

The nations of the World.

CHAPTER 2

MICRO IRRIGATION TECHNOLOGY: GLOBAL SCENARIO*

FELIX B. REINDERS

CONTENTS

2.1 Introduction 22
2.2 Irrigated Agriculture in the World 22
2.3 Agricultural Water Usage 24
2.4 Early History of Micro Irrigation 25
2.5 Summary 29
Keywords 29
References 30

*Modified from, "F. B. Reinders, Micro-irrigation: World overview on technology and utilization. International Commission on Irrigation and Draiage, www.icid.org. Keynote address at the opening of the 7th International Micro-Irrigation Congress in Kuala Lumpur, Malaysia, 13–15 September 2006."

2.1 INTRODUCTION

Water gives life and is crucial to development all over the world. It waters the fields; nurtures the crops and stock; provides recreation; it support mines, industry; electricity generation and it provide life for plants and animals that make up ecosystems. Irrigated agriculture plays a major role in the livelihoods of nations all over the world and although irrigation is one of the oldest known agricultural techniques, improvements are still being made in irrigation methods and practices [2]. During the last three decades, micro irrigation systems made mayor advances in technology development and the uptake of the technology increased from 3 Mha in 2000 to more than 6 Mha in 2006. Micro-irrigation is an irrigation method that applies water slowly to the roots of plants, by depositing the water either on the soil surface or directly to the root zone, through a network of valves, pipes, tubing, and emitters (see Fig. 1).

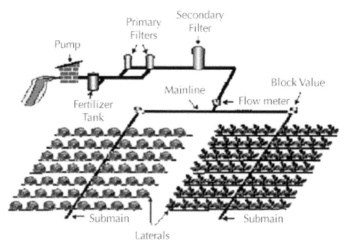

FIGURE 1   Components of a micro irrigation system.

2.2 IRRIGATED AGRICULTURE IN THE WORLD

Irrigation is practiced on more than 299 million ha [3]. As it can be seen in Fig. 2, the top ten countries irrigate 68.5% of the total irrigated area in the world. Four countries (i.e., India, China, USA and Pakistan) have the largest irrigated areas.

Irrigation is an age-old art and in the words of N.D. Gulati of India: "Irrigation in many countries is an old art, as old as civilization, but for the whole world it is a modern science—the science of survival [2]." Although irrigation is of first importance in the more arid regions, yet it is becoming increasingly important in humid regions. Although irrigation is one of the oldest known agricultural techniques, improvements are still being made in irrigation methods and practices. Everyone involved in irrigation has a certain responsibility:

1. the researcher, developer, and supplier for providing practical and useful technology;
2. the designer, who must adapt the design so that it is technically and agriculturally appropriate; and

3. the producer, who must use the system properly and exercise sound irrigation
    practices.

As the population of the world continues to increase, the demand for food and fiber
for the people has also increased. Food production (cereals) in the world is 2,200 million tons (ICID 2011, 3; see Fig. 3) and it is produced on arable and permanent cropped
area of 1,533 × 10⁶ ha (see Fig. 4).

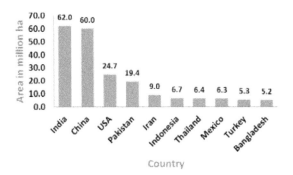

FIGURE 2   Irrigated areas: Top ten countries in the world.

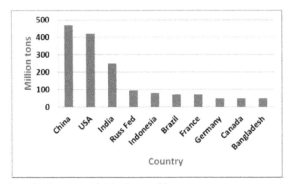

FIGURE 3   Food production (cereals) in the world.

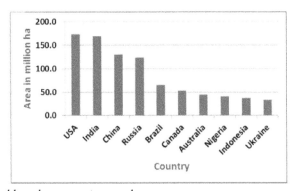

FIGURE 4   Arable and permanent cropped area.

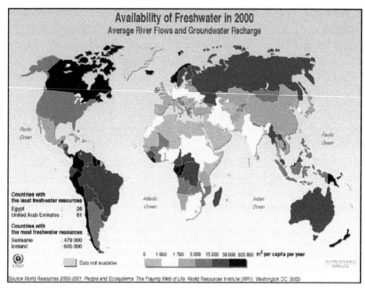

FIGURE 5   Availability of freshwater in 2000 [7].

2.3   AGRICULTURAL WATER USAGE

All life depends on water. The earth is covered with 70% water of which 97% is seawater, 2% the polar ice caps and only 1% is freshwater available for use. In Fig. 5, the availability of freshwater for 2000 is shown in m³ per capita per year. On a global scale, agriculture, industrial and domestic freshwater withdrawal is shown in Fig. 6.

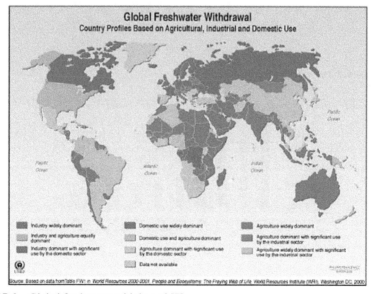

FIGURE 6   Global freshwater withdrawal [7].

The agricultural sector is by far the biggest user of water [1]:

1. In 2000, agriculture accounted for 67% of the world's total water withdrawal and 86% of consumption [6].
2. In Africa and Asia, 85–90% of all the freshwater used is for agriculture.
3. To satisfy global demand for food, by 2025, agriculture expected to increase water requirements by 1.2 times, (industry 1.5 and domestic consumption 1.8).
4. By 2000, 15% of the world's cultivated lands were irrigated, accounting for almost half of the value of global crop production [6].

Today 31 countries in the world face chronic freshwater shortages. Among the countries likely to run short of water in the next 25 years are Ethiopia, India, Kenya, Nigeria and Peru. Parts of other large countries (e.g., China) already face chronic water problems.

2.4   EARLY HISTORY OF MICRO IRRIGATION

Drip irrigation has been used since ancient times by filling buried clay pots with water and allowing the water to gradually seep into the soil. Modern drip irrigation began its development in Germany in 1860 when researchers began experimenting with sub irrigation using clay pipe to create combination irrigation and drainage systems. In 1913, E.B. House at Colorado State University succeeded in applying water to the root zone of plants without raising the water table. Perforated pipe was introduced in Germany in the 1920s; and in 1934, O.E. Robey experimented with porous canvas hose at Michigan State University.

With the advent of modern plastics during and after World War II, major improvements in drip irrigation became possible. Plastic drip tubing and various types of emitters began to be used in the greenhouses of Europe and the United States. A new technology of drip irrigation was then introduced in Israel by Simcha Blass and his son Yeshayahu. Instead of releasing water through tiny holes, blocked easily by tiny particles, water was released through larger and longer passageways by using friction to slow the water flow rate inside a plastic emitter. The first experimental system of this type was established in 1959 in Israel by Blass, where he developed and patented the first practical surface drip irrigation emitter.

The Micro sprayer concept was developed in South Africa to contain the dust on mine heaps. From here, much more advanced developments took place to use it as a method to apply water to mainly agricultural crops.

2.4.1   ADVANTAGES OF MICRO IRRIGATION

1. Sophisticated technology.
2. Maximum production per mega liter of water.
3. Increased crop yields and profits.
4. Improved quality of production.
5. Less fertilizer and weed control costs.
6. Environmentally responsible, with reduced leaching and run-off.
7. Labor saving.
8. Application of small amounts of water more frequent.

2.4.2   DISADVANTAGES OF MICRO IRRIGATION

1. High initial cost: Expensive.
2. Need managerial and technical skills.
3. Waste: The plastic tubing and "tapes" generally last 3–8seasons before being replaced.
4. Clogging.
5. Plant performance: Studies indicate that many plants grow better when leaves are wetted as well.

2.4.3   TECHNOLOGY TRANSFER

With rapid developments in research, it was inevitable that the first micro irrigation congress took place in September 1971 to exchange ideas and principles. Since then, eight congresses have been held as listed in Table 1.

TABLE 1   International Micro Irrigation Congresses over the years [3].

Date	Place and Country	Details
15–23 October 2011	Tehran, Iran	8th International Micro Irrigation Congress **Theme:** Innovation in Technology and Management of Micro Irrigation for Crop Production Enhancement
13–15 September 2006	Kuala Lumpur Malaysia	7th International Micro Irrigation Congress **Theme:** Advances in Micro Irrigation for Optimum Crop Production and Resource Conservation
22–27 October 2000	Cape Town South Africa	6th International Micro Irrigation Congress, **Theme:** Micro Irrigation Technology for Developing Agriculture
2–6 April 1995	Orlando, Florida, USA	5th International Micro Irrigation Congress, **Theme:** Micro Irrigation for a Changing World: Conserving Resources/Preserving the Environment
23–28 October 1988	Albury–Wodonga, Australia	4th International Micro Irrigation Congress
18–21 November 1985	Fresno, California, USA	3rd International Micro Irrigation Congress
7–14 July 1974	San Diego, California, USA	2nd International Micro Irrigation Congress
6–13 September 1971	Tel Aviv, Israel	1st International Micro Irrigation Congress

2.4.4   UTILIZATION OF MICRO IRRIGATION

On-farm water application systems are normally classified in four basic groups (Table 2) and under micro irrigation there are three main irrigation systems: Drip, microjet and microsprinkler.

The large-scale use of drip irrigation system started in 1970s in Australia, Israel, Mexico, New Zealand, South Africa, and USA to irrigate vegetables, orchards and its coverage was reported as 56,000 ha. The micro irrigated area grew slowly but steadily and it was 0.44 Mha in 1981, 1.03 Mha in 1986, 1.83 Mha in 1991, 3.20 Mha in 2000 up to 10.31 Mha in 2012. Over the past 31 years (1981–2012) an increase in the usage of micro irrigation of 2262% took place. The utilization data has been compiled by

ICID [3, 4] mainly from National Committees of ICID, FAO-AQUASTAT, websites and other published documents.

TABLE 2 Classification of irrigation systems.

Irrigation Group	Type of Irrigation System	
Flood (Gravity)	Furrow Border Basin	
Sprinkler: Non-mechanized	Dragline Portable Permanent	Hop-along Big gun floppy
Sprinkler: Mechanized	Side-roll Traveling gun Boom	Centre pivot Linear traveling boom
Micro:	Drip (Surface and Subsurface) Micro jet Micro sprinkler	

TABLE 3 Utilization of micro irrigation.

Year	1981	1986	1991	2000	2006	2010	2012
Area (ha)	436 590	1 030 578	1 826 287	3 201 300	6 089 534	8 400 000	10 310 441
% increase		136.1	77.2	75.3	90.2	37,9	22.7

The decade 1990–2000, witnessed a quantum leap in expansion of micro irrigation technology (See Table 3), both in developed and developing countries. In case of micro irrigation, the highest coverage was in Americas (1.9 Mha) followed by Europe and Asia (1.8 Mha each), Africa (0.4 Mha), and Oceania (0.2 Mha).

The area under micro irrigation increased almost six fold during last 20 years, from 1.03 million ha in 1986 to 6.1 million-ha in 2006 (see Fig. 7).

2.4.5 MICRO IRRIGATION TECHNOLOGY

During the last three decades, micro irrigation systems made significant advances in technology development, such as: From drippers (see Fig. 8), microjets (Fig. 9), filters (Fig. 10) to control equipment (Fig. 11).

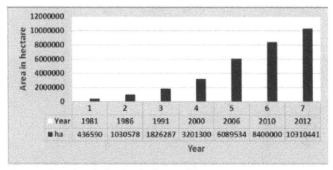

FIGURE 7 Area under micro irrigation in the world.

FIGURE 8   Types of drippers and emitters: On-line and in-line.

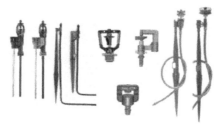

FIGURE 9   Types of mini-sprinklers: Micro and sprayers.

FIGURE 10   Types of filters.

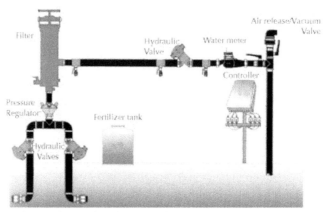

FIGURE 11   Valves and control equipments [5].

> MICRO-IRRIGATION succeeds
> not because it is big
> or because it has been long established
> but because there are people
> building it
> who live it
> sleep it
> dream it
> believe in it and
> build great future plans
> for it.

FIGURE 12   Life must continue trickling on Ref. [3].

2.5 SUMMARY

Micro irrigation (drip/ trickle or localized irrigation) was introduced on commercial scale in the world in the seventies. Micro irrigation is the most efficient method of water application to crops. However, owing to its technicalities involved in design, operation and maintenance, the pace of its adoption has been rather slow. In order to promote the use of micro irrigation on a large-scale, irrigation community worldwide, especially in developed countries, has been hosting "International Micro Irrigation Congress" since 1971. Subsequently, ICID volunteered to organize the event commencing from 5th International Micro Irrigation Congress held at South Africa in 2000 with an objective of creating awareness among its members about latest developments in micro irrigation technology to enhance crop production. Since 2012, the event has been renamed as International Micro Irrigation Symposium.

The success or failure of a micro irrigation system depends to a large extent on careful selection, thorough planning, accurate design and effective management. One thing we can be certain of, the demands of irrigated agriculture will certainly not diminish, they will indeed increase almost exponentially. Surface irrigation will still dominate as the primary irrigation method, but with the current trends, the area under micro irrigation will continue to expand.

KEYWORDS

- **ASABE**
- **drip irrigation**
- **dripper**
- **emitter**
- **gravity irrigation**
- **ICID**
- **irrigation**

- **micro irrigation**
- **micro sprinkler**
- **pressurized irrigation**
- **sprinkler irrigation**
- **subsurface drip irrigation**
- **surface drip irrigation**
- **surface irrigation**

REFERENCES

1. Cook, 2005. Challenge Program on Water and Food. Personal communication.
2. Hagen, R. M., 1967. *Irrigation of Agricultural Lands*. American Society of Agronomy.
3. International Commission on Irrigation and Drainage, New Delhi, India. <http://www.icid.org/index.html>.
4. Kulkarni, S.A., Reinders, F.B, and F. Ligetvari, 2006. Global scenario of sprinkler and micro irrigated areas. 7th International Micro Irrigation Congress in Kuala Lumpur, Malaysia, 13–15 September 2006.
5. Sne, Moshe, 2006. Micro Irrigation in Arid and Semi-arid Regions: Guidelines for Planning and Design. International Commission on Irrigation and Drainage, ISBN. 81-896-10-09-0, pp. 126.
6. UNESCO, 2000.
7. World Resources Institute, Washington D.C. World resources, People and Ecosystems: 2000.

CHAPTER 3

AFFORDABLE MICRO IRRIGATION FOR SMALL FARMS*

JACK KELLER and ANDREW A. KELLER

CONTENTS

3.1   Introduction 32
3.2   Systems for Small Landholders.. 32
3.3   Farmers' Experiences 33
3.4   Expected Longevity of "L-*Cm*-Drip Systems"....................................... 35
3.5   Technical Development and Marketing.. 35
3.6   General Design Strategy.. 36
3.7   Developing the L-*Cm*-Drip System Market Demand.............................. 36
3.8   Specifications and Costs.. 37
8.2   Costs 39
3.9   Performance Characteristics 40
3.10 Design Tools 41
3.11 Summary.. 46
Keywords 46
References.. 47

*This chapter is based on the site visits by the authors to the developing countries. In addition, it covers the status of affordable micro irrigation technology in these countries. The prices of drip irrigation equipment and materials are in Indian Rupees (Rs.) and for the years of their visit. It was presented as paper number 0415 at the 2003 meeting of Irrigation Association, USA.

3.1   INTRODUCTION

It is estimated that three-quarters of the farmers in developing countries cultivate less than 2 hectares (5 acres) of land. For example, a typical farm in Bangladesh supports six people on what they can earn and eat from one acre of land. Typically the family income is only $200 to $300 a year, far too little to afford the modern irrigation devices that are often promoted by development experts. However, without improved irrigation, they cannot gain full access to green revolution inputs. Furthermore, many development experts expect that in an open marketplace, small inefficient farms will be taken over by larger and more efficient farms. But in the face of rapid population growth, actual farm size in developing countries is steadily decreasing! The failure of the development community to take these simple facts into account is a major factor constraining emergence of practical solutions to both improved irrigation performance and to hunger and poverty.

About 20 percent of the 6 billion persons in the world live in families with incomes of less than a dollar a day, and 800 million people regularly go hungry. Roughly 80 percent of this core group of 800 million hungry and poor people live in rural areas in developing countries and earn their livelihoods from agriculture. The green revolution, with its high yielding varieties of seeds combined with access to improved irrigation and fertilizer, tripled global grain production and tripled the incomes derived by farmers with sufficient water supplies and relatively large land holdings. But for the most part it left farmers who only have access to small plots of land and limited water supplies standing on the sidelines [5].

An area of land that can be fully irrigated from a given volume of applied water can be significantly increased by converting from traditional surface irrigation to micro irrigation. Of even greater importance from a basin-wide water resources perspective, the production per unit of water depleted by evaporation, transpiration and salt loading is often increased by 30 to 50%. Furthermore, the availability of micro irrigation systems in small affordable packages unlocks these potential benefits for literally millions of resource-poor farmers.

It opens the potential benefits of irrigation even where water supplies were considered insufficient or too costly to acquire for traditional irrigation methods to be practical. The drip irrigation systems described in this chapter are low-cost, available in small packages, operate at very low pressure, and are easy to understand and maintain. These features are what distinguish them from commercial "state-of-the-art" micro irrigation systems. Compromises are made in operational convenience, manufacturing tolerances, and the uniformity of irrigation applications to achieve these advantages. However, the water conservation and productivity gains from converting from traditional small-scale surface irrigation to low-cost micro irrigation may even be greater than the comparative gains from converting large-scale commercial surface irrigation to state-of-the-art micro irrigation systems.

3.2   SYSTEMS FOR SMALL LANDHOLDERS

A potential micro irrigation customer who had limited access to capital can purchase an expandable drip system capable of irrigating a garden plot of 20 to 100 m^2 at a cost of

$2.00 to $10.00. Poor farmers, who only have small plots, can afford to invest in them [4]. After they gain technical competence and sufficient financial capacity, they can use the profit generated to expand the system. These systems are also affordable enough to be attractive to home gardeners with access to small patches of land adjacent to their dwellings (or elsewhere) to invest in them.

The "drip kits" that irrigate 20 to 40 m² only need a 20-liter (5-gallon) water supply bucket or tank. For intermediate sized drip-kits that irrigate 100 m² to 400 m² (1/10th of an acre) with a 50-to-200 L water supply tank supported about 1.0 meter (3.3 feet) above the ground is sufficient (see Fig. 1). It is practical to manually fill the water supply tank of such drip kits. However, for larger systems a pumped or gravity fed water supply is needed to either periodically fill the tanks, or be directly connected to them. Still, these low-cost systems are only suitable for irrigating plots of 2 hectares (5 acres) or less. **Pressure-type manually operated treadle-pumps** at a cost of $50.00 are easily capable of providing enough water to irrigate up to 2,000 m² (half an acre) of vegetable crops when operated about four hours per day during peak water use periods

As a Board member (Jack Keller) and a Professional Associate (Andrew Keller) of International Development Enterprises (IDE), we periodically provided technical assistance to IDE-India. IDE and IDE-India are nongovernment development organizations (NGOs). IDE with headquarters in Colorado, USA is the international entity and IDE-India is an associated organization that is managed separately and operates mainly in India. A principal objective of IDE-India is to refine, test, and promote *generic* low-cost irrigation technologies. These include low-cost microtube-drip kits (see Fig. 1) and customized microtube-drip systems supplied directly from wells.

This chapter is focused primarily on the customized low-cost microtube-drip (*L-Cm-drip*) systems. We have used a case study approach that covers the following aspects related to them: farmers' experiences and profitability; technology development and marketing; specifications including manufacturing requirements and costs; performance characteristics; and design tools and procedures.

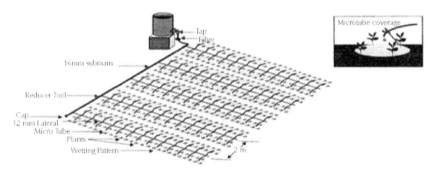

FIGURE 1   Schematic of a low-cost microtubedrip irrigation system.

3.3 FARMERS' EXPERIENCES

In the semiarid region of the state of Maharashtra, India, the average small land holding is less than one hectare (2.47 acres). The following comments are based on discussions that occurred with over 25 purchasers of the low-cost microtube drip (*L-Cm-drip*) systems promoted by IDE-India.

Most of the farmers had previous experience producing vegetable crops (such as tomatoes, eggplant, okra, squash, etc.) using traditional surface irrigation supplied from hand-dug, open wells fitted with electric or diesel-powered pumps. Many smallholders were using portable small diameter (63 to 75 mm) flexible or rigid recycled-plastic pipe to convey the water to the vegetable plots. During the dry season, their wells produced very little water. A typical hand-dug-open well might only produce 5 to 20 m³ per day. This will last for one or two hours of pumping at a rate of 2 to 4 L per second (30 to 60 gpm). The size of vegetable plot ranged from roughly 200 m² (1/20 of an acre) to 2,000 m² (half an acre). All of the farmers, who were interviewed, commented that the conversion to micro irrigation was very cost-effective and many of them had increased annual net return from additional vegetable production that were several times the investment cost.

Farmers reported yield increases of roughly 50 to 100 percent and decreases in water use of 40 to 80 percent compared to experiences with traditional surface irrigation systems. Their gross returns from the typical vegetable crops grown under "*L-Cm-drip* systems" ranged between $0.25 and $1.00 per m² for each crop season, with a typical annual return of roughly $0.50 per m² for a single crop and $1.00 per m² per year for double cropping. The net returns from a given area under double cropping were roughly $0.50 per m² greater under the "*L-Cm-drip* systems" than with the traditional surface irrigation systems. However, in most cases water was a limiting resource and they have been able to double or even triple the irrigated area by converting from surface to "*L-Cm-micro irrigation*," and generate increased net returns of $1.00 per m² from the newly irrigated land. In addition to the increased crop production, they found the *L-Cm-drip* systems to be much easier and less time consuming to operate than traditional surface systems, particularly where water supplies were very limited.

However, the reader is cautioned that by increasing the irrigated area, the total water consumed by crop evapotranspiration will increase proportionally. From a basin-wide water resource perspective, this will not increase the production per unit of water consumed if the so called "losses" from the less efficient traditional surface irrigation were being reused. The losses are only "real losses" if the water is discharged to salt sinks or consumed by salinization or unwanted evaporation and transpiration.

Farmer, B. Harat of Jalina, Maharashtra, had some regular commercial drip-tape with built in emitters. He commented that the uniformity of application from his newly purchased "*L-Cm-drip* system" was not as good. However, he had observed that the water savings and productivity were almost the same. He also noted the sophisticated drip-tape required relatively high pressure heads with careful filtration and special acid treatment to keep it operating properly; However, the "*L-Cm-drip* system" can be operated at much lower pressure heads and it does not need such special care and a specialized filtration system. Therefore, considering its ease of use and low-cost (less than 20% of a sophisticated system cost), he intends to continue buying "*L-Cm-drip* systems" to expand the irrigated area. He also noted that the sophisticated drip-tape would probably last three times as long but it was still not worth the difference in cost.

We observed that microtube clogging was not a problem. (A few farmers elected to simply punch holes in the thin wall drip-tape with a needle or thorn instead of using microtubes as emitters, which did clog.) They were not even experiencing severe problems with clogging with any of the "*L-Cm-drip* systems" with short microtubes of internal diameters between 1.2 and 1.5 mm and with no filtering system. Though the water supply was directly from the open wells, (and good quality water), most of them did not even have simple in-line screens for safety purposes. The few microtube emitters that clogged were simply replaced if flushing did not unclog them.

3.4   EXPECTED LONGEVITY OF "*L-CM-DRIP* SYSTEMS"

We estimate that the longevity of the lay-flat lateral tubing (called drip-tape in this chapter) that IDE-India recommends for "*L-Cm-drip* systems" serving row and vegetable crops is up to four vegetable cropping seasons or roughly two years. The estimated longevity of the heavier wall tubing recommended for horticultural crops is roughly four years. These estimates are based on discussions with IDE-India field agents and the following observations in the field:

- A farmer [The farm is near Jalgon, Maharashtra and it is owned and operated by Mr. Uttam Digambar Bari] with a 1.6-hectare (4-acre) field, has used an *L-Cm-drip* system for three different crop seasons beginning in 2001 (cotton followed by two crops of watermelon). The system was actually operated for a total of 12 months. It had 16 mm diameter drip-tape with a wall thickness of 110 microns (4+ mils). In March 2003, he replaced it with the recommended drip-tape, which has 125-microns (5-mil) wall thickness because the original lateral tubing was mechanically damaged (not because of weathering). He is confident that the new drip-tape will last for four cropping seasons or at least two years. He also installed an *L-Cm-drip* system in a banana planting using the recommended heavier 250-micron (10-mil) drip-tape and is confident it will last for four to five years.
- A farmer (Mr. Narayan Bahi Jeram Bahi Dogariya, whose farm is near Rajkot, Gujarat) used 140-micron drip-tape with hot-punched holes instead of microtubes. He only had a lateral for every six rows to irrigate his one-hectare (2.5-acre) lentil field. The drip-tape was shifted six times per irrigation and he applied four irrigations to the lentils. Thus each lateral was moved dozens of times, but the drip-tape is still in excellent condition with practically no signs of deterioration and he was planning to use it in a similar manner next cropping season.

3.5   TECHNICAL DEVELOPMENT AND MARKETING

The marketing approach to development used by IDE and others is beautifully presented and explained in the Swiss Agency for Development and Cooperation's Report by Heierli [1]. This report includes program descriptions with examples of success along with the performance indicators. This and other reports describing the marketing approach to development as well as smallholder irrigation technologies are available at: http://www.siminet.org/fs_start.htm, the Small Irrigation Market Network (SIMI-net).

The strategy of subsidizing the cost of conventional micro irrigation systems, so farmers with small plots can afford and use them, has generally proven to be unsustainable. It has not been a very efficient mechanism for addressing the needs of smallholders, nor has it resulted in the expected improvements in irrigated agricultural performance. For economically sustainable success, the uptake of micro irrigation systems by smallholders should be demand driven and without direct subsidies. Thus the systems must be financially feasible (or affordable), and farmers should be willing to pay the full ongoing cost (including reasonable profit margins) associated with producing and marketing them once the market demand is well established. There are circumstances resulting from extreme poverty, disasters, and socio/economic/political situations when farmers are unable to pay the "full ongoing costs" and are thus subsidized. But the experience has generally been that system uptake and continued use are not sustained in such circumstances when the subsidies are discontinued.

3.6   GENERAL DESIGN STRATEGY

L-Cm-drip systems provide subsistence farmers with an affordable means for irrigating their small plots in order to reap the associated potential crop production increases and water savings. The general technical and economic criteria IDE-India employed in this successful development effort are:

- The systems are designed around the best available components, with preference given to local manufacturing that only require relatively unsophisticated facilities, but not at the expense of affordability and functionality.
- The assemblage, sales, and service tactics required for the systems are compatible with local microenterprises and require limited skill and capital to design, service, and maintain.
- The systems are designed to optimize economic returns based on the availability and opportunity costs of both local capital (which may be higher than 100% per year) and labor (which is often ≥ $1.00/day).
- The income generation potential of the systems (compared to the systems they replace) at least covers the investment cost in one irrigation season.
- The systems are available in a range of small packages (from as little as 20square meters to a couple of hectares). They are also expandable so that the area served can be enlarged as farmers gain confidence in the technology and become more financially capable.
- The systems are simple and easily understood, operated, and maintained by unsophisticated users.
- The required inlet pressure head for *L-Cm-drip* systems ranges from 1 to 4 meters.
- They provide the potential for high irrigation efficiencies and superior crop yields.

3.7   DEVELOPING THE *L-CM-DRIP* SYSTEM MARKET DEMAND

Rather than direct subsidies, IDE-India provides what are in effect indirect subsidies to farmers by covering the costs of developing and promoting *L-Cm-drip* systems and establishing the market demand for them. The strategy they used prior to establishing the production capacity and implementing the marketing program included the following:

- Identifying promising technologies that had the potential to improve productivity if packaged for small plots of land and affordable to the potential smallholders. While the *L-Cm-drip* systems may appear simple, developing this affordable and user-friendly product line required inventive concepts followed by talented engineering.
- Promising low-cost drip systems were then field-tested and modified to trim cost, increase functionality, and better address field requirements so they would be more acceptable to smallholders.
- The most promising systems were again field-tested and then market-tested prior to initiating supply chain development and promotional programs for them.
- The design strategy is described above, but it essentially never ends because of remaining possibilities for reducing costs and increasing the functionality of *L-Cm-drip* systems in response to suggestions and insights gained from the manufacturers, dealers, assemblers, and farmers who work with them.

This role played by IDE-India using donor funding was necessary since the *L-Cm-drip* system was designed around more or less generic (without patent protection) components. Thus microenterprise manufacturers (or importers and assemblers) and vendors could not have afforded to invest in the necessary product and market development activities because these costs would not be recoverable after the market demand was established. If the innovators attempted to maintain sufficient profits to cover these development costs, competitors would arrive on the scene and undercut their prices. However, this competition is actually a positive aspect of the market creation approach that led to development of the *L-Cm-drip* system. It continuously stimulates inventiveness by attempting to increase cost-effectiveness and functionality.

3.8 SPECIFICATIONS AND COSTS

One might consider the *L-Cm-drip* systems recommended and promoted by IDE-India to be a regression back to systems previously used in the US and elsewhere. This is because the laterals are simple plastic tubing and microtubes are used for emitters. However, this is not the whole story of *L-Cm-drip* systems. They represent refinements by using modern plastic technologies, generic off-the-shelf auxiliary components that have been developed for modern drip systems, and various innovations that use simple manufacturing techniques and increase layout flexibility. The ideas underlying the development of *L-Cm-drip* systems resulted from a blend of:

a) IDE-India's experience using standard wall polyethylene (PE) tubing with long microtube emitters so that a single lateral could serve two crop rows (see Fig. 1); and

b) Some innovative farmers' experiences using drip laterals made from very thin-wall clear plastic tubing produced for packaging a confectionary treat called "Pepcee or Pepsi." Instead of using emitters, holes are punched in the "Pepcee" tubing with a needle. The resulting water application is not uniform and the tubing begins to disintegrate in a few weeks; furthermore, algae grow in it. But these systems are successful because they only cost about \$0.01 per m^2 (\$40 per acre) and last long enough to germinate a cotton crop six weeks before the monsoon rains begin, which increases yields by 25 to 50 percent.

IDE-India recognized the cost advantages of the "Pepcee" systems but they were not suitable for irrigating a vegetable crop for a full season. So IDE-India focused on blending the cost effectiveness of the thin-wall tubing using modern plastic technologies and microtube emitters with relatively large internal diameters in developing the *L-Cm-drip* systems. The lay-flat tubing is a mixture of linear low density PE (LLDPE) and low density PE (LDPE) with carbon black so it is strong and resists stress cracking, deterioration from ultraviolet light, and internal algae buildup. Under low operating pressure heads, the discharge rates from the microtube emitters are about ideal, clogging problems are minimal, and on relatively level small fields the application uniformity is high. Furthermore, the systems are very affordable, with the laterals plus submain (see Fig. 1) costing less than $0.04/m² ($400/hectare or $160/acre) installed.

2.8.1   RECOMMENDED SPECIFICATIONS

The general specifications recommended by IDE-India for *L-Cm-drip* system components are:

- Lay-flat lateral tubing.
- The recommended composition is 80 units of LLDPE, 20 units of LDPE and 2.5 units of Master Batch containing 50% carbon black. Virgin film grade plastic should be used and the tubing scrap should not be recycled in the process, but used for making other products.
- Recommended wall thicknesses:
    — For regular row and vegetable crops use 125 ± 5 microns (5 ± 0.2 mils); and
    — For horticultural crops (such as banana, vines and fruit trees) use 250 ± 5 microns (10 ± 0.2 mils).
- The width of the tubing when flat should be 26 ± 0.5 mm, which gives a minimum inflated inside diameter of 16 mm.

8.1.2   MICROTUBE EMITTERS

- Recommended internal diameters (IDs):
    — For regular row and vegetable crops use I.2 mm ID microtubes; and
    — For horticultural crops, use I.5 mm ID microtubes.
- Recommended microtube lengths:
    — For regular row and vegetable crops use 20-cm (8-inch) long precut microtubes with a tight overhand knot [The overhand knots are tied so about 5 cm of the microtubes are inserted inside of the drip-tape pointing downstream. The knots produce an angular bend in the microtubes so they lay flat along the drip-tape] when a lateral is used for each crop row. The microtubes should be I0-cm (4-inches) longer than half the row-width where two rows are served by each lateral as shown in Fig. 1.
    — For horticultural tree crops use I.0 to I.5 m (3.3 to 5 feet) long tubes. For bananas and papayas use 0.75 to I.5 m (2.5 to 5 foot) long tubes, depending on the plant and drip-tape spacing and microtube layout. (Typically, there are four microtubes per tree for citrus and deciduous tree fruit and only one per plant for bananas and papayas.

- Microtubes must be installed in the field when the drip-tape is inflated and inserted so that 2.5 to 5 cm (1 to 2 inches) are inside the drip-tape with their inlet ends pointing downstream.

8.2 COSTS

On March 2003, the conversion rate for $US to Indian Rupees was: $1.00 = Rs. 46.00. However, in India's rural sector, when the purchasing price parity of a Rupee is considered, Rs. 46.00 will buy roughly $4.00 worth of services. The cost of *L-Cm-drip* systems is very low. Some insights into how this low cost is achieved in India are:

- *L-Cm-drip* tape is a simple continuous tube and with microtube emitters installed in the field rather than standard drip-tape with integral emitters that requires expensive manufacturing machinery. Although installing *L-Cm-drip* is labor intensive, in India a skilled installer can install about 1000 m² (one-fourth of an acre) of *L-Cm-drip* per day at a labor cost of $2.00/day.
- The *L-Cm-drip* tape without emitters and the simple emission devices are economical because they can be manufactured using inexpensive machinery. Furthermore, the rolls of drip-tape and microtubes are very compact and easy to transport, even on a motorbike. The drip-tape is manufactured using the same blow extrusion process used to make plastic films and bags. In India, the locally made extruders range in price from $3,200 to $6,400 and produce from 1.25 to 5 Kg of drip-tape per hour, respectively. They require two operators, each earning roughly $0.25 per hour: one to manually maintain the required width and wall thickness of the drip-tape, and the other to manage and change the take-up reels.
- The cost and profit margins in the supply chain are very low in India. For example, the cost of the virgin plastic in a kilogram (Kg) of the drip-tape is about $1.30 and the cost to the farmer including installation is about $2.60/Kg, only twice the raw material cost.
- Approximately 160 m of 125-micron (5-mil) drip-tape can be extruded form one Kg of plastic, therefore the drip-tape costs 2.60/160 = $1.60/100 m ($0.50/100 ft.). Bulk microtube stock costs about $15.00 per 1,000 meters, which when cut into 20-cm (8-inch) lengths makes 5,000 microtube emitters. Thus the installed cost of drip-tape with 30-cm (12-inch) between microtube emitters is only $2.65/100 m ($0.80/100 ft.).
- The smallholders' clients of IDE-India seldom have fields over one acre and the typical row length ranges from 30 m to 100 m. (Row lengths up to 50 m can be served from one end, but for longer rows, pairs of drip-tape laterals must be fed from a submain laid across and near the middle of the rows). The drip-tape has 16 mm ID and the friction head loss at a given flow rate is only half that of regular 1 mm thick, 16 mm drip tubing with an ID of 14 mm, that is, tubing friction head loss is a function of $1/(ID)^{4.75}$ and $(14/16)^{4.75} = 0.5$.
- The required inlet pressure head is usually only 1 to 3 meters so very simple pipefittings and connections can be used. For example, the submain pipes can be made using recycled plastic and the pipe sections do not need to be glued, which

further reduces costs and makes it easy to rearrange the system after each crop cycle. Some other advantages of low pressure operation include:

— There are no problems with leakage or splitting of the drip-tape at the sub-main connections or where the microtubes are inserted.

— Microtube with wide flow paths are used for the emitters, thus minimizing screening and filtration requirements that are required to eliminate clogging.

— Low quality standard sized irrigation fittings and pipe extruded from recycled plastic can be used for the submains and drip-tape connections to them. For example:

(a) Both flexible semilay-flat and rigid pipes made from recycled plastic are suitable for submains. The 75-mm semilay-flat tubing only costs $0.20/m but only lasts for 2 to 4 years, while 63-mm rigid pipe cost $0.40/m and lasts from 6 to 8 years.

(b) Regular 16-mm grommet connectors that cost $0.03 each are used for connecting the drip-tape to the submains.

3.9 PERFORMANCE CHARACTERISTICS

We recommend that the micro irrigation performance standards, such as those proposed by the American Society of Agricultural Engineers, be relaxed for *L-Cm-drip* systems designed for smallholders in developing countries. Thus we propose the following performance standards be institutionalized and "officially accepted" so smallholders are eligible for bank loans and government assistance to finance *L-Cm-drip* systems:

3.9.1 UNIFORMITY STANDARDS FOR SMALLHOLDER DRIP SYSTEMS

The uniformity of water distribution or Emission Uniformity (EU) has been defined by Keller and Bliesner [3] and it is typically used as a primary measure of the potential performance of micro irrigation systems. EU is dependent on the combined effects of the water supply head available; the elevation differences throughout the irrigated area; the friction losses in the pipe distribution network; and the discharge characteristics and manufacturer's (and assembler's) coefficient of uniformity of the water emission devices. Usually systems serving small plots can be laid out so that elevation differences throughout the irrigated area are relatively small and both ground slopes and flows are in the same direction. Thus "elevation decreases," that increase the available pressure head, can be used to offset pipe friction losses. The "rule of thumb" criteria for designing a subunit of a *L-Cm-drip* system is to try to maintain the pressure head difference due to pipe friction losses and elevation differences between −25% and +50% of the average microtube emitter pressure head, *Ha*.

3.9.2 COEFFICIENT OF VARIATION UNIFORMITY, CVU

We recommend using the *coefficient of variation*, *v*, of the individual emitter discharges from the field test data as the measure of uniformity for postinstallation evaluation of *L-Cm-drip* systems:

$$v = [sd]/[q_a]$$ (1)

where, q_a is the average rate (lph) of catch (i.e., discharge) for the population of emitters; and sd is the estimated standard deviation of the catch rates (lph) of the population. The v is easy to calculate using a hand calculator or a computer spreadsheet program. Furthermore, it is a useful parameter with consistent physical significance for populations of normally distributed data, and it is a generally known and accepted measure of the variability within a population. The physical significance of v is derived from the classic bell-shaped normal distribution curve in which approximately 68% of the catch rates fall within $(1 \pm v)qa$; approximately 95% of the catch rates fall within $(1 \pm 2v)qa$; essentially all of the observed catch rates fall within $(1 \pm 3v)qa$; and the average of the low one-quarter of the catch rates is approximately equal to $(1-1.27v)qa$.

Wu and Barragan [6] have also proposed using the equivalence of CvU for micro irrigation systems and recognized the above relationship between EU and Cv. However, we subscribe to the use of a term we refer to as the "Coefficient of Variation Uniformity (CvU)," as the standard measure of application uniformity for smallholder micro irrigation systems:

$$CvU = 100(1.0-v) \tag{2}$$

The relationship between the field emission uniformity, EU' [3] and CvU for relatively normally distributed field catch data is:

$$EU' \approx 100(1.0-1.27v) \tag{3}$$

Combining Eqs. (2) and (3) gives:

$$EU' \approx [100-\{1.27(100-CvU)\}] \tag{4}$$

3.9.3 RECOMMENDED VALUES OF *CVU* FOR *L-CM-DRIP* SYSTEMS

According to Keller and Bliesner [3], conventional micro irrigation systems serving relatively small fields with uniform topography should be designed to produce EU' values above 85%, which is equivalent to a CvU of 88%. However, they also suggest that for systems serving relatively large fields with undulating topography design EU values as low as 70%, which is equivalent to a CvU of 76%, are acceptable. Therefore, it is clear that system acceptability is dependent on site, equipment selection, and cropping conditions rather than a rigid adherence to fixed EU values. Based on these observations, we recommend adapting the following general performance criteria [4] for field evaluation of *L-Cm-drip* systems serving smallholder plots:

- CvU above 88% is excellent;
- CvU between 88% and 80% is good;
- CvU between 80% and 72% is fair; and
- CvU between 72% and 62% is marginally acceptable.

3.10 DESIGN TOOLS

Many of the people in the supply chain for *L-Cm-drip* systems have little knowledge of pipeline hydraulics and the use of the typical equations engineers use for designing irrigation systems. In view of this, we have developing preengineered design tables that are intuitive and convenient to use. We anticipate that by using these design tools, it will be relatively easy to train inexperienced IDE-India staff and other field personnel as well as assemblers and dealers so they can provide rather expert designs for their farmer clients. The purpose of this section is to provide a sample of types of design tools we are developing for *L-Cm-drip* systems.

3.10.1 *L-CM-DRIP* LATERAL DESIGN TABLES

We developed a unique program for designing *L-Cm-drip* lateral design tables. To use these tables, user needs the input data: The lateral inlet pressure head, *HL;* lateral length, *LL*; and microtube emitter spacing, Se. Development of these tables requires an iterative process to compute the total lateral flow rate, *QL*, for a given *HL, LL,* and Se. Starting with *HL* values, is most convenient for designing *L-Cm-drip* systems since the design strategy is to begin with the system operating pressure head, *HS,* that satisfies the desired system flow rate, *QS*, and uniformity, *EUL,* under the given field conditions. Entering design tables with *HL* may seem unusual for designers in the US because we usually begin our design by computing the required *qa* to meet peak crop water requirements assuming some desired *EU.* Also developing design tables for different *qa* values does not require an iterative solution because if *qa, LL, and Se* are given, then *QL*, and the remaining hydraulic characteristics can be computed directly by assuming $q = qa$ at all emitters. In addition to the lateral design tables for level rows (*see* Tables 1 and 2), we are also developing tabular tools for the design of submains and for positioning submains on sloping fields.

The preengineered *L-Cm-drip* system design tables we have developed for IDE-India are based on the following input variables:

- Lateral inlet pressure head, *HL*, such as: 0.50 m, 0.75 m; I.0 m, I.5 m, 2.0 m; 2.5 m, and 3.0 m;
- Microtube emitter spacing, *Se*, such as: 45 cm (I.5 ft.), 60 cm (2.0 ft.), 75 cm (2.5 ft.), and 90 cm (3 ft.);
- Lateral length, *LL*, such as: 20 m, 30 m; 40 m, and 50 m;
- Lateral drip-tape inside diameter, which is 16.00 mm for I25-micron drip-tape and 15.75 mm for 250-micron drip-tape;
- Minor loss due to the insertion of the microtube emitters into the drip-tape, which is expressed as an equivalent length of 0.1 m;
- Microtube emitter pressure head/discharge relation based on bench tests and entered either as a curve or   equation; and
- Coefficient of variation of the microtube emitters, *v*, based on bench test data.

The following information is developed for each of the above lateral configurations assuming the *L-Cm-drip* laterals are lying along crop rows that are nearly level, i.e. zero slope:

- Total lateral discharge, *QL*, in liters per minute (lpm);
- Average emission device discharge, qa, in liters per hour (lph);
- Lateral friction head loss, *hf*, in meters (m); and
- Design emission uniformity of the *L-Cm-drip* lateral, *EUL*, as a percentage, %.

The *EUL* is computed using the same equation for the design emission uniformity developed by Keller et al. [2] and Keller and Bliesner [3], which is commonly used for micro irrigation system design purposes in the US and elsewhere. However, *EUL* is a metric for lateral uniformity rather than for the whole system or subunit of a system, which would include pressure variations due to elevation differences along the laterals as well as pressure differences along the submain supplying them. The equation for computing *EUL* is:

$$EU_L = [100(1.00-1.27\ v)]\ [q_n/q_a] \tag{5}$$

where, EU_L is the design emission uniformity, %; v is the microtube emitter coefficient of variation; q_n is the minimum microtube emission rate computed from the minimum pressure along the lateral line based on the emitter's nominal flow rate versus pressure curve, lph; and q_a is the average microtube emission rate, lph.

TABLE 1   Row and vegetable microtube drip-tape hydraulic design tables for laterals on zero slope: Lateral/row length of 20 meters and 30 meters.

Inlet Head, HL	Microtube Spacing, Se	Lateral/row length, 20 m				Lateral/row length, 30 m			
		QL	qa	hf	EUL	QL	qa	hf	EUL
m	cm	lpm	lph	m	%	lpm	lph	m	%
0.50	45	2.17	2.96	0.04	94	3.04	2.72	0.11	91
	60	1.66	3.01	0.02	95	2.38	2.86	0.07	93
	75	1.37	3.04	0.02	95	1.96	2.93	0.05	94
	90	1.12	3.06	0.01	96	1.64	2.98	0.03	95
0.75	45	2.91	3.97	0.07	94	4.05	3.63	0.18	91
	60	2.23	4.06	0.04	95	3.19	3.83	0.11	93
	75	1.84	4.09	0.03	95	2.63	3.94	0.08	94
	90	1.51	4.12	0.02	96	2.20	4.01	0.06	95
1.00	45	3.59	4.90	0.10	94	4.96	4.44	0.25	90
	60	2.76	5.01	0.06	95	3.93	4.71	0.16	93
	75	2.28	5.06	0.04	95	3.23	4.85	0.11	94
	90	1.87	5.10	0.03	96	2.72	4.94	0.08	94
1.50	45	4.82	6.58	0.16	94	6.60	5.91	0.41	90
	60	3.71	6.74	0.10	95	5.25	6.30	0.27	92
	75	3.07	6.81	0.07	95	4.34	6.50	0.19	94
	90	2.52	6.87	0.05	95	3.65	6.64	0.14	94
2.00	45	5.94	8.10	0.23	94	8.07	7.23	0.58	89
	60	4.58	8.32	0.14	95	6.44	7.73	0.38	92
	75	3.79	8.42	0.10	95	5.34	8.01	0.27	93
	90	3.11	8.49	0.07	95	4.50	8.19	0.20	94
2.50	45	6.98	9.52	0.31	94	9.43	8.45	0.76	89
	60	5.39	9.79	0.19	95	7.55	9.06	0.50	92
	75	4.46	9.91	0.14	95	6.27	9.40	0.36	93
	90	3.67	10.01	0.10	95	5.29	9.62	0.26	94

TABLE 1   *(Continued)*

Inlet Head, HL	Microtube Spacing, Se	Lateral/row length, 20 m				Lateral/row length, 30 m			
		QL	qa	hf	EUL	QL	qa	hf	EUL
m	cm	lpm	lph	m	%	lpm	lph	m	%
3.00	45	7.96	10.86	0.39	93	10.71	9.59	0.94	89
	60	6.15	11.19	0.24	95	8.60	10.32	0.63	92
	75	5.10	11.33	0.17	95	7.15	10.72	0.45	93
	90	4.20	11.45	0.12	95	6.04	10.99	0.33	94
4.00	45	9.80	13.36	0.56	93	13.08	11.71	1.33	88
	60	7.59	13.80	0.35	94	10.54	12.65	0.90	91
	75	6.30	13.99	0.25	95	8.78	13.18	0.65	93
	90	5.19	14.15	0.18	95	7.44	13.53	0.48	94
5.00	45	11.51	15.69	0.74	93	15.26	13.67	1.74	88

TABLE 2   Row and vegetable microtube drip-tape hydraulic design tables for laterals on zero slope: Lateral/row length of 40 meters and 50 meters.

Inlet Head, HL	Microtube Spacing, Se	Lateral/row length, 40 m				Lateral/row length, 50 m			
		QL	qa	hf	EUL	QL	qa	hf	EUL
m	cm	lpm	lph	m	%	lpm	lph	m	%
0.50	45	3.62	2.44	0.18	87	3.98	2.15	0.26	81
	60	2.95	2.65	0.13	90	3.34	2.41	0.19	87
	75	2.45	2.78	0.09	92	2.88	2.58	0.15	89
	90	2.10	2.86	0.07	93	2.51	2.69	0.12	91
0.75	45	4.78	3.22	0.30	86	5.22	2.82	0.41	80
	60	3.93	3.52	0.21	90	4.41	3.19	0.30	86
	75	3.28	3.71	0.15	92	3.81	3.42	0.24	89
	90	2.81	3.83	0.11	93	3.34	3.58	0.19	90
1.00	45	5.82	3.92	0.41	85	6.32	3.42	0.56	79
	60	4.80	4.30	0.30	89	5.37	3.88	0.43	85
	75	4.02	4.55	0.21	91	4.66	4.17	0.34	88
	90	3.45	4.71	0.16	93	4.09	4.38	0.27	90
1.50	45	7.67	5.17	0.67	84	8.28	4.47	0.89	77
	60	6.37	5.70	0.48	88	7.07	5.11	0.68	84
	75	5.36	6.07	0.35	91	6.16	5.52	0.55	87
	90	4.61	6.29	0.27	92	5.43	5.82	0.44	89

TABLE 1   *(Continued)*

Inlet Head, HL	Microtube Spacing, Se	Lateral/row length, 40 m				Lateral/row length, 50 m			
		QL	qa	hf	EUL	QL	qa	hf	EUL
m	cm	lpm	lph	m	%	lpm	lph	m	%
2.00	45	9.32	6.28	0.93	83	10.01	5.41	1.22	76
	60	7.77	6.96	0.68	88	8.59	6.21	0.95	83
	75	6.56	7.43	0.50	90	7.51	6.73	0.77	86
	90	5.66	7.72	0.39	92	6.64	7.11	0.62	89
2.50	45	10.83	7.30	1.20	83	11.60	6.27	1.57	75
	60	9.07	8.12	0.89	87	9.98	7.22	1.23	82
	75	7.68	8.69	0.66	90	8.75	7.84	1.00	86
	90	6.64	9.05	0.51	92	7.75	8.30	0.81	88
3.00	45	12.25	8.26	1.48	82	13.08	7.07	1.92	75
	60	10.29	9.21	1.10	87	11.29	8.16	1.52	82
	75	8.73	9.88	0.82	90	9.91	8.88	1.24	85
	90	7.56	10.30	0.64	92	8.79	9.42	1.01	88
4.00	45	14.86	10.02	2.06	81	15.80	8.54	2.64	73
	60	12.53	11.22	1.55	86	13.69	9.90	2.11	81
	75	10.67	12.08	1.16	89	12.06	10.80	1.73	85
	90	9.27	12.64	0.90	91	10.73	11.49	1.43	87
5.00	45	17.26	11.63	2.66	80	18.28	9.88	3.37	72

Tables 1 and 2 show preengineered lateral for an *L-Cm-drip* systems based on an average $v = 0.031$. The pressure head versus emitter discharge curves were determined for a series of bench tests at pressure heads ranging from 0.5 to 2.0 m for 20-cm (8-inch) long microtube emitters with an internal diameter (ID) of 1.2 mm installed in drip-tape. The ID used for the drip-tape laterals was 16.0 mm and the microtube/drip-tape connection loss was assumed to be equal to an equivalent length of 0.1 m of drip-tape.

To use these tables for designing an *L-Cm-drip* system the designer enters them with the plant spacing and lateral length that fits the field size and shape. Then the designer searches for the combination of *HL, QL,* and *EUL* that appears most reasonable for the field layout while considering the following:

- Microtube emitters with large unobstructed passageways have relatively high discharges even under very low operating pressure heads. Therefore, it is not practical to use laterals much longer than 50 m for *L-Cm-drip* systems. If rows are longer than 50 m the submain must be laid out to bisect the field rather than being placed at the head of the rows. If rows are longer than 100 m, a second submain is needed.
- The ideal inlet pressure head for *L-Cm-drip* systems is between 1 and 3 m.

- The system discharge, QS, is equal to the lateral discharge, QL, times the number of laterals along the submain. Thus to find a reasonable QL, divide the available QS by the number of laterals that will be required to irrigate the field. If the available QS is insufficient irrigate one-half, one-third, or one-quarter, etc. of the field at a time.
- It is desirable to have the design emission uniformity, EUL, as high as practical. Assume that in view of minor elevation variations and losses in the submain, the CvU of field test data will be close to the EUL. Thus the designer can use EUL values from the tables in place of recommended CvU presented earlier as a guide to evaluating the anticipated system performance with the lateral configuration selected.

3.11 SUMMARY

Many farmers in developing countries use conventional (surface or gravity) irrigation to irrigate their small fields with water from wells that have small average discharges. They are being assisted in converting to very low-cost drip systems that are nonpropri-etary and manufactured locally. These drip systems are affordable and payback is quick in view of the reported water savings of 50 to 80 percent and yield increases of 30 to 50 percent. The systems operate at pressure heads of less than 3 meters (10 feet), require minimal filtration, use cheap recycled plastic submains (manifolds), and use simple drip-tape with short lengths of microtubes for emitters. This presentation covers the following aspects of these low-cost drip systems: a) farmers' experiences and profit-ability; b) technology development and marketing assistance; c) system specifications and component costs; d) local manufacturing requirements and costs; e) system perfor-mance characteristics; and f) design tools and procedures using design tables.

KEYWORDS

- affordable micro irrigation
- coefficient of variation
- design emission uniformity
- drip irrigation
- drip irrigation potential
- drip irrigation technology, low cost
- emitter
- emitter flow rate
- emitter spacing
- Food and Agriculture Organization, FAO
- India
- Indian Rupees, Rs.
- International Development Enterprises, IDE

- **Irrigation**
- **hydraulic tables**
- **micro irrigation, MI**
- **microtube**
- **million-hectares, Mha**
- **poor-friendly technology**
- **smallholder community**
- **surface irrigation**
- *uniformity coefficient*

REFERENCES

1. Heierli, U. 2000. Poverty alleviation as a business: the market creation approach to development. *Swiss Agency for Development and Cooperation.* Berne, Switzerland.
2. Keller, J. and D. Karmeli. 1974. Trickle irrigation design parameters. *Transactions of the ASCE.* Vol. 17, No. 4: 678–684.
3. Keller, J., and R.D. Bliesner. 1990. *Sprinkler and Trickle Irrigation.* New York, N. Y.: Van-Nostrand Reinhold. (Reprinted. Caldwell N.J. (The book has been reprinted and is available at: www.BlackburnPress.com
4. Keller, J., D. L. Adhikari, M. R. Petersen, and S. Suryawanshi. 2001. Engineering low-cost micro irrigation for small plots. Report No. 5. Journal of International Development Enterprises, Longmont, Colorado (This can be downloaded from: http://www.siminet.org/fs_start.htm).
5. Postel, S., P. Polak, F. Gonzales, and J. Keller. 2001. Micro irrigation for small farmers: a new initiative to alleviate hunger and poverty. *Water International* Journal of the International Water Research Association. Vol. 26, No.1: 3–13.
6. Wu, I.P., and J. Barragan. 2000. Design criteria for micro irrigation systems. *Transactions of the ASAE.* Vol. 43 No. 5. pp. 1145–1154.

CHAPTER 4

MICRO IRRIGATION POTENTIAL IN THE DEVELOPING COUNTRIES*

TUSHAAR SHAH and JACK KELLER

CONTENTS

4.1   Introduction 50

4.2   Evolution of Micro-Irrigation Program at Four Locations 51

4.3   Impact on Livelihoods, Water Productivity, Environment: Early Impressions 53

4.4   Adopters and Adoption Behavior 57

4.5   Market Dynamics of Micro Irrigation 65

4.6   Assessment and Future Challenges 68

4.7   Summary 71

Keywords 72

References 72

Appendix I 73

Appendix II 75

Appendix III 75

Appendix IV 76

---

*This chapter is based on the site visits by authors to: Indian States: Gujarat, Karnataka and Madhya Pradesh; and three hill districts in Nepal and Kenya. In addition, it covers the status of micro irrigation technology in these countries during the years of my visit. The prices of drip irrigation equipment and materials are in Indian Rupees (Rs.) and for the years of their visit.

Modified from, "*Shah, Tushaar and Jack Keller. Micro-Irrigation and the Poor: A Marketing Challenge in Smallholder Irrigation Development. Technical report prepared for International Development Enterprises (IDE),* http://www.siminet.org/fs_start.htm"

4.1 INTRODUCTION

Micro-irrigation technologies—drip and sprinkler based systems—were first perfected in Israel during the 1960s and have spread to many other parts of the world, especially in the USA. These seem particularly suited to conditions in water-scarce regions such as western and southern India and North China. However, introduced first in the 1970s, the total area under drip irrigation in India has expanded to just around 60,000 ha against the ultimate potential of 145 million-hectares [Mha]. Of these, 40,000 ha are in Maharashtra where it is extensively used in grape and orange orchards; the bulk of the rest are in Tamil Nadu and Karnataka (Sivanappan, Chapter 6). Drip irrigation is a big success for citrus and orange orchards and grape in Maharashtra; for coconut in Coimbatore (Tamil Nadu); and for mulberry in Kolar (Karnataka). However, despite proactive promotion by a growing private irrigation equipment industry and subsidies provided by governments, the appeal of these technologies has remained confined to 'gentlemen farmers [Large farm holders]. A common perception that has held sway over the popular mind is that drip and sprinkler irrigation require: High initial capital; high maintenance cost; technical/managerial skills; high skilled labor; and sophisticated scientific knowledge.

In recent years, there have been efforts to promote and mainstream a nearly opposite notion that these technologies are particularly suited to very small farms and poor farmers; that, for small plots, they require surprisingly little capital; they are easy to manage and, in fact, save labor; and most importantly, these can significantly enhance productivity of land and water, quality of the produce and the farm income of the adopter household. Pioneering efforts made in this direction are by: US based Chapin drip industry; International Development Enterprises [8, 9]; Jain Irrigation; Netafim, the Israeli irrigation equipment major and some other players. All these have developed and launched 'miniaturized' versions of drip and sprinkler systems adapted to small vegetable gardens. Best known are bucket and drum kits promoted by Chapin mostly in Southern Africa and by IDE in India, Nepal and several African countries. Particularly with the IDE, the focus has been on cutting the cost of the technology so that poor men, marginal farmers and women farmers can afford it without subsidy. One estimate indicates that some 13,000 IDE bucket and drip kits were already in use by small holders in Asia and Africa; and the potential seems great. A larger global initiative is already in the making for "*scaling up poverty-oriented micro irrigation by creating a global dissemination network:* http://www.siminet.org."

This chapter includes our views based on one-month fieldwork in Gujarat, Karnataka and Madhya Pradesh in India; and three hill districts in Nepal and Kenya. This chapter attempts a first-cut assessment of the potential of the drip irrigation technology, its social impacts, and issues in 'scaling it up.' Although this chapter does not intend to make a definitive statement, yet presents our impressions in the manner of 'field-notes.'

IDE's micro irrigation program is barely five years old; and in many regions, adopters have not tried the technology long enough to realize its full benefits and constraints [8, 9]. There is therefore an element of speculation even in the broad qualitative assessment we offer. It will take some more years for the technology to 'sink' before it is ready for a proper and full-scale assessment.

4.2 EVOLUTION OF MICRO-IRRIGATION PROGRAM AT FOUR LOCATIONS

In all the four locations, the marketing environment—and therefore, the IDE approach—have evolved differently. In Gujarat and Nepal, the micro irrigation (MI) program is operating in a developmental mode with IDE being the only player in small-scale MI market in selling the 'concept' of micro irrigation to small farmers. In Maikaal (Madhya Pradesh, India) and Kolar (Karnataka, India), the scene is different; here, IDE and its partners are among the several mainstream players in the MI business; and since they are all in marketing custom-built systems, the distinctive aspect of poverty-focused MI is somewhat diluted.

The history of the program has also had some impact on the current status. In Gujarat, for example, IDE's MI program is barely two years old; and we found it difficult to find farmers who had completed one full cropping cycle using MI technology. It is implemented in the dry region of Saurashtra and Baroda/Panchmahal districts of North-Eastern Gujarat, India. Some 600 bucket and drum kits were in use here. Kits have been moving at a rate of 150–200/year; but the pace has been accelerating. IDE's marketing organization in Gujarat is simple and thin. Until recently, the IDE itself played distributor and got supplies from a Nasik-based manufacturer. A distributor was appointed just before our visit. He caters to some eight 'Assemblers' who are the dealers. The assembler sells ready drum and bucket kits as well as custom-built systems for specific requirements of each former. Within Gujarat too, we found that the focus of IDE effort in Saurashtra (Gujarat) is to promote low-cost, mostly custom-built drip irrigation through the assembler who does most of the extension, promotion and custom designing. In Chhota Udepur (Gujarat), the target market is dominated by poor tribal women; and standardized MI kits are marketed to them for kitchen gardens.

Nepal's MI program is focused squarely on the poorest segments and on standardized drum and bucket kits. An early assumption of the program managers here was that farmers shifting the pipes around could cut costs of drip technology drastically. However, the promotional work with farmers suggested that they did not quite like shifting the pipes around. Indeed, a Unique Selling Proposition (USP) of the MI technology is that it saves labor and on-farm water management effort. If a drip system is designed so that it has to be frequently shifted around, this USP is lost. So ultimately, IDE Nepal designed and put on the market a proper drip system in three sizes. IDE Nepal has grounded some 3200 kits in around 450 villages in the Nepal hills. They have also launched the microsprinkler, which is probably getting more popular in Nepal hills as well as in Himachal Pradesh on the Indian side of the Himalaya's.

In Kolar town—Karnataka, we met Pragathi Enterprises, a dealer who doubles up for Primere Irrigation as well as for KB (Krishak Bandhu) Micro-irrigation. He uses Primere material for building drip systems sold under the KB brand name. He sold 180 acres for drip irrigation systems for mulberry to 80 customers and 400 acres of 'vegetable systems' to 50–100 farmers essentially for mango and coconut orchards. In his 80 mulberry customers, 5–6 are large; but 75 are *relatively* small farmers with 1 to 1.5 acre under drip-irrigated mulberry. In Kolar town, IDE has been promoting MI for nearly a decade. The focus of IDE's promotional effort in this town is on custom-built drip systems mostly for mulberry farmers but also for commercial orchards.

There is hardly any sale of bucket or drum kits; nor has horticulture emerged yet as a major customer (as probably it has in Andhra Pradesh, which we could not visit). So in Karnataka, IDE is in a primarily promotional role for the drip irrigation industry as a whole. The costs of IDE products too are comparable to those of the mainstream players though they vary hugely; for horticulture, the cost of laying a drip system is Rs. 7,000 to 8,000/acre; for mulberry, it is Rs. 20,000–25,000 for paired row system and Rs. 20,000/acre for the pit system. Costs also vary according to make; KB systems made with Jain Irrigation material cost Rs. 2,000–3,000 more per acre compared to Pioneer, Krishi, Telecom and other brands. Micro-tube technology has been popular and is now becoming increasingly so. Jain and Pioneer, two leading suppliers aggressively have promoted microtube systems for decades, in any case, long before IDE came in to promote them. Kolar is a major center for promoting drip irrigation. It has 70,000 acres under mulberry within 40 km radius of Kolar town. In principle, it can be a major thrust region for IDE's Micro-irrigation program. The issue is: to what end. There is some confusion about what exactly is the role of the IDE here. It does not market cheaper systems; it does not market smaller systems; it does not market primarily to the poor; and it is not the only one to promote the microtube system. So, 'what business are we in' is the key strategy issue for IDE here?

In Madhya Pradesh, the micro irrigation program has evolved still differently. In late 1990s, IDE began to work with Mikaal Cotton Spinning Company, an Indo-Swiss Collaborative Company, and its development NGO BioRe promoting bio-cotton cultivation around the Maheshwar area (the site of one of the Narmada dams) in the Maikaal region of Madhya Pradesh. In this dry, hilly terrain, cotton has been cultivated mostly with well irrigation. However, with mushroom growth in wells and pumps, well yields are dropping and in dry months of summer, most wells turn very dry. Some dynamic farmers had already begun trying out the drip irrigation technology in cotton. In a short cooperation, IDE encouraged Maikaal's member farmers to experiment with the microtube technology for drip irrigation on 25 acres. For some reason, IDE was moved out of the region soon thereafter; however, the seed of drip irrigation it has sown here has blossomed and borne fruit.

Some 1500 acres of Maikaal Cotton's bio-cotton area is already under drip. BioRe initiated a scheme to install drip systems on farmers' fields: the advantage to the farmer is that BioRe buys tubes and laterals in bulk to get a good price; second, farmer gets an interest-free 3-year loan. Many small farmers are taking up the offer by BioRe. There are indications all around that the drip technology is being rapidly internalized by farmers and is on the verge of taking off in a big way in this region through commercial channels. The best indicator of this is that the farmers have begun to play around with the material as well as the design on their own.

Therefore, we can affirm that the MI program in "Chhota Udaipur in Gujarat" and Nepal has evolved quite differently than in Saurashtra, Kolar and Maikaal as shown below:

MI program in "Chhota Udaipur	MI program in Saurashtra, Kolar and Maikaal
Has been engaging primarily with very small-holder, mostly women farmers.	Has primarily reached the middle-peasantry.
Is heavily into promoting 2–3 standard configurations of bucket and drum kits.	Is primarily into promoting the custom-built systems.
IDE is primarily playing a development NGO with little or no sign of other market players on the horizon trying to get a cut in the business.	The playing field is dominated by mainstream players, and the distinctive role of IDE as well as of micro irrigation technology awaits sharper definition.

4.3 IMPACT ON LIVELIHOODS, WATER PRODUCTIVITY, ENVIRONMENT: EARLY IMPRESSIONS

The beneficial impacts of drip and sprinkler irrigation in water-stressed regions have been widely studied in Israel, US and many other countries where commercial farmers have taken to it on a large scale. Even in India, several researchers have highlighted the benefits of the technology. For example, Sivanappan (in Chapter 6) suggests that based on field trials conducted at various Indian agricultural universities, MI reduces water application by 40–70 percent and raises crop yields by 200 percent for most of the crops. It permits efficient saline irrigation since salt gets accumulated only at the surface of the periphery of the wetting zone without affecting crop growth. Like many other researchers have, in a survey of 160 farmers in Maharashtra, Narayanamoorthy [13] found that drip irrigation cuts cost of cultivation especially in inputs like fertilizers, labor, tilling and weeding. The yield of drip irrigated banana and grapes was estimated to be 52 and 23 percent higher compared to flood irrigation. Benefit-cost ratio of investment in MI is estimated to be 13 without taking into account the value of water saved and 32 with water saving accounted for in the benefit-cost calculation. Net profit per hectare of MI over conventional irrigation is 100,000 rupees for grapes compared to 87,000 rupees for banana crop. Unlike flood irrigation, drip irrigation works in undulating topography. Despite these advantages, the moot question is: Why is MI technology spreading so slowly? According to Narayanamoorthy [14], it is because of high capital cost, absent or inadequate subsidy, poor product quality and lack of awareness and knowledge of the farmer. Above all, the notion which holds powerful sway over the industry—that MI is appropriate only for large commercial farmers with resources and farm management skills—had led industry leaders to offer relatively expensive products designed only for large commercial farmers. IDE's MI program is a major breakthrough because it has down-sized, simplified and demystified the drip and sprinkler irrigation technology for targeting it to the ultra-poor.

To the commercial mulberry farmers in Kolar and cotton farmers in Maikaal, the impact of productivity of MI—in particular, producing quality crops under extreme moisture stress—was of paramount interest. In Kolar, we interviewed the mulberry farmers who listed a number of advantages of the mulberry drip-irrigated versus the mulberry flood-irrigated:

- Water needed for half acre of flood irrigation will suffice for 2 acres of drip irrigation;
- Labor requirement is drastically reduced due to low weed growth;

- Drip irrigation itself requires far less labor and management than flood irrigation;
- The plant population and health are better; and
- Larger area that can be irrigated from available power supply.

In many parts of India, shortage of power is the binding constraint rather than water availability or the cost of pumping, which at the margin is zero for bore well owners under flat system of electricity tariff. In Kolar, for instance, farmers get 4 h of power during the day and 4 h in the night; they use night power to fill up their farm ponds and tanks that are used for irrigation during the day time.

Besides these direct, private benefits to adopters, Professor Sundar at the University of Agriculture, Karnataka (www.uasbangalore.edu.in) enumerated several other indirect, social benefits of MI:

- It reduces soil erosion and nonpoint pollution because MI water percolates only to 45–60 cm;
- Fertilizers and pesticide residues do not mix with the water table;
- It promotes more efficient use of nutrients;
- It ensures better and longer moisture retention in the root zone; and
- MI is a powerful instrument of drought proofing.

Overall, then, in Maikaal and Kolar, the gains from MI technology seemed convincingly established; here, the ground is ready for major up-scaling. However, we could not make a firm assessment of many low-end adopters to whom the IDE program is targeted. It was only in Nepal that we could make a firm assessment of the livelihood impact of the MI program; and the evidence we gathered here validated the high expectations from the program in terms of livelihood impacts.

In the IDE parlance, the term MI implies drip and sprinkler irrigation technologies down-sized in scale and costs to suit very small and marginal farmers' needs and financial capacity. Studies are beginning to show that all the benefits that commercial drip and sprinkler irrigation confer on their users accrue to small and marginal farmers who take to MI. In a study of IDE's MI program for poor women vegetable farmers in Aurangabad and Bijapur districts of Maharashtra, Bilgi [2] concluded that a typical MI kit resulted in 55% savings in water applied, 58% decline in labor-days, 16% savings in fertilizer and pesticide use, 97% increase in output and 142% increase in gross income. We wanted to explore if gains of this scale were experienced by women microirrigators we met in Gujarat and Nepal. Gujarat, as we mentioned earlier, offered little understanding since most of the MI kits were nonoperational because of heavy out-migration of tribal families due to drought. However, our experience in Nepal suggested livelihood grains of the order Bilgi [2] found in Maharashtra. Women we met in Nepal hills had all been growing some vegetables earlier; but they took only one crop during the rainy season. Many households ate meals without vegetables for days; they grew a few plants mostly for family consumption; they seldom or never had vegetables to sell on the market; instead, most spent Rs. 900–1,200/year on the purchase of vegetables earlier; and the quality and size of their rain-fed crops were far from satisfying. The drip-kit changed all these in significant ways. Adopters began growing drip-irrigated vegetables in winter and summer while continuing to grow rain-fed vegetables during the rainy season; they all grew a variety of vegetables (Okra, bottle gourd, sponge gourd, snake gourd, pumpkin, tomato, and chili), they grew vegetables on a larger *net* area; their crop was better in size as well as quality. Eating vegetables daily became the habit of most families. Before the drip came, only four confessed to selling any vegetables; now, they all became net sellers of vegetables; while their purchase of vegetables declined sharply, their sales increased to Rs. 2000–15000/year. In sum, the 30 odd adopters whom we met at Kahun, Nepal have been enjoying an IRR of well over 300–500% on their original investment of Nepal of Rs. 320 on the purchase of the *Saral Thopa Sinchai* (Local Nepali name), the name given by IDEN to the bucket kit system.

Elsewhere in Nepal, we found gross income from sale of vegetables to be Rs. 1500–20,000 per MI system with the modal value around Rs. 3500–4500. In Tansen (Nepal), we met farmer representatives from six Village Development Committees and NGO representatives from LISP (Local Initiatives Support Program, See Appendix I) project of Halvitas, besides a dealer and the District Agricultural Development Officer (http://www.helvetasnepal.org.np/lisp.htm). Together, the dozen or so farmer representatives present reflected the experience of over 200 drip adopters in the neighboring areas. The overall patterns showed little variation. The technology has met with uniform success. The MI program is having a good run in Nepal hills. Many people believe that this run will soon be affected by water scarcity. However, it is likely that marketing limitations may do this earlier than water scarcity does. IDE therefore needs to keep working on these second-generation issues that will soon begin to affect the spread of the technology. (*Note*: Some farmer groups have already begun to work on this. The organized women of Darham Danda, for example, first agreed on a staggered harvesting program among themselves to avoid self-inflicted glut and then had their president enter into a smart tacit agreement with two local vegetable traders who supply to a large workforce working on a local dam project. The women agreed to offer stable supply of cabbage and cauliflower at Rs. 11 and Rs. 13 per kg respectively; they could sell initially at much higher prices but as the glut builds up, prices plummet. Therefore, instead of taking a myopic view, they made a stable long-term arrangement, and in the process ensured stable market. The drip irrigator women of Darkham Danda (Nepal) were lucky in having a far-sighted president who has figured out that market bottlenecks and water scarcity may seal the fate of her members especially in a remote location like hers; so she is already planning a diversification strategy; she would like, on the one hand to grow coffee and ginger, both of which are easier to market. To fight water scarcity, she hopes to get support for a rainwater-harvesting project that can help them build a storage tank with a capacity of 100,000 L.).

The same technology can produce significantly different livelihood impacts in two different communities. *Ramadi* and *Aaboo Khaiseni Yekle Phat*, two other villages we visited in Nepal hills, followed the same broad general pattern as several other hill communities we visited, but heightened the contextual variations. In both, we met groups of 15–20 drip users—MI communities—who were introduced to the technology by IDE; and benefited very significantly from the adoption. (*Note*: Ramadi would probably not have qualified for IDEN's drip kit program but for the fact that it is covered under another project on development of 'Mountain Marketshed.' The village has only 16 users of which 10 had collected to meet us. These women seemed markedly poorer; and their adoption was perhaps aided in some measure by the fact that Social Welfare Center, a local NGO, offered 25% capital cost subsidy to the first group of adopters. They had used the drip kit only for one season; and already there was great interest among others to adopt. In fact, eight nonusers had showed up just out of curiosity; they had not joined the adopters so far because either they did not know or were not sure about whether it will work; and/or because they had trouble raising the cash. Some women felt, correctly, that although there are significant benefits, it takes a higher overall level of effort and engagement in the vegetable enterprise. All of these were now keen to take to drip irrigated vegetables. In addition, the adopters all wanted to increase their area under vegetables and plant numbers by shifting the pipes around a little more).

We met 18 women and 4 men in Ramadi (Nepal), who were significantly poorer; and before they took to drip irrigation, none or few of them grew vegetables to sell in the market. They also experienced extreme water stress; and after the first season of drip irrigated vegetables, a majority of them sold Rs. 500–1000 worth vegetables. The women from Ramadi were concerned that as their vegetable enterprise reaches a serious scale, water scarcity may catch up with them. In contrast, " *Aaboo Khaiseni Yekle Phat* [Local Nepali name]" consisted of none too small professional vegetable sellers, who expanded the vegetable business very significantly after the adoption of the drip kits. Water is not a problem at all with "Aaboo Khaiseni Yekle Phat," which has plenty of it, and slightly more land than Ramadi. Besides, this village is right on the highway and ideally suited for vegetable cultivation for the market. We were told that IDEN has worked with farmers in Aaboo Khaiseni Yekle Phat for nearly 4 years; and the earliest adopter of the drip kit here undertook dramatic expansion in this area under drip and sold Rs. 100,000 worth of vegetables last year. Here, every one (of the 20 odd adopters, we met) doubled the vegetable area after drip, and a third of them tripled it; over half of the drip users sold Rs. 10–15 thousand worth vegetables. Several bought multiple kits or went for upgrades; the original pioneer installed five large drip kits; in addition, he has to shift his tubes once every day. The booming growth in the vegetable production and incomes that the drip technology catalyzed have been due to: Tradition of vegetable cultivation for the market; abundance of water; IDE's low cost storage tank program under which these adopters have built their private water storage ranging from 1000 to 14,000 L; and proximity to markets.

What happens to the additional income from sale of drip-irrigated vegetables? In Darkham Danda (Nepal), many women adopters have to manage their households in the absence of their husbands who are away working in India. The first charge on the earnings then is sugar, tea and other daily necessities, and school fees. Often, the remittances from husbands are delayed; so these women heads of households are always in need of cash to keep the household going. Clearly, the MI program in Nepal is attacking IDE's target segments; even so, in one of our meetings. Tulsi Neupane and Dr Adhikari of the LISP project of Helvetas shared their major concerns: The low-cost drip technology was penetrating only the middle-poor; it is still not easily accessible to the very poor who have some land to grow vegetables. According to them, Rs. 900 is not much for a middle-poor household but it is a good deal for a very poor household to spend on a technology, and they are not certain about the success. Their second concern was about sustainability of an irrigation technology whose success depends so critically on a high quality, intensive technical support in drip irrigation technology as well as horticulture that IDEN have so far provided.

As of now, then, Nepal's powerful positive experience is the prime leading indicator we have of the vast potential of MI technology for poverty alleviation. In Gujarat, it is still early days for even adopters to experience the full range of benefits of the technology. The experience with the technology in Maikaal and Kolar is very interesting but in a different way; in both these sites, we saw little adoption by the poor vegetable growers; but the aggressive adoption by the middle peasantry—and the subsequent spurt in market development—opens up unforeseen opportunities for large-scale propagation of the technology to the poor as well.

4.4 ADOPTERS AND ADOPTION BEHAVIOR

In the larger backdrop of the subject of 'scaling up through market development,' one aspect of the program we explored throughout our fieldwork was:
- The profile of the adopters;
- The 'adoption behavior' of MI customers;
  a. What triggered the first trial of the product by early pioneers?
  b. How did the by-standers process their experience with the technology?
  c. How did the word spread around?
  d. Where early experience with the technology was happy and satisfactory;
  e. At what stage does the technology 'take off' and begin to spread all by itself like wild-fire?

From the past experience and research in drip and sprinkler irrigation in India and elsewhere, there exist stylized propositions about factors that promote and inhibit the adoption of this technology by farmers. In general, it is considered to be the technology for well-off, commercial farmers; farmers take to these not so much to save water but to increase output and incomes and save labor and inputs. For example, Shreshtha and Gopalakrishnan [18] estimated that over 80% of Hawaii's sugarcane farms came under drip irrigation during the 1970s not because it saved over 20 inches (12%) in water application but because it raised cane yield/acre by 1.7 metric tons valued at US $578 at 1987 prices.

Likewise, we also know that major barriers to adoption are high capital cost, unfamiliarity, and the high risk of failure; and that adoption tends to build up as early adopters' successful experience gets confirmed and widely known, and as technology becomes simpler and cheaper. In a survey of some 160 farmers in Nagpur district, Puranik et al. [15] found that all the farmers interviewed—adopters as well as nonadopters—find the high initial capital cost the major barrier to adoption of drip irrigation technology. Interestingly, nearly as many thought difficulty of accessing the subsidy equally important barriers.

In their Hawaii study of the wild-fire spread of drip irrigation for sugarcane cultivation during the 1970s, Shreshtha and Gopalakrishnan [18] concluded that '… continued improvements in the technology have made it more applicable and affordable, thus reducing the risk involved with new technology as well as reducing the cost of information over time….' To what extent are these stylized propositions playing themselves out in the MI scene in India and Nepal?

In Gujarat, we were convinced that the ongoing drought has been the principal 'trigger' for the adoption by pioneers. Most adopters we met took to MI to cut potentially big crop and capital losses from water stress. Veerjibhai Metalia of village Lalavadar installed a MI system 6 months ago at a cost of Rs. 2500 to save a plantation of 90 papaya, guava, lemon trees which is 3 years old but would surely perish due to moisture stress during the current drought. He assembled a MI system with the help of the assembler; he pumps water into a *pucca* tank from his open well some 100 meters away. The tank is connected to the well through a buried pipe; and the drip system is hooked on to the tank. The well can be pumped only once in 2 weeks, and yields just enough water to fill the tank. However, these 15,000 L tank apparently saved his plantation. Veerjibhai appeared sold on the technology. And having adopted the technology for one reason, he has now discovered many other reasons why he should stick to it; he found moisture retention is better with MI than with the flood irrigation system; and his plants were now healthier. Panabhai in Jasdan Taluka too installed a custom-built MI system at a cost of Rs. 1100 to protect his small plantation of 30sapota (*Manilkara zapota*), lemon and other plants. His experience too was similar.

In Vinchhia village, we met a community of professional small-scale horticulturists who raised lemon gardens. These were under tremendous moisture stress during the current (2001 summer) drought spell as their wells dried up. So one of them installed a drip system and found he could make his plants survive with very little water. He pumped his well 12 h daily for flood-irrigating his plantation; now he uses 4 drums of 350 L each—that is, about 1400 L of water—to irrigate his 50 lemon trees. Following this experiment, 11 lemon farmers in the neighborhood all installed MI systems. They made a new group-managed bore well to fill up their tanks. In Saurashtra, then, the current MI buying spree is triggered by the drought. The experience has been good; but it will be interesting to see what these adopters do if there is a good monsoon in 2001. Many will probably keep using it if at all because they see the significant productivity impact of MI systems. The common motivation among Saurashtra adopters is that they are early in their learning curve about what the technology can deliver besides saving their plantations during the current drought.

If drought triggered MI adoption in one part of Gujarat, it induced adopters to fold up their kits and shelve it in another part. In Chhota Udaipur in Gujarat area, another pocket of MI marketing thrust we visited in Baroda district, the IDE assembler is Anand Niketan Ashram, Rangpur, and a local NGO with high credibility with the tribal communities here. Rangpur Ashram has been aggressively promoting the MI technology; and the prime purchase motive here was irrigating vegetable gardens in the homesteads. The most popular product here was the bucket kit; the promotional message is: it can ensure steady supply of 500 grams daily of vegetables per household for three months a year. Some 450 bucket kits are grounded in four talukas; in the Rangpur area itself, some 200 have been sold through the NGO.

This is a predominantly tribal area: Bhil tribals who live here are first generation farmers. The Ashram has been popularizing modern agricultural methods here for 50 years. Its experience with promoting some technologies followed the trajectory we expect the MI technology too will follow. In the 1960s, it installed scores of lift irrigation schemes to promote irrigated farming. It took 8–10 years for the new technology to sink among these communities used to rain-fed, slash-and-burn farming. The Ashram's lift irrigation schemes faced endemic problems of economic viability; and the program was ultimately folded up; but its purpose of popularizing lift irrigation and irrigated farming was achieved. Fed up with the unreliability of the community lift irrigation systems, farmers took to private wells and diesel pumps in a big way as benefits of irrigation got internalized. Now groundwater markets are booming; and pump irrigation is widely used.

Total drought for the second year in a row has, however, put the tribal agrarian economy under great stress, resulting in massive out-migration. MI kits purchased are mostly out of use since wells have no water. We could see some kits in operation in Bhekhadia village where hand pumps as well as dug wells had some water. Mostly, MI kits are used to sustain small kitchen gardens; however, one farmer also raised a somewhat larger garden with a custom-built kit. There is a tradition of vegetable gardens besides the homesteads in the Bhil households; this is good augury for the MI kit program. However, domestic water supply systems are traditionally designed to channelize domestic wastewater into the kitchen gardens. No special effort is made to irrigate the garden. Therefore the MI kit does not offer a significant water-saving advantage over the traditional system of wastewater irrigation. Our overall sense was that poverty-focused MI as a concept is yet to be well established in Gujarat; however, in many ways, this water-stressed state offers opportune conditions for it.

In Nepal hills, on the other hand, MI concept is already firmly established among poor women vegetable growers. The trigger for new purchase decisions is not so much water stress but generating significant household income. Some very interesting work has been done here by IDE in adapting the product to the customer need. IDE, Nepal has been steadfast in pursuing the original mission of introducing the micro irrigation intervention: of designing a product appropriate to the needs of the small farmer household and promoting it aggressively to that target group with intensive after-sales support system. An impressive aspect of the way it has gone about doing it is the adaptive design response to farmer feedback. IDE began with a set of assumptions about what might cut costs best and yet find favor with the target households; as it went ahead testing out those assumptions, it cast aside those that were not supported and developed new ones based on feedback from users. This resulted in much ingenious experimentation in design; and all of it seemed driven by user feedback and functionality. A new, improved product has been launched almost every year since inception.

Thus, for example, the 1998 *Saral Thopa Sinchai* (Local Nepali name for drip-kit) Kit they introduced had a very simple common household filter on the neck of the tank. The 1998 kit was also made available in 'very small' size for 40 plants. These were both changed in the 1999 model, which incorporated several new design features. Similarly, in the early models, IDEN used black recycled rubber laterals; but these were found too hard and nondurable; so they used 8 mm green PVC lateral, which is better in quality and image. Finally, IDEN has avoided the use of microtubes; instead, they have punched fine holes in the lateral itself and fitted it with raffles which when fitted over the holes ensure the water is delivered in trickle rather than in a sprinkle. This has made frequent shifting around of pipes a major requirement; it has also imposed a tough planting discipline on users; if they do not maintain the same distance as between the holes, the system will mis-deliver water.

And the 2000 model of *Saral Thopa Sinchai* kit has fixed nearly all problems the feedback on earlier models pointed out—except, of course, the propensity for clogging. However we observed that farmers have come to terms with it: some problems have to be just lived with.

IDE Nepal has closely followed the development NGO model in promoting the MI technology among the poor. By supplying MI kits to close-knit groups of vegetable growers along with intensive after-sales support, it has created MI communities, and has actively discouraged its dealers from selling kits to isolated buyers, lest they should fail and damage the product image. Nepal hills then have some major clusters of drip kit users; and we saw and interacted with several of these. On our first day of field visit, we went to Kahun near Pokhra and Bhimad in Tanahu. Kahun VDC was a village of some 600 (including 56 drip-kit user) households with 9 wards; Bhimad is a trifle larger. In Bhimad, we met a sizeable group of some 35 women and 8 men adopters of drip kits. In Kahun, we interacted with a group of women in ward 1; this had 70 households; 40 of these have adopted drip kits; of the remaining, 6 have already placed their orders. So it will not be long before this village becomes a 100% drip user village. However, such examples must be few; for, if 50 drip kits are grounded per village, IDE Nepal's total kits should be in 60 villages instead of 450–500. Therefore, there must be many villages, which have isolated adopters of drip kits.

IDEN's distinctive approach emphasizing MI communities supplied with intensive technical support in both drip system use and maintenance as well as in horticulture has produced major impacts. Vegetable production increased manifold, and generally surpassed the wildest expectations of the adopters; average gross income from sale was less in the first year but averaged Rs. 4,000–6,000 in the second year. Once the women adopters were convinced about the profit, they began to learn fast. Soon, IDE found that farmers with two years of experience can be easily weaned away from the IDEN support system; they have enough experience to carry on their own, and even guide new adopters. IDEN is now developing a Lead Farmer Concept to multiply its technical support capability, because intelligent and dynamic farmers with two years of experience with MI of vegetables offer ideal candidates for such appointments. Many of them have already upgraded their drip-systems.

The demonstration effect of MI communities is already strong; some of the women we met came to know first about the drip system not from IDE but from the gardens of some early adopters; but they faced tough time laying their hands on the kit because of IDEN's policy of not selling to isolated buyers. Many prospective farmers have to request the existing groups so that these will accept the prospective farmers as members to obtain the kit and covered by IDEN's technical support. Many keen potential adopters also mobilize 15–20 others to form an MI community so that IDEN would work with. In Darham Danda VDC (ward 1 and 8), this organizing role was performed by the dynamic chairwoman of the Jagriti Mahila Samuha, a local CBO. She visited IDE office several times but could not connect with the staff who were mostly in the field; fed up, she slid a hand-written application for support under the closed door and returned to her village. Sure enough, a marketing officer from IDE turned up a week later to 'process' the application of the women of Darham Danda. This opened a new chapter in the lives of these women.

One drawback of this success was that it attracted attention of subsidy-providers. Subsidies to the tune of 25–33% are already available from local NGOs and even from the office of the Nepal Government Agriculture Development Officer (ADO). One representative of a Nepal Federation of Women Group was in our meeting canvassing for a regular subsidy program. The ADO, who has already been offering 25% subsidy to 30–40 women so far, has offered to expand the program to cover 300 women; he has been asking ready-to-buy potential adopters to wait for next year so that he can oblige them. If this subsidy menace grows, it must hit the program in ominous ways.

Constraints are showing up for wider propagation of MI kits in Nepal hills from two directions: water scarcity and output market glut. Using drip irrigation is not easy for many of these women farmers since their only source of water in the dry season is the public drinking water taps. One such tap is available for 15–20 households; and they share the water equally. Most fill buckets and fetch it to fill the drip tank manually; a few lucky ones are close enough to the tap so that a hose can be connected to the storage tank. For many, however, filling the drum may involve 10–30 min of fetching. One of the women in our meeting mentioned acquiring a 14,000 L tank under trial by IDE; that is her water insurance; her plan is to fill it up with rain water and seal it; it is to be used to save her vegetable crop during the summer days of acute water scarcity.

Water scarcity is a major constraint in the Pulpa district of Nepal, which is mostly dry. Some of the users in our meeting collected surplus overflow from the drinking water system during the night and used it for drip irrigation. In Darham Danda, the remote village in Palpa's mountains, women have to make 14 turns to fetch water for domestic, livestock and drip irrigation requirements. If 3–4 people help in fetching water, they can do the household's water-fetching in 4–5 turns; however it takes half a day since each turn takes 1 h for a slow walker and 30–40 min for a fast walker.

Then, the output market is rapidly emerging as a constraint, too. Members of MI communities tend to grow the same vegetables and their products end up in the same limited local market at around the same time. This results in a glut, and prices go down crashing. In Aaboo Khaiseni Yekle Phat, some women vegetable farmers sold the cauliflower and cabbage products at rock bottom prices; and even then, they have to dump some in the drain. Trucking vegetables to distant towns individually is a dicey business, as some have found out from daily worst experience; so now most depend on buyers to lift vegetables ex-farm.

Much lip service is paid to organizing for marketing; but nothing concrete has happened. Even at its early stage of development, drip users in Ramadi are concerned about the limited market. IDEN helped them to meet local vendors from Bhesisahar and Bhotowodar in a workshop to create better understanding between the both parties. The growers urged vendors to stop buying vegetables from terai. The vendors force-fully argued their position that women producers do their best to sell door-to-door, and come to them only to sell their left-overs; moreover, if they want vendors to sell their produce, growers must ensure a wide variety of vegetable crops; consumers cannot be expected to buy only what they grow. Apparently, both the parties have agreed to play ball, at least for now.

However, it is clear that limited and shallow local vegetable markets may nullify some of the benefits. Smart growers anticipate the glut and can prepare for it. And along with agricultural support, perhaps IDEN may also need to think of some training in vegetable marketing. We met a stray MI adopter at lunch in Tanahun who has been using a drip-kit for 3 years to regularly earn Rs. 7,500–10,000 from a single crop of cucumber. He probably gets 10 kg of cucumber per plant (100 metric-tons/ha) on his tiny plot; and markets two-third of it by weight in the retail market and the rest as a snack food to travelers. He spends 2 months marketing his crop.

For, as the vegetable market becomes a buyers' market, drip-users will need to innovate to keep their incomes stable or to even increase them. Alternatively, IDEN might want to reconsider its present approach of creating concentrated MI communities which glut the shallow local vegetable markets, and instead, spread the kits more thinly over a wider area by letting the dealers loose.

In Maikaal (Madhya Pradesh) and Kolar (Karnataka): the IDE program was in direct competition with mainstream players; and hence, we found here a very different and an interesting dynamic approach. It played a pioneering role in introducing MI among cotton growers in Maikaal and mulberry growers in Kolar; however the adoption is confined largely to middle peasantry. It is an open question whether IDE does not need to redefine its role, now that the concept is established. In Maikaal, we met a group of 15–20 cotton growers from 2–3 villages who had gathered in Mohna village. They were all Patidars, and had 5–15 acres of land, mostly under bio-cotton. All of these were drip irrigators and good cotton farmers. And all of them were using the microtube system although the government subsidy scheme allows only drippers. Only a few large, influential farmers got access to subsidies; most others purchased the material from the open market and built their own microtube based system. One farmer had built a microtube drip system with micro tubes only for one row of plants; this required more lateral but offered the advantage that he can weed and intercultivate without having to shift the pipes around.

The gray market of unbranded products offers limitless opportunities for economizing on capital investment here. Although BioRe has been collecting tube and lateral prices from several prominent market centers in MP, Maharashtra and Gujarat, the best deal it can offer to farmers is Rs. 12,500/acre. However most farmers we met laid their drip systems at Rs. 6,000–7,000/acre by assembling them with gray market material. True, BioRe offers only ISI-approved products; and farmers buy mostly gray products. However the group we met saw absolutely no quality difference. One farmer quipped: 'Big brands charge exorbitant prices and uncertain quality; gray market charges rock-bottom prices and uncertain quality. So who wants big brands?" The gray market deal-ers also offer them written guarantee of 5 years that they believe would be honored if invoked. Some farmers, who have been using gray products since 1996, were quite happy.

As the drip technology is internalized here, the name of the game is cutting its cost down to the minimum. And the farmer's main partner here is the private gray sector. The business has probably recognized that many first time users will try out drip technology only in a drought to save their crops with limited water. They also recognize that the demand is highly price elastic. To encourage such small farmers to try out drip irrigation, one innovative manufacturer has just introduced a new product labeled 'Pepsy'—no pun intended to the soft-drink giant who just introduced dispos-able plastic bottles in Madhya Pradesh—which is basically a disposable drip irrigation system consisting of a lateral with holes. At Rs. 1,500/acre, Pepsy costs a small fraction of the more enduring systems that Maikaal offers to its members at Rs. 12,500/acre. However for small farmers who are trying out the technology for the first time, it offers an important alternative. As one Patidar farmer mentioned, "if I can buy a system at the cost of the interest amount, why should I invest capital? Why spend Rs. 1,200 on a filter when a piece of cloth can serve the same purpose as effectively?" The boom in the private gray trade in laterals and microtubes—and the falling prices of parts—suggests that IDE's ultimate goal of market development is likely to be achieved in this region rather effortlessly.

In Kolar district, the mulberry heartland of India, we met a similarly dynamic and resourceful group of 20–25 mulberry farmers of all classes and social groups in Nayatharahally village. We took a quick inventory of our sample that yielded the in-formation in Table 1. Although small by western standards, these were certainly not the smallest farmers, one could find in the area. This group felt that the kind of drip ir-rigation systems they use are beyond the resources of small and marginal farmers. The farmers face several barriers to adoption: capital requirement is one; lack of education and awareness is another; but the most important is that small and marginal farmers do not have bore-wells. Moreover, a majority of farmers are too small and poor to take to professional sericulture. The group we met represented only the upper crust. We figured that Nayatharahally has some 300 farmers of which 275 probably raise some silk worms. But 7–10 households, each having >7 acres, have all taken to drip and sericulture as sole or primary enterprise. At the other end of the spectrum, over 100 households with <2 acres do some sericulture but only one has a drip system. This is because only 1–2 of the marginal farmer households have their own bore-wells. In-deed all the 60–70 bore-wells in the village were owned by large and medium farmers. And the ownership of a bore-well seems a precondition to adoption of micro irrigation for mulberry. Most poor sericulturists without their own bore-well depend upon larger farmers for the supply of mulberry leaves, which has catalyzed a vibrant exchange institution in mulberry leaves. Small silk farmers buy leaves on a regular basis at Rs. 100–150/bag; some also buy water from big farmers on one-third share cropping basis; the seller provides the water and claims one-third of the mulberry leaf output.

TABLE 1   A profile of Mulbery farmers using drip irrigation in Kolar (Karnataka, India).

Farmer	Village	Total farm land acres	Area under mulberry acres	Area under drip acres	Type of Drip system	Experience with drip irrigation
Govinda Gawda	Nayatharahally	5	5	0	0	0
Muniappa	Thondala	5	5	3	Micro-tube	4 years
Nanjudappa Gawda	Nayatharahally	10	10	2	Integral	4 years
				2	Online	
				3	Micro-tube	
Narayana Gawda	Nayatharahally	10	8	6	Micro-tube	2 years
Narayanappa	Thondala	7	2	2	Micro-tube	2 months
Ramappa	Thondala	10	7	7	Micro-tube	4 years
Ramappa	Thondala	1	0	0	0	0
Ranganath	Nayatharahally	5	5	1	Micro-tube	2 months
Ravakrishna Ppa	Thondala	20	9	9	Micro-tube	4 years
Siva Reddy	Chikapannahally	3	1	0	0	0
Sonappa	Gujjarahally	15	8	0	0	0
Srirama Reddy	Pumbarahally	12	2	1.5	Micro-tube	3 years
Venkatarama ppa	Nayatharahally	5	4	0	0	0

The Kolar group of drip irrigators we met then were a totally different class than the poor women microirrigators we interviewed in Chhota Udaipur in Gujarat and Nepal. These were well-off farmers; but more importantly, they had a dynamism, enterprise and awareness of technology and market conditions we did not expect to find in the poor women vegetable farmers. For instance, the Kolar group's assessment of the pros and cons of alternative drip technologies reflected their knowledge and experience with drip irrigation. We were told that integral systems have higher chance of clogging; microtubes clog less easily but these make intercultivation difficult; they are also more prone to damage; women weeders pull out microtubes to tie the bundles of forage. However in a summary, microtube technology is the best and least-cost option especially for a paired row planting of mulberry: it provides greater aeration and sunlight to plants; it provides greater moisture retention and better root penetration, making the plants more tolerant to dry spells. Paired row system also yields more plants—5,300 per acre compared to 4,600 in the pit system. As paired row system becomes popular, so does the microtube technology that the IDE is promoting. All in all, IDE's Kolar story so far has been the affluent farmer story. However it seems poised at a point where the small mulberry farmer too may take to drip irrigation if he had the right options.

Overall, it seems that the drip sales are set to take off in a big way; and a challenge for IDE is to increase its penetration in the small-holder market segment.

4.5 MARKET DYNAMICS OF MICRO IRRIGATION

An extraordinary aspect of the MI intervention in the four sites was the emergence and nature of the market dynamic. In Gujarat and Nepal, we found little evidence of competition to IDE in the MI market. In Gujarat, the intervention itself is very young; the benefits of the technology are yet to be discovered by the adopters; and a potential for profitable business is yet to emerge. In Nepal, there are signs of such potential emerging; but it is not clear to us if IDEN is doing much to egg this process along. Our impression is that IDEN's approach of providing intensive support to MI communities and of discouraging dealers from selling MI kits to isolated buyers may in fact hamper the market development process.

In **Kolar and Maikaal**, however, we witnessed highly charged market dynamics in micro irrigation material. We saw earlier that Maikaal farmers have begun to experiment with the technology and the gray market has emerged to help them do it at much lower cost than leading suppliers of branded drip products. Products like 'Pepsy' are likely to be welcome by first-time adopters, especially by the poor farmers who want to avoid undue risk of technology failure. However, BioRe's approach is somewhat agnostic—if not suspicious—of the gray market activity since it continues to sell only ISI marked branded material that more than doubles the cost of MI systems. Karnataka has a similar market dynamics; and IDE's posture here is pretty similar to BioRe's in Maikaal. 100,000

Against the national cake of Rs. 2,000–2,500 million/year (One crore = 10 million), the Karnataka drip irrigation business is estimated at Rs. 400–500 million/year. Fifteen years ago when drip irrigation was commercially marketed for the first time, some of the leading players—especially, Jain Irrigation—invested heavily in market development and were beginning to reap the benefits. However in the 1990s, Government of India introduced subsidy in drip systems. For sericulture, subsidy was fixed at 50% for general farmers, 70% for women and 90% for the farmers from scheduled castes and scheduled tribes; for horticulture, it was 30% for general category farmers and 50% for the farmers from scheduled castes and scheduled tribes. Subsidies were available only on systems larger than one acre. The major industry players—like Jain irrigation—are frustrated by the distortions caused by the subsidy. In reality, it also increased competition for them. The subsidy has attracted a large number (40–50 companies are registered) of shady players in the drip business who peddle low quality products, and often claim subsidy without selling systems. Getting the ISI registration involves a one-time bribe of Rs. 60,000–80,000; but then the manufacturer was entitled to market his products under the subsidy scheme. This made big players uncompetitive; it has also created quality problems and impeded market growth due to diminishing farmer faith in the technology.

Even today, the drip irrigation industry does not see much promise in the small farmer segment. According to the Jain Irrigation dealer, a successful adopter is typically a large commercial farmer with some education. However, since such farmers are few in number, the potential in a district is exhausted fairly soon. Moreover, with such farmers—who maintain their systems well—there is little replacement demand; so the market gets easily saturated. These farmers integrate drip technology into their farming enterprise very well; so they buy it for its long-term productivity and economic benefits, not for the expedient goal of tiding over a drought season.

Government of India are now cutting subsidies on drip irrigation; and this is creating a new generation of problems for the industry main-stream which has got hooked onto the opiate of subsidies over several years. Until last year when the subsidy was as high as 90%, the marketing dynamics of the drip system was fired by the subsidy culture. Indeed, the manufacturers and dealers—including the leading brands—were after the 'unearned profit' in the form of subsidies than manufacturing and marketing margins from serving satisfied customers. Since ISI-marked product enjoyed a degree of monopoly in the form of subsidy access, their manufacturers hiked their prices pretty much to levels where they and the bureaucrats empowered to approve subsidies claimed the bulk of the subsidy. However, since claiming the subsidy involved between 1–3 years and 15–20% bribe money, there was always a market for nonsubsidy drip system and products.

Now that the subsidy was reduced to 30%, the profits in ISI marked drip systems have taken a plunge. All players with major names in the ISI-sector are facing declining fortunes; they have been progressively cutting their prices to stimulate the nonsubsidy sales. However, they face stiff competition from the non-ISI players who sell unbranded products at rock bottom prices. We met two dealers in Bangalore who deal in the "cash and carry market" for ISI as well as non-ISI products. Jai Kisan Irrigation and SN Pipe Products were two such examples. Saiyad from SN Pipe was of the view that "ISI mark + subsidy = fraud." He stocked best as well as second quality material from ISI as well as non-ISI sectors; he himself was a manufacturer and sells ready-made products as well as executes orders for material of required quality with a 24-hour lead time. Saiyad asserted, as did the other dealers we met, that there is no real difference between the quality of average ISI and non-ISI products; under ISI-marked branded products, farmers often get cheated with poor quality; at the same time, many non-ISI products are of excellent quality. In general, then ISI-mark is at best a poor indicator and guarantee of quality. The company brand name is a much better indicator; for example, the brand name Jain Irrigation conveys assured quality; but companies with such respected brands exact a commensurately high price. However for discerning consumers, there are non-ISI marked products, which are nearly as good as the best available in the market, that are sold at 60–70% lower price. The prices are compared in Table 2 (Saiyad, SN Pipe Products).

For a majority of potential adopters, however, high-perceived risk in drip irrigation investment is a major barrier to adoption. And this perception is not unfounded; even reputed suppliers grant that the failure rate in the drip system is as high as 50–60%. Many farmers invest in the technology but then dump it because of poor experience with it. As a result of the uncertainty about how well it will work, many first time buyers of drip products view the investment decision more as an expenditure decision (like buying a bag of fertilizer) than as a long-term capital investment decision. In turn, this means that most first-time buyers are highly price-sensitive and search for lowest-priced products available. This tendency is also strengthened by the lack of faith in the quality assurance of ISI-marked products. Because of all these, very little nonsubsidy demand goes to ISI-marked branded products; and the gray market has a field day.

TABLE 2 Price comparison (Rs. per m) for micro irrigation products.

Product type	12 mm lateral	16 mm lateral	micro tubes	Total system
	Rs. per m	Rs. per m	Rs. per m	Rs.
Non-ISI top quality	1.90	2.80	0.50	2,500
Non-ISI II quality	1.40	2.30	—	—
ISI Branded	2.80–3.50	4.60	0.60	10,000
Jain irrigation	4.35	5.50	1.10	12,500

The industry representatives we met did not seem to take IDE and its MI venture very seriously. Most thought there is a better fit between commercial farming and drip irrigation technology than between low-input subsistence farming system and the MI technology. One of them explained to us his viewpoint: "When a Nasik farmer makes Rs. 200,000/acre from grape orchards a year, he does not mind investing Rs. 2000/acre on installing a drip irrigation system. Similarly, coconut, areca-nut farmers internalize the drip technology easily; but vegetable growers on small scale find it more difficult to do so. Vegetable prices fluctuate heavily, and growers need to deal with output as well as price risks; so they are lukewarm to capital intensive farming." The industry had thought similarly about mulberry growers too; but IDE's breakthrough has begun to change the thinking somewhat. Problems with the availability of spare parts, insufficient and erratic power supply are other factors that impede wide acceptance of drip irrigation among small holders. While the industry respected IDE's marketing ethos, it betrayed its doubt about IDE's propensity to down-scale and simplify the MI technology. The Jain dealer had quipped: "drip irrigation technology involves more than just joining tubes with laterals—microtubes are an obsolete technology—besides the kits overlook the importance of custom design."

Arguably, IDE in Karnataka could have carved out a strategic role for itself in Karnataka's fluid market environment. Now that the subsidy is cut down, the business in the non-ISI brands is booming. According to Saiyad, a drip product dealer we interviewed in Bangalore, for every 100 meters of ISI-marked branded laterals, the off-take of non-ISI laterals is 1000 meters. When we tried to cross check this data, Sundar, one of IDE's friendly ISI-marked manufacturers, suggested this ratio of 1:10 (= 1000/100) is hugely exaggerated. According to Sundar, the actual ratio is probably 1:5 or 1:6.

By tying up with top brands in the ISI-sector, IDE has ensured that it promotes drip systems in the highest price range without commensurate quality assurance, and by doing that, it has virtually excluded from its ambit the low-end customers who are its target segment. Even if these had an intent to purchase, the poor farmers are likely to be far more readily drawn to the non-ISI market than to Jain Irrigation and Primere or

KB of IDE which is—and is perceived to be—in the same league. Since the company brand name has stronger association with consumer's perception of quality than mere ISI mark, IDE could use the KB brand name to develop and market a range of low-cost, high quality drip products in the non-ISI sector that can not only achieve quick penetration in small farmer segment however make KB a leading non-ISI brand. Indeed, IDE can develop the bucket and drum kit market by introducing trial kits at rock bottom prices: bucket kit at Rs. 100 and Drum kit at Rs. 250. These can be made using recycled but good quality material under a minimalist IDE quality control mechanism.

In general, then, Paul Pollak's original insight that drip kits made from recycled plastic should sell at rock bottom prices still remains valid and unfulfilled. IDE-India is once again falling into the same trap that has kept its treadle pump market from expanding to its full potential: of offering a high quality product at a high price to a target market that is extremely price sensitive. If MI is to take off in a big way, it seems to us that this will need to change, and marketing elitism will need to make way for some street-smart market maneuvering.

4.6   ASSESSMENT AND FUTURE CHALLENGES

A critical strategic issue for IDE is what exactly is it marketing, or indeed, what business is it in the field of MI? As a product, bucket and drum kits hardly offer a USP, since all parts are available in the gray market and assembling a kit is no big deal. Therefore, if the technology had some special benefits to offer to small holders, the more dynamic approach would have surely taken to it since, besides upper-end brands like Jain Irrigation, they also have access to a whole range of tubes. Only microtube cannot easily be accessed (is this true?) The drip systems are marketed in India for over 15 years now; and the microtube is considered by the industry an obsolete technology when compared to dripper. One concept that the IDE is trying to market is a completely new farming system. In Gujarat, IDE found growing lucrative market in drumstick in Padra taluka; and drum stick is specially amenable to drip irrigation. In Saurashtra, IDE is marketing 'drum- stick' MI as a concept. In Karnataka, MI of mulberry has been a big hit, and can be considered as a significant IDE break through.

The exciting aspect of the IDE's Micro-irrigation program—as, indeed its organizational philosophy—then is the implicit vision about how market development takes place for new products and technologies with potential for livelihoods creation. Figure 1 sets out the roadmap outlining IDE's entry into a new domain with a new technology concept. Typically, it spends a good deal of time and energy initially in establishing a new technology-concept, adapting it to the local conditions and demonstrating its potential benefits to its target customers.

The best example of 'concept establishment' work is to be found in IDE Nepal's micro irrigation program. IDE Nepal has by far the clearest strategic position: it is in the business of marketing low cost MI technologies to 'selected' small holder communities along with an intensive pre and after-sales support system. IDEN actively discourages direct sale of drip kits by its dealers without its recommendation because it believes that without adequate technical support, adoption may neither be beneficial nor sustainable. Since IDE Nepal believes that such support can be best provided to groups of adopters organized by IDE, therefore the potential adopters outside the IDE group may find it very difficult to get the IDE MI-kits. We met the Pokhra assembler and a dealer. Both of them suggested that there is a direct demand for drip-kits without IDEN recommendation. It is not clear how dealers respond; the assembler said they service the demand; the dealer said he does not.

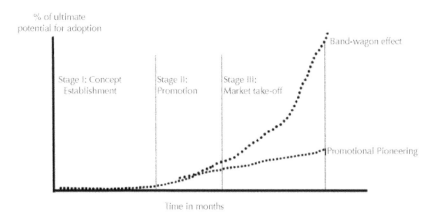

FIGURE 1   The roadmap outlining IDE's entry into a new domain with a new technology concept.

IDEN follows an elaborate process—and has invested significant organizational resources—in achieving its strategic goal. This intensive support and back-stopping make IDEN's drip program virtually failure-proof. We could see this in course of our visits in five days. We met over 200 adopters, mostly women, a few men; and we did not find anyone who was disappointed with the system. Everyone was happy, some more, some less; clogging bothered everyone. However, we did not hear in any case adoption failure. The process of introducing the MI technology in a new community involves following steps:

Step 1: The Marketing Supervisor makes an exploratory visit to the village to under-take a rough feasibility analysis. He explores a range of questions: Does the village have a tradition of vegetable cultivation? Is there some water available? Is there access to a market nearby? Are farmers open to new ideas?

Step 2: If the village passes this test, a meeting is planned and organized, if possible with the entire community. At this meeting, the technology is demonstrated, a sales speech is delivered with the idea of generating interest in it. Invariably, 10–15 farmers show readiness to try it out.

Step 3: A training workshop is conducted for the 'pioneers'—who showed interest in the trial in two aspects: (a) agronomic-seed preparation, common nursery, spacing, etc.; and (b) drip kit purchase, installation, operation and maintenance. After this, a common nursery is raised in 4–6 weeks; as it gets ready, pioneers are asked to approach the dealers and obtain the kits.

Step 4: Marketing Supervisor, Agricultural Technician and Installer visit the community again to train in the proper-installation process of the system, its uses, its operation and repairs. Further, some agronomic training is also given on planting and spacing.

Step 5: After this, the Marketing Supervisor and Agriculture Technician give follow-up visits to the community alternatively at an interval of one week; this interval grows longer as the community becomes at familiar with the system; but IDE support is available virtually on demand.

One issue, which might become important as drip sales grow, is the capacity of IDEN to sustain such a support system. We probably saw some of IDEN's best-performing drip-irrigation communities [9]. One wonders if it is easy to provide such a cover to all 3200 adopters so far—or even to the 1200 odd who will buy the kits this year. IDEN's challenge then is to find innovative ways to extend its technical support cover in a cost-effective manner—through collaboration with NGOs or through enlisting successful and enterprising adopters in the task of supporting new ones.

This first stage involves hard, patience and often frustrating work of pioneering a new concept, support and 'hand-holding' for early adopters, developing manufacturers, setting up supply chains. The adoption is slow and restricted to a small number of risk-loving customers. Many potential customers—the by-standers—closely watch the trials with the new concept by early adopters, gathering the evidence and drawing their own inferences. It takes time for this 'evidence gathering and analytical process' to mature, since each 'by-stander' sets up and works with his own mental model. If the technology delivers against the expectations of early adopters and by-standers, the market development process enters the second phase when IDE's promotional efforts begin to deliver results in rapid growth in technology adoption and sales volumes. In this phase, promotion and marketing acquire a critical role; sales begin to build up; and awareness about the technology spreads. If the product/technology is capable of sustaining on its own—without subsidy and other external support—then we begin to see inkling of interest in it from other players in the market who want to jump onto the bandwagon and build profitable line of business on the groundwork of pioneering and promotion done by the IDE. It is here that IDE differs from other NGOs; whereas most NGOs would view this growing interest of private players in their product with a sense of concern and insecurity, IDE views it as the sign of its success: the fruit of its arduous labor through stage I and II. The role of IDE might become extremely complex at this stage; as the pioneer and the oldest player, it could set standards for others, become a rallying point, actively assist its competitors to take on its own brand; for its ultimate aim in stage III is not the gains from promotional pioneering it did laboriously but to also capitalize on the 'bandwagon effect' produced by the entry of other players—which, in the ultimate analysis, is the market development role it claims to be playing.

Against this model, we found that Gujarat and Nepal are still pretty much in stage I of the market development process for MI; however, Kolar and Maikaal are somewhere in stage II or even III. In that sense, the market dynamics here is different from Gujarat and Nepal, and offers interesting insights. After five years of stage I labor by IDE as well as BioRe, some 1500 acres of Maikaal Cotton's bio-cotton area is now under drip. BioRe initiated a scheme to install drip system on farmers' fields: the advantage to the farmer is that BioRe buys tubes and laterals in bulk to get a good price; second, farmers gets an interest-free 3-year loan. Many small farmers are taking up the

BioRe offer. There are indications all around that drip technology is being rapidly internalized by farmers and is on the verge of taking off in a big way in this region. The best indicator of this is that the farmers have begun to play around with the material as well as the design on their own. In Maikaal, the MI market is already in stage III of Fig. 1. There are strong indications that private business is doing far more to cut the costs and reach the technology to poor farmers than IDE and BioRe, because the private business understand the mindset and the behavior of the poor.

4.7 SUMMARY

Ever since they became popular in Israel and the US, drip and sprinkler irrigation technologies have appealed to large, commercial, technology—sauvy farmers. In recent years, attempts have been made—by NGOs like IDE and corporates like Netafim and Chapin—to adapt these technologies and promote them as livelihood-creators for the poor of Asia and Africa. IDE, which has simplified and demystified the technology, has focused on cutting its cost to the minimum and on promoting it massively among the poor.

Does micro irrigation offer promise as a poor-friendly technology? Based on a month's fieldwork by the authors in Gujarat, Karnataka and Madhya Pradesh in India, three hill districts in Nepal and Kenya, this chapter attempts a first-cut answer to this question. It offers a preliminary, impressionistic assessment of the potential of micro irrigation technology, its social impacts, and issues involved in 'scaling it up.' We conclude that:

1.   In South Asia, IDE micro irrigation program has responded to two critical but distinct needs: of the poor women to create a new means of income and livelihood; and of farmers in water scarce areas to cope with extremes of water scarcity;

2.   The best example of the improved livelihood is found in Nepal hills, where MI Communities—mostly of poor women vegetable growers—created by IDEN have experienced major improvements in cash income and household food and nutrition security;

3.   The best examples to cope up with water scarcity are found among: organic cotton farmers in Maikaal region near the site of the Maheshwar dam in Madhya Pradesh; mulberry farmers of Kolar district in Karnataka; and lemon growers in Saurashtra in Gujarat;

4.   The strategic issues in marketing micro irrigation bucket-kits and MI drum-kits to the poor women vegetable growers are totally different from promoting MI to farmers coping with extreme water scarcity;

5.   In terms of sheer scale of outreach, promoting micro irrigation as a means to coping with water scarcity offers much greater potential than promoting it to poor women vegetable growers;

6.   In doing both, *prima facie*, it seems that the IDE operating philosophy of paring the cost of the technology down to the minimum and of using normal market processes to mainstream it holds great promise.

KEYWORDS

- Chapin drip irrigation system
- drip irrigation
- drip irrigation potential
- drip irrigation technology, High cost
- drip irrigation technology, Low cost
- Food and Agriculture Organization, FAO
- India
- Indian Rupees, Rs.
- Indian Standards Institute, ISI
- International Development Enterprises Nepal, IDEN
- International Development Enterprises, IDE
- irrigation
- Jain Irrigation
- Krishak Bandhu (India), KB
- Kenya
- Local Initiatives Support Program in Nepal, LISP
- micro irrigation, MI
- million-hectares, mha
- Nepal
- Netafim micro irrigation system
- poor-friendly technology
- small holder community
- Symbiotic Research Associates India, SRA
- TARU Research and Information Network, TRIN
- Unique Selling Proposition, USP

REFERENCES

1. Bandur, C., T. Nataraj, K. R. Prasad, S. N. Srinivas, 1997. IDE drip irrigation system for micro irrigation: A case study. Bangalore, India: Department of Management Studies, Indian Institute of Science.
2. Bilgi, M. Socio-economic study of the IDE-promoted micro irrigation systems in Aurangabad and Bijapur. New Delhi, India: International Development Enterprises.
3. Catalyst Management Service, 1999. A preliminary study on role and effects of subsidy in Mulberry micro irrigation systems. Bangalore, India: Catalyst Management Systems.
4. Devkota, B. M. 1999. Linking local initiatives to new know-how: Low-cost Drip Irrigation Pilot Project. Kathmandu, Nepal: Helvetas Nepal.
5. HURDEC, 1999. Impact assessment of drip irrigation system. Kathmandu, Nepal: Human Resource Development Center.

6. Gurung, J. B., 2000. Preliminary impact assessment of low-cost water storage structures in Ta-nahun and Kaski districts. Kathmandu, Nepak: International Development Enterprises.
7. KARI, 2000. Transfer of micro/drip irrigation technology: Business plan report by Nairobi: Ke-nya Agricultural Research Institute. 2000 Scaling-up Plan of Affordable Micro-Irrigation Tech-nology with Complimentary Small Farm Intensification and Micro-Enterprise Development, New Delhi, India: International Development Enterprises.
8. IDE and SRA, 1999. Study on performance of micro irrigation systems in Mulberry in Karnata-ka and Andhra Pradesh. Bangalore, India: International Development Enterprises and Symbiotic Research Associates.
9. IDE, Nepal. 2001. A Nepalese model of simplified and low-cost drip irrigation technology. Kathmandu, Nepal: International Development Enterprises.
10. IPTREAD, 2000. Affordable irrigation technologies for smallholders: opportunities for technol-ogy adaptation and capacity building. IPTRID Secretariat, Food and Agriculture Organization of the United Nations, Rome.
11. MBL, 1999. Brand Image of KB Reshme Drip-A Report. Bangalore: MBL Research and Con-sultancy Group Pvt. Ltd.
12. Naik, G. and V. K. Dixit, 2001. A market development approach study of affordable micro irrigation technology program of IDE. (with contributions from Sobhan Sinha), New Delhi, India: International Development Enterprises.
13. Narayanamoorthy, A., 1996a. *Evaluation of Drip Irrigation System in Maharashtra*. Pune, In-dia: Gokhale Institute of Politics and Economics.
14. Narayanamoorthy, A., 1996b. Micro-irrigation. *Kisan World*, January, 23(1): 51–53.
15. Puranik, R. P., S. R. Khonde and P. L. Ganorkar, 1992. Constraints and problems as perceived by nonadopters and adopters of drip irrigation systems. *Agricultural Situation in India*, April, 47(1): 33–34.
16. Rajshekar, S.C., 1997. Case study of mulberry farmers using micro irrigation systems. Banga-lore, India: Symbiotic Research Associates.
17. Shankar, M. A. and H. R. Shivakumar, 2000. Drip and fertigation to Mulberry. Bangalore, India: University of Agricultural Sciences, Technical Bulletin.
18. Shreshtha R. B. and C. Gopalakrishnan, 1993. Adoption and diffusion of drip irrigation technol-ogy: An econometric analysis. *Economic Development and Cultural Change*, 41(2): 407–418.
19. TARU, 1997. Market potential for affordable micro irrigation systems for Mulberry. Bangalore, India: TARU Research and Information Network (TRIN, info@taru.org).

APPENDIX I

The Local Initiatives Support Program (LISP)

http://www.helvetasnepal.org.np/lisp.htm: GP.O. Box 688, Kathmandu, Nepal. Phone: 075–520075

Email: lisp@helvetas.org.np

Development can succeed if it is demand based, led and owned by local com-munities. This is one of the key learning Helvetas has gained from its approximately 28 years of involvement in the development activities in Palpa District. It requires strengthening of institutional mechanisms at community level on the demand side. On the supply side, it needs to develop demand-responsive services, organizations and funding modalities. However, development efforts may still fail unless the needs, interests and voices of socially and economically excluded people are well-regarded.

LOCAL INITIATIVES SUPPORT PROGRAM, PALPA (LISP)

The Local Initiatives Support Program (LISP) has strengthened local organizations for institutional capacity building in Palpa for the last 10 years. This was accompanied by supporting income generation initiatives at local level. It is now in its final phase of localizing the entire planning, implementation and financing modality.

OUTCOMES

- Planning and managerial skills enhanced at village and district level Local institutions (local NGOs, farmer associations, federations) tackle the socioeconomic problems and enhance opportunities of people to achieve a higher level of self-reliance.
- Local economic initiatives (e.g., vegetable, coffee, ginger, NTFP, goat raising) generate cash income of small holders after adopting improved technology and through better marketing mechanisms.

CURRENT STATUS

- Planning and managerial capacity enhanced to 1 DDC and 31 VDCs through capacity building activities (preparation/implementation of District Transport Master Plan; District Periodic Plan; Village Development Periodic Plan).
- More than 20,800 households (41% of the district's total households) participate in project activities.
- Proportion of participating Dalit (15%) and Janajati (64%) households above district average.
- Local funding system initiated which finances demands from groups for technical support from Local Resources Persons ("drSP fund"); more than 5700 households served through this system in 2005.
- Farmers generated a gross income of NRs. 2.0 million from coffee, 42.2 million from ginger, 18.5 million from vegetables and 12.48 million from NTFP (for 2005).
- 77 microhardware activities (including small drinking water/sanitation schemes, irrigation, agro product collection Center, Picohydro power and plastic ponds) supported on cost sharing basis (over 3 years).
- 26.72 km. of village roads under construction following the green road concept.

PARTNERSHIP

- All works are implemented by local partner organizations or directly by self-help groups.
- Self-help groups demand services from the drSP-fund and are supported accordingly.
- District level association of coffee farmers and forest users.
- Saving/credit cooperatives, milk cooperatives.
- Five local NGOs, two local community radios.
- DDC, 31 VDCs and line agencies for coordination and joint implementation of development activities (e.g., road, microhardware).

APPENDIX II

Map of India.

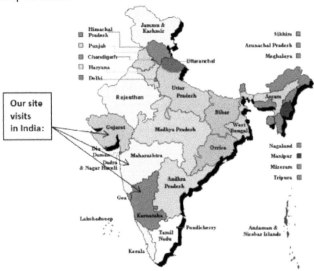

APPENDIX III

Map of Municipalities in Nepal.

APPENDIX IV

Map of Kenya.

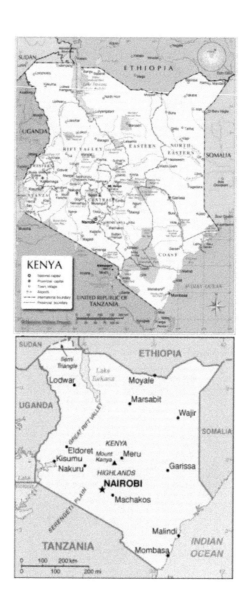

CHAPTER 5

LOW COST MICRO IRRIGATION IN KENYA*

JACK KELLER

CONTENTS

5.1 Introduction 78
5.2 Promotors of Low-Cost Drip Systems in Kenya 79
5.3 Technical and Socio-Economic Issues 82
5.4 Promotion and Marketing 83
Keywords 84
Appendix I 85
Appendix II 86

*In this chapter, cost of drip system is based on 1999–2000 prices. Information in this chapter is based on the visits by the author to Kenya in 1999–2000. Author acknowledges the consultancy opportunity in Kenya by Winrock International and the International Water Management Institute (IWMI). During February 17–18 of 2000, he also attended the workshop on "the state of the low-cost drip irrigation situation in Kenya" at the National Agricultural Research Laboratories, Nairobi. Proceedings for this workshop were printed. This chapter has been modified from: *"Keller, Jack, 2000. Gardening with low-cost drip irrigation in Kenya: For health and profit. Technical Report prepared for International Development Enterprises (IDE), http://www.siminet.org/fs_start.htm."*

5.1 INTRODUCTION

Like other sub-Saharan countries, land and water resources are becoming even scarcer and arable land in Kenya has decreased from about 5 ha per person in 1960 to 0.2 ha per person in 1998. The high population growth has contributed to this decline, but it is evident that agricultural production has also kept pace perhaps because it was based on intensification policies, which are no longer feasible. Coupled with lack of appropriate technologies, this decline in per capita arable land is the main cause of declining food security and increasing poverty, which the country has been experiencing over the last decade.

According to information I obtained from various sources (Ministry of Agriculture, Kenya Agricultural Research Institute, and the University of Nairobi) the estimated irrigation potential of Kenya is somewhere between 250,000 and 540,000 ha depending on who made the estimate, and somewhere between 65,000 and 78,000 ha have been put under irrigation to date. Of this about 40% is privately managed by large firms, 20% is managed by the government, and 40% is managed by smallholders. However, because of inadequate maintenance it is estimated that less than 55,000 ha are now being used. The main constraints to irrigation development are water availability and poor water management. While there are several efficient water use irrigation systems, those used in Kenya are mainly traditional furrow and overhead sprinkler systems, which tend to use too much of the limited water.

According to the Permanent Secretary, Ministry of Agriculture and Rural Development in his remarks at *Low-head Drip Irrigation Review Workshop in February 2000*, the root cause of the poor economic performance, poverty and food insecurity in Kenya is because of unsound development policies. He stated that everybody who has been attending seminars on public policy questions agree that there is presently a lack of new ideas and policy options. It seems that currently, everybody believes in the market logic, in privatization and in the primacy of the private sector. But the policy debates should shift from discussing privatization, retrenchment and incentives to the private sector, to finding ways to assist small-scale rural agriculture. Poverty has entrenched itself in Kenya because the government had ignored the small-scale rural farmer and supported the large-scale farmers. He believes that Kenya achieved economic growth in the Sixties and the Seventies because the government implemented land redistribution policies. In other words, policies that gave the majority of the population improved access to economic assets and social services. He expressed his conviction that such simple and adapted technologies as small-scale drip irrigation give the rural population improved access to economic asset (technologies).

He continued by pointing out that increased production in the 60s and 70s was achieved mainly due to expansion of agriculture (extensification) into new areas in addition to new technologies. However, over the last 20 years, because of the rapid increases in population the option of extensification has dwindled. The Kenyan government is now convinced: a) that intensification of agriculture including use of irrigation is the only solution to the problems of increasing agricultural productivity and hopefully food sufficiency and the reduction of poverty; and b) that appropriate technologies, in both improved crop varieties and soil and water management technologies, are the key to intensification.

5.2   PROMOTORS OF LOW-COST DRIP SYSTEMS IN KENYA

5.2.1   CHAPIN LIVING WATERS FOUNDATION

In the early 1970s widespread droughts and famines in Africa were reported in the news media. At the request of the Catholic Relief Services, Mr. Richard D. Chapin went to Senegal in 1974 to provide a small-scale drip system that could operate without a pump. This system operated from a bucket or small drum supported about a meter above the ground. It worked well and produced excellent vegetables where there was little or no rain. Since then Mr. Chapin has developed and promoted a simple, low-cost, and efficient drip irrigation system, called a "drip-kit."

To make it possible for low-income families to afford these drip-kits for their subsistence needs in areas that faced long dry seasons, Chapin Watermatics produced and sold the kits on a nonprofit basis and promoted their use through the Chapin Living Waters Foundation (CLWF). Mr. Richard D. Chapin is the founding executive director of CLWF (He celebrated his 96th birthday in August 2013). The vision of CLWF was to promote micro irrigation for small plots in developing countries. The weblink of CLWF states that "*Chapin Living Waters was founded as a means to express Christian love to needy people in third world nations through small-scale drip irrigation technology. While much of our effort is directed toward missionaries working among these suffering peoples, Chapin Living Waters stands ready to assist any group, regardless of race, color, or creed. With 1.1 billion people going to bed hungry tonight, we are dedicated to working in partnership with as many organizations as possible. Today, Chapin Living Waters exists to help poor people in third world countries grow vegetables when there is insufficient rain. We offer simple, sustainable technologies like drip irrigation with the use of "Bucket Kits" for subsistence farming, bringing hope to the poorest people in over 150 countries and the opportunity to solve their own problems* (http://www.chapin-livingwaters.org/)." The reader is advised to visit this website for updated information.

The CLWF has in place a training program on drip-kit use as well as training in dry season vegetable production. In 1996, it assisted the Kenya Agricultural Research Institute (KARI) in introducing the Chapin drip-kits in Kenya with the objective of improving the economic conditions, nutrition level and enhanced food self-sufficiency of small-scale resource poor farmers (Fig. 1).

FIGURE 1   Low cost bucket type drip irrigation system promoted (See Appendices I and II) by http://www.chapinlivingwaters.org/Pages/Contact.htm.

5.2.2 ROLE OF KENYA AGRICULTURAL RESEARCH INSTITUTE (KARI) IN DRIP IRRIGATION

Since 1996, KARI has made use of grants from USAID and the World Bank and has imported, assembled, and sold/distributed over 3,500 Chapin bucket-kits and roughly 400–1/8-acre (500 m²) drip-systems. In addition KARI has established drip-kit and low-cost drip irrigation demonstration sites at the National Agricultural Research Laboratories (NARL) in Nairobi, the National Dry Land Farming Research Center in Makueni, and the Regional Research Centers in Mtwapa and Perkerra. The Agricultural Engineering Department of the University of Nairobi has also conducted research on the effectiveness of the bucket-kit drip irrigation systems. Initially, for demonstration and promotional purposes, KARI freely distributed bucket-kits to various women's groups and NGOs working with small-scale poor farmers.

A bucket-kit has a 20-liter bucket that is supported 1-meter above the ground and supplies two 15-meter (or four 7.5-meter) long drip limes. The drip lines are lay-flat 16 mm tubing (0.25 mm wall thickness) with internal long-tortuous-path emitters spaced at 0.3-meter intervals. A bucket-kit supplies irrigation to a small vegetable plot of 15 to 25 m², depending on row spacing, and costs about $15. This is usually sufficient for a family's self-sufficiency in vegetables and many families even have some extra vegetable, which they market. Women own most of the bucket-kits and they also do the vegetable gardening. KARI researchers informed me that farmers report gross incomes of $40 to $60 per crop and grow two crops per year where there is sufficient water.

KARI has recently scaled up the Chapin bucket-kit to what they call a "drum-kit." It is comprised of 10 of the 15-meter long drip lines connected to a 200-liter drum to serve a plot of from 75 to 125 m² (depending on row spacing) at a cost of about $100. They also provide a double drum-kit with 20 of the 15-meter long drip lines connected to a single 200-liter drum at a cost of about $135.

KARI's 1/8-acre (500 m²) drip systems have a 50 mm diameter 15-meter long header that serves 20 drip lines that are each 30 meters long. The drip lines come in a 600- meter roll and are cut in the field. This system actually irrigates a 450 m² plot and it costs about $200, which does not include the water supply or a head tank. If a "Super MoneyMaker treadle pump" supplied through "Appropriate Technologies for Enterprise Creation's (ApproTEC, now called KickStart) marketing system" were used for the water supply, then this would add about $75 for the pump plus about $25 for the necessary hoses and head tanks. They have also developed a one cylinder pressurized treadle pump that they intend to sell for about $45.

Currently KARI sells drip-kits to farmers and Projects. Farmers and donor-funded project personnel visit the demonstration sites and purchase the kits there. Some bucket-kits are still being donated to farmers through various charitable organizations but farmers are now directly purchasing roughly half of the kits that KARI sells.

While survey results have indicated that many of the bucket-kits that have been given to farmers are not being fully used, most farmers who have purchased their kits use them for two crops per year except where the water supply has failed.

5.2.3 OTHER PROMOTERS OF LOW-COST DRIP SYSTEMS

Besides KARI there are a number of other institutions and groups that have begun marketing drip-kits. Following are some brief comments about some of these other promoters:

FPEAK: The Fresh Produce Exporters Association of Kenya (FPEAK), which was established through a grant from USAID, has purchased a few container loads of bucket-kits and 1/8-acre drip-kits plus some roles of drip tape. They sell the systems as an aid to their growers and as an income generating activity to help fund the activities of the association.

KickStart: KickStart, which was formerly Appropriate Technologies for Enterprise Creation (ApproTEC), is in the process of developing a 1/16-acre and a 1/8-acre drip-kit to be used in conjunction with their "Super MoneyMaker pressurized treadle pump." They intend to sell the 1/16-acre kit for approximately $130. Unlike KARI they plan to promote the drip-kits through a mass media advertising campaign and sell them through the network of roughly 100 dealers they have established to market their treadle pumps.

Dream Drip-Kit: Dream Drip Kit was established by Water Resources Management (WAREM) Consultants who wanted to use locally manufactured products as much as possible. The Dream Drip Kits use a high quality locally extruded light-weight 16 mm outside diameter hose with inline emitters. The Victoria Driplines are extruded by a local company called Shade-Net, that uses a modern Italian manufactured extruder and high quality LLDPE plastic. They also make the inline emitters locally and currently extrude over 7 million meters of drip line for sale to the commercial irrigation sector in Kenya. The drip line is sold under the brand name "Victoria Driplines." The available systems are similar to the Chapin drip-kits, and cost about the same although the drip lines are made of thicker (0.5 mm vs. 0.25 mm wall thickness) plastic.

NRMMM Project: The Natural Resources Monitoring, Modeling, and Management (NRM[3]) Project is located in the rain shadow of Mt. Kenya in the vicinity of Nayuki, which is a very drought prone and impoverished area. This is a research project activity related to comprehensive rural development that is being carried out by the University of Nairobi with Swiss Development Corporation funding. Bucket-kits are a part of the water management program. The project personnel have assembled bucket-kits that are similar to the KARI Chapin kits, but they have three 15-meter long drip lines instead of two and cost about $16, which is only $1 more. They innovated their own system using basically locally manufactured components; in fact they even have a local plumber craft elbows and other fittings from straight PVC tubing. They have sold or donated about 100 bucket-kits to local farmers and women's groups, and have set up a nursery to provide fruit, vine, and tree plants that are adapted to the area, and a demonstration plot to promote them. They have also worked with water storage tanks that capture rainfall runoff for use in the drip systems during the dry season. Another innovation they have tried is the use of donkey-carts to deliver water for the bucket-kits.

Typical incomes from the kits are in the neighborhood of $25 per vegetable crop plus enough small vegetables for the family's consumption. One women's group we visited had their first bucket-kit donated to them, had purchased a second kit, and were intending to purchase a third kit for the next crop season.

ALIN: The Arid Lands Information Network (ALIN) has been an active prompter of bucket-kits. They have either purchased kits directly form Chapin or from KARI. They have promoted and distributed over 400 bucket-kits, about half of which have been sold. They periodically place articles in their tri-annual magazine *Baobab* to promote bucket-kits throughout Africa.

MISCELLANEOUS: There are several other groups promoting bucket-kits in addition to the above. These include Catholic relief Services, the Mennonite Church, the Christian Children Fund, German Agro-Action, and many others.

5.3   TECHNICAL AND SOCIO-ECONOMIC ISSUES

This section focuses on the technical and socioeconomic issues related to the low-cost drip systems that were available in Kenya during my trip. The comments in this section are based on my personal observations and the conclusions agreed at the Low-head Drip Irrigation Review Workshop, held at the National Agricultural Research Laboratories [NARL], Nairobi, 17–18 February 2000.

5.3.1   TECHNICAL ISSUES

1. The filters used are too small so they clog very quickly and they also break.
2. The drip lines are fragile and emitter clogging is a problem so the lines only seem to last for a couple of years.
3. The drip lines are vulnerable to attack by rodents in the field and in storage.
4. There is a general lack of skills in installation, operation, maintenance and repair of the bucket-kits.
5. Both the kits and the spares are in short supply nationally as well as at the user level.

5.3.2   SOCIO-ECONOMIC ISSUES

At the workshop, a number of socioeconomic issues emerged. The bucket-kit is very popular with users because the benefits are easily recognizable. They have potential for growing high value crops and provide an opportunity to create employment and better nutrition in rural areas. However, about 80% of the users are women, and they receive little support from men at present.

Although the $15 bucket-kits are fairly inexpensive, the prevailing high poverty levels in rural areas make them unaffordable to many potential beneficiaries. It would appear that there are two ways to alleviate this problem. First more attention should be given to reducing the cost of the bucket-kits. For example, in India and Nepal bucket-kits of the sizes sold by KARI and others cost less than $10. Secondly, access to credit should be investigated, including mobilization of local resources through self-help groups, revolving funds and merry-go-rounds.

5.3.3   CONCLUSIONS AND RECOMMENDATIONS

Both drip irrigation systems serving large commercial farms and low-cost drip-kits designed for resource poor farmers with small plots are well established in Kenya. I would estimate that commercial drip systems are serving somewhere between 4,000 and 7,000 ha of land. In total there are probably somewhere between 5,000 and 6,000 bucket-kits, 300 to 500 drum-kits and perhaps as many as 200 to 300–1/16 to 1/8-acre (250 to 500 m²) drip systems serving poor farmers who have limited financial and land resources.

A rough estimate of the potential market for low-cost micro irrigation systems for small plots in Kenya is in the order of 1 million bucket-kits, 100,000 drum-kits, and 100,000–1/8- to 1/2-acre drip and mini-sprinkler systems. These systems would irrigate roughly 20,000 ha and serve roughly 1/3 of Kenya's 4 million rural households.

It appears that the demand for low-cost drip irrigation for small plots is becoming well established in Kenya. However, there has not yet been a concerted effort to develop a comprehensive marketing approach for low-cost micro irrigation for small plots.

5.4 PROMOTION AND MARKETING

Limited promotion, availability, and accessibility of drip-kits to farmers at the local rural levels are a major constraint to the rapid expansion of low-cost micro irrigation systems for small plots. At this point a true marketing approach directed at selling drip-kits to the poor rural farmers with small holding has not been used. This can be improved by developing a marketing approach like International Development Enterprises (IDE) uses to discriminate the technology.

The marketing approach to development used by IDE and others is diligently presented and explained in the "Swiss Agency for Development and Cooperation's Report by Urs Heierli (2000)": *Poverty Alleviation as a Business: The Market Creation Approach to Development*. This report includes program descriptions with examples of successes along with the performance indicators used. It covers treadle pumps in Bangladesh, the agro-forestry and sanitary water-trap latrines in Bangladesh, maize storage-silos in Central America, and concrete-roofing tiles globally. The Report contains Annexes that indicates: *The Setup of the Marketing Approaches for Each of the Above Mentioned Projects*; *The Practical Steps of How IDE Installed 1.3 Million Treadle Pumps in Bangladesh by Activating the Private Sector*; and *The Design Process for the IDE Low-Cost Drip Irrigation Systems*.

There has not yet been a concerted effort to develop more robust drip-kits and produce them locally at a reduced cost. Furthermore, I saw no evidence of any efforts at developing an appropriate mini-sprinkler kit to accompany excellent pressurized treadle pumps by KickStart.

In view of these findings I recommend that efforts be made to further refine and reduce the cost of the various drip-kits and to develop a mini-sprinkler kit, and market test them in accordance with the following strategies:

5.4.1 MICRO-IRRIGATION DEVELOPMENT

Developing suitable low-cost micro irrigation systems and establishing the market demand for them through private sector microenterprises is the role that organizations like IDE or ApproTEC must perform. Donors are needed to provide what are in effect indirect subsidies to farmers by covering the costs of developing the affordable technologies that improve productivity and establishing the market demand for them. Funding the development of low-cost systems and establishing the demand driven markets for them has proven to be a very cost-effective role for donors.

The strategy that has been successfully used by IDE to develop suitable low-cost micro irrigation systems prior to establishing the production capacity and implementing the marketing program for them includes the following:

1. The initial challenge is to identify promising technologies that have the potential to improve productivity if they can be packaged in a manner that is suitable for small plots of land and affordable to the potential farmers. While the final products may appear simple, developing affordable and user-friendly products requires inventive and talented engineering.

2. Promising designs must be market-tested and those that are not rejected must usually undergo considerable change to trim cost, increase functionality, and better address field requirements to gain farmer acceptability.

3. The most cost-effective way to provide systems for the initial market testing and after that to provide larger but still limited numbers of systems for the early stages of marketing may be to purchase imported or locally available components that are relatively costly. Then discount the selling price of the systems to more affordable levels based on the anticipated ongoing cost (including reasonable profit margins) after the market has matured.

4. The design stage essentially never ends because of remaining possibilities for reducing cost and increasing functionality. However, at some point further design changes should be restricted to those that are necessary to address significant problems uncovered by users or take advantage of new insights that have the potential to significantly reduce costs and or increase functionality.

The roles described above for organizations like IDE and the donors are necessary since low-cost systems should be designed around more or less generic (without patent protection) components. Thus microenterprise manufacturers and vendors cannot afford to invest in the necessary product and market development activities since these costs would not be recoverable after the market demand is established. If they attempt to maintain sufficient profits to cover these development costs, competitors will arrive on the scene and undercut their prices. However, this competition is actually a positive aspect of the donor supported market creation approach. It continuously stimulates inventiveness to gain market share by increasing cost-effectiveness and functionality.

KEYWORDS

- **Chapin Living Waters Foundation, CLWF**
- **Chapin Watermatics**
- **drip irrigation**
- **drip irrigation, low cost**
- **drip system**
- **drip-kits**
- **International Development Enterprises – Kenya, IDEK**
- **International Development Enterprises, IDE**

- **International Water Management Institute, IWMI**
- **Kenya**
- **Kenya Agricultural Research Institute, KARI**
- **micro irrigation, MI**
- **microenterprise**
- **National Agricultural Research Laboratories, NARL**
- **Natural Resources Monitoring, Modeling, and Management – Kenya, NRM[3]**
- **treadle pump**
- **University of Nairobi**
- **Winrock International**

APPENDIX I

Seminar in Uganda	Rotary Display in El Salvador	Ready to Transplant in Nepal
Completed Installation—Mexico	Laying Drip Hose in Burkina Faso	Bucket Kit Irrigation in Kenya
Field in Honduras	Group Work in Haiti	Seminar Participants in Cambodia
Seminar Participants in Thailand	Rotary Project in Guyana	A Group Photo in Guinea

Bucket-kits distributed by:

APPENDIX II

Chapin Living Waters (CLW)
Bucket Kit News
August 2013

Board Member Dr. J.L. Williams has set up Super Bucket Kits for Haitians !

Dr. J.L.   Patt
Williams   Williams

Mr. VanWingerden helping to roll out Drip Tape for the new garden.

Dr. Williams helped set up a drip system for the Haitians living in Abaco, Bahamas.

Workers Planting Crop in Abaco, Bahamas

Bucket-Kit Activities by Chapin Living Waters Foundation.

CHAPTER 6

MICRO IRRIGATION POTENTIAL IN INDIA*

R. K. SIVANAPPAN

CONTENTS

6.1   Micro Irrigation: Macro Future in India 88
6.2   Microirrigation Technology for Marginal Farmers 94
6.3   Advances in Microirrigation and Fertigation iIn India 100
6.4   Microirrigation for Closely Spaced Crops 109
6.5   Micro Irrigation in Fruit Crops 110
6.6   Summary 115
Keywords 117
References 119
Appendix I 120

*In this chapter, conversion rate for an Indian currency is Rs. 64.00 = US$ 1.00 on September 30, 2013. The prices of drip irrigation system/crops are based on 1999–2000 prices. Information in this chapter is based on the site visits by the author and interviews with farmers/scientists in India. Author acknowledges the consultancy opportunity by Water Technology Center at TNAU, Coimbatore; Central and State Governments in India; NGOs in India; and the International Development Enterprises.

6.1 MICRO IRRIGATION: MACRO FUTURE IN INDIA

India is blessed with abundant water resources. However, due to various physiographic conditions, existing legal constraints and the current available technology of water development, the utilizable water for irrigation is very limited. This is happening at a time when the demand for various agricultural commodities, Industries and drinking purposes are steeply rising due to population growth. The emerging balance is to increase the agricultural productivity per unit area, per unit water and per unit time. In this section, the readers will be able to appreciate the macro future of micro/drip/ trickle (surface and subsurface) irrigation systems under Indian conditions. Data presented in this section pertains to the period 1970 to 2002 A.D.

Technological innovations are to be exploited to achieve the twin objectives of higher productivity and better water use efficiency. Micro or drip or trickle irrigation has revolutionized cultivation of agricultural crops under irrigation in many countries in the world. India is embarking on a massive program for popularizing this method. Although the economic and social advantages are substantial, yet the expected coverage of micro irrigation system has not happened due to institutional, political, financial and technical problems. The area under micro irrigation in India is only about 3.5×10^5 to 4.0×10^5 hectares (= ha), of which more than 50% is in Maharashtra state alone followed by Karnataka, Andhra Pradesh and Tamil Nadu states of India.

Although irrigation is as old as civilization, yet it is a modern science—the science of survival. Many changes have taken place in all walks of life, but the age old practice of irrigation is still followed without much change. In the traditional irrigation method, the water use efficiency is very poor and in some cases it causes water logging and salinity. Therefore, we must adopt advanced methods of irrigation like sprinkler and micro irrigation in large areas to cope up with the demand of water. By introducing pipe conveyance of water coupled with sprinkler and micro irrigation systems for the crops wherever it is possible and necessary. All the cultivable land in India (i.e., 145 million ha) can be brought under irrigation.

Micro irrigation is an accepted method of irrigation even in the developing countries. The farmers are aware of the advantages of the system, but the adoption is very slow due to high initial investment cost for the system. More than 3 million hectares are under drip irrigation for about 60 crops in the world; whereas in India, it is only about 0.4 million-ha. Research and demonstration efforts for this system have been in progress since 1970 onwards, but large-scale adoption has been only during the nineties. The Government of India has constituted a National Committee on the use of Plastics in Agriculture (NCPA), that has helped in popularizing the MI system and to provide government incentives to the farmers. There are numerous successful micro irrigation farms and some failed farms due to the defects in design and poor quality of water and maintenance of the systems. The success story of Kuppam in Andhra Pradesh, India will enhance the adoptions of the system in the coming years.

Micro irrigation (MI) system consists of: pump head, main, submain, lateral, drippers, fertigation equipment, filtration system, and other accessories to deliver the required quantity of water near the root zone of crop. The main accessories are filters and fertilizer tanks or venturi. Filters are necessary to prevent clogging of the drippers. Fertilizers can be applied through venturi or fertilizer tanks to avoid wastage of this costly commodity. In India, biwall or canewall system is used for closely spaced crops: Sugarcane, vegetables, cotton etc. Micro sprinklers and micro sprayers are also available to sprinkle the water around the root zone of the tree crops (fruit/chards). There are many other types MI: Typhoon, T-tape, Twin wall, etc. The advantages and disadvantages of micro irrigation are:

ADVANTAGES	DISADVANTAGES
Controlled root zone environment.	High initial investment.
Energy conservation.	Highly technical.
Enhanced plant growth and yield.	More care and maintenance of the system.
Improved fertilizer and other chemical application.	Persistent maintenance requirements.
Improved quality of production.	Rats/Rodents damage the line/drippers.
Judicious use of water and uniform application of water at low application rate.	
Reduced operation and labor cost.	
Reduced salinity hazards to crops.	
Reduced weed growth.	
Uninterrupted cultural operations.	
Used in difficult terrain and problem soils.	

Since 1970, research work on MI has been in progress in the various agricultural universities, research institutions/centers, manufacturing firms and farmers. The institutions in India are: the Tamil Nadu Agricultural University—Coimbatore, Indian Agricultural Research Institute—New Delhi, University of Udaipur and Jobner, Haryana Agricultural University—Hisar, CAZRI Jodhpur, University of Agricultural Sciences—Bangalore, Agricultural University Kalyani, Centre for Water Resources Development and Management—Calicut, Punjab Agriculture University—Ludhiana, Indian Council of Agricultural Researches Complex for North Eastern Hill region—Shillong, Indian Institute of Horticultural Research—Bangalore, PKV Akola, MPAU—Rahuri, Vasantdada Sugar Institute Manjari (Pune), Directorate of Irrigation Research and Development—Pune, Indian Petrochemicals Limited—Vadodara, etc.

The studies conducted by various institutions have revealed that the water saving in the MI method compared to surface irrigation is about 40–70% and increased yield up to 100% for some crops. In addition, the saline water can also be used in this system and the salt is accumulated only at the surface of the periphery of wetting zone and hence does not affect the crop growth. The water consumption and crop yield for MI and conventional methods is given in Table 1. Therefore, the potential/prospects of the MI system is very high in India.

TABLE 1   Water used and yield for various crops in drip and conventional irrigation methods in India (1990).

Crop	Yield (100 Kg per ha)			Water depth applied (cm)		
	Con-ven-tional	Drip	Increase in yield (%)	Conven-tional	Drip	Water Saving (%)
Banana	575.00	875.00	52	176.00	97.00	45
Grapes	264.00	325.00	23	53.20	27.80	48
Mosambi, 000	100.00	150.00	50	166.00	64.00	61
Pomegranate, 000	55.00	109.00	98	144.00	78.50	45
Sugarcane	1280.00	1700.00	33	215.00	94.00	65
Tomato	320.00	480.00	50	30.00	18.40	39
Watermelon	240.00	450.00	88	33.00	21.00	36
Cotton	23.30	29.50	27	89.53	42.00	53
Lady finger	152.61	177.24	16	53.68	32.44	40
Egg plant	280.00	320.00	14	90.00	42.00	53
Bitter gourd	154.34	214.71	39	24.50	11.55	53
Ridge gourd	171.30	200.00	17	42.00	17.20	59
Cabbage	195.80	200.00	2	66.00	26.67	60
Papaya	13.40	23.48	75	228.00	73.30	68
Radish	70.45	71.86	2	46.41	10.81	77
Beet root	45.71	48.87	7	88.71	17.73	79
Chilly	42.33	60.88	44	109.71	41.77	62
Sweet potato	42.44	58.88	39	63.14	25.50	60

Source: National Committee on the use of Plastics in Agriculture (NCPA), Status, potential and approach for Adoption of Drip and Sprinkler Irrigation System, Pune, 1990.

The development of the system has been slow and that too it is adopted only in the South and North Western States due to water scarcity and maintenance of sustainable agriculture. The area has increased from almost zero in 1970 to 1,000 ha in 1985; and 60,000 ha in 1995; 225,000 ha in 1998; 400,000 ha in 2002 covering about 30 crops in India. The experience of numerous farmers in the new method reveals many interesting results. The old coconut farms are not possible to irrigate as the ground water is depleting. Numerous farmers in Coimbatore district has taken up this irrigation for their coconut trees and it has been successful for them. Now water is not available even for drip systems. The development of micro irrigation is very spectacular in Maharashtra after 1987. This is due to the encouragement provided by the government and the publicity given by the drip system manufacturers to the farmers. The farmers are also forced to take up since water has become scarce commodity in Nashik, Jalgoan, Pune, Ahmednagar, and other districts. In Nashik, the grape farmers are purchasing truck-load of water during summer (April–June) to give water through micro irrigation system to their grape vine. It was informed that more than 90% of grape vine is irrigated by drip method. Although the cost for the system is about Rs. 30,000–40,000/ha, yet the BCR is very high. Similarly many farmers are using micro irrigation for: Banana, pomegranate, *mosambi* (sweet lime), tomato, cotton, sugarcane and various other crops in Maharashtra state.

In Kerala state, the coconut and other plantation crops need water during January to May, and the farmers are introducing micro irrigation due to the shortage of water. The experiences in Karnataka and Andhra Pradesh states are also encouraging, especially for grapes, coconuts and other fruit crops. It is slowly catching up in Gujarat, Rajasthan and Madhya Pradesh states. This system is not familiar (not adopted) in Northern and North Eastern States. It may be due to abundance of water potential in these states. However, this method is very well suited to the undulated terrain and for commercial and plantation crops like tea, cardamom, rubber etc., to get more yield and income to the farmers. Therefore, the farmers should be educated to adopt this system: to get more income, to save fertilizer and labor.

If you ask a farmer in USA, why he is practicing micro irrigation, his answer will be: 1st more yield, 2nd less labor, 3rd less inputs (fertilizer/pesticides) and 4th saving of water.

Although we have not come to the level in USA and Israel, but farmers are convinced that this system is beneficial and useful to get more income for sustainability especially where water is very scare—in South and North Western States of India. However, a farmer Mr. S.V. Kulkarni of Vivrekhand village in Jalgoan district, Maharashtra, when asked why he has introduced drip for the banana crop, his answer was: Increased production, decreased labor, early maturity and good quality of banana, and water saving.

Micro irrigation is thought to be an expensive venture. The experimental results are supportive of micro irrigation and its technical feasibility has also been established. The social acceptability is also decision-making factor in all the states. The large-scale adoption involves a crucial question of economic viability. The cost of the system depends on: the crop type, row spacing, crop water requirements, location of water source, etc. It varies from Rs. 15,000/ha for coconut/mango—wide spaced crops—to Rs. 50,000–60,000/Ha for sugarcane, vegetables, cotton—closely spaced row crops.

The benefit-cost ratio (BCR) for drip system was worked out by interviewing the farmers in Maharashtra and Tamil Nadu states. The range of BCR excluding the proposition of water saving was from 1.31 to 2.60 for various crops excluding grapes, and for grape it was about 13.35. If water saving is considered, the BCR range goes up from 2.78 to 11.05 for various crops and 30.00 for grapes, respectively (Table 2). This accounts for economic logic of entrepreneurial grape farmers to go in for the drip system in an expanded scale throughout India.

TABLE 2   Benefit-cost ratio (BCR) for various crops under drip irrigation in India.

Crop	Row spacing m × m (ft. × ft.)	Cost of the MI system Rs/acre	Benefit-cost ratio, BCR	
			Excluding water saving	Including water saving
Coconut	7.62 × 7.62(25' × 25')	7,000	1.41	5.14
Grapes	3.04 × 3.04(10' × 10')	12,000	13.35	32.32
Grapes	2.44 × 2.44(8' × 8')	16,000	11.50	27.08
Banana	1.52 × 1.52(5' × 5')	18,000	1.52	3.02
Orange	4.57 × 4.57(15' × 15')	9,000	2.60	11.05
Acid lime (citrusSp.)	4.57 × 4.57(15' × 15')	9,000	1.76	6.01
Pome-granate Mango	3.04 × 3.04(10' × 10')	12,000	1.31	4.04
Papaya	7.62 × 7.62(25' × 25')	7,000	1.35	8.02
Sugarcane	2.13 × 2.13(6' × 6')	18,000	1.54	4.01
Vegetables	Between lateral1.83 (6')	20,000	1.31	2.78
	Between lateral1.83 (6')	20,000	1.35	3.09

To encourage the use of MI system, the central and state Governments in India are giving subsidy to the farmers with a matching contribution of 50% from the State Governments. However, in practice not all the State Governments are implementing this. Apart from this, some State Governments are operating subsidy schemes to install MI. The rate of subsidy varies from state to state and also differ from one category of farmers to others. However, this is still in the initial stage, and we have to go a long way in achieving the objectives to popularize the MI system. Although NABARD has approved to the commercial banks to provide loans for the drip farmers, yet it is very cumbersome procedure for the farmers to get loans from the bank especially by the small and marginal farmers. There are more than 40 manufacturing company/firms in MI system in the country.

The MI system is technically viable and socially acceptable. Therefore, this method can be adopted under Indian conditions in the following situations:

a. Well irrigated areas which constitutes about 35% of the irrigated area in the country.

b. It is ideally suited to all row crops especially coconut, fruit trees, vegetables, flowers, commercial crops like cotton and sugarcane, plantation crops etc., which occupy more than 18 Mha of land.

c. Waste lands by planting trees including fruit trees. More than 25 Mha of land is available under this category.

d. Hills and semiarid areas.

e. Coastal sandy belts.

f. Water scarcity areas.

g. Command area of the community wells.

Drip irrigation in the world is under varying degree of development. In Israel, the entire area is now irrigated by MI. In USA, large areas of crop like grape, sugarcane, cotton and other fruit orchards are irrigated through MI. Similar trend is noticed in Australia, Southern Europe and many other countries. Drip irrigation has increased at a faster rate, due to: The demand of water for various needs, and to increase the yield and quality of the produce.

The farmers have been compelled to opt for the advanced method of irrigation, due to stepping up of irrigation, limited available water resources and high demand of water from all sectors (industries and drinking). The awareness of the farmers to increase the production and income has kindled them to use the water more efficiently. It is reported that drip irrigation farmers in Maharashtra and other states are able to get a net profit of Rs. 50,000 to 100,000 per ha (0.40 ha = one acre) by growing grapes, orange, pomegranate and other fruit and vegetable crops. Similarly, the yield of tea crop has increased by about 30% with drip irrigation in the summer dry months. In spite of above facts, the area under drip is very meager in India. Therefore, it is projected to have 1 million ha in 2005 that is 1% of the irrigated area, and about 10 million ha in 2025.

The cost of MI for 1 million ha will be about Rs. 40 billion (average Rs. 40,000 per ha). This financial investment is not much compared to the benefits achieved by increasing the crop yield. It will also generate employment opportunity on large scale. The following constraints are evident to bring such large area under micro irrigation.

- High initial cost.
- Quality of drip material—clogging of drippers and cracking of lateral pipes and drippers.
- Poor awareness by the farmers
- Lack of adequate technical inputs.
- Damage due to rats and rodents.
- High cost of spare parts and components.
- Difficulty in getting subsidy and loans.
- Insufficient extension and promotional activities.

To exploit the full potential of the MI system, the constraints need be relaxed by appropriate policy instruments, financial support and technical guidance. This calls for an integrated approach and endeavor on the part of the Central and state Governments, implementing agencies, manufacturing companies and farmers. The technology is to be perfected. Therefore, more field oriented research needs to be initiated by the Universities and Research institutions. Training should be imparted to the officials and farmers to learn about the installation, operation and maintenance of MI. Seminars and workshops can be organized at the site, block, district, state and central levels to learn about: Potential, management, operation, service and maintenance, fertigation and automation procedures, potential problems and other details related to drip irrigation.

1. The main policy should encourage and motivate all categories of farmers, since the critical issue is saving of water and augment productivity.
2. The cost of the system should be brought down.
3. There should be a separate agency in-charge with the responsibility for popularizing and introducing the MI system.

4. The manufacturing companies should commit themselves for the supply of standard materials, equipment, and after-sale services.
5. Farmers should develop conviction and confidence in the system.
6. The research institution should integrate their activities with Research and Development unit's studies in potential problems in micro irrigation.

The Governments should prepare action plan for the coming years to bring one Mha i.e. 1% of the irrigated area within the next 5 years and 10 Mha in 2025. Although land/water/input resources are constant and limited, yet it is possible to increase the gross area by 2–3 times to the net area due to water availability by introducing drip irrigation. The water constraint can be overcome to some extent by introducing micro or drip irrigation, wherever it is possible and suitable. This will solve many problems in India, such as: Unemployment and foreign exchange by exporting vegetables, fruits, and flowers, etc. in the coming years.

6.2 MICROIRRIGATION TECHNOLOGY FOR MARGINAL FARMERS

6.2.1 INTRODUCTION

The net cultivated/sown area in India is about 145 Mha. The gross sown area will be about 210–220 Mha in another 10 to 15 years. The irrigation potential is about 85 Mha at the present, though the net irrigated is only about 82.0 Mha. The potential utilizable water is about 112 Mhm (surface water 69 Mhm + ground water 43 Mhm) which may be sufficient to irrigate about 135–140 Mha. Therefore, even with full exploitation, about 40–50% of the cultivated area will remain as rain-fed agriculture.

The per hectare investment cost for an irrigation project has increased enormously from about Rs. 1500 during first five year plan (1951–56) to more than Rs. 100,000 during 10th plan period (2002–2007). Therefore, it is necessary to economize on the use of water for agriculture to bring more under irrigation and reduce the cost of irrigation per hectare.

6.2.2 MICRO IRRIGATION

Micro irrigation can be practiced for all row crops in all soils. In this method, the required quantity of water is given to each plant daily at the root zone through net-work of pipes. Therefore, there is minimum loss of water in the conveyance and distribution in this system. Since the required quantity of water is applied daily, soil moisture will be always available more or less at the field capacity. It means the crop can take moisture from the soil without water-stress. Therefore, the growth of the crop is uniform and optimum yield can be obtained by MI system.

Micro irrigation system is very well suited for undulated terrain, shallow soils and water scarce areas. Saline/brackish water can also be used to some extent, since water is applied daily, which keeps the salt stress at minimum and the salt will be pushed to the periphery of the moisture regime which is away from the root zone of the crop. Therefore, it does not affect the crop growth. The main advantage of micro irrigation as compared to gravity irrigation system are:
• Increased water use efficiency (90–95%).
• Higher yield (40–100%).
• Decreased tillage requirements.

- High quality products.
- Higher fertilizer use efficiency (30% saving in fertilizer).
- Less weed growth.
- All operations can be done at all times.
- Less labor requirements.

6.2.3   NEED FOR MICRO IRRIGATION

The total rainfall is about 400 Mhm in India, but the utilizable quantity is only about 112 Mhm both from surface and ground water. Even with full exploitation of the potential, nearly 40–50% of the cultivated area will remain under rain-fed. Further, the investment per hectare for an irrigation system has increased from Rs. 1500 during first five year plan (1951–1956) to about Rs.45,000 during seventh plan and it is more than Rs. 100,000 during tenth plan (2002–2007). In addition, the ground water table in many parts of India is depleting year after year at a rate of about one meter per year. In many parts India, thousands of wells are abandoned in view of the alarming depletion of water in the wells. In Tamil Nadu state, it is reported that more than 150,000 open wells are abandoned for lack of water. The crop yield per hectare is not comparable with other developed and even in some developing countries for all crops including paddy, vegetable, fruit, etc. Therefore, it is necessary: to economize the use of water for agriculture to bring more area under irrigation; to reduce the cost of irrigation per hectare; and to increase the productivity per unit area from unit quantity of water. This can be achieved by introducing advanced irrigation methods like micro irrigation with improved water management practices.

The experiments/ studies conducted in various research institutions in India have indicated that the water saving for any crop is about 40–70% and the crop yield has increased up to 100% (i.e., double the yield). In spite of high installation cost of MI system, the economics were worked out by the author for various crops and it is viable. The pay-back period varies from 6–24 months and the BCR ratio is about 1.77 to 7.03 (Table 3). Therefore, the advanced method can play an important role in improving the living condition of the farmers in the country. Further to popularize the system, the Central and State Governments in India are giving subsidy. It was up to 70–90% but now it has come down to 25%. In addition, 90% of the subsidy is borne by the Central Government and only 10% is borne by the State Government. However, it is not available for all crops, which are suitable for drip irrigation. Another constraint is that the system and its usefulness is not known among farmers and even to Government officials who are in charge of popularizing drip irrigation, that is, agriculture and irrigation department officials.

TABLE 3 Benefit-cost ratios (BCR) and payback periods for selected crops under micro irrigation in India (1996).

Variable	Crops									
	Sugarcane 2.75'x 2.75' x5.5'	Banana 5'x5' 3'x5' x6'	Cotton 3'x 5'x 6'	Papaya 6'x6' 8'x8'	Grapes 10'x 6'	Pomegranate 14'x 14'	Ber 15'x 15'	Tomato 45'x 45'x 165 cm	Strawberry 9'x 12'x 9'	Rose 3'x 2' x5'
System Cost, Rs/acre	19,000 Drip 10,000 Canewall	19,000	19,000	16,000	17,000	12,000	12,000	19,000 Drip 12,000 Canewall	75,000	10,000 Micro tube
System Cost, Rs. per hectare	47,500 Drip 25,000 Canewall	47,500	47,500	40,000	44,000	30,000	30,000	47,500 Drip 30,000 Canewall	188,000	25,000 Micro tube
Water used	20,000 Lpd/ acre	15–20 Lpd/ acre	8–10 Lpd/P	15 Lpd/P	12–20 Lpd/P	50–60–60 Lpd/P Lpd/P 12,000–15,000 Lpd/acre		20,000 to 24,000 Lpd/ acre	2 Lpd/pl	10 Lpd/pl
Yield tons/ acre	80 tons	30 tons	14 (100 Kg)	750 kg Latex 60 tons fruit	20 tons	9 tons	10 tons	30 tons	3 tons	1000/ day
Payback period, months	12	12	18 three crops	18 one crop season	<12	<12	<12	6 or one season	24 or two season	<12
B.C.R.	3.44	3.08	1.77	4.09	3.64	7.03	6.51	1.91	2.34	2.71
Extra Income* Rs	31,620	49,320	14,360	72,040	2,64,200	1,51,280	1,33,280	49,280	67,000	1,19,190

Note: *Due to drip irrigation over conventional method. B.C.R. = Benefit-cost ratio.

Source: Sivanappan R K., Strengths and weakness of growth of drip irrigation in India, Paper presented in the national seminar organized by the National Water Development agency at Bhopal, 1996.

Micro irrigation is an accepted method of irrigation. Most of the farmers are convinced of the usefulness of the system, but the adoption is very slow due to its high initial investment cost. The cost of the drip system for wide spaced crops like coconut, mango, pomegranate, oil palm is about Rs. 15,000–25,000 per hectare. The cost for closely spaced crops like vegetables, cotton, sugarcane, mulberry etc. is about Rs. 50,000–60,000 per hectare for MI system using 4 or 5 drippers per tree or one dripper at 50–75 cm spacing for crops like sugarcane, cotton, vegetables, etc.

The area under drip irrigation in India was only about 1000 Ha in 1985 and increased to 60,000 Ha in 1993. The present area under micro irrigation in the country is about 500,000 ha covering about 20 different crops like coconut, mango, oil palm, guava, sapota, pomegranate, *ber*(Indian jujube, Zizyphus spp.), lime, orange, grapes, cotton, sugarcane, vegetables (potato, tomato, onion, etc.), plantation crops (tea, coffee, cardamom, pepper, chilly) and flowers like rose, and mulberry (Morus spp.) etc. The projected area is estimated about 1 million hectare (1% of the irrigated area) in the year 2000–2005 AD and about 10 Mha (10% of the irrigated area) by 2020, though the crop area which is suited for the MI system is about 27 Mha as recorded by the government task force on the micro system. But the main constraint is its cost especially to the small and marginal farmers in India who comprise about 83% of total number of farmers and get only about 35% by income

6.2.4   NEED FOR AFFORDABLE LOW COST MICRO IRRIGATION SYSTEM

The drip irrigation system is capital intensive and involves the sophisticated technology. Therefore, it is beyond the reach of the majority of small and marginal farmers in India. Therefore, if the drip system is made affordable and suitable to the crop needs of small and marginal farmers in India, then we can help to increase the productivity and income of the farmers; and help to conserve the scarce water. This will also improve the capability of these farmers to manage and manipulate the soil, water and crop to obtain maximum yields, because it is suitable for all topographical and agro-climatic conditions, and different crops.

To make the MI technology accessible to small and marginal farmers, International Development Enterprises (IDE: 4,6,7) has developed a low cost drip system and is now extensively field testing this affordable drip system in India, Nepal, Bangladesh and Kenya. Costs are being reduced by eliminating the use of sophisticated parts, replacing with low cost substitutes without sacrificing the quality and performance. The IDE has reduced the cost by 80% by making the system portable and doing away with emitters/drippers but using holes and sockets or microtubes to deliver water. In case, the system cannot be portable, the reduction is about 50–60%, which is affordable for the marginal farmers even without government subsides. Because it is very difficult and cumbersome for the farmers to get subsidy from the Government and it is not available throughout the year and for all crops. This technology was used in the early seventies by the author in College of Agricultural Engineering of Tamil Nadu Agricultural University (TNAU) – Coimbatore, when there was no drip irrigation manufacturers in the country and most of the research work was done using the material available in the market. The system will catch up very well if the required drip irrigation quality material and accessories are available in the market.

Affordable or low cost MI system can be used for all row crops and by all farm-ers, since the cost is reduced to an extent of 40–50%. As an example of closely spaced crops with row spacing of 5 cm or 7.5 cm (2″ or 3″), each lateral pipe can serve 4 to 6 rows of crop by providing suitable microtubes on both sides of the lateral. The uni-formity of water application can be attained by adjusting the lateral lengths/providing appropriate lengths of microtubes in the outlets to get uniform water to all plants. The farmers can be educated about the installation and operation of the MI system to get the uniformity in water application. For in depth study on "*Low Cost Micro Irrigation System for Marginal Farmers,*" the reader may consult:

1. Small holder Irrigation Market Initiative: <http://www.siminet.org/fs_start.htm>
2. <www.ideorg.org>
3. Polak, Paul and R. K. Sivanappan, 2001. The potential contribution of low cost drip irrigation to the improvement of irrigation productivity in India. Technical Report prepared for International Development Enterprises (IDE).
4. Polak, Paul, 2008. *Out of Poverty: What Works When Traditional Approaches Fail.* San Francisco, CA, USA: Barrett-Koehler Publishers.

6.2.4.1   LOW COST MICRO IRRIGATION SYSTEM: CASE STUDIES
The International Development Enterprises (IDE) is popularizing the low cost MI sys-tem in hill areas of Nepal for vegetable crops, mulberry cultivations in South India (Karnataka, Andhra Pradesh and Tamil Nadu), sugarcane in Rajasthan and cotton crop in Madhya Pradesh, etc. (Fig. 1). The mission of IDE is to introduce the affordable MI system on large scale in most places in India to all small/marginal farmers to all crops except paddy. This venture will help to save water and to improve the living condi-tions of these farmers. "Smallholder irrigation technologies can make a difference for the poor. Access to irrigation is a limiting factor to the productivity and profitability of small farms in many parts of the world. Low-cost micro irrigation and a series of other low-cost technologies related to small-scale irrigation like treadle pump, rope and similar pumps, small-scale water storage technologies etc. have a good potential to allow a large number of small farm households to escape the most severe poverty. These technologies can enable them to produce high value cash crops for local and more distant markets, or to produce food during the dry season, and thus to increase household incomes and improve livelihood security. However, smallholders can only exploit the potential that these technologies offer if they have access to the technolo-gies themselves including spare-parts and to adequate means of production, know-how and markets." The demonstration and case studies in Nepal and Karnataka state in India are briefly discussed below:

6.2.4.1.1   NEPAL—VEGETABLES

There is more than adequate moisture during the rainy season (June–September), and the crop suffers for lack of moisture for rest of the year. There are water in the springs throughout the year and these sources are tapped for providing water through drip ir-rigation. The systems were installed on small farms for vegetable crops in hill areas

Western Katmandu. The results of these trials were extremely encouraging. All farmers in the project doubled the irrigated area as a result of introducing the system with 50% of the workers. The uniformity of flow from the drip holes within terrains was found to be 84% and 73% across all terrain varying from 2–4 meters.

FIGURE 1   Examples of low cost affordable systems (IDE).

6.2.4.1.1   *KARNATAKA—MULBERRY*

The International Development Enterprises has installed low cost drip irrigation on 15small farms using microtubes in place of emitters/drippers. Each lateral line can deliver water to 4/6 rows of mulberry depending upon the spacing both in pit method and row methods. The cost is about Rs. 18,000/acre for row system and Rs. 10,000–12,000/acre for pit system (3'x3'/2'x2'). The traditional drip system costs about Rs. 30,000–35,000/acre. After onsite visits and interviews with the farmers, the following observations were made:

   i)   The yield is more for drip irrigation compared to conventional surface irrigation.

ii)   The water used in MI is about 50% of the surface method.

iii) It is possible to maintain the water table in the area if the entire area is brought under drip irrigation.

6.2.5   CONCLUSIONS

The target of 10 Mha of drip-irrigated area by 2020 can be achieved only if all the farmers take up drip irrigation for all row crops. This will be possible only if the cost of MI system is affordable for small and marginal farmers without sacrificing the quality and uniformity of water application.

In areas with a high concentration of small and marginal farmer operations and scarcity of water, the low cost drip irrigation system will provide access to water saving irrigation technology, which is affordable on a small scale and divisible. This system is likely to be applicable to areas where both water and crops have relatively high value and where the high capital cost of existing drip irrigation technology constitute a significant barrier to its adoption.

6.3   ADVANCES IN MICROIRRIGATION AND FERTIGATION IN INDIA

6.3.1   INTRODUCTION

There has been a great deal of intensive and serious concerns about the limited availability of earth's fresh resources and management since the "1972 International Conference on Water and the Environment in Dublin" and the "June 1992 United Nation Conference on Environment and Development in Rio de Janeiro, Brazil." This has bought out the worldwide reforms in the water sectors. The Agenda 21 of the conference indicates: "Sustainability of food production increasingly depends on sound and efficient water use and conservation practices consisting primarily of irrigation development and management with respect to rain-fed areas, livestock, water supply, inland fisheries and agroforestry. Achieving food security is a high priority in many countries, and agriculture must not only provide food for rising population, but also save water for other uses. The challenge is to develop and apply water saving technology and management methods; and through capacity building enable communities to adopt new approaches, for both rain-fed and irrigated agriculture."

According to a recent survey, 30 four countries will face water scarcity by 2025 AD indicating that the per capita availability of fresh water supplies will be less than 1000 M³/person/year. The threshold renewable water availability on an annual per capita basis is about 1700 M³. A country, with a threshold value will suffer only occasional or local water problem. Below this threshold, countries begin to experience periodic or regular water stress. According to a study by the World Resource Institute, the countries with semiarid regions will be in a critical situation. India (1400 M³/person/year) and China (1700 M³/person/year) will come into this category in the year 2025 AD, but USA will have more than 7000 M³/person/year.

Raising demand for urban and industrial water supplies in the world, poses a serious threat to irrigated agriculture i.e., the allocation for agriculture will come down to 50–55% from the present 66–68% on global base (Table 4). In India, it is estimated that the allocation of water for agriculture will be reduced to 71% from the present 85% in the next 15–20 years.

TABLE 4   Percentage of total water demand in the world.

Various uses	% of total demand in the world	
	In 1993	In 2025
Domestic	6–8	15–20
Industry	26	30
Agriculture	66–68	50–55

Land and water are the basic needs for agriculture and economic development of any country. The demand for these resources is continuously increasing. Further, the per capita availability of water resources in India is much less compared to many other countries. Indian experts have assessed that water supply will be a major resource constraint to limit the economic development. The potential utilizable volume of water is estimated at 110 million hectare meters (Mhm). Even with full exploitation of this potential, nearly 40–50% of the cultivated area will remain under rain-fed. Further, the per hectare investment in irrigation projects has increased from Rs. 1500 during the first five year plan (1951–1956) to about Rs. 40,000 to 45,000 during the seventh plan (1985–1990) according to Naralawala (1992) and is about Rs. 100,000 during 9th plan (1997–2002). Therefore, it is necessary to economize the use of water for agriculture: To bring more area under irrigation; to reduce the cost per hectare of the irrigation system; and increase the crop productivity. This can be achieved by introducing advanced methods of irrigation like micro irrigation and sprinkler irrigation.

6.3.2   CONVENTIONAL IRRIGATION METHOD

Gravity irrigation is the dominant method of irrigation in India and it is going to be so in the coming years also. Generally, under open canal conveyance and surface irrigation methods, less than one-half of the water released reaches the plants. In major irrigation projects, the overall efficiency ranges from 30–35%. The low efficiency may be accounted for, in part, by conveyance losses due to seepage, evaporation and non-beneficial use by phreotophytes. The losses are also partly the result of poor farm distribution of water due to inadequate land preparation and lack of know-how of farmer in application of water, with consequent excess applications and deep percolation.

Higher efficiency in gravity irrigation can be obtained by improved facilities and skilled irrigators. These facilities and skills include: Proper land leveling and preparation; the planning of advanced techniques in the determination of irrigation frequencies; quantity in stream size and duration of irrigation; the installation of water measurement and regulation systems; the adaptation of surface irrigation technique to suit the crops; and the laying down of systems for the removal of excess waters and reusing the same again for irrigation. The preparation of land for gravity irrigation, in addition to the skills and knowledge of local conditions that are required, is costly and time-consuming. Furthermore, losses in productivity through the removal of the upper fertile soil layers due to runoff may be considerable.

Presently, water is applied once in every 7 to 10 days in surface (gravity) irrigation depending on the soil. Hence water stress will be noticed just before irrigation and the crop growth is affected. Furthermore, it is difficult to constantly supply the required quantity of water to the root zone using surface irrigation, thus resulting in yields less than the optimum. However, in micro irrigation water is given daily and hence moisture is available always equal to the field capacity to the plants.

6.3.3 MICRO IRRIGATION

Of the various ways of irrigating the crops, drip/micro irrigation ranks the top in water economy and productivity. In a MI, the water is directly delivered to the root zone of individual plants by network of tubing. The tubing can be laid permanently or part of it can be moved to different locations to reduce the capital cost of the system. The water is delivered at low pressure through emitters/microtubes. Drip irrigation techniques were developed commercially in Israel. Australia, Israel, Mexico, New Zealand, South Africa and the USA have been using drip irrigation for the various crops. Due to numerous advantages and benefits, this technology has been adopted by many countries in the world. The drip irrigated in the world is rapidly increasing (Table 5).

TABLE 5   Drip irrigated area (hectares) in the world.

Year	Area in ha	Year	Area in ha
1970	56,000	1999	2,800,000
1988	1,055,000	2001	>3,000,000
1991	1,600,000		

Although the drip-irrigated area is more than 3 Mha in the world, it represents only about 1% of the total irrigated area (270 Mha) in the world. In view of worsening water scarcity and raising water costs, there is tremendous scope of increasing the drip irrigated in the world. It is worthwhile to note that already six international conferences on drip irrigation have been organized, including in South Africa in 2000.

Experiments using drip irrigation have been conducted at various Research Institutions and Agricultural Universities in India, Israel, Jordan, Spain, USA and Australia. These studies have indicated that the water saving was about 40–70% and crop yield for all crops increased by 10–100% (Table 1) with micro irrigation. In addition to this, this system saves labor/ fertilizer to about 30% and produces high quality of fruit/ vegetables and other commodities. Table 6 indicates water productivity gains due to shifting from surface to drip irrigation in India. Drip system costs about Rs. 20,000–75,000/ha depending upon the crop, topography, source of water supply etc. Most farmers think it to be expensive. However, the economic studies by author indicate that the system is economically viable (Table 7).

TABLE 6   Water productivity gains from shifting to drip irrigation from surface irrigation in India (1994).

Crop	Change in yield	Change in water use, %	Change in water productivity
Banana	+52	-45	+173
Cabbage	+2	-60	+150
Cotton	+27	-53	+169

TABLE 6   *(Continued)*

Crop	Change in yield	Change in water use, %	Change in water productivity
Cotton	+25	-60	+255
Grapes	+23	-48	+134
Potato	+46	-0	+46
Sugar cane	+6	-60	+163
Sugar cane	+20	-30	+70
Sugar cane	+29	-47	+91
Sugar cane	+33	-65	+205
Sweet Potato	+39	-60	+243
Tomato	+5	-27	+49
Tomato	+50	-39	+14

Source: Sandra Postel—Pillar of sand—1999
Adapted from data Indian National Committee on Irrigation and Drainage, Drip Irrigation in India, Compiled by R.K. Sivanappan (New Delhi, 1994)
R K Sivanappan 'Prospects of Micro Irrigation in India' Journal of Irrigation and Drainage Systems, Volume 8, pp. 49–58 (1994).

TABLE 7   Benefit-cost ratio (BCR) and payback period for various crops under micro irrigation (1993).

Crops	Crop spacing	Cost of the system	Water used	Yield	Payback period	BCR
	m	Rs./ha	Lpd/plant	Tons/ha	months	
Banana	0.91x1.5x1.8 pair row	47,500	15–20	75	12	3.00
Grape	3.03x1.8	44,000	15–20	45	<12	3.28
Pomegranate	4.3x4.3	30,000	50–60	25	<12	5.16
Ber	4.5x4.5	30,000	60	25	12	4.56
Tomato	0.45x.45x1.65 paired row	30,000 Cane- wall	40,000 (Lpd/ha)	75	one season, 6 Months	1.09
Papaya	1.81x1.81	40,000	15	60	12	4.09
Cotton	0.9x1.5x1.8 pair row	47,500	8–10	1.5	18	1.83
Sugarcane	0.83x1.66 pair row	47.500	30,000 Lpd/ha	200	12	3.45

Source: Case studies conducted by the author with numerous farmers in Maharashtra State, November 1993

In India, the drip area has increased from 1000 ha in 1985 to about 400,000 ha in 2002, which is very meager (<0.5% of the irrigated area). The Drip irrigation is increasing gradually as shown in Table 8.

TABLE 8   Area under drip irrigation in India.

Year	Area in ha	Year	Area in ha
1970	—	1996	1,62,000
1985	1,000	1998	2,25,000
1993	55,000	1999	2,54,000
1994	71,000	2001	3,10,000
		2002	>400,000

It is believed that there is a great potential in India for MI and it is possible to bring about 1 Mha (i.e., about 1% of the countries irrigated area) in 2004/2005, and about 10 Mha (10%) in the year 2020/2025. The case studies conducted by the author in different parts of India have indicated that all farmers are convinced the usefulness of the system. Their main complaints are:

- Though the Government is giving subsidy in some states (not in all states) farmers are experiencing innumerable problem to obtain the subsidy because of the procedural difficulties.
- Subsidy is not given for all crops
- Subsidy is not available throughout the year and only one or two months before the end of financial year
- The cost of the system is increased by the manufacturers when subsidy is increased.
- The field staff at various level are not fully conversant with the basic philosophy and concept of drip/sprinkler system.

This may be the case in many developing countries. Scientists/NGO's are researching to bring down the cost of the system so that the poor and small farmers can afford the system. It is possible to reduce the cost by about 40–50% by proper geometry of the crop planting, irrigating 2 to 4 rows by providing suitable micro tubes on both sides of the laterals and by proper design and layout of the system. For popularizing this low cost/low energy system in India, field trails are being conducted in farmer's field in large scale for crops like vegetables, cotton, sugarcane, mulberry and orchard crops by Universities and NGO, etc. and the response are very much encouraging.

6.3.4   SCOPE OF MICRO IRRIGATION

The experience of numerous farmers in this new method revealed many interesting results. The old coconut farms are not possible to irrigate as the water availability in the wells has reduced and water table is depleting every year. Numerous farmers in Coimbatore District in Tamil Nadu state have adopted micro irrigation for the coconut trees and it has proved successful. Due to drought this year, thousand of coconut trees died for lack of water even for drip irrigation. The development of micro irrigation is very spectacular in Maharashtra state after 1987. This is due to the encouragement provided by the government and promotional efforts given by the manufacturers. The farmers are also forced to take up since water has become scarce commodity in many districts. In Kerala, the coconut and other plantation crops need water during the dry period of January through May and the farmers are installing micro irrigation to manage the shortage of water. It is picking up very well in Andhra Pradesh, Karnataka, Madhya Pradesh, Gujarat and Rajasthan states. The farmers are now convinced that MI system helps them to get more yield, with less inputs apart from water saving and get more income. The arecanut farmers in Karnataka (Shimoga) are going in a big way to use drip irrigation with 4/5 multi crop-layers (Inter crops) along with arecanut such as coconut, black pepper, banana, beans etc. About 20,000 ha of crops are under drip in and around Shimoga and farmers get more profit due to Drip Irrigation.

In Andhra Pradesh, the Drip Irrigation has been introduced for vegetable crops with the technical support of Israel and Financial help of Govt. of India and Japan Government. The farmers are taking 3 or 4 crops in a year. Although the cost of the system is about Rs. 30, 000 to 40,000 per acre including bore well with pump sets, yet the farmers are happy because they get more income after shifting to drip Irrigation.

In Maharashtra, the story is different, the grape wine farmers of Nasik, the banana farmers of Jalgoan, pomegranate farmers of Pune were encouraged by going for drip Irrigation and now more than 85% of the above crops in these districts are under drip Irrigation.

In Tamil Nadu, the Government has decided to develop the entire waste land/fallow land of 2.0 Mha under tree crops/horticulture crops using drip irrigation wherever sufficient water is available for drip system by giving subsides/incentives to the farmers. The same analogy can be followed throughout India to bring all the wasteland under production and to generate employment opportunity.

6.3.5   ADVANCES IN MICRO IRRIGATION

Numerous innovations and developments have taken place in a short period of 20–25 years in Micro irrigation system. The research was started using available pipes in the market and making holes and socket/micro tubes in 1969/70 by the author at TNAU Coimbatore and development has been tremendous: The advances in irrigation pipes, especially the LLDPE for laterals in drippers/emitters, micro sprayers/Sprinklers and jet/self-compensated, pulsation and bubbler etc. The inline/online drippers of different categories have been developed by various companies. Many types of filters are available to suit the quality of water/source of water. In fittings also the company's developed easy fittings and quick connections. Fertigation equipment having quick injection tanks/venturi to suit the farmers and for the demand have been introduced. Now many farmers are going to automation with use of computers. All the developments are taking place in a quick successions, but the extent of drip area is progressing very slowly in the country.

6.3.6   CONSTRAINTS

The constraints experienced in bringing large area under this method are: High initial cost, clogging the drippers and cracking of pipes, lack of adequate technical inputs, damages due to rats and rodents, high cost of spares and components and insufficient extension efforts. To exploit the full potential of MI system, the constraints are to be removed by appropriate policy instruments, financial supports and technical guidance. This calls for an integrated approach and endeavor on the part of the federal and state governments, implementing agencies, manufacturing companies and the users i.e., farmers. The technology is to be perfected and hence more field oriented research are to be taken by the universities to reduce the cost and problem free drip system. Training should be imparted to the officials and farmers to learn about the system and its maintenance for good performance.

Seminars and workshops can be organized in the village, block, taluk (Indian word for *tehsil*) and district levels to popularize the system to understand the problems and socioeconomic factors. The main policy should be to encourage and to motivate all categories of farmers, since the critical issue is saving of water and augment productivity. The government can prepare time bound action plan for the coming years.

6.3.7   FUTURE PROSPECTS

The studies conducted and information gathered from various farmers have revealed that drip irrigation is technically feasible, economically viable and socially acceptable. Drip irrigation can be implemented in most of the areas irrigated by open/ tube wells, which make about 35% of the total irrigated area in the country. The drip irrigation can be extended to the following category of lands:

1.   Waste lands after planting tree crops including fruit trees
2.   Hilly area
3.   Semi arid Zones
4.   Coastal sandy belts
5.   Water scarcity areas
6.   Command area of the community wells
7.   Wind mill (farm) areas.

At present, on an average about Rs. 100,000 is invested to bring one hectare of land under irrigation in the new irrigation project if water is available. As water is becoming increasingly scarcer, adoption of micro irrigation system offers potential for bringing nearly double the area under irrigation with the same quantity of water.

It has been considered as a boom for wide spaced perennial crops namely, mango, coconut, banana, grapes, pomegranate, ber, citrus, tea, coffee, cardamom and the like. It is also suited for vegetables, flowers and other commercial crops like cotton, tobacco and sugarcane. The details of cultivated and irrigated areas under fruits, vegetables, plantations crops at present and in the year 2020/25 are given in Table 9, which gives an idea of large potential area to bring under drip irrigation system in the coming years.

TABLE 9   Areas (Mha) sown and irrigated: Suitable for drip irrigation in India (2000).

Crop	Present area, 2000		Expected area, 2020/25	
	Sown	Irrigated	Sown	Irrigated
Coconut/Arecanut	1.5	0.9	2.0	1.0
Cotton	11.5	7.5	12.0	8.8
Flowers	0.8	0.3	1.6	0.90
Fruits	4.0	1.2	4.2	2.2
Plantation crops	2.8	1.0	3.0	1.6

TABLE 9   *(Continued)*

Sugarcane	5.5	5.5	6.0	6.0
Tobacco	0.6	0.4	0.6	0.6
Vegetable	8.3	4.2	10.0	7.4
Total	35.0	21.0	39.4 = 40	28.5 = 29

The area under micro irrigation at present is only about Four Lakh Hectare. Therefore there is a great future for the rapid expansion of the area under micro irrigation in India in the coming years. It is expected that the projected area of one Mha (i.e., 1% of the irrigated area) will be brought under micro irrigation in the next 2–3 years and about 10 Mha by the year 2020/2025 AD.

6.3.8   FERTIGATION

Fertigation is an effective tool in precision farming. Application of required fertilizers to the crops through drip irrigation water is known as fertigation. It offers potential for more accurate and timely crop nutrition leading to increase in yield, quality of the products and early maturity of crops. In fertigation water and nutrients are applied in effective root zone simultaneously and ensures application uniformly. The draw backs in conventional method are volatilization, leaching and fixation get minimized. When using water soluble fertilizer with drip irrigation, it saves fertilizer dozes, time, labor etc. It also minimizes Soil deterioration. This technology is adopted by farmers on large scale in Maharashtra state for high value crops like grapes, pomegranate, tomato, roses etc. and catching up on other states also.

6.3.8.1   ADVANTAGES OF WATER SOLUBLE FERTILIZER

- Application of fertilizers through drip system will result in uniform distribution at effective root zone and prevention of losses from leaching and run-off during heavy rainfall. This results in substantial saving in quantity of fertilizers (30% to 50%).
- By combining liquid fertilizers with insecticides and herbicides, labor and machinery saving occur.
- Allow crops to be grown on marginal lands, such as sandy or rocky soils, where accurate control on water and fertilizers in the plant's root environment is critical.
- Since these fertilizers can be applied through drip irrigation through venturi or fertilizer tank or injection pump, the application becomes very simple and thus results in saving labor.
- Higher water and fertilizer use efficiency and greater recovery of them.
- Uniform nutrients application to all plants.
- Optimizing the nutritional balance by supplying nutrients directly into the root zone in readily available form.
- Better availability of nutrient especially P&K due to maintenance of proper moisture level.
- Improvement in Physio-chemical properties of soil.

- Accurate place of fertilizer in available form.
- High osmotic potential.
- Less mechanical damage to soil and plants.
- Minimize pollution in soil.

6.3.8.2   LIMITATIONS OF WATER SOLUBLE FERTILIZERS
- Limitations for the use of water soluble fertilizers, drip irrigation is essential.
- Investment in fertigation equipments.
- High cost of suitable fertilizers.
- Lack of training, knowledge about crop and concepts.
- Continuous availability of these fertilizers.

The equipment required for fertigation are venturi injector, fertilizer tank (flow by pass system) and fertilizer injector pump. Both fluid and solid fertilizers are now available in India. Fluid fertilizers are manufactured in India and while solid fertilizers are imported. Several studies have been carried out at the Department of Horticulture, UAS, Bangalore in fertigation in fruits, vegetables, flowers and plantation crops. Studies on the Bangalore blue grapes have revealed that application of 80%. WSF is superior over all other treatments. Similarly studies in Sapota revealed that application of 80% WSF through fertigation was superior over all treatments including conventional method of application of 100% Natural fertilizer. Studies conducted for other crops like oil palm, arecanut, coffee, cardamom, cashew, potato etc. have indicated that fertigation has advantage of over soil application apart from saving of about 25–35% of the fertilizer.

In Cyprus, the fertigation experiments have revealed that the yield improvements for crops like potato, carrot, tomato, watermelon are substantial compared to conventional method of fertilization. Yield improvements (Kg per hectare) from fertilizer experiments in Cyprus are in given Table 10.

TABLE 10   Yield improvements due to fertigation in Cyprus.

Crop	Crop yield	
	Fertigation	**Conventional**
	Kg/ha	
Carrot	54,000	42,000
Potato	70,000	37,000
Tomato	1,80,000	55,000
Water Melon	1,15,000	60,000

6.3.9   SUMMARY
In the age of increasing scarcity of water resources, the agricultural sector as the World's largest water consumer has to find ways and means to use the water very efficiently and judiciously especially when the demand is increasing for feeding the increasing population and at the same the allocation will be diminished as the need for water for industry and drinking water will be increasing especially in the developing countries. Figures are quoted which do not show exactly the careful use by agriculture

of water resources. In addition irrigation has absorbed over 50% of public investment in Agriculture. According to FAO, some 60% of the water supplied in surface irrigation goes unused and leads to water logging and salinization. Therefore the use of water saving technology like micro irrigation is very necessary in the coming years. This will not only save the scarce water, but also increase the productivity by at least 2 to 3 times per unit of water with various other advantages.

The development of micro irrigation in India is slow covering only about 400,000ha, but it should go up to 10 Mha by 2020/25. This is possible, as India must increase area of irrigation for agricultural production.

Fertigation along with drip irrigation will enhance the production (Table 1) and will reduce the fertilizer requirements by about 30%. Efforts should be taken up to achieve the goal by educating/training the officials and farmers on micro irrigation and fertigation.

6.4 MICROIRRIGATION FOR CLOSELY SPACED CROPS

The irrigated agriculture accounts for about four fifth of water in India and is increasingly required to produce more food from limited land area using less water. Over a period of 15 to 20 years, the water for agriculture will be reduced to 70 percent (of India's water resource), and at the same time increase the production from 210 to 400 million tons. Surface irrigation is dominant to grow all types of crops. Though advanced method of irrigation like sprinkler, mini-sprinkler and drip irrigation systems are available, the spread of the MI systems are very slow. Out of 97 Mha of irrigated area in India, only about 600,000 ha is under sprinkler and about 350,000 ha is covered by drip irrigation, which is only a fraction of percentage of the total irrigated area in the country. The drip irrigation technology is the latest and it is suited to almost all crops except rice crop. Research in drip irrigation was started in 1969 in India and useful findings have been emerged from various Agricultural universities and Research Institutions. The studies revealed that the water saving for all type of crops vary from 40–70%, and crop yield is increased up to 100%. In other words from the same quantity of water, the area of irrigation can be increased to 2 to 3 times and the productivity per unit quantity can be increased to 3 to 4 times (Table 1).

Drip irrigation is introduced for many crops: Widely spaced as well as closely spaced crops with increased productivity per unit quantity of water. This system is used mainly for widely spaced commercial crops as the cost of the system is within the reach of the farmers, whereas the MI is costly for closely spaced—vegetables, cotton, groundnut, pulses, sunflower, etc. Further, the widely spaced orchard crops are perennial in nature and if the system is laid out, it will serve the purpose for many years: 20–30 years. In case of closely spaced crops like vegetables, flowers, pulses, cotton, sugarcane, groundnut, etc., the cropping system should be changed once in 4–6 months or one year. Further the layout of the drip system also may vary with reference to the spacing of the crops. The approximate costs of the drip system for various crops are given in Table 2. A typical layout is shown in Fig. 2. Here a raised bed is formed having two furrows on both sides. The distance between the middle of the furrows is about 1.92 meters. In the raised bed, the lateral pipe is placed at center of the bed and the dripper/ emitter spacing is 50 cm. The two rows of cotton with a row spacing of

1.92 meters are raised at the end of the bed. The bed is compacted in order to have uniform flow in all directions to wet the soils laterally and longitudinally. Some examples are shown in Appendix I at the end of this chapter.

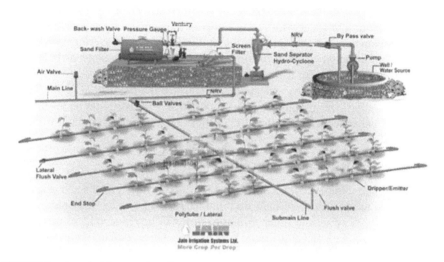

FIGURE 2   A typical layout of a micro irrigation with necessary accessories and equipment.

The same bed and layout of drip system is used to grow other crops after harvesting the cotton. In this way, the layout made for cotton is used for growing any closely spaced crops like maize, groundnut, sunflower, pulses, sugarcane, chili etc. Hence the farmers can grow any crop and have any cropping pattern if the layout as described above is formed and maintained. The cost for this drip (layout) system is about Rs. 18,000 to 21,000 per acre depending upon the quality of water, the location of water source etc.

Drip irrigation is very popular in the Southern states—Tamil Nadu, Andhra Pradesh, Karnataka, Maharashtra not only for widely spaced crops and closely spaced crops (Table 3). The author conducted a detailed study on the benefit-cost ratio for various crops in the southern states mainly in Maharashtra, Tamil Nadu (Tables 2, 3, 7, (14 and 15 in Section V)). He concluded that the system is technically feasible, economically viable and socially acceptable one. The main problems in popularizing the system are high initial investment and the need for different field layout of drip system for each crop. The cost of the system can lowered by suitable changes in the geometry of the crop layout and the layout is designed in such a way the same layout can be used for all closely spaced crops.

6.5   MICRO IRRIGATION IN FRUIT CROPS

6.5.1   INTRODUCTION

Water is a prime natural resource to achieve number of significant functions. Unlike most other natural resources, water does not have a substitute in its main uses. It can be used more or less lavishly or efficiently, but it cannot be replaced. It is indispensable, finite and vulnerable resource. Virtually no activity in society or process in the landscape or in the environment is possible in the absence of water. Water is one of the plentiful resources. Covering more than 2/3 of the earth, water travels from the sea into the air to the land and back to the sea in a seemingly endless cycle of renewal (Hydrologic cycle).

Yet water is a finite resource, and the tiny fraction suitable for drinking or irrigating crops is distributed unevenly throughout the regions. At the same time, the human needs for water is escalating because of rapid population and industrialization, especially in the regions where water is a most scarce. Between 1940 and 1990, world population has more than doubled, from 2.3 to 5.3 billion and the per capita use of water has also doubled from 400 to 800 m^3/year. Hence, the global water use has increased by more than 4 times during this period. In many of the regions of the world, population is growing more rapidly; the needed water is simply unavailable. The critical limits are not at the global level but at regional, national and local levels. The area of irrigated land worldwide nearly doubled in the first half of the twentieth century (from 48 Mha to 94 Mha). Land under irrigation has nearly trippled (260 Mha).

Worldwide agriculture is a single biggest user of water supply, accounting for about 69% of all use. About 23% of water is used to meet the demands of industry and just 8% to domestic use. Pattern of use varies greatly from country to country, depending on factors such as: Economic development, climate and population. As an example, India and Africa consume about 90% of water for agriculture, (irrigation) while highly industrialized countries in Europe allot more than half the water for industry and energy production. Only in the recent years, the growth has slowed down. In California and some parts in India, farmers are selling their land and the accompanying water rights to the metropolitan area with huge demand. The proportion of water used for industrial purposes is often seen as an indicator of economic development.

According to the data available on the global water supply, there seems to be no lack of fresh water worldwide (Table 11). However, it must be taken into account that in individual region, because of the spatial and temporal variations in precipitation, the potential usable water supply is very small. The most frequently used criterion for assessing the availability of the renewable water supply is the per capita supply, a supply of less than 500 M^3 per capita per annum being regarded as the critical lower limit, 1000 M^3 per capita as very low, 2000 M^3 as critical. According to a study carried out by the World Resource Institute it is especially the countries, which are in or near semiarid regions that are in a critical situation. Based on a supply of less than 500 M^3 per capita per annum, the most endangered regions are North Africa with 3 and the Middle East with 8 countries as threat. Twenty-two of the countries in the world currently [2] have renewable water supplies of less than 1000 M^3 per capita per year. The World Bank estimates that by the year 2025 one person in three, in other words 3.25 billion people, in 52 countries will live in conditions of water shortage.

TABLE 11   Global water supply for earth (1999).

Region	Km³/a	mm	M³/p/year
Europe	3110	319	4410
Asia	13190	293	4130
Africa	4225	139	6581
North America	5960	287	13925
South America	10380	583	34949
Australia	1965	225	75577
	38830 (Total)	294 (Avg)	7337 (Avg)

Source: National Resource and Development, Focus—Water the life line of our future, Vol. 49/50, Institute of Scientific Co Operation, Tubingen, Germany.

6.5.2   CURRENT STATUS OF IRRIGATION METHODS
Reader is referred to Section 6.3.2 in this chapter.

6.5.3   CURRENT STATUS OF MICRO IRRIGATION
Reader is referred to Section 6.3.3 in this chapter. Drip system costs $400–$1500/Ha depending upon the crop, topography, source of water supply etc., which most farmers feel to be too expensive though the economic worked out will prove that the system is economically viable. In India, the drip area is increased from 1000 Ha in 1985 to about 400,000Ha in 2002, which is very meager (less than 0.5% of the irrigated area). The drip-irrigated area in India is increasing gradually as given below (Table 12):

TABLE 12   Drip irrigated area in India.

Year	Area in ha	Year	Area in ha
1985	1,000	1998	2,25,000
1993	55,000	1999	2,54,000
1994	71,000	2002	4,00,000
1996	1,62,000		

It is believed that there is a great potential in India for this advanced method and it is possible to bring about one Mha, that is, about 1% of the countries irrigated area in 2004/2005, and about 10 Mha in the year 2020/2025.

6.5.4   FRUIT CROPS
The major potential of drip irrigation is for fruit crops where the system can provide a substantial water economy and better productivity. Further the cost of the system will be reasonable, economical, and viable. In fact, MI is well suited for all widely spaced and commercial crops. India has a total area of about 3.5 Mha of fruit crops producing about 42 metric tons per year. The major fruit crops are mango, apple, guava, pineapple, grapes, papaya, etc., where good water economy can be affected if drip irrigation

is used with technical and scientific recommendations these fruit crops. The area of different fruit crops and their production in India is given in Table 13.

TABLE 13 Area and production of fruit crops in India.

Crop	Area	Production
	$\times 10^5$ha	$\times 10^5$ tons
Mango	12	110
Banana	4.45	130
Citrus	4.5	38.0
Apple	2.2	12
Guava	1.3	15
Pineapple	0.71	10.7
Grapes	0.35	6.0
Papaya	No data	3
Other fruits:	5	65

Ber, Pomegranate, Custard apple, Strawberry, Sapota, etc.

Source: NCPA, Perspective plan for Drip & Sprinkler Irrigation (1990–2000).

6.5.5 RESEARCH ADVANCES IN MICRO IRRIGATION FOR FRUIT CROPS

The results of the survey conducted on micro irrigation for fruit crops has revealed that in Europe, the yield increase was in the range of 10–50% as a result of switching to micro irrigation and the water saving was significant (20–25%) compared to sprinkler irrigation, and was 40–60% compared to surface methods. France reported a similar trend from mini-sprinkler to drippers in Orchards.

However, in Australia and South Africa the trend was opposite with an expanding use of microsprinkler/sprayer in fruit orchards. The major problem reported by most of the countries has been clogging of drippers which was overcome in most cases by installation of efficient filters. In some cases, injection of chemicals was necessary to overcome buildup of algae or carbonate/iron compounds in the lines and drippers. The experiments conducted in India for fruit crops at various Agricultural Universities/ Research Institution have revealed that the water savings is about 40–70% and yield increase varied from 20–100% (Table 1).

The author has analyzed economics of micro irrigation system, BCR ratios, etc., by interviewing farmers/scientists in Maharashtra, Tamil Nadu, Karnataka at two different occasions (1990 and 1993–1994). The results are summarized in Tables 14 and 15.

TABLE 14   Benefit cost ratio for various fruit crops under micro irrigation (1990).

Crops	Spacing m × m	Benefit-cost ratio, BCR	
		Excluding water saving	Including water saving
Grapes	3 × 3	13.35	32.32
	8 × 8	11.50	27.08
Acid lime	4.57 × 4.57	1.76	6.01
Banana	1.52 × 1.52	1.52	3.02
Mango	7.62 × 7.62	1.35	8.02
Orange	4.57 × 4.57	2.60	11.05
Papaya	1.84 × 1.84	1.54	4.01
Pomegranate	3.04 × 3.04	1.31	4.04

Source: Constraints and potential in popularizing Drip Irrigation, R. K. Sivanappan & Associates, 1990.

TABLE 15   Benefit–cost ratio and payback period for fruit crops under micro irrigation (1993–1994).

Crop	Spacing, m × m	Payback period	Benefit–cost ratio
Banana	0.91 × 1.5 × 1.8	One year	3.00
Grapes	3.03 × 1.8	One year	3.28
Pomegranate	4.3 × 4.3	One year	5.16
Ber	4.5 × 4.5	One year	4.56
Papaya	1.80 × 1.80	One year	4.09

Source: R K. Sivanappan, Case study with number of farmers in Maharashtra, 1993–1994.

At Gujarat Agricultural University in a 14 ha farm having various fruit crops namely: Mango, sapota, ber, guava, pomegranate, acid lime, sonala, phalsa (or falsa, Indian word for *Grewia asiatica*), sweet orange, mandarin, coconut, the research studies indicated: Higher yield; reduction in farm labor; weed/pest; huge water saving; and a better fruit quality. Experiments conducted at Dapoli has revealed that irrigation for mango applied at 60 L/tree/week through MI produced 152.7% higher yield compared to manual watering with equivalent amount of water. For Pomegranate, MI gave comparable yield (6.84 T/Ha) and saved 45% water over check basin method of irrigation as indicated by the scientists of Agricultural University at Rahuri. Several studies carried out at the UAS, Bangalore on fertigation of fruit crops (grapes/sapota) have revealed that application of 80% of water soluble fertilizers were superior over the conventional method of application of 100% normal fertilizer. The experiment conducted at Marathwada Agricultural University, Parbhani has indicated that drip irrigation for banana crop gave better yield than surface/conventional method of irrigation.

6.5.6 FUTURE OF MICRO IRRIGATION FOR FRUIT CROPS

The area under fruit crops in India will increase to 5–5.5 Mha in the year 2020/25 from the present area of about 4 Mha. The irrigated area at present is about 12×10^5 ha of which only about 1.35×10^5 ha is under drip irrigation (Table 16).

Since it has been proved that micro irrigation helps in increasing the productivity and saving of water, it is possible that about 2 Mha of fruit crop will be brought under micro irrigation by 2020/25. This will require detailed and phrased plans by the Governments, NGO sand manufacturers, and determination on the part of farmers. Furthermore, micro irrigation should be supported by the suppliers and extension staff to help farmers to maintain and operate their system properly.

TABLE 16 Area under drip for fruit crops (1999).

Crop	Area ha	Crop	Area ha
Amla	220	Guava	4,930
Banana	24,565	Mango	21,863
Ber	4,700	Papaya	2,115
Citrus	22,210	Pomegranate	19,250
Custard apple	810	Sapota	5,125
Grapes	29,630	Strawberry	170
Total for all crops		**1,34,588 Ha**	

Source: Proceedings of the All India Seminar on Micro Irrigation. Prospects and potential in India held in June 1999 at Hyderabad.

6.5.7 CONCLUSIONS

Micro irrigation is well suited for fruit crops but it has not been fully exploited. In India, the area of irrigation of fruit crops is only about 30% and the average productivity is not even in sufficient quantity prescribed for the population, though there is tremendous scope in extending the area of irrigation for fruit crops. Drip irrigation can increase productivity and also the quality of fruits. Therefore, it is planned to bring at least 2 Mha under micro irrigation in the year 2020/25. This will not only meet the demand of the population and at the same time, it will fetch the much required foreign exchange by exporting the fruits.

6.6 SUMMARY

The micro irrigation system is technically viable and socially acceptable under Indian situations, such as:
1. Well irrigated areas, which constitutes about 35% of the irrigated area in the country;
2. It is ideally suited to all row and commercial crops;
3. Waste lands by planting trees including fruit trees;
4. Hills and semiarid areas;
5. Coastal sandy belts;

6.   Water scarcity areas; and

7.   Command area of the community wells.

The constraints to bring such large area under micro irrigation are: High initial cost; quality of drip material—clogging of drippers and cracking of lateral pipes and drippers; poor awareness by the farmers; lack of adequate technical inputs; damage due to rats and rodents; high cost of spare parts and components; difficulty in getting subsidy and loans; and insufficient extension and promotional activities. To exploit the full potential of the MI system, the constraints need be relaxed by appropriate policy instruments, financial support and technical guidance. This calls for an integrated approach and endeavor on the part of the Central and state Governments, implementing agencies, manufacturing companies and farmers.

The target of 10 Mha of drip irrigated area by 2020 can only be achieved if all the farmers (including marginal farmers) take up drip irrigation for all row crops. This will be possible only if the cost of the system is affordable by small and marginal farmers without sacrificing the quality and uniformity of water application. In areas with a high concentration of small and marginal farmer operations and scarcity of water, the low cost drip irrigation system will provide access to water saving irrigation technology, which is affordable on a small scale and divisible.

At present, on an average about Rs. 100,000 is invested to bring one hectare of land under irrigation in the new irrigation projects if water is available. The development of micro irrigation in India is slow covering only about 400,000 ha, but it should go up to 10 Mha by 2020/25. This is possible, as India has to increase area of irrigation for agricultural production.

Fertigation along with drip irrigation will enhance the production and will reduce the fertilizer requirements by about 30%. Efforts should be taken up to achieve the goal by educating/training the officials and farmers on micro irrigation management.

The cost of the system has been brought down by suitable changes in the geometry of the crop layout and the layout is designed in such a way the same layout can be used for all closely spaced crops like cotton, sugarcane, groundnut, flowers, pulses etc., as detailed above.

Micro irrigation is very well suited for fruit crops but it has not been fully exploited. In India, the area of irrigation of fruit crops is only about 30% and the average productivity is minimum. The fruit is not even sufficient; the meager quantity prescribed for the population, though there is tremendous scope in extending the area of irrigation for fruit crops. Drip irrigation can increase productivity and also the quality of fruits. Therefore, it is planned to bring at least 2 Mha under micro irrigation in the year 2020/25. This will not only meet the demand of the population and at the same time, it will fetch the much-required foreign exchange by exporting the fruits.

KEYWORDS

- accepted method
- action plan
- advanced method
- advantages
- affordable
- availability
- clogging
- closely spaced crops
- command area
- commercial crops
- compact
- confidence
- conventional methods
- conveyance
- conveyance and distribution
- cost benefit
- cost, drip system
- cultivable land
- development
- drippers
- economic development
- education and training
- emitter
- emitter spacing
- evaporation
- expensive
- exploit
- fertigation
- filter
- food security
- foreign exchange
- fruit crops
- great potential
- increased yield
- India
- inefficient extension
- investment
- investment cost
- irrigation investment
- Jain irrigation
- lay out
- low cost drip system
- maintenance

- major constraints
- marginal farmers
- Nepal
- new approaches
- overall efficiency
- pay back
- plantation crops
- plastics in agriculture
- population growth
- potential
- precipitation
- precision farming
- productivity
- raising demand
- renewable water
- root zone
- row crop
- salinity
- scarce commodity
- science of survival
- scope
- seminar
- shallow soil
- stress, crop
- subsidy
- sustainability
- Tamil nadu—India
- target area
- technical feasibility
- technology
- uniformity
- usefulness
- utilizable water
- water economy
- water management
- water problems
- water saving
- water table
- water use efficiency
- weed growth
- workshop
- yield improvements

REFERENCES

1. Abbot, J.S., Micro Irrigation: World Wide Usage, (Personal communication). Section 5.
2. Engelman, Robert and Pamela Levoyt (1993). Sustaining water—Population and the future of renewable water supply. Population Action International. Section 3.
3. Institute for Scientific Co-operation. Natural Resources and Development, Focus: Water—the life line of our future. Volume 49/50, Tubingen, Germany, 1999. Section 5 and 3.
4. Keller, Jack, 2000. Gardening with low-cost drip irrigation in Kenya: For health and profit. Technical Report prepared for International Development Enterprises (IDE), http://www.simi-net.org/fs_start.htm."
5. National Committee on the Use of Plastics in Agriculture (NCPA), 1990. Status, Potential and Approach for Adoption of Drip and Sprinkler Irrigation Systems. Pune, India. Section 1.
6. Polak, Paul, Nanes, Bob, and Adhikari, Deepak. A Low Cost Drip System for Small Farmers in Developing Countries. Journal of the American Water Resources Association, February 1997, Vol 33, No. 1.
7. Polak, Polak and R. K. Sivanappan, 2001. The potential contribution of low cost drip irrigation to the improvement of irrigation productivity in India Technical Report prepared for International Development Enterprises (IDE), <http://www.siminet.org/fs_start.htm>."
8. Postel, Sandra (1999). Pillar of Sand: Can the Irrigation Miracle Last. WW Norton & Co, New York. Section 3.
9. Proceedings of the XI International Conference on the Use of Plastics in Agriculture, New Delhi—India, 1990. Section 5.
10. Proceedings of VIII IWRA World Congress on Water Resources (1994). National water resources Center, Cairo—Egypt. Section 3.
11. Proceedings of the National Seminar on Micro Irrigation and Sprinkler Irrigation Systems, Delhi, India, April 28–30, 1998.
12. Sivanappan, R.K., 1990. Constraints and potential in popularizing drip and sprinkler irrigation system. National committee in the Use of Plastics in Agriculture, New Delhi, India. Section 5.
13. Sivanappan, R.K., 1994. Prospects of Micro Irrigation in India. Journal of Irrigation and Drainage System. 8: 49–58. Section 5.
14. Sivanappan R.K., 1994. Status of drip irrigation in India. Journal Irrigation and Drainage Systems, 8: 49–58. Section 3.
15. Sivanappan R.K., 1996. Strengths and weakness of growth of drip irrigation in India, Paper presented at the National seminar organized by the National Water Development agency at Bhopal. Section 2.
16. Sivanappan R.K., 1999. Scope for micro irrigation in India. Proceedings of the National seminar on Problems and prospects of Micro Irrigation—A critical Approach, IE(I) Bangalore, pp. 12–23. Section 3.
17. Sivanappan R.K., 2000. Micro irrigation including low energy precision applications. International Conference on Management of Water Resources for twenty-first century, New Delhi, India. Section 3.
18. Sivanappan, R. K., Rao, A. S., and Dikshit, N.K., 1994. Drip Irrigation in India. Indian National Committee on Irrigation and Drainage, Jolly Reprographics, New Delhi, 110 008.

APPENDIX I
EXAMPLES OF MICRO IRRIGATION

CHAPTER 7

ADVANCED TECHNOLOGIES FOR SUGARCANE CULTIVATION UNDER MICRO IRRIGATION: TAMIL NADU*

R. K. SIVANAPPAN

CONTENTS

7.1   Technology to Increase Sugarcane Yield.. 122
7.2   Advanced Technologies For Sugercane Cultivation Using Micro Irrigation
with Fertigation in Tamil Nadu (To Double Cane Yield, Save Water and
Mitigate Power Shortage by Producing Ethanol and Electricity) 135
7.3   Conclusions 142
7.4   Summary... 143
Acknowlededements.. 144
Keywords 144
Appendix I 145

*This chapter is a combined version of two unpublished reports by the author: *"Technology to Produce 100 Tons per Acre in Sugarcane"; and "Advanced Technologies in Sugarcane Cultivation using Drip-Fertigation in Tamil Nadu, India: (To Double Cane Yield, Save Water and Mitigate Power Shortage by Producing Ethanol and Electricity in Tamil Nadu)."* December 2012.In this chapter, conversion rate for an Indian currency is Rs. 64.00 = US$ 1.00 on September 30, 2013.

7.1   TECHNOLOGY TO INCREASE SUGARCANE YIELD

7.1.1   INTRODUCTION

Sugarcane is one of the most important commercial crops in India and plays a vital role in Indian agriculture in general and rural economy in particular. India ranks second after Brazil among important sugar producing countries in the World and contributes 22% and 25% in area and production of sugarcane respectively. In the recent years, though our Nation is registering a record production of sugarcane and sugar, the productivity and sugar recovery have not been increasing significantly. The potential yield of sugarcane is 485 tons/ha, whereas the average cane productivity in India is only 65 to 70 tons/ha (Table 1). The sugarcane is grown in 3.20×10^5 ha in Tamil Nadu. The yield is the highest in Tamil Nadu (more than 100 tons/ha) and the lowest in Assam (less than 40 tons/ha).

7.1.2   SUGARCANE CULTIVATION

The demand of water is increasing in all sectors. Already, more than 95% of surface water and about 85% of ground water are exploited in Tamil Nadu. After paddy (which consumes 72% of water), sugarcane crop consumes about 14 to 15% of water in this state. The sugarcane crop is cultivated in about 3×10^5 to 3.20×10^5 ha in Tamil Nadu. The productivity of the cane is about 106 to 110 tons/ha, whereas the average yield in the country is about 70 tons/ha. Continuous drought for 3 to 4 years and the resultant declined ground water levels lead to a reduction in cane area and productivity. In addition, yield decline is also due to inefficient water and fertilizer management practices.

TABLE 1   State wise productivity (tons/ha) of sugarcane in India, during 2009–2010.

States	Production tons/ha	State	Production tons/ha
Andhra Pradesh	74.1	Orissa	61.2
Assam	39.1	Punjab	61.6
Bihar	43.4	Rajasthan	57.4
Gujarat	80.5	Tamil Nadu	101.4
Haryana	72.1	Uttar Pradesh	59.2
Karnataka	90.3	UttaraKhand	60.8
Madhya Pradesh	40.8	Other states	33.8
Maharashtra	84.8	Average, all India	70.0

Source: Indian Sugar, 2011.

Sugarcane is a high water consuming crop, and requires about 2000 mm, that is, about 200 tons of water to produce 1 ton of cane. Sugarcane is the major commercial crop under contract farming (under each sugar factory) and pays rich dividends to the farmers, industries and to the Governments. Further, there is more demand of sugar from all walks of life. Hence there is a need to cultivate sugarcane in large areas with high productivity per unit of water. This will not only create more employment opportunities in the rural areas, but also increase the overall economy of the rural areas which will in turn help the sugar industry and Government to get more returns by way of revenues and taxes. But to bring more land area under sugarcane, large quantity of water is needed. As mentioned earlier, the available water is almost fully harnessed, and hence to bring more area, better water management practices including practicing the newer technology, namely, micro irrigation are only possible alternative. *By introducing micro irrigation with fertigation, it is possible to: Increase the yield potential by three times with the same quantity of water; to save about 45–50% of irrigation water; and increase the productivity of cane by about 40%.* When fertilizer is applied through micro irrigation (fertigation), the tonnage increases and about 30% of the fertilizer can also be saved. Unlike the surface method (flood or gravity irrigation), the overall water use efficiency is increased by more than two times (Table 2) by drip irrigation because the daily water need is applied to the root zone. Also evaporation/ conveyance and distribution losses are reduced with drip irrigation. Further, the soil moisture in micro irrigation is always near the field capacity (Fig. 1).

TABLE 2   Irrigation efficiency of three irrigation systems.

Details	Irrigation efficiency (%)		
	Surface	**Sprinkler**	**Micro**
Conveyance	65–76	100	100
Application of water	65–75	80–85	90–95
Surface water evaporation/soil moisture evaporation	20–30	20–30	5–10
Total	35–45	60–70	85–90

Source: R. K. Sivanappan, 1998, cited from Report by "Task Force on Micro Irrigation," Ministry of Agriculture, Govt. of India, 2004.

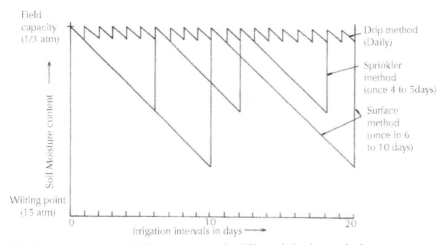

FIGURE 1   Moisture availability for crops under different irrigation methods.

TABLE 3 Yield potential of sugarcane.

Details	tons/ha	tons/acre
Genetic potential yield	485	194
Farm potential yield in Maharashtra (Mr. Sanjeev Monae)	298	120
Farm potential yield in	257	103
Tamil Nadu (Mr. Selvaraj)		
Average for Tamil Nadu	106–110	42

However, the average sugarcane productivity in India is low compared to many other sugarcane producing countries in the world. To meet the sugar requirement of increasing population, it is possible only through introduction of new sugarcane varieties, which contribute to 60% of the cane yield and adoption of scientific methods including irrigation in cane cultivation. *Since, there is no scope to bring additional area and more water for sugarcane cultivation; the targets have to be achieved only through increase in sugarcane productivity.* The genetic potential yield of sugarcane and realized maximum potential yield of sugarcane are mentioned in Table 3. The reasons for low productivity in India are:
1. Soil fertility is not properly enriched;
2. Improper water management practices;
3. Imbalanced nutrition management;
4. Not planting in the proper season; and
5. Lack of scientific methods of sugarcane cultivation.

In 2009–2010 planting season, the Sakthi Sugars Ltd., Sakthi Nagar in the Erode District of India have initiated 100 tons *UzhavarMandram* (Indian word for "Farmers Club") with a goal to increase the sugarcane productivity by scientific methods of cane cultivation, increasing the net profit of the sugarcane farmers and to sustain cane areas. Sixty one model plots in 135.16 acres extent were selected. The farmers of the model plot were trained with 100-ton cultivation package of practices.

7.1.3 EXPERIENCES FROM MAHARASHTRA STATE
During 1990s, the author extensively toured Maharashtra State in India to study the micro irrigation practices for sugarcane crop and interacted with numerous cane farmers. Mr. PopatHadawala Aliephata, Jinnar Talukin of Pune District, introduced micro irrigation system for sugarcane in large areas. His observations are detailed as follows:

The cost of the micro irrigation system (1993) for sugarcane crop was about Rs. 19,000 per acre. The present cost is about Rs. 30,000 to 35,000 per acre including fertigation equipment. The rainfall in the area was 600 mm. The amount of irrigation water through micro irrigation was about 8,000–16,000 L/day/acre, that is, about 2 to 4 mm per day for sugarcane cultivation. He obtained 80 tons/acre of sugarcane with micro irrigation compared to 50tons/acre with surface irrigation earlier. His aim was to get about 120 tons/acre using drip fertigation method. According to him, it was possible, as he knew some farmers in the area producing 135 tons/acre. He obtained net extra income of Rs. 35,000 taking into account the savings in water because of adoption to micro irrigation. *The Benefit-cost (BC) ratio was 3.45 for micro irrigation compared to 2.85 for surface irrigation method.* He also informed that the sugar content of the cane under micro irrigation was about one percent more than the cane with surface irrigation. The duration of the crop period was also reduced by about one month. Further, it was reported that he took four ratoon crops in case of micro irrigated field without affecting the yield compared to maximum of two ratoon crops in case of surface irrigated field earlier.

The demand of sugar is increasing, but availability of land and water are main constraints. Further, the groundwater is intensively mined. In some pockets in Tamil Nadu, the ground water level has gone beyond 700–1,000 feet. To overcome the water scarcity problem, the only solution is opting for micro irrigation with fertigation method for sugarcane for which the initial investment cost is about Rs. 30,000 to 35,000/ acre including fertigation equipment.

In addition, the water used for producing one ton of sugarcane under micro irrigation is only about one-third quantity of the water used in the surface method, besides giving 40% higher yield. The study conducted in Tamil Nadu has revealed that micro irrigation consumes only 12.8 horsepower (HP) hours of water to produce one ton of sugarcane against 28.28 HP-hours of water under surface irrigation. It was observed that under micro irrigation, the electricity saving was about 40% compared to surface method (flood or gravity irrigation).

According to Mr. Nerkar of Vasandada Sugar Institute at Pune—Maharashtra State, the average sugarcane yield in India is only 70 tons/ha (it is 106 to 110 tons/ha in Tamil Nadu). Being a C4 plant, sugarcane is physiologically one of the most efficient plants. As per agro-biological calculations, it is possible to harvest 600 tons per hectare. Some of the progressive cane farmers have achieved a yield up to 350 tons/ha (or 140tons/acre). This can also be achieved in Tamil Nadu, by using paired-row method and pit-method of sugarcane cultivation using micro irrigation with fertigation and also sustainable sugarcane initiative (SSI) methods using subsurface drip irrigation (SSDI) with fertigation and by adopting mechanization in harvesting and following all recommendations of package of practices and good agricultural practices.

7.1.4 EXPERIENCES IN TAMIL NADU

The author made extensive field trips in Coimbatore, Erode and Sivaganga districts in Tamil Nadu to study potential of micro irrigation for sugarcane crop. He interacted with progressive individual farmers and group of farmers who are growing cane under micro irrigation with fertigation, using paired row and pit method of planting by using drip fertigation, Subsurface drip fertigation with dual row of crops and Sustainable Sugarcane Initiative (SSI) method of cultivation using sub surface drip fertigation. A progressive farmer in Gobichettipalayam in Erode district obtained 98.3 tons/acre using subsurface drip fertigation. His initiative motivated to organize an "exclusive farmers" club, viz "100 tons/acre club" in Sakthi Sugar Mills at Sakthi Nagar in Erode district. Many farmers in Erode district can get 70–75 tons/acre of sugarcane using the latest methods. The following are the observations and recommendations for future implementation of the sugarcane cultivation in the State.

When implemented, this method of cultivation will enable farmers to get the required sugar in the coming years without diverting any extra land or water and get about 100 tons/acre cane yield. This will help to get the required sugar and also to solve the energy problem by providing ethanol and cogeneration of power in the State.

The progressive farmers are using paired-row method (2 ½' ′ 4 ½' ′ 2 ½') as recommended by the scientists of TNAU. Recently the Sustainable Sugarcane Initiative (SSI) method using subsurface drip with fertigation is being used in many sugar factory areas. To suit the mechanized harvesting, cane is planted at 5 feet interval between rows and proper spacing within the rows. Micro irrigation with fertigation equipments are installed to give water and fertilizer as per the requirements of the plant. The normal method of sugarcane cultivation under surface irrigation is by providing furrows at 3 feet interval with only one row of sugarcane (Fig. 2). The layout of the field for paired-row method with micro irrigation is given in (Fig. 3). The layout of the field for the pit-method of cultivation is shown in (Fig. 4). The subsurface drip fertigation method is shown in (Fig. 5). The details of these methods are discussed below.

7.1.4.1   PAIRED ROW METHOD

The paired-row method with micro irrigation is practiced by many farmers in the State and they obtain about 60 to 75 tons/acre at present. For the sugarcane crop 30 cm (75 cm) wide shallow furrows are formed at an interval of 4 ½ to 5 feet (135 to 150 cm) and two rows of plantings are done on either side of the furrow. The lateral drip line is laid with drippers/emitters at 15 to 18 inches (38 to 45 cm) interval and the required water is given daily through drip irrigation. In this method, water saving is about 40–45% and the yield has increased by 30 to 40% (Fig. 3).

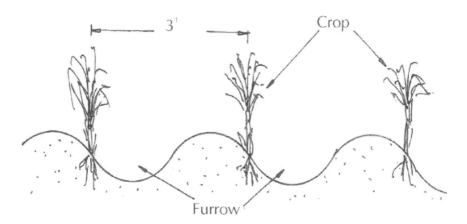

FIGURE 2   Furrow method of sugarcane cultivation with surface irrigation.

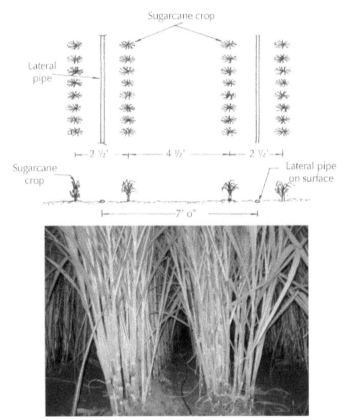

FIGURE 3 Drip irrigation with fertigation and paired row method.

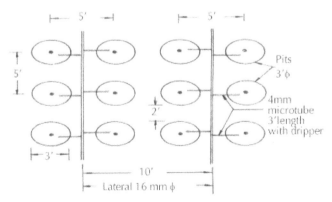

FIGURE 4 Pit method of cultivation with drip irrigation and fertigation.

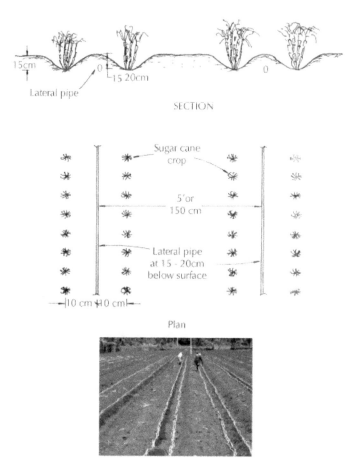

FIGURE 5 Subsurface drip irrigation with twin or dual rows at 20 cm away from the lateral and SSI method of cultivation.

7.1.4.2 PIT METHOD

There are about 1750 pits per acre at 5´5 feet spacing as shown in Fig. 4. In each pit, 16 numbers of double budded or 32 numbers of single budded sets are planted. The distance between laterals is 10 feet and 4 mm micro tubes of 3 feet length with dripper are provided to supply the required quantity of water to each pit. The quantity of water to be supplied in paired-row method and pit method can be analyzed based on crop stage, climate, and soil, etc. The information gathered in the field such as cost, yield in the conventional and drip methods are discussed in detail in this chapter.

Onsite meetings were conducted (during 2006–2007) with many farmers individually and in groups who are traditionally sugarcane growers. According to them, 3 or 4 years ago, the cane was cultivated in furrows at 3 feet interval and surface method of irrigation (flooding) was followed, as shown in Fig. 2. They used to irrigate once in 6 to 8 days by flooding the furrows. The total water used for 12 months period was about 1800 to 2000 mm and the yield of cane varied from 40 to 50tons/acre. In the last 2 to 3 years, sugar factories have been canvassing the farmers to adopt micro irrigation with fertigation using paired row method in order to reduce the cost of the micro irrigation system, since each lateral in this method caters to two rows of crop. About 1000 acres of sugarcane was irrigated through drip fertigation in Coimbatore district and another 1000 acres in Sivaganga district during 2006–2007. The data collected from the farmers have revealed that the water saving was about 45 to 50% and the yield of the crop varied from 60 to 75 tons/acre indicating that the yield increase was about 15 to 20 tons/acre or 30% more yield. It should be noted that in 2011, micro irrigation was adopted in about 75,000 ha of sugarcane in Tamil Nadu.

The latest method of cultivation of sugarcane to get more yield and save water are (a) Subsurface drip fertigation with twin/dual rows method and (b) Sustainable Sugarcane Initiative (SSI) using subsurface drip fertigation method.

(a) Subsurface drip fertigation with twin or dual rows method

In this method, drip lateral lines are placed about 15 to 20 cm (6 to 8 inches) below the surface. The spacing of lateral line is 150 to 165 cm. The drippers or emitters are spaced in the laterals at 20 to 30 cm for light soils (sandy soils), at 40 to 50 cm for loamy/medium soils, and 60 to 70 cm for heavy clayey soils. The pressure-compensating drippers with a provision for flushing at the end of the laterals are used. Also provision of antivacuum valves at the submain and main lines is made and daily irrigation avoids root intrusion into the laterals. Irrigation and fertilizer application are given through micro irrigation system. The sugarcane (twin or dual rows) is planted on either side of the lateral at 10 cm distance from the laterals. This is placed at the same level of the drip lines with a seed rate of 20.000 two-budded-setts per acre (Fig. 5).

As the laterals and emitters are located below the soil surface, this system is called subsurface micro irrigation system. The subsurface micro irrigation system with fertigation system is the "triple wonder" technology comprising irrigation, fertigation and preventing evaporation of water. Even though the cost of installations for the subsurface micro irrigation system is slightly high when compared with surface micro irrigation system with fertigation equipments (Rs. 40,000/acre), cane growers are very much prompted to adopt this technique on account of various benefits in sugarcane cultivation because the harvesting can be done by machines since the interval between the crop row is 5 feet (Fig. 5). Though the system has been introduced only recently, the sugarcane area under subsurface drip fertigation system is around 10,000 ha out of 75,000 ha for surface micro irrigation and there is increasing trend in its adoption. Though the area under cane or other crops using subsurface irrigation in India is low, but this method is widely adopted in many countries in the world like Australia, Brazil, South Africa, and USA.

(b) Sustainable Sugarcane Initiative using subsurface drip fertigation method

Sustainable Sugarcane Initiative (SSI) method is the latest cane cultivation method, which is more or less similar to subsurface drip fertigation, except in planting material. Here, instead of sugarcane-setts, seedlings raised in specialized nurseries are used.

In SSI, single-budded-chips, carefully removed from healthy canes 7 to 9 months old which have good internode length (15–20 cm or 7 to 8 inches), are used for raising nursery. The selected buds are placed in trays half filled with well decomposed cocopith (coconut coir waste) and then fully covered with cocopith and watered. By raising nursery, high percentage of germination can be achieved within a week under shade-net conditions.

Under proper conditions (especially, warm temperature) within 3–5 days, white roots (primodia) will come out and shoots will also appear in the next 2 to 3 days. Based on the moisture content of cocopith, watering to the trays (seedlings) is initiated in the evenings for the next 15 days using rose cans. Shoots will start growing strong and leaves will start unfolding. During six leaf stage (about 20 days old seedling), grading of the plants is done. Plants of similar height are lifted up and placed in one tray, eliminating the damaged or dead seedlings. The seedlings raised in such a way at the age of 25–35 days are transplanted in the main field at 60 cm plant to plant spacing in a row just above the lateral line (Fig. 5).

It is important to observe that this one month growth of seedlings with SSI method cannot be achieved even after two months with the conventional method. They offer synchronous tillering leading to uniform growth and maturity of stalk population, which usually gives better yield and sugar recovery.

7.1.4.3   IRRIGATION

In many cases, unfortunately, the exact irrigation requirements and recommendations for cane crop are not followed by the farmers. Neither the irrigation company nor the sugar factories have given the recommendations for irrigation scheduling to the farmers but only have given arbitrary instructions to give irrigation by micro irrigation for 3 or 4 h per day irrespective of the weather conditions and stage of the crop. Based on enquiry, it is noticed that the farmers are giving water varying from 4 to 8.5 mm per day. However, if the water is given based on the climate, stage of the crop using pan evaporimeter and crop factor, it is possible to provide the exact quantity of water need and fertilizer through the drip water. If micro irrigation with fertigation is used, it is possible to obtain 100 tons/acre, using only 45–50% of water under traditional surface irrigation.

If the farmers are educated and class A pan is installed at least one in each revenue village or block, it is possible to give the exact quantity of water daily and it can increase the cane yield more than 100 tons/acre. Therefore, the extra cane output can be used to produce ethanol and for cogeneration. The water thus saved (45 to 50%) can be used to raise other crops under irrigation (through saved water) or divert the water for municipal/other uses.

7.1.5   IRRIGATION AND FERTIGATION SCHEDULING

7.1.5.1   IRRIGATION SCHEDULING

Irrigation is needed to meet daily evapotranspiration (ET) requirements of the crop that can be calculated by the following formula:

$$WR = PE \times K_c \times K_p \times Area \tag{1}$$

$$Time\ required(hr) = \frac{WR(litres\ /\ day)}{Dripper\ disch \arg e(litres\ /\ hr) \times No.\ of\ Drippers} \tag{2}$$

where, WR = Water requirement in liters/day; PE = Pan evaporation (mm); K_c = Crop coefficient; K_p = Pan factor (0.90); and A = Area in square meters. Using Eq. (1), the time required for irrigation can be calculated using Eq. (2). The crop coefficient (K_c) values for the sugarcane crop from the beginning to the end of the crop growth stages are given in the Table 4.

TABLE 4   The crop efficient (K_c) values for the sugarcane crop.

Age of the crop	Kc value
1 to 31 days	0.40
1 to 2 months	0.75
2 to 2.5 months	0.95
2.5 to 4 months	1.10
4 to 10 months	1.25
10 to 11 months	0.95
11 to 12 months	0.70

Source: TNAU, Coimbatore

7.1.5.2   FERTIGATION SCHEDULING

Studies, conducted at Sugarcane Breeding Institute (SBI) – Coimbatore, showed that an average crop of sugarcane yielding 100tons/ha removes 208 kg of N, 53 kg of P, 280 kg of K, 30 kg of sulfur, 3.4 kg of iron and 0.6 kg of copper. For sustained high yields in sugarcane, integrated nutrient management is the best option. Farmyard manure or compost at the rate of 25 tons per ha is applied at the time of land preparation. Soil test based on addition of inorganic fertilizers is recommended to reduce the cost of fertilizers. For sugarcane, a general NPK recommendation of 275:63:115 kg/ha is followed. The fertilizer dose when applied through micro irrigation ensures higher nutrient use efficiency, where both water and fertilizers are delivered to crop simultaneously. Subsurface drip fertigation ensures that essential nutrients are supplied precisely at the area of most intensive root activity according to the specific requirements of sugarcane crop and soil type resulting in higher cane yield and sugar recovery. The fertigation scheduling for sugarcane crop under drip fertigation method of cultivation is indicated in Table 5. Fertigation recommendations adopted by different sugar mills vary from these general recommendations to suit the local conditions.

TABLE 5   Recommended dose and fertigation scheduling for sugarcane (RDF), 275:63:115 of NPK kg/ha.

Days after planting	N	P	K	Days after planning	N	P	K
20	13	0	0	130	20	5	5.5
30	13	0	0	140	10	0	6.5
40	13	0	0	150	10	0	6.5
50	16	9	3	160	10	0	6.5
60	16	9	3	170	10	0	6.5
70	16	9	3	180	10	0	6.5
80	19	7	5	190	10	0	6.5
90		7	5	200	3	0	9
19							
100	19	7	5	210	3	0	9
110	20	5	5.5	220	3	0	9
120	20	5	5.5	230	3	0	9

Source: TNAU, Coimbatore-3.

7.1.6   BENEFIT-COST ANALYSIS OF SUGARCANE CROP UNDER CONVENTIONAL AND DRIP FERTIGATION METHODS

The author evaluated the Benefit-Cost (B.C.) ratio for various crops including sugarcane in Maharashtra State by contacting many farmers in 1994. Based on the study, it was found that the B.C. ratio for sugarcane with drip fertigation method was 2.66 compared to conventional (surface) method of 2.21. The water use efficiency was 180.85 kg/ha-mm with drip fertigation method compared to 59.53 kg/ha-mm for surface method (Table 6 and Fig. 6).

TABLE 6   Benefit-cost of Sugarcane (Maharashtra State, India).

	Cost economics	Micro irrigation system	Conventional system
1.	Fixed cost, Rs. per ha	30,000	NIL
		(65,000 to 70,000 at present)	
	Life, years	5	NIL
	Depreciation, Rs.	6,000	NIL
	Interest, @ 12%	1,800	NIL
	Repairs and Maintenance	600	NIL
	Total = (b)+(c)+(d); Rs.	8,400	NIL
2.	Cost of cultivation, Rs./ha	11,445	17,375
3.	Seasonal total cost = (1e) + (2); Rs/ha	19,845	17,375
4.	Water used, mm	940	2,150
5.	Yield of produce, 100 Kg/ha or tons/acre	1,700 or 68	1,280 or 51

TABLE 6   *(Continued)*

	Cost economics	Micro irrigation system	Conventional system
6.	Selling Price, Rs. /100 Kg	31	30
7.	Income from Produce = (5)×(6); Rs.	52,700	38,400
8.	Net seasonal income= (7)–(3); Rs.	32,855	21,025
9.	Additional area cultivated due to saving of water, ha	2	NIL
10.	Additional Expenditure due to additional area = (3) × (9)	39,690	NIL
11.	Additional Income due to additional area = (7) × (9)	105,400	NIL
12.	Additional Net income = (11)– (10); (Rs.)	65,710	NIL
13.	Gross cost of Production = (3) + (10); Rs	59,535	17,375
14.	Gross income = (7) + (11); Rs.	158,100	38,400
15.	Gross benefit cost ratio = (14) ÷ (13)	2.66	2.21
16.	Net extra income due to micro irrigation system over conventional = (12) + (8, drip) – (8, conventional)	77,540	NIL
17.	Net profit per mm of water used = (8) ÷ (4)	34.95	9.78
18.	Water use efficiency [(5) ÷ (4)] × 100; Kg/ha-mm	180.85	59.53

Source: Sivanappan, R.K., 1994. Micro irrigation in India. Indian National Committee on Irrigation and Drainage (INCID), New Delhi, July.

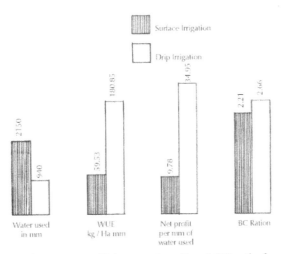

FIGURE 6   Water used, water use efficiency, net profit and BC ratio for surface and drip irrigation methods.

Besides the water savings and yield advantages, the micro irrigation system has following advantages:

a. Labor saving in irrigation and in weed removal.
b. Diseases and pest problems are less.
c. Frequent watering eliminates water stress and uniform growth and hence yield is increased.
d. Prevents soil erosion due to controlled wetting.

The water used, water use efficiency, net profit and BC ratio for surface and micro irrigation methods are shown in Table 6 and Fig. 6. Similar studies were carried out in Tamil Nadu recently, comparing the cost of cultivation for sugarcane under surface irrigation, subsurface drip fertigation (SSDF) and SSI under SSDF system (Table 7).

TABLE 7   Benefit-cost analysis of sugarcane (Tamil Nadu).

S. No.	Particulars	Surface irrigation	SSDF	SSIDF
		Indian Rupees, Rs.		
1	Micro irrigation system cost (discounted cost), Rs. per year per ha	0	12,000	12,000
2	Preparatory cultivation	8,800	10,950	10,950
3	Setts and plating	23,640	16,160	15,500
4	Crop maintenance	8,400	8,400	8,400
5	Fertilizers cost	5,250	5,250	5,250
6	Irrigation/Drip fertigation	4,200	3,000	3,000
7	Weeding	3,000	3,000	3,000
8	Plant protection	2,470	5,060	5,060
9	Herbigation	0	1,257	1,257
10	Chlorine treatment	0	600	600
11	Acid treatment	0	800	800
12	Micronutrients	900	900	900
13	Harvesting	42,000	48,750	48,750
		Manual	Mechanical	Mechanical
14	Cane yield, **tons/ha**	98	175	195
Economics				
15	Gross income, @ Rs.1,950/ton	19,1100	34,1250	390,000
16	Cost of cultivation, Rs.	98,660	116,127	115,467
17	Net income, Rs.	92,440	225,123	264,783
18	Benefit—cost ratio, BCR	1.93	2.93	3.29

Source: TNAU (2012) SSDF = Subsurface drip fertigation, and SSIDF = SSI under SSDF system.

7.2 ADVANCED TECHNOLOGIES FOR SUGERCANE CULTIVATION USING MICRO IRRIGATION WITH FERTIGATION IN TAMIL NADU (TO DOUBLE CANE YIELD, SAVE WATER AND MITIGATE POWER SHORTAGE BY PRODUCING ETHANOL AND ELECTRICITY)

7.2.1 INTRODUCTION

Tamil Nadu has a geographical area of 13.0 Mha (million hectares), which is about 4% of India's land area of 329 Mha. The water resource is only about 2% and the population share is about 6% of the country. The annual per capita water availability in India is about 1970 m^3, whereas it is only about 650 m^3 in Tamil Nadu. If the water availability is about 1000 m^3/person/year, this means that the state or the country is under water scarcity or stress condition, as accepted by world bodies. Therefore, Tamil Nadu can be classified as water scarcity zone.

Water is extremely valuable and a critical resource without which sustainable development is impossible. Therefore, it calls for everything humanly possible to be done to satisfy the need of the growing population in a sustained way. This chapter discusses how the water being used at present for growing sugarcane crop in Tamil Nadu can be profitably used for doubling the cane production in the same area of land by using only 50–55% of the water used at present, and irrigation can be done for another 4×10^5 to 5×10^5 hectares with the saved water. At the same time, the spirit and ethanol (gasohol-mixing up to 20% with petrol) can be produced that is required to generate about 4,000 million units of energy for Tamil Nadu from the bi-products (bagasse). This will not only mitigate the growing water scarcity problem to some extent, but also address the energy problem in Tamil Nadu in the years to come if implemented in a phased manner within next 3 to 5 years.

7.2.2 WATER RESOURCES AND IRRIGATION STATUS IN TAMIL NADU

The total surface water potential of Tamil Nadu is estimated as 2.42 Mha-m (million hectare-meter) (853 TMC-ft.: TMC-ft. is an abbreviation for "thousand million cubic feet," and is commonly used in water management in India. One TMC-ft. = 28.317 million cubic meters = 22,956.8 acre-feet) including the contribution of 0.74 Mha-m from the neighboring states. The ground water potential is assessed as 2.24 Mha-m (in 1998). Hence, the total water potential in Tamil Nadu is about "2.42 + 2.24 = 4.66 or 4.70" Mha-m, of which, it is estimated that about 95 to 97% of the surface water is used and nearly 85% of ground water is extracted. Hence, the available water for future development is very limited and meager. Under these circumstances, how to manage the water required for the escalating population in the coming years for all sectors especially in agriculture is the question to be addressed by all of us.

Of the total water used in Tamil Nadu at present, about 80 to 82% is diverted for irrigation use. The remaining quantity is used for human population, animal population, industries, aqua culture, power, and maintaining the environment, etc. The net area for irrigation varies from 2.70 to 2.80 Mha and the gross irrigated area varies from 3.3 to 3.5 Mha. However, the Government of India, while assessing the ultimate irrigation potential of various states in the country, observed that the ultimate irrigation potential of Tamil Nadu is estimated as 5.53 Mha (Water and Related Statistics, CWC, New

Delhi, 2000). Since agriculture draws more than 80% of water, the future demand of water for increasing the area of irrigation and for other sectors can mainly be obtained by saving water in irrigation i.e. by better water management practices in irrigation and in other sectors.

Therefore, it is mentioned that agricultural scientist have evolved technologies for increasing productivity and advanced irrigation management practices to save substantial quantity of irrigation water for various crops. But only about 5% to 7% of the research findings have reached so far from the laboratories (office) to field.

7.2.3   WATER UTILIZATION AND PRODUCTIVITY

In Tamil Nadu, paddy crop alone takes more than 72% of water diverted for agriculture (irrigation) and the next crop, which uses more water is sugarcane (14 to 14.5%). The details of the gross area irrigated, water used for different crops, and percentage of water used for various crops at present in Tamil Nadu are shown in Table 8.

TABLE 8   Gross area irrigated, water used for different crops, and percentage of water use for various crops at present in Tamil Nadu.

S. No.	Crop	Gross irrigated area, Mha	Water use, Mha-m	Percentage of water used, %	
1	Paddy	2.10	2.52	72–73	
2	Sugarcane	0.30	0.48	14–14.5	90-91%
3	Banana	0.09	0.14	4–4.5	
4	Coconut	0.30	0.12		
5	Cotton	0.08	0.03		
6	Groundnut	0.25	0.08		
7	Millets	0.11	0.05	9-10%	
8	All other crops (vegetables, fruits, flowers, fodder etc.)	0.17	0.02		
9	**Total**	**3.40**	**3.44/3.45** (80% of water used at present)	100	

Mha = million hectare; Mha-m = million hectare meter.
Source: R K Sivanappan.

We must increase the irrigated area from 3.40 to about 5–5.2 Mha or more by adopting appropriate water technologies to save water and to increase the productivity per unit area and unit water to feed the growing population and provide water for all other sectors. As mentioned earlier, this chapter gives the details on how to achieve the goals in sugarcane crop, which uses about 15% of current water allotment for agriculture.

7.2.4 SUGARCANE CULTIVATION AND USE OF BI-PRODUCTS

The sugarcane crop is cultivated in about 50×10^5 ha in India while it is cultivated in about 3 to 3.2×10^5 ha in Tamil Nadu. The average productivity of cane in India and Tamil Nadu are about 67 to 68 and 100 to 106 tons/ha, respectively. Being a C4 plant, sugarcane is physiologically one of the most efficient plants. As per the agro-biological calculations and considering 50% use of solar radiation and 30% transpiration loss in sugarcane, it is possible to harvest 600 tons of cane/ha (240 tons/acre). Therefore, there is a great scope to bridge the gap between the potential and actual cane yields. Similarly, the water requirement of the crop varies from 1,800 to 2,500 mm depending upon the method of cultivation of the crop, method of irrigation, agro-climate region and the farmer. In Tamil Nadu, the water used for the crop in surface irrigation varies from 1,800 to 2,000 mm. Scientists have shown that by introducing micro irrigation and by adopting some crop geometry techniques, the irrigation water use can be reduced by above 40–50% and by giving nutrients at frequent intervals through fertigation in drip method, not only 30 to 35% of fertilizers can be saved but also the yield can be increased to 100 to 120 tons/acre (250–300 tons/ha). This is only 50% of the potential yield of the sugarcane in the country (600 tons/ha).

Presently, sugarcane is used mainly to make sugar (white) and the bi-products namely bagasse and molasses are not used to the maximum benefit to the cultivators, industries and to the state (Nation). Further, only 75–80% of the cane is used to make sugar and the rest is used to manufacture *gur* (Indian word, jaggery or gur is a traditional uncentrifuged sugar) in the South and *khandsari* (Indian word, khand is known as either gur or jaggery or less refined sugar-which is a raw cane sugar) in North India. In order to increase the yield and save water and fertilizer, it is necessary to introduce drip with fertigation. In addition, the bi-products, namely molasses and bagasse can be scientifically processed to produce spirit/ethanol and power by providing the needed infrastructure. In Tamil Nadu, by adopting the available technology in crop production and water management from the existing cropped area of 3×10^5 ha, apart from obtaining sugar (20×10^5 tons from 200×10^5 tons of sugarcane), there is possibility to crush another 200×10^5 to 250×10^5 tons sugarcane directly to produce ethanol, which is also economically viable and from each ton of cane, it is possible to take 100 units of power (1 unit = 1 KWH, kilo watt hour). The details of water used, yield obtained, cost of micro irrigation system with fertigation to be installed, quantities of sprit and ethanol production and power generation and the economics of all these items are provided in the following sections.

7.2.4.1 WATER USE AND YIELD OF SUGARCANE WITH PRESENT METHOD OF CULTIVATION

TABLE 9   Data for water use and sugarcane yield, Tamil Nadu.

Area under sugarcane in Tamil Nadu at present	$\times 10^5$ ha	3
Water requirement for sugarcane	mm/year	1,800 to 2,000
Effective or useful rain for the crop (in 12 months)	mm	350 to 400
	(average)	

TABLE 9   *(Continued)*

Therefore irrigation water requirement	mm	2,000–400 = 1,600
		= 1,600/1,000 = 1.6 meter
Total water requirement both ground water and surface water for 3×10^5 ha (0.3 Mha)	Mha-m	$0.3 \times 1.6 = 0.48$
Average yield of sugarcane	tons/ha	100
Therefore total yield of sugarcane from 3×10^5 ha	$\times 10^5$ tons	300
Of this 70–80% (75% average) is used for sugar factory	$\times 10^5$ tons	225
The balance cane is used for jaggery and other uses	$\times 10^5$ tons	75

7.2.4.2 WATER REQUIREMENT AND ANTICIPATED YIELD WITH IMPROVED AGRICULTURAL TECHNOLOGY AND IRRIGATION PRACTICES

7.2.4.2.1 WATER REQUIREMENT

Paired row with drip fertigation/subsurface drip with fertigation (SSI method)	0.3 Mha (3×10^5 ha)
Water requirement (irrigation) in micro irrigation (50 to 60%)/ take 60%	800 to 900 mm (or) 850 mm (average)
Total water needed	$[0.3 \times 850]/1000 = 0.25$ Mha-m
Water requirement is approximately 50% of the water used in the present practices (0.48 Mha-m in surface irrigation)	0.48 Mha-m
Water saving	$0.48 - 0.25 = 0.23$ Mha-m
	(or 2.3×10^5 ha-m)

7.2.4.2.2 ANTICIPATED YIELD OF SUGARCANE

0.3 Mha $\times$ 175 tons/ha (70 tons/acre), although we can get 250 tons/ha (100 tons/acre)	525×10^5 tons

Of the above 525×10^5 tons, 425×10^5 tons will be available for crushing and the remaining 100×10^5 tons for jaggery and other uses.

With the saved water, area that can be irrigated by cultivating dry crop such as groundnut, millets, oil seeds, cotton, etc.: Water requirement for one hectare under surface irrigation with improved water management practices is 0.50 ha-m (500 mm of water excluding rainfall during crop period). From 2.3×10^5 ha-m saved water, area that can be irrigated = $(2.3 \times 10^5)/0.50 = 4.60 \times 10^5$ ha (or 5.0×10^5 ha)

From the water used to grow sugarcane at present, by introducing drip, fertigation and other agricultural technology from the same 3×10^5 ha: We can produce:

1. About 525×10^5 tons of sugarcane with using only about 50% of water used at present;
2. From the saved water, it is possible to bring 4.5 to 5.0×10^5 ha under irrigation of dry crops of 4 to 5 months duration.

To implement the project, the fund requirements (i.e. drip and fertigation equipments) are:

Area	3×10^5 ha
Cost, Rs./ha: For drip with fertigation equipments	Rs.30,000/acre (or) Rs.75,000/ha
Therefore, total cost	Rs. 75,000 × 300,000 = Rs. 2,250 crores
Note: One crore = 10,000,000 = 10 million	= Rs. 22,500 million
	= Rs. 22.5 billion

Of which 75% (or Rs. 22,500/acre) is by the Government as subsidy and about Rs. 5000/acre or about 16.7% is paid by the factory (as being done by M/s. Sakthi Sugars) and the balance 8.3% or Rs. 2,500/acre is met by the farmer. Government can help the farmers to get loans from the banks to adopt micro irrigation and fertigationat low interest rate (say 4 to 5%).

7.2.4.3   THE DEMAND OF SUGAR IN TAMIL NADU IN 2010 AND IN 2020–2025

The present demand of sugar in Tamil Nadu is 12×10^5 to 15×10^5 tons/year. The estimated demand in 2020/25 will be about 16×10^5 to 20×10^5 tons. The present production from 200×10^5 tons of cane is 20×10^5 tons of sugar. Hence, the cane requirements for production of sugar, 150×10^5 tons of sugarcane at present and 200×10^5 tons of sugarcane in 2020/2025 will be sufficient. This indicates that the present production of sugarcane, 225×10^5 tons excluding *gur* making, is more than sufficient for the production of sugar even after 15 to 20 years (i.e., 2020/2025).

7.2.4.4   ADOPTION OF ADVANCED TECHNOLOGIES

By using advanced technology, from the same area of cultivation (3×10^5 ha) using about 50% of water, we can produce 525×10^5 tons of sugarcane (in another 5–7 years) from which more than 425×10^5 tons of sugarcane will be available for sugar industries (100×10^5 tons may be used for *gur* making etc.). The cane requirement for sugar preparation in 2020 will be about 200×10^5 tons as stated above and additional 225×10^5 tons of sugarcane is also available for other uses namely ethanol making/ production.

From sugarcane, after making sugar, we get two bi-products namely: (1) Molasses, and (2) Bagasse. Molasses can be used to prepare Spirit/alcohol, and Ethanol (mixing with petrol/diesel as fuel/alcohol-gasohol). Bagasse is used to produce electricity (energy).

7.2.4.4.1   MOLASSES

- One ton of sugarcane can produce 45 kg of molasses when it is crushed for sugar making and it can produce 10.3 liters of ethanol

- The 225×10^5 tons of sugarcane can produce $225 \times 10^5 \times 45 = 10,125 \times 10^5$ kg or 10.13×10^5 tons of molasses
- One ton of molasses can produce 230 L of ethanol/spirit
- Therefore 10.13×10^5 tons of molasses can produce $230 \times 10.13 \times 10^5 = 2,330 \times 10^{5 L}$ of ethanol/spirit.
- For 5% ethanol mix with petrol/diesel:
  (a) Tamil Nadu requirement at present $= 1,100 \times 10^5$ liters $=11$ crore liters $= 110$ million
  (b) Indian requirement at present $= 10,000 \times 10^{5 L} = 100$ crore liters $= 1000$ million
- For 10% ethanol mix with petrol/diesel:
  (a) Tamil Nadu requirement $= 2,200 \times 105$ liters $= 22$ crore liters $= 220$ million liters
  (b) India requirement $= 20,000 \times 10^5$ liters $= 2,000$ million liters.

In 2020:

   a) Alcohol/spirit from 200×10^5 tons of sugarcane.
   b) Ethanol form 225×10^5 tons of sugarcane is sufficient for 10% mix of 2,200 $\times 10^{5 \text{ liters}}$s $(2,330 \times 10^5$ liters) for Tamil Nadu.

The present production of spirit/ethanol is sufficient for industries till 2020/25.

7.2.4.4.2 BAGASSE

- 1 ton sugarcane yields 280 kg of bagasse.
- 1 ton bagasse can produce 2.3 tons of high pressure steam.
- 1 ton sugarcane (i.e., 280 kg bagasse) produces 100 units of power (under high pressure).
- 225×10^5 tons of cane produces $22,500 \times 10^5$ units or 2,250 million units of electricity.
- 425×10^5 tons of cane can produce (with drip and advanced methods of cultivation) $42,500 \times 10^5$ units of electricity or 4,250 MU (million units).

At present, only 27 plants out of 43 are producing energy (Co-generation). Action is to be taken for cogeneration in the remaining 16 plants to produce 4,250 million units/year.

7.2.5   STATUS OF POWER GENERATION IN TAMIL NADU

Item	2004–2005	2006–2007
Installed capacity	9531.70 MW	10237 MW
Ownership by States	5401.04 MW	7717.98 MW
Hydro	1987.40 MW	2191.00 MW
Thermal	2970.00 MW	2970.00 MW
Wind mill, private	19.36 MW	2040.00 MW
Gas, harnessed up to	424.28 MW	516.08 MW
31–03–2005.		

Item	2004–2005	2006–2007
Total	5401.04 MW	7717.98 MW
Maximum Demand	7473.00 MW	10859 to
		11500 MW
Power Generation (MU, million units) in Tamil Nadu, India		
Hydro	4426.00	6292.00
Thermal	20004.00	21228.00
Wind mill	17.00	17.00
Gas	2003.00	1944.00
Total	26450.00	29481
Power purchased (MU, million units) by Tamil Nadu State		
Source from where purchased		
Neyveli I and II	7688 MU	—
Madras Atomic Plant	913 MU	—
National Thermal Power	—	—
Corp./Ramagunda	4010 MU	—
Manali and others	13284 MU	—
Total	25895 MU	52345 MU
Gross power available	52345 MU	63,038 MU
Total Power Consumption	40298 MU	77,218 MU
Per capita consumption of power KWH/Unit	815 units	1040 units
	(2004–05)	(2011–12)

This indicates that the power that can be generated from sugarcane is approximately equal to that of hydroelectric power produced in the state in 2005 and 70% in 2007.

From 3×10^5 ha, at present, only about 225×10^5 tons of cane is available for crushing (75% of average production). The ethanol produced from this cane will be sufficient for spirit/alcohol for the state even up to 2020/25. After introduction of pit method/paired row micro irrigation and fertigation/subsurface drip fertigation/SSI method, it is possible to produce about 525×10^5 tons of cane in the same area with 50–55% of water, saving about 45–50% of water (i.e., 0.23 Mha-m). The cane available after using for *gur* making, may be 425×10^5 tons. Of which 200×10^5 tons can be crushed for sugar and preparation of spirit. By this, the required sugar and spirit can be obtained. The balance 225×10^5 tons of sugarcane can be used to produce only ethanol directly from sugarcane. About 75 liters of ethanol can be produced from one ton of sugarcane without going to sugar. Therefore, 225×10^5 tons of sugarcane will produce about $(225 \times 10^5) \times 75 = 16,87.5$ million liters of ethanol, which can be used to mix with petrol (gasohol). The requirement for Tamil Nadu is about 2.20 million liters only at 10% mix and India's requirement is about 20.00 million liters at 10% mix. By introducing micro irrigation to irrigate 3×10^5 ha of sugarcane crop, the water requirement will be reduced by 40–50%, which means electricity will also be reduced to that extent. It is estimated that the total electricity saving, when 3.0×10^5 ha of cane being cultivated under micro irrigation/fertigation system, will be around Rs. 1,350 million/year, which again is a saving for the Government.

In addition, many benefits are available for the State and Government of India. Sugarcane is one of the most important commercial crops, which is being taxed by both Central and State governments and by increasing the cane production will ultimately result in higher return for the government. It is understood that the government of Tamil Nadu and India gets revenue of around Rs.450/ton of cane. At this rate, the income for the government (525×10^5 tons) will be Rs. 23,625 million per annum. By deducing the subsidy of Rs. 16,875 million, a net balance of Rs. 6,750 million is available. After one year, the entire Rs. 23,625 million per year can be obtained by the Government. Hence, it is recommended that this venture/profit can be taken up without any further delay.

7.3 CONCLUSIONS

Dr. K. Ramasamy, Vice Chancellor at TNAU, indicates that *"Demand for sugar will rise by 50% more by 2030. Climate change is likely to impact productivity irrespective of the region where sugar cane is grown. As much as 20% reduction is predicted for every degree rise in temperature. It is not possible to find new lands and more water to cultivate sugarcane for satisfying the demand. So, we need to increase productivity through new technologies with less water to achieve it. The number of sugar mills increased from 30 in 1930 to 560 in 2012 in India, producing about 26 million tons of sugar in 2012. Sugarcane researchers, 5 million farming communities and 560 sugar industries played a major role in making India, the second largest producer of sugar in the World. However, there is a need for improving the productivity in the coming years. The sugar industry should take an active part in sugarcane research and development activities so that the country can achieve a newer height in sugarcane production and increase profits to all stakeholders. Further sugarcane is also viewed as the most important raw material for bio-fuel production. To meet these demands, there is a need for enhancing the cane production. This is possible only by increasing productivity per unit area and per unit water since the scope for further expansion of cane area and to get more water for the additional areas are highly limited. The technology described in this chapter by the author is urgently needed for the sugar industry in India. The technology can increase the cane yield from the existing 40 tons/acre to 100 tons/ acre and by using the same area of land and 50% less water. With this technology, the production of sugarcane can be increased from 300×10^5 tones to 525×10^5 tons in Tamil Nadu. In addition, the required ethanol for mixing with petrol and generating about 4,000 million units of electricity can be produced, which is the need of the State. Apart from the above direct benefits, the Central and State Governments will get more income by way of taxes. The State and Central Governments should take necessary action to implement the two findings of this technology, namely: (a) helping the farmers to obtain 100 tons of sugar cane per acre and produce ethanol for mixing with petrol;*

and (b) to cogenerate in all sugar mills in the State to meet the sugar demand and spirit/ethanol requirement for energy, and produce substantial amount of electricity for Tamil Nadu as early as possible to solve some critical problems faced by the State."

The available technologies, namely, pit method and paired row method of cultivation using micro irrigation with fertigation/subsurface drip fertigation/SSI method should be transferred quickly and efficiently using modern techniques. By these methods, sugarcane production can be doubled and about 40–50% of water (i.e., 0.23 Mha-m) can be saved, which can bring an additional area of about 4×10^5 to 5×10^5 ha without any extra water or funds. If all the bi-products are used, the ethanol requirement of the State at 10% mix (gasohol) can be produced. Furthermore, from the cane produced, (bagasse) above 4,000 million units of electricity can be produced from this 3×10^5 ha of land which is equivalent to the power generation with hydro units in Tamil Nadu. A detailed action plan can be prepared by the Govt. (i.e., Directorate of Sugar) involving all sugar factories, experts and Planning Commission to take up the project as soon as possible for the benefit of Tamil Nadu State.

7.4 SUMMARY

Sugarcane is the second largest commercial crop cultivated in about 5 Mha in India and about 3×10^5 to 3.20×10^5 ha in Tamil Nadu. In India, there are, about, 560 sugar factories producing nearly 26 million tons of sugar. In Tamil Nadu, there are about 37sugar factories producing about 20×10^5 to 22×10^5 tons of sugar. The average production of sugarcane is about 70 tons/ha in India, whereas the average yield in Tamil Nadu is about 106 to 110 tons/ha. It is possible to harvest 600 tons of cane/ha (i.e., 240 tons/ac). Water requirement of the crop varies from 1,800 to 2,500 mm depending upon the method of cultivation, the method of irrigation, the agro climatic region and the farmer. Scientists have shown that by introducing drip with fertigation and adopting required crop geometry techniques, the irrigation water used can be reduced by 45 to 50% fertilizer application can be saved (by giving nutrients at frequent intervals) by 30 to 35% and the yield can be increased to 100 to 120 tons/acre. In the Pune District of the Maharashtra State and in the Coimbatore District of the Tamil Nadu, farmers have obtained 100 to 120 tons/acre of cane.

About 15% of water used for irrigation in Tamil Nadu and 60% of water used for irrigation in Maharashtra are diverted for sugarcane. In the future, it may not be possible to allot more water for irrigating sugarcane. In fact, the future water allocation should be reduced. However, more cane is required to meet the demands of the growing population. The author has provided detailed information on how to obtain 100 tons/acre sugarcane based on the research and experience. It is heartening to note that farmers have already formed 100 tons/acre Clubs in some sugar factory areas in Tamil Nadu.

When sugar cane is irrigated by the drip fertigation method, it is possible to double the crop yield without any extra land or water, thus enabling us to meet our future sugar demand. At the same time, by using molasses/bagasse from the cane, the energy problem of the State can also be solved to some extent. All these actions will result in extra income to the farmers.

ACKNOWLEDEMENTS

Author is thankful to: (1) Dr. M. V. Ranghaswami, former Director of Water Technology Centre at Tamil Nadu Agricultural University (TNAU), for estimating the daily water requirement for the sugarcane crop and fertigation application scheduling; (2) Mr. P. Natarajan, Senior President of Sakthi Sugars for many useful suggestions, comments and providing more information about cane production; (3) Mr. K. T. Varadharajan, retired Chief Engineer of TNEB for providing information about energy production in Tamil Nadu; Dr. S. R. Sree Rangasamy, former Dean at TNAU; and Dr. Arumugam Kandiah former visiting Professor at TNAU and FAO Expert—Rome for providing many useful suggestions and editing the manuscript. Author also acknowledges the support, for preparing the two unpublished reports that were used to prepare this chapter, from: Dr. N. Mahalingam, Chairman of Sakthi Group of Companies; and Dr. K. Ramasamy, Vice Chancellor at TNAU.

KEYWORDS

- advantages
- bagasse
- Benefit-cost
- bi-product
- cogeneration
- commercial crop
- demand of sugar
- ethanol
- experiences
- fertigation
- fertilizer schedule
- fund requirement
- high water consuming crop
- irrigation
- micro irrigation
- million hectare, meters, Mha-m
- million hectares, Mha
- molasses
- paired row
- pit method

- productivity
- profit
- spirit/ alcohol
- sustainable sugarcane initiative, SSI
- Tamil Nadu Agricultural University, TNAU
- water scarcity
- yield potential

APPENDIX I

(Sugarcane Cultivation in Tamil Nadu)

Shade net nursery

Sustainable Sugarcane initiative (SSI) with single row planted above the lateral

CHAPTER 8

DESIGN AND MANAGEMENT OF IRRIGATION SYSTEMS: CHILE

EDUARDO A. HOLZAPFEL, ALEJANDRO PANNUNZIO,
IGNACIO LORITE, AUREO S. SILVA DE OLIVEIRA,
and ISTVÁN FARKAS

CONTENTS

8.1 Introduction 148
8.2 Irrigation Systems.. 149
8.3 Water Management for Irrigation Systems.................................... 153
8.4 Conclusions 155
8.5 Summary... 156
Keywords 156
References.. 157

8.1 INTRODUCTION

Water required by crops is supplied by nature in the form of precipitation, but when it becomes scarce or its distribution does not coincide with demand peaks, it is then necessary to supply it artificially, by irrigation. Several irrigation methods are available, and the selection of one depends on factors such as water availability, crop, soil characteristics, land topography, and associated cost.

In the near future, irrigated agriculture will need to produce two-thirds of the increase in food products required by a larger population [16]. The growing dependence on irrigated agriculture coincides with an accelerated competition for water and increased awareness of unintended negative consequences of poor design and management [11].

Optimum management of available water resources at farm level is needed because of increasing demands, limited resources, water table variation in space and time, and soil contamination [44]. Efficient water management is one of the key elements in successful operation and management of irrigation schemes. Irrigation water management involves determining when to irrigate, the amount of water to apply at each irrigation event and during each stage of plant, and operating and maintaining the irrigation system. The main management objective is to manage the production system for profit without compromising environment and in agreement with water availability. A major management activity involves irrigation scheduling or determining when and how much water to apply, considering the irrigation method and other field characteristics.

Under limitations in water availability, it is required to develop new irrigation scheduling approaches focused on to ensure optimal use of available water, and not based on full crop water requirements. The determination of these efficient and effective irrigation schedules (including deficit irrigation strategies) requires the accurate determination of water requirements for the main crops, in order to assist the farmers in deciding when and how much to irrigate their crops. New technologies such as remote sensing, modeling, and lysimeters are contributing to obtain a better knowledge of the crop agronomy and irrigation, detecting water stress, determining crop water requirements and obtaining accurate irrigation schedules.

Irrigation technology has made significant advances in recent years. Criteria and procedures have been developed to improve and rationalize practices to apply water, through soil leveling, irrigation system design, discharge regulations, adduction structures, and control equipment. However, in many regions these advances are not yet available at the farm stage.

Irrigation systems are selected, designed and operated to supply the irrigation requirements of each crop on the farm while controlling deep percolation, runoff, evaporation, and operational losses, to establish a sustainable production process. Playán and Mateos [58] mentioned that modernized irrigation systems at farm level implies selecting the appropriate irrigation system and strategy according to the water availability, the characteristics of climate, soil and crop, the economic and social circumstances, and the constraints of the distribution system.

8.2 IRRIGATION SYSTEMS

8.2.1 DESIGN AND MANAGEMENT

Design of irrigation systems is a very important topic in the process to improve irrigation application, efficiency and economical return in the production process [53–55].

Irrigation system design substantially affects application efficiency and involves numerous variables and restrictions, whose principal objective is to maximize benefits and minimize costs. In a successful irrigation system, a set of resources produces maximum returns. To achieve this, an optimization process supporting the design and operation of water application systems in agriculture is required due to many possible combinations of design variables that satisfy irrigation conditions.

Irrigation systems have specific applications that are based in several factors, among which the most relevant are the crop, soil type, topography, and water availability and quality. The application efficiency of the different surface and pressurized irrigation methods varies and depends on design, management, and operation [33].

Undoubtedly, well-designed and correctly used irrigation systems will have the highest efficiency and water distribution levels, which can result in a good production and high product quality [34, 35]. For an adequate management and operation of the surface irrigation systems, a series of support elements have been developed, including simulation models and control and derivation structures, such as adduction systems. In the case of pressurized irrigation systems, enormous advances have resulted in greater automation in its operation, greater precision in its application, and the incorporation of chemical elements for nutrition and disease control into the irrigation process [33, 62].

In order to increase agriculture sustainability, an important aspect that has been considered by several researchers and studies is to design efficient irrigation systems at farm level [31, 37, 43, 50, 51]. These appear to be a very crucial aspect for the irrigated agriculture and a key factor due to the competition for water resources. In recent years, several irrigation systems have improved significantly the application efficiency at farm level, improving the irrigation water management. For instance, in the main irrigation districts of Mexico, introducing new technologies and more efficient associated with real-time irrigation scheduling, demonstrated water savings in the order of at least 20%, without any appreciable decrease in crop yields [60].

Some studies have analyzed the differences in irrigation efficiency for several irrigation systems, as Al-Jamal et al. [2], where sprinkler, trickle and furrow irrigation systems were analyzed in Southern New Mexico, USA, with onion production. Thus, the maximum irrigation water use efficiency (IWUE) was obtained using the sprinkler system. The lower IWUE values were obtained under subsurface drip and furrow irrigation systems compared with sprinkler irrigation were due to excessive irrigation under subsurface drip and higher evaporation rates from fields using furrow irrigation.

8.2.2 COMPARISON OF IRRIGATION SYSTEMS

Ibragimov et al. [39] compared drip and furrow irrigation, obtaining that 18–42% of the irrigation water was saved with drip systems in comparison with furrow, and the IWUE increased by 35–103% compared with furrow irrigation. Same comparisons were made by Maisiri et al. [46] in a semiarid agro tropical climate of Zimbabwe; in this study, drip irrigation used about 35% of the water used by the surface irrigation systems, providing higher IWUE. The gross margin level for surface irrigation was lower than for drip irrigation.

Other studies analyzed surface and subsurface drip irrigation systems in Turkey [76]. Both methods had similar yield results but surface drip had more advantages due to difficulty in replacement and higher cost for subsurface systems. Additionally, they recommended the surface drip in early potato under Mediterranean conditions.

Tognetti et al. [71] determined that drip irrigation influenced positively many of the physiological process and technological parameters in semiarid conditions, as compared to low-pressure sprinkler irrigation. Hanson and May [28] obtained yield increases when drip system was used compared to the sprinkler systems with similar amounts of applied water; additionally, drips systems reduced percolation below the root zone. Another study analyzed low-energy precision application (LEPA) and trickle irrigation for cotton in Turkey, concluding that both irrigation systems could be used successfully under the arid climatic conditions of this country [76].

8.2.3   PRESSURIZED IRRIGATION SYSTEMS

According to many papers, drip irrigation is widely regarded as the most promising irrigation system, and is specially recommended in combination with saline water. Karlberg et al. [42] indicate that two low-cost drip irrigation systems with different emitter discharge rates were used to irrigate tomatoes (*Lycopersicon esculentum*), concluding that the combination of drip systems with plastic mulch increased the yield.

In relation with drip systems, some analyzes have been made to determine the optimum lateral spacing for drip-irrigated corn (*Zea mays*) in Turkey [6]. Lateral spacing of 0.7, 1.40 and 2.10 m were compared, concluding that the optimum lateral spacing for corn was 1.40 m (one drip lateral per two crop rows). Ravindra et al. [62] developed a linear programming model for optimal design of pressurized irrigation system subunit, considering implementation cost and energy. Holzapfel et al. [32–36] analyzed the optimal selection of emitters and sprinkler taking into account the total cost, operation and implementation. Rodríguez et al. [64, 65] studied the selection of emitter on the basis of hydraulic characteristics. On the other hand, annual water application cost per unit area has been analyzed at the subunit level designed with pipes of different materials by Ortiz Romero et al. [49].

Other important aspects are related with the analysis of strategies for water management in pressurized irrigation systems [61]. The timing or frequency of water applications will depend on soil texture (e.g., sand *vs.* clay), the irrigation system used (e.g., drip *vs.* sprinkler), the rate at which the plant is using water, and the overall development of the plants root system. Frequent water applications are especially important when using drip systems, which tend to restrict soil wetting and thus produce a smaller root system. When done properly, frequent irrigations are beneficial and often increase growth and yield in many orchards and crops. For example, frequent irrigation by drip in peach (*Prunus persica*) increased fruit size and yield compared to other irrigation methods by maintaining higher tree water status between irrigations [8]. It is also important to apply water to both sides of the plant and to maintain an adequate volume of soil wetted, as was observed in citrus (*Citrus* sp.) and avocados (*Persea americana*) [30, 33]. Abbott and Gough [1] found that when water was applied to only one side, growth and production were severely restricted on the other. However, irrigators, particularly those using drip, should be careful to avoid the temptation to over irrigate. Over-irrigation depletes the root zone of much-needed oxygen, thus reducing both root growth and nutrient uptake, leading to a host of potential root disease problems as well as aquifer contamination. Yadav and Mathur [75] found with tested model that the water extraction is closely related with soil moisture availability in addition to the root density, which occurs normally from the upper root dense soil profile.

Considering the relevant support that means a good automation of the irrigation systems to reduce management time and simplify design, it is necessary to use the modern methods of surveying and analysis tools [72]. Remote sensing and Geographic Information Systems (GIS) with their capability of data collection and analysis are viewed as efficient and effective tools for pressurized irrigation system design and management [64]. The capability of GIS to analyze the information across space and time would help in managing such dynamic systems as pressurized irrigation systems. Hazrat et al. [29] found that the GIS are an important tool that can be used to optimal allocation of water resources of an irrigation project. Mean water balance components results for different months were stored in GIS databases, analyzed, and displayed as the monthly crop water requirements maps.

Intensive irrigated agriculture under pressurized systems involves water application, fertilizers and other chemicals, applied in varying amounts to the crop and orchards areas. Yet, like any enhanced productive activity, it leads to environmental contamination if it is improperly managed [74]. When utilization of chemicals is incomplete or inefficient, or when water is applied in excess, the resultant seepage ends up in drainage systems or in recharging the aquifers beneath the cropped land [27].

8.2.4   SURFACE IRRIGATION SYSTEMS

During the past two decades procedures and criteria of surface irrigation have become available, which have established the optimum design and efficient management and operation of the water applied to crops to obtain maximum yield and avoid environmental effect due to leaching and runoff.

The most frequently used surface irrigation systems are: contour irrigation, border irrigation, and furrow irrigation. The latter is more frequently used to irrigate row crops and orchards. This method possesses various factors and elements of design and management, which have been extensively analyzed.

Furrow irrigation is the oldest and more commonly used irrigation system. Lately, it has become important because of the high cost of energy in pressurized irrigation methods and the incorporation of automation in its operation [33]. Eldeiry et al. [15] found in a study of design of furrow irrigation in clay soil that the furrow length and application discharge are the main factor affecting application efficiency. On the other hand, environmental impacts of furrow irrigation has been studied by different research-ers [45] and has been found that in irrigation, risks of nitrate leaching depends of water and fertilizer application [59]. Burguete et al. [9, 10] developed and calibrated a model and methodology for fertigation in furrow irrigation, which allow adequate fertilizer uniformity and avoid contamination problems. The automation of fertilization as well as water application are very important advances for the development of furrow irriga-tion and the improvement of surface irrigation systems.

The amounts of water and leachates vary with the type of irrigation system used, irrigation efficiency, crop or orchard, utilization of water and fertilizers, decomposi-tion of the added organic materials, and absorption of the decomposed fractions. The leachates contain the chemicals mostly added in irrigation water, but nitrate is the domi-nant anion, derived from fertilizers and decomposing organic materials. In addi-tion, pesticides are other chemicals incorporated in irrigation and it has been shown that leaching increased in direct relation with the increment of the amount of percolating water produced by irrigation [3]. The close association between leaching and amount of deep-percolating water produced by irrigation is expected because leaching occurs through dissolution of solute in soil solution and subsequently moves with soil water. These results can show the importance of using an adequate design of surface and pressurized irrigation systems and management techniques to reduce chemical leach-ing by decreasing deep percolation.

In many irrigated regions of the western United States, commercial growers that use furrow irrigation systems are facing serious challenges to improve irrigation efficiency and reduce contamination of water supplies [63]. However, the most important points are to adequately select the irrigation variables in each system with appropriate de-sign criteria, improving the irrigation scheduling and water management of the field that will also potentially reduce over irrigation and deep leaching of applied water and chemicals.

8.2.5   DESIGN OF IRRIGATION SYSTEMS: GENERAL CONSIDERATIONS

Design of irrigation systems is a very important topic in the process to improve water application, efficiency and economical return in the production process. To develop adequate engineering design, the use of well-known criteria it is a basic component; this can only be obtained with deep knowledge on irrigation, and the technical pa-rameters associated with crop, soil water characteristics, energy and environment. Pereira [57] reported that to improve the irrigation systems requires the consider-ation of the factors influencing the hydraulic processes, the water infiltration and the uniformity of water application to the entire field. The consideration of all these aspects makes irrigation management a complex decision-making and field practice process.

Playán and Mateos [58] reported that in general all irrigation systems can attain approximately the same level of efficiency, when they are well designed and appro-priated selected for the specific condition, due that irrigation is site specific. How-ever, differences among irrigation systems appear in many areas as a consequence of design, management and maintenance. There are several reports indicating that trickle irrigation systems increased yields in Washington Navel oranges (*Citrus sinensis*) in humid regions [51, 52], determining the convenience of wetting a minimum percentage of the shaded area by the crop to maximize yields, finding the mini-sprinkler system as the most efficient.

An adequate selection and distribution of number of emitters in drip irrigation may give substantial different results [65]. Colombo and Or [12] found that in a drip irrigated field, local crop yield patterns are determined by spatial patterns in water accessibility to plants that could be quantified and used for optimal irrigation system design; the results demonstrated that around each emitter, the spatial distribution of above- ground dry mass per plant followed a general pattern. Number and location of drippers in a blueberry (*Vaccinium corimbosum*) crop was tested, finding out that the shallow and spread roots of the crop were well wetted by two laterals per each row of crop compared with a single line of drippers per row of crop, resulting in greater yields [53, 54]. Henríquez [30] found that with a correct number and distribution of emitter on the base of soil and plant characteristic, there are differences of 100% in yield for citrus orchards.

8.3 WATER MANAGEMENT FOR IRRIGATION SYSTEMS

8.3.1 DEFICIT IRRIGATION IN FRUIT TREES

The concept of regulated deficit irrigation (RDI) was initially proposed to control vegetative growth in peach orchards, determining in addition savings in irrigation water without reducing yield. Thus, deficit irrigation has been demonstrated as a useful tool to improve the irrigation management at field scale for arid and semiarid conditions.

Deficit irrigation has been used extensively for annual crops in the past. Thus, was applied to corn [19, 56]; cotton (*Gossypium hirsutum*) [18, 41]; wheat (*Triticum aestivum*) [41]; sorghum (*Sorghum bicolor*) [19]; onion (*Allium cepa*) [5] and sugar beet (*Beta vulgaris*) [17].

However deficit irrigation has had significantly more success in tree crops and vines than in field crops [20]. Analyzes of deficit irrigation in trees have been related with peaches [21]; almonds (*Prunus dulcis*) [22, 25, 66, 67]; pistachio (*Pistacia vera*) [24]; plum (*Prunus domestica*) [40]; olives (*Olea europaea*) [48, 69]; citrus [13, 26, 73] and wine grapes (*Vitis vinifera*) [7, 47]. Analyzing all these studies, the conclusion would be that supplying full ET requirements could not be the optimal irrigation scheduling for management of irrigation systems.

A correct water management at field scale is highly influenced by the irrigation method and the irrigation operating system. Traditionally deficit irrigation strategies were carried out on-demand irrigation systems, where the date and rate of the irrigation was decided by the farmer or technician, and the irrigation scheduling was not limited by the distribution system. Equally, for fruit trees drip and microsprinkler systems are the most common irrigation methods, although surface irrigation methods under an appropriate water management could provide similar efficiency and performance results.

Girona et al. [21] analyzed the response to single and combined deficit irrigation regimes for peach trees with an automated drip system with four compensating emitters per tree. Thus, it was obtained that deficit irrigation during Stage II of fruit development and/ or during postharvest significantly reduced vegetative growth of the trees. However, fruit production was not affected by any deficit irrigation strategy until the fourth year when yield decreased slightly with combined deficit irrigation [21].

Similar studies were made in almond, applying RDI during the kernel-filling phase [22]. In the study during the first two experimental years, kernel dry matter accumulation did not decrease, however, similar to previous works, cropping and kernel growth were reduced during the third and fourth years of experiment due to a hypothetical depletion of the reservoirs. Water savings obtained with these deficit irrigation strategies may help to promote the adoption of regulated deficit irrigation in areas with water availability reduced. The main difference among this experiment and previous analyzes was the irrigation system; in this case microsprinklers were used. This fact could affect the wetted soil volume. Hutmacher et al. [38] suggested the possibility to obtain better results using microsprinklers, confirmed by Girona et al. [23] who indicated that wet soil volume was a very important factor for almond. Thus, specific studies comparing drip and microsprinklers systems indicated that also for others crops as grapevines irrigated by micro sprinklers, showed greater root presence as the distance from the trunk increased [4].

Also in almond with microsprinkler irrigation, Goldhamer et al. [25] analyzed the impacts of three different water stress timing patterns. The most successful stress timing pattern in terms of yield (considering fruit size and load) was the pattern that imposed sustained deficit irrigation by applying water a given percentage of full ETc over the entire season. Also Romero et al. [66] analyzed the influence of several RDI strategies under subsurface and surface drip irrigation. Thus, RDI, with a severe irrigation deprivation during kernel-filling (20% ETc), and a recovery postharvest at 75% ETc or up to 50% ETc under subsurface drip irrigation, can be adequate in almonds under semiarid conditions, saving a significant amount of irrigation water. The comparison between surface and subsurface irrigation indicated higher levels of volumetric soil water content for sub superficial systems [66].

Comparisons among surface drip, subsurface drip and microjet on almonds indicated that microjet irrigated trees tended to increase yield about 10% in some years although no consistent yield differences were found [14]. Considering growth under deficit irrigation, microjet irrigated trees induced slightly higher growth, similar results that when root development was analyzed. However, as negative aspect, microjet irrigation increased maintenance and required higher use of herbicide.

Deficit irrigation effects on yield and vegetative development also have been analyzed for drip-irrigated olives. Thus, Tognetti et al. [69] determined that water availability might affect fruit weight before flowering or during the early stages of fruit growth rather than later in summer season. Thus, irrigation of olive trees with drip systems from the beginning of pit hardening could be recommended, at least in the experimental conditions of the study. Comparing different treatments, deficit irrigation during the whole summer resulted in improved plant water relations with respect to other watering regimes, while severe RDI differentiated only slightly treatments from rain fed plants [70, 71]. Nevertheless, regulated deficit irrigation of olive trees after pit hardening could be recommended. Citrus also has been studied under deficit irrigation strategies. Thus, deficit irrigation treatments compared with the control drip irrigated by six pressure compensated emitters by tree allowed seasonal water saving between 12 and 18% [73].

8.3.2 CURRENT TECHNOLOGIES FOR WATER MANAGEMENT AND STRESS DETECTION

Several studies about deficit irrigation have used new technologies as remote sensing techniques. The use of infrared thermometry and thermal imaging is considered an interesting option in order to monitoring stress in trees, as Sepulcre-Cantó et al. [68] described for stress detection in olive trees from infrared imagery. Also Falkenberg et al. [18] with an infrared camera distinguished between biotic (root rot) and abiotic (drought) stress with the assistance of ground truthing, determining that deficit irrigation around 75% ETc had no impact on yield, indicating that water savings were possible without reducing yield.

In addition to remote sensing technology, lysimeter also has been used to characterize deficit irrigation and stress for orchard trees as peaches [21]. In this work it was established threshold levels of available soil water content for ET, leaf conductance and photosynthesis during deficit irrigation periods, demonstrating the importance of considering wetting patterns, soil depth and root exploration in high frequency irrigation management, especially for tree crops.

8.4 CONCLUSIONS

Irrigation system design substantially affects application efficiency and involves numerous variables and restrictions whose principal objectives are to maximize benefits and minimize costs. Design of irrigation systems is a very important topic to improve irrigation application, efficiency and economical return in the production process. The design must be on the basis of sound criteria that are mainly related with knowledge on irrigation, hydraulic, economic, energy, environmental and agronomics aspects.

Efficient irrigation systems design at farm level appear to be a very important aspect for the irrigated agriculture and a key factor due to the competition for water resources with other sectors and to permit the economical and environmental sustainability of agriculture. In general, surface and pressurized irrigation systems can attain a reasonable level of efficiency, when they are well designed, adequately operated and appropriately selected for specific conditions; because irrigation is site specific.

Automation of the irrigation systems (surface and pressurized) coupled with sensor and equipment to register the information of the irrigation process are important topic that must be developed in order to use the systems with their total potentiality.

Regulated deficit irrigation proposed to control vegetative growth in orchards produce, in addition, savings in irrigation water without reducing yield. Thus, deficit irrigation has been demonstrated as a useful tool to improve irrigation management at field scale for arid and semiarid conditions.

8.5 SUMMARY

Irrigation systems should be a relevant agent to give solutions to the increasing demand of food, and to the development, sustainability and productivity of the agricultural sector. The design, managing, and operation of irrigation systems are crucial factors to achieve an efficient use of the water resources and the success in the production of crops and orchards. The aim of this chapter is to analyze knowledge and investigations that enable to identify the principal criteria and processes that allow improving the design and managing of the irrigation systems, based on the basic concept that they facilitate to develop agriculture more efficient and sustainable. The design and managing of irrigation systems must have its base in criteria that are relevant, which implies to take into account agronomic, soil, hydraulic, economic, energetic, and environmental factors. The optimal design and managing of irrigation systems at farm level is a factor of the first importance for a rational use of water, economic development of the agriculture and its environmental sustainability.

KEYWORDS

- almond
- annual crops
- arid
- Chile
- deficit irrigation
- environmental sustainability
- evaporation
- evapotranspiration
- irrigation
- irrigation design
- irrigation, application efficiency
- irrigation, gravity
- irrigation, high pressure
- irrigation, low pressure
- irrigation, sprinkler
- irrigation, surface
- operation systems
- rational use of water
- semiarid
- sustainable agriculture
- water management
- water use efficiency, WUE

REFERENCES

1. Abbott, J.D., and R.E. Gough. 1986. Split-root water application to highbush blueberry plants. HortScience 21:997–998.
2. Al-Jamal, M.S., S. Ball, and T.W. Sammis. 2001. Comparison of sprinkler, trickle and furrow irrigation efficiencies for onion production. Agric. Water Manage. 46:253–266.
3. Asare, D.K., T. Sammis, D. Smeal, H. Zhang, and D.O. Sitze. 2001. Modeling an irrigation strategy for minimizing the leaching of atrazine. Agric. Water Manage. 48:225–238.
4. Bassoi, L.H, J.W. Hopmans, L.A. de C. Jorge, C.M. de Alencar, and J. Silva. 2003. Grapevine root distribution in drip and microsprinkler irrigation. Sci. Agric. (Piracicaba, Braz.) 60:377–387.
5. Bekele, S., and K. Tilahun. 2007. Regulated deficit irrigation scheduling of onion in a semiarid region of Ethiopia. Agric. Water Manage. 89:148–152.
6. Bozkurt, Y., A. Yazar, B. Gencel, and M.S. Sezen. 2006. Optimum lateral spacing for drip-irrigated corn in the Mediterranean Region of Turkey. Agric. Water Manage. 85:113–120.
7. Bravdo, B., and A. Naor. 1996. Effects of water regime on productivity and quality of fruit and wine. Acta Hort. (ISHS) 427:15–26.
8. Bryla, D.R., E. Dickson, R. Shenk, R.S. Johnson, C.H. Crisosto, and T.J. Trout. 2005. Influence of irrigation method and scheduling on patterns of soil and tree water status and its relation to yield and fruit quality in peach. HortScience 40:2118–2124.
9. Burguete, J., N. Zapata, P. García-Navarro, M. Maïkaka, E. Playán, and J. Murillo. 2009a. Fertigation in furrows and level furrows systems. I. Model description and numerical test. J. Irrig. Drain. Eng. 135:401–412.
10. Burguete, J., N. Zapata, P. García-Navarro, M. Maïkaka, E. Playán, and J. Murillo. 2009b. Fertigation in furrows and level furrows systems. II. Field experiments, model calibration, and practical applications. J. Irrig. Drain. Eng. 135:413–420.
11. Cai, X., D.C. McKinney, and M.W. Rosegrant. 2003. Sustainability analysis for irrigation water management in the Aral Sea region. Agric. Syst. 76:1043–1066.
12. Colombo, A., and D. Or. 2006. Plant water accessibility function: A design and management tool for trickle irrigation. Agric. Water Manage. 82:45–62.
13. Domingo, R., M.C. Ruiz-Sánchez, N.J. Sánchez-Blanco, and A. Torrecillas. 1996. Water relations, growth and yield of Fino lemon trees under regulated deficit irrigation. Irrig. Sci. 16:115–123.
14. Edstrom, J.P., and L.J. Schwankl. 2002. Microirrigation of almonds. Acta Hort. (ISHS) 591:505–508.
15. Eldeiry, A., L.A. Garcia, A.S. El-Zaher, and M.E. Kiwan. 2005. Furrow irrigation system design for clay soils in arid regions. Appl. Eng. Agric. 21:411–420.
16. English, M.J., K.H. Solomon, and G.J. Hoffman. 2002. A paradigm shift in irrigation management. J. Irrig. Drain. Eng. 128:267–277.
17. Fabeiro, C., F. Martín de Santa Olalla, R. López, and A. Domínguez. 2003. Production and quality of the sugar beet (Beta vulgaris L.) cultivated under controlled deficit irrigation conditions in a semiarid climate. Agric. Water Manage. 62:251–227.
18. Falkenberg, N., G. Piccinni, J.T. Cothren, D.I. Leskovar, and C.M. Rush. 2007. Remote sensing of biotic and abiotic stress for irrigation management of cotton. Agric. Water Manage. 87:23–31.
19. Farré, I., and J.M. Faci. 2006. Comparative response of maize (Zea mays L.) and sorghum (Sorghum bicolor L. Moench) to deficit irrigation in a Mediterranean environment. Agric. Water Manage. 83:135–143.
20. Fereres, E., and M.A. Soriano. 2007. Deficit irrigation for reducing agricultural water use. J. Exp. Bot. 58:147–159.
21. Girona, J., M. Gelly, M. Mata, A. Arbonés, J. Rufat, and J. Marsal. 2005a. Peach tree response to single and combined deficit irrigation regimes in deep soils. Agric. Water Manage. 72:97–108.

22. Girona, J., M. Mata, E. Fereres, D.A. Goldhamer, and M. Cohen. 2004. Evapotranspiration and soil water dynamics of peach trees under water deficits. Agric. Water Manage. 54:107–122.

23. Girona, J., M. Mata, and J. Marsal. 2005b. Regulated deficit irrigation during the kernel-filling period and optimal irrigation rates in almond. Agric. Water Manage. 75:152–167.

24. Goldhamer, D.A., and R.H. Beede. 2004. Regulated deficit irrigation effects on yield, nut quality and water-use efficiency of mature pistachio trees. J. Hort. Sci. Biotechnol. 79:538–545.

25. Goldhamer, D.A., M. Viveros, and M. Salinas. 2006. Regulated deficit irrigation in almonds: effects of variations in applied water and stress timing on yield and yield components. Irrig. Sci. 24:101–114.

26. González-Altozano, P., and J.R. Castel. 1999. Regulated deficit irrigation in 'Clementina de Nules' citrus trees. I. Yield and fruit quality effects. J. Hort. Sci. Biotechnol. 74:706–713.

27. Hadas, A., A. Hadas, B. Sagiv, and N. Haruvy. 1999. Agricultural practices, soil fertility management modes and resultant nitrogen leaching rates under semiarid conditions. Agric. Water Manage. 42:81–95.

28. Hanson, B., and D. May. 2004. Effects of subsurface drip irrigation on processing tomato yield, water table depth, soil salinity, and profitability. Agric. Water Manage. 68:1–17.

29. Hazrat, A., L. Teang Shui, and W.R. Walker. 2003. Optimal water management for reservoir based irrigation projects using Geographic Information System. J. Irrig. Drain. Eng. 129:1–10.

30. Henríquez, K. 2005. Efecto de la localización y aplicación de agua y fertilizante en la producción y calidad de naranjas cv. Valencia. 154 p. Tesis Ingeniero Civil Agrícola. Universidad de Concepción, Facultad de Ingeniería Agrícola, Chillán, Chile.

31. Hillel, D., and P. Vlek. 2005. The sustainability of irrigation. Adv. Agron. 87:55–84.

32. Holzapfel, E.A., W.A. Abarca, V.P. Paz, A. Rodríguez, X. Orrego, y M.A. López. 2007a. Selección técnico económico de emisores. Rev. Bras. Eng. Agric. Ambiental 11:456–463.

33. Holzapfel, E., y J.L. Arumí. 2006. Interim Report. Tecnología de manejo de agua para una agricultura intensiva sustentable. 70 p. Proyecto D02I-1146. Universidad de Concepción, Facultad de Ingeniería Agrícola, Chillán, Chile.

34. Holzapfel, E.A., R.F. Hepp, and M.A. Mariño. 2004. Effect of irrigation on fruit production in blueberry. Agric. Water Manage. 67:173–184.

35. Holzapfel, E., R. Merino, M. Mariño, and R. Matta. 2000. Water production function in kiwi. Irrig. Sci. 19:73–80.

36. Holzapfel, E.A., X. Pardo, V.P. Paz, A. Rodríguez, X. Orrego, y M.A. López. 2007b. Análisis técnico económico para selección de aspersores. Rev. Bras. Eng. Agric. Ambiental 11:464–470.

37. Hsiao, T.C., P. Steduto, and E. Fereres. 2007. A systematic and quantitative approach to improve water use efficiency in agriculture. Irrig. Sci. 25:209–231.

38. Hutmacher, R.B., H.I. Nightingale, D.E. Rolston, J.W. Biggar, F. Dale, S.S. Vail, and D. Peters. 1994. Growth and yield responses of almond (*Prunus amygdalus*) to trickle irrigation. Irrig. Sci. 14:117–126.

39. Ibragimov, N., S.R. Evett, Y. Esanbekov, B.S. Kamilov, L. Mirzaev, and J.P.A. Lamers. 2007. Water use efficiency of irrigated cotton in Uzbekistan under drip and furrow irrigation. Agric. Water Manage. 90:112–120.

40. Intrigliolo, D.S., and J.R. Castel. 2006. Performance of various water stress indicators for prediction of fruit size response to deficit irrigation in plum. Agric. Water Manage. 83:173–180.

41. Jalota, S.K., A. Sood, G.B.S. Chahal, and B.U. Choudhury. 2006. Crop water productivity of cotton (*Gossypium hirsutum* L.)—wheat (*Triticum aestivum* L.) system as influenced by deficit irrigation, soil texture and precipitation. Agric. Water Manage. 84:37–146.

42. Karlberg, L., J. Rockstrom, J.G. Annandale, and J.M. Steyn. 2007. Low-cost drip irrigation: A suitable technology for southern Africa? An example with tomatoes using saline irrigation water. Agric. Water Manage. 89:59–70.

43. Khan, S., R. Tariq, C. Yuanlai, and J. Blackwell. 2006. Can irrigation be sustainable? Agric. Water Manage. 80:87–99.

44. Kumar, R., and J. Singh. 2003. Regional water management modeling for decision support in irrigated agriculture. J. Irrig. Drain. Eng. 129:432–439.
45. Lehrsch, G.A., R.E. Sojka, and D.T. Westermann. 2000. Nitrogen placement, row, spacing, and furrow irrigation water positioning effects on corn yield. Agron. J. 92:1266–1275.
46. Maisiri, N., A. Sanzanje, J. Rockstrom, and S.J. Twomlow. 2005. On farm evaluation of the effect of low cost drip irrigation on water and crop productivity compared to conventional surface irrigation system. Phys. Chem. Earth 30:783–791.
47. McCarthy, M.G., B.R. Loveys, P.R. Dry, and M. Stoll. 2002. Regulated deficit irrigation and partial rootzone drying as irrigation management techniques for grapevines. In Deficit irrigation practices. Water Reports N°. 22. p. 79–87. FAO, Rome, Italy (Available at http://www.fao.org/ docrep/004/y3655e/y3655e11.htm).
48. Moriana, A., F. Orgaz, M. Pastos, and E. Fereres. 2003. Yield responses of mature olive orchard to water deficits. J. Amer. Soc. Hort. Sci. 123:425–431.
49. Ortiz Romero, J.N., J. Martínez, R. Martínez, and J.M Tarjuelo. 2006. Set sprinkler irrigation and its cost. J. Irrig. Drain. Eng. 132:445–452.
50. Pannunzio, A. 2008. Efectos de sustentabilidad de los sistemas de riego por goteo en arándanos. 113 p. Tesis Mg.Sc. Universidad de Buenos Aires, Facultad de Ciencias Veterinarias, Buenos Aires, Argentina.
51. Pannunzio, A., and J. Genoud. 2000. Four trickle irrigation treatments in four varieties of oranges. p. 629–635. Proceedings of the IV Decennial National Irrigation Symposium, Phoenix, Arizona, USA. 14–16 November. American Society of Agricultural Engineers (ASAE), St. Joseph, Michigan, USA.
52. Pannunzio, A., J. Genoud, F. Covatta, T. Barilari, and A. Agulla. 2000. Trickle irrigation assessments with orange varieties in Argentina. p. 34–35. Proceedings of the VI International Microirrigation, Cape Town, South Africa.
53. Pannunzio, A., M. Román, J. Brenner, and A. Wölfle. 2004a. Economic overview of drip and micro irrigation systems in humid regions. p. 52. Proceedings of the VII World Citriculture Congress, Agadir, Marruecos. International Society of Citriculture (ISC), Riverside, California, USA.
54. Pannunzio, A., M. Román, A. Wölfle, and J. Brenner. 2004b. Design and operation decisions in drip and micro irrigation systems, as a tool to achieve better quality and profitability in citriculture. p. 51. Proceedings of International Congress on Citriculture, Agadir, Marruecos. International Society of Citriculture (ISC), Riverside, California, USA.
55. Pannunzio, A., P. Texeira, D. Pérez, G. Sbarra, y A. Grondona. 2008. Efectos de sustentabilidad de los sistemas de riego de arándanos en la Pampa Húmeda.
paginas 24–25. Libro de Resúmenes del I Congreso Latinoamericano de Arándanos y otros berries. Agosto 2008. Universidad de Buenos Aires, Facultad de Agronomía (FAUBA), Buenos Aires, Argentina.
56. Payero, J.O., S.R. Melvin, S. Irmak, and D. Tarkalson. 2006. Yield response of corn to deficit irrigation in a semiarid climate. Agric. Water Manage. 84:101–112.
57. Pereira, L. 1999. Higher performance through combined improvements in irrigation methods and scheduling: a discussion. Agric. Water Manage. 40:153–169.
58. Playán, E., and L. Mateos. 2006. Modernization and optimization of irrigation systems to increase water productivity. Agric. Water Manage. 80:100–116.
59. Popova, Z., D. Crevoisier, P. Ruelle, and J.C. Mailhol. 2005. Application of Hydrus2D model for simulating water transfer under furrow irrigation—Bulgarian case study in cropped lysimeters on Chromic Luvisol. Pp. 1–13. ICID 21th European Regional Conference, Frankfurt, Germany. International Commission on Irrigation and Drainage (ICID), New Delhi, India.
60. Quiñones, P.H, H. Unland, W. Ojeda, and E. Cifuentes. 1999. Transfer of irrigation scheduling technology in Mexico. Agric. Water Manage. 40:333–339.
61. Ramalan, A.A., and R.W. Hill. 2000. Strategies for water management in gravity sprinkle irrigation systems. Agric. Water Manage. 43:51–74.

62. Ravindra, V.K., R.P. Singh, and P.S Mahar. 2008. Optimal design of pressurized irrigation sub-unit. J. Irrig. Drain. Eng. 134:137–146.
63. Rice, R.C., D.J. Hunsaker, F.J. Adamsen, and A.J. Clemmens. 2001. Irrigation and nitrate movement evaluation in conventional and alternate–furrow irrigated cotton. Trans. ASAE 44:555–568.
64. Rodríguez, A. 2001. Uso de GIS para diseño de riego. Tesis Ingeniero Civil Agrícola. Universidad de Concepción, Facultad de Ingeniería Agrícola, Chillán, Chile.
65. Rodríguez, J., L.R. Sánchez, and A. Losada. 2007. Evaluation of drip irrigation: Selection of emitters and hydraulic characterization of trapezoidal units. Agric. Water Manage. 90:13–26.
66. Romero, P., P. Botia, and F. Garcia. 2004. Effects of regulated deficit irrigation under subsurface drip irrigation conditions on water relations of mature almond trees. Plant Soil 260:155–168.
67. Romero, P., J. Garcia, and P. Botia. 2006. Cost-benefit analysis of a regulated deficit-irrigated almond orchard under subsurface drip irrigation conditions in South-eastern Spain. Irrig. Sci. 24:175–184.
68. Sepulcre-Cantó, G., P.J. ZarcoTejada, J.C. Jiménez- Muñoz, J.A. Sobrino, E. de Miguel, and F.J. Villalobos. 2006. Detection of water stress in an olive orchard with thermal remote sensing imagery. Agric. For. Meteorol. 136:31–44.
69. Tognetti, R., R. d'Andria, A. Lavini, and G. Morelli. 2006. The effect of deficit irrigation on crop yield and vegetative development of *Olea europaea* L. (cvs. Frantoio and Leccino). Eur. J. Agron. 25:356–364.
70. Tognetti, R., R. d'Andria, G. Morelli, and A. Alvino. 2005. The effect of deficit irrigation on seasonal variations of plant water use in *Olea europaea* L. Plant Soil 273:139–155.
71. Tognetti, R., M. Palladino, A. Minnocci, S. Delfine, and A. Alvino. 2003. The response of sugar beet to drip and low-pressure sprinkler irrigation in southern Italy. Agric. Water Manage. 60:135–155.
72. Troy, Peters, R., and S.R. Evett. 2008. Automation of center pivot using the temperature-time-threshold method of irrigation schedule. J. Irrig. Drain. Eng. 134:286–291.
73. Velez, J.E., D.S. Intrigliolo, and J.R. Castel. 2007. Scheduling deficit irrigation of citrus trees with maximum daily trunk shrinkage. Agric. Water Manage. 90:197–204.
74. Wichelns, D., and J.D. Oster. 2006. Sustainable irrigation is necessary and achievable, but direct costs and environmental impacts can be substantial. Agric. Water Manage. 86:114–127.
75. Yadav, B.K., and S. Mathur. 2008. Modeling soil water uptake by plants using nonlinear dynamic root density distribution function. J. Irrig. Drain. Eng. 134:430–436.
76. Yazar, A., S.M. Sezen, and S. Sesveren. 2002. LEPA and trickle irrigation of cotton in the South-east Anatolia Project (GAP) area in Turkey. Agric. Water Manage. 54:189–203.

CHAPTER 9

SUBSURFACE MICRO IRRIGATION: REVIEW*

LEONOR RODRÍGUEZ-SINOBAS and MARÍA GIL-RODRÍGUEZ

CONTENTS

9.1 Introduction 162

9.2 History 162

9.3 Advantages and Disadvantages of SDI 163

9.4 Components of SDI Units 164

9.5 Field Installation of SDI Units 165

9.6 Effect of Soil Characteristics on Emitter Discharge.................. 165

9.7 Field Performance of SDI... 169

9.8 Simulating Subsurface Drip Irrigation 171

9.9 Criteria for the Design and Management of SDI........................ 177

9.10 Perspectives for Subsurface Drip Irrigation 180

9.11 Summary... 181

Keywords 181

References.. 182

*Authors thank the Spanish Ministry of Science and Technology (CICYT) for the financial support under project No. AGL2008-00153/AGR.

Modified versión of, *"Leonor Rodríguez Sinobas and María Gil Rodríguez (2012). A Review of Subsurface Drip Irrigation and Its Management. In: Water Quality, Soil and Managing Irrigation of Crops, Dr. Teang Shui Lee (Ed.), ISBN: 978-953-51-0426-1, InTech, Available from: http://www.intechopen.com/books/water-quality-soil-and-managing- irrigation-of-crops/a-review-of-subsurface-drip-irrigation-and-its-management."*

9.1 INTRODUCTION

According to the ASAE Standards [1], subsurface drip irrigation (SDI) is "the application of water below the soil surface though emitters with discharge rates generally in the same range as drip irrigation." Thus, aside from the specific details pointed out in this chapter, an SDI unit is simply a micro irrigation network buried at a certain depth. In this chapter, the word emitter is used instead of dripper since the dripper has a lack of meaning, and water does not drip in SDI as it does in drip irrigation.

There are wide varieties of plants irrigated with SDI all over the world such as herbaceous crops (lettuce, celery, asparagus and garlic), woody crops (citrus, apple trees and olive trees) and others such as alfalfa, corn, cotton, grass, pepper, broccoli, melon, onion, potato, tomato, etc. Over a quarter of a billion hectares of the planet are irrigated and entire countries depend on irrigation for their survival and existence. Growing pressure on the world's available water resources has led to an increase in the efficiency and productivity of water-use of irrigation systems as well as the efficiency in their management and operation. The efficiency of subsurface drip irrigation SDI could be similar to drip irrigation but it uses less water. It could save up to 25–50% of water regarding to surface irrigation. Throughout this chapter, the specific characteristics of SDI will be presented and some criteria for its design and management will be highlighted.

9.2 HISTORY

The first known reference of a subsurface irrigation comes from China more than 2000 years ago [4] where clay vessels were buried in the soil and filled with water. The water moved slowly across the soil wetting the plants' roots. SDI, as we know it nowadays, developed around 1959 in the US [38], especially in California and Hawaii, as a drip irrigation variation. In the 60s, SDI laterals consisted of polyethylene PE or polyvinyl chloride PVC plastic pipes with punched holes or with punched emitters. These systems used to work at low pressures, depending on water quality and filtration systems. By the 70s, this method extended to crops as citrus, sugar cane, pineapple, cotton, fruit trees, corn, potato, grass and avocado. Its main disadvantages were: the difficult maintenance, the low water application uniformity and the emitters' clogging due to, mainly, oxide and soil particles. Alternatively, the equipment for SDI installation in the field developed [18], and the fertilizers' injection with the irrigation systems started in Israel [15]. Further, the quality of commercial emitters and laterals improved and SDI gained respect over other irrigation methods mainly due to the decrease of clogging.

At the beginning of the 80s, factors such as: the reduction of the costs in pipes and emitters, the improvement of fertilizers application and the maintenance of field units for several years promoted the interest in SDI. This has been extended since then.

FIGURE 1 Layout of the lateral of subsurface drip irrigation.

9.2.1 SUBSURFACE DRIP IRRIGATION IN THE WORLD

There is a general agreement about the spread of SDI. However, it is difficult to obtain data to confirm this trend since the areas under SDI is counted as a drip irrigation in most surveys. In the US, the USDA Farm and Ranch Irrigation Survey [37] indicates that SDI comprises only about 27% of the land area devoted to drip and subsurface drip irrigation. Nonetheless, this percentage is continuously increasing over drip irrigation as farmers substitute their conventional drip irrigation methods by SDI systems. If this framework continues in the future, SDI would still have a potential increase.

In underdeveloped countries, there are many references about the so called low-cost SDI with rudimentary laterals and low emitter pressures that highlight even more, the SDI potential increase (See Chapters 2 to 6 in this book). Likewise, this is enhanced by the savings on water and energy.

9.3 ADVANTAGES AND DISADVANTAGES OF SDI

Experimental evidences of SDI advantages over other irrigation methods, specifically drip irrigation, are vast. Some advantages and drawbacks of this method, compiled by Lamm [19] and Payero [24], are shown below.

- The efficiency of water use is high since soil evaporation, surface runoff, and deep percolation are greatly reduced or eliminated. In addition, the risk of aquifer contamination is decreased since the movement of fertilizers and other chemical compounds by deep percolation is reduced.
- The use of degraded water. Subsurface wastewater application can reduce pathogen drift and reduce human and animal contact with such waters.
- The efficiency in water application is improved since fertilizers and pesticides can be applied with accuracy. In widely spaced crops, a smaller fraction of the soil

volume can be wetted, thus further reducing unnecessary irrigation water losses. Reductions in weed germination and weed growth often occur in drier regions.

• Hand laborers benefit from drier soils by having reduced manual exertion and injuries.

• Likewise, double cropping opportunities are improved. Crop timing may be enhanced since the system need not be removed at harvesting nor reinstalled prior to planting the second crop. On the other hand, laterals and submains can experience less damage and the potential for vandalism is also reduced.

• Operating pressures are often less than in drip irrigation. Thus, reducing energy costs. Drawbacks:

• Water applications may be largely unseen, and it is more difficult to evaluate system operation and water application uniformity. System mismanagement can lead to under irrigation, less crop yield quality reductions, and over irrigation. The last may result in poor soil aeration and deep percolation problems.

• If emitter discharge exceeds soil infiltration, a soil overpressure develops around emitter outlet, enhancing surfacing and causing undesirable wet spots in the field.

• Timely and consistent maintenance and repairs are a requirement. Leaks caused by rodents can be more difficult to locate and repair, particularly for deeper SDI systems. The drip lines must be monitored for root intrusion, and system operational and design procedures must employ safeguards to limit or prevent further intrusion. Roots from some perennial crops may pinch drip lines, eliminating or reducing flows. Periodically, the drip lines need to be flushed to remove accumulations of silt and other precipitates that may occur in the laterals. Likewise, operation and management requires more consistent oversight than some alternative irrigation systems. There is the possibility of soil ingestion at system shutdown if a vacuum occurs, so air relief/vacuum breaker devices must be present and operating correctly. Compression of laterals by soil overburden can occur in some soils and at some depths, causing adverse effects on the flow.

9.4 COMPONENTS OF SDI UNITS

The typical SDI unit, contrarily to drip irrigation units, is a looped network. It is composed by a submain or feeding pipe, and a flushing pipe that connects all lateral ends (Fig. 1). The looped network enhances the hydraulic variability and thus, the flow variability within the unit is also improved. The pipes and laterals are buried at a certain depth, generally from 7 to 40 cm, supplying the water directly in the crop root area.

The flushing pipe eliminates possible sediments or other water suspended elements. Two relief valves may be placed at the head and tail of the feeding and flushing pipes to release the air at the beginning of the irrigation, and to let it pass when irrigation stops. Thus, soil particles are not introduced into the laterals. Alternatively, some flushing valves for the lateral end are also developed for that purpose (Fig. 3).

A manometer and a flow meter are placed at the unit head to measure the inlet unit pressure and the total unit discharge, respectively. Filters are also installed at the unit head and additional containers with fertilizers and other chemical products are located there, as well.

Emitters in SDI are similar to drip irrigation, but they generally are impregnated with a weed killer to reduce root intrusion at the emitter outlet. In countries where herbicides are forbidden, commercial emitter models are developed with specific geometries to reduce the risk of root intrusion. Likewise, there are some models with elements such as diaphragm and geometries inducing vertical movement, to avoid the entrance and deposition of soil particles (Fig. 3).

9.5   FIELD INSTALLATION OF SDI UNITS

The installation of SDI units in the field shows differences with conventional drip irrigation. In most cases, the soil is chiseled to a depth close to the crop roots length prior the layout of laterals. This will also favor the horizontal water movement. Then, the feeding and flushing pipes are laid on trenches dug following the lateral ends keeping an extra space for the connections between pipes and laterals. Once these are done, the other elements such as: valves, relief valves, flow meters can be installed. Finally, all the subsurface elements are buried.

Among all the installation tasks, the difficult one corresponds to the deployment of laterals. These are introduced into the soil by one or several plows connected to a tractor (Fig. 4). Care must be taking to ensure the laterals are placed following a straight line and at a proper depth, and also that the lateral spacing stays constant.

9.6   EFFECT OF SOIL CHARACTERISTICS ON EMITTER DISCHARGE

One of the key differences between SDI and surface drip irrigation is that soil properties in fine-pore soils affect emitter discharge. A spherical-shaped saturated region of positive pressure (h_s) develops at the emitter outlet. Consequently, the hydraulic gradient across the emitter decreases reducing emitter flow rate according to the following equation:

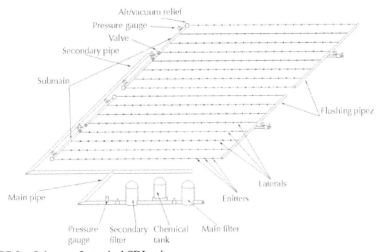

FIGURE 2   Scheme of a typical SDI unit.

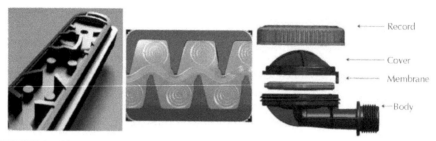

FIGURE 3   Elements of SDI units: (a) Emitter models and (b) Flushing bulb. (Courtesy of NaaDan Jain Irrigation).

$$q = k \times (h - h_s)^x \tag{1}$$

where: q is the emitter flow rate, h is the working pressure head, k is the emitter co-efficient, and x is an exponent, respectively. The constants k and x define the emitter characteristics.

In fine-pore soils, the overpressure hs increases rapidly at the beginning of irrigation until it stabilizes after, approximately, 10 to 15 min. Values of hs from 3 m up to 8 m have been recorded on single emitters in the field within a wide range of emitter discharges [20, 33]. However for similar flow rates and soils, these values were smaller, within an interval from 1 to 2 m, in laboratory tests since the soil structure increases the soil mechanical resistance to water pressure under field conditions [12]. Emitter flow reaches a permanent value coinciding with the stabilization of hs. Sometimes, the overpressure produces what it is called a chimney effect or surfacing (Fig. 2). Overpressures develop locally in the soil displacing the soil components and creating preferential paths. Soil surface is wetted and big puddles are also observed.

a   b

FIGURE 5   Surfacing in SDI units: (a) Wetted soil surface above the laterals, and (b) Details of the wetted area.

Pressure-compensating and noncompensating emitters exhibit different performance. For noncompensating emitters, discharge reductions from 7% up to 50% have been measured in controlled conditions on single emitters in fine-coarse soils. However, the decrease is less than 4% in sandy soils. For pressure-compensating emitters, flow variation is negligible if hs stays below the lower limit of the emitter compensation range.

Water movement in SDI can be considered as a buried point source (Fig. 6a). Philip [25] developed an analytical expression to determine the pressure at the discharge point in a steady-state conditions that was applied by Shani and Or [33] to relate h_s with the soil hydrophysical properties and the emitter flow rate q as shown below:

$$h_s = \{[(2-\alpha.r_0) \div (8\pi.K_s.r_0)](q)-[1/\alpha]\} \tag{2}$$

where: r_0 is the spherical cavity radius around emitter outlet, K_s is the hydraulic conductivity of the saturated soil and α is the fitting parameter of Gardner's [11] of the nonsaturated hydraulic conductivity expression. In moderate flows, h_s is linear and q follows a straight line whose slope depends on r_0, K_s and α. For a given discharge q and radius r_0, the pressure h_s is more sensitive to K_s and less sensitive to α.

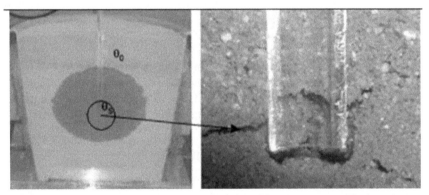

FIGURE 6   Water movement of SDI emitters in loamy soils: Left—wetting bulb, and Right—detail of horizontal cracks in the cavity. Note: θ_o = Initial soil water content, and θ_s = Saturated soil.

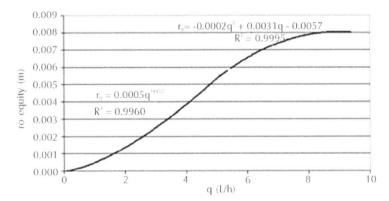

FIGURE 7   Variation of the radius spherical cavity with emitter discharge in uniform soils [14].

Experiments carried out in uniform soil samples in pots show that the cavity shape tended to be spherical at small emitter discharges [13]. At higher emitter discharges, horizontal cracks initially appear in the cavity, but slowly they fill with soil and, ultimately resulting in a spherical cavity (Fig. 6b). The radius of the cavity linearly increases with small emitter discharges and stabilizes at higher discharges (Fig. 7).

9.6.1 WATER DISTRIBUTION VARIABILITY

The main causes that affect flow rate variability in buried emitters are: hydraulic variation, manufacturing variation, root intrusion, deposition of soil particles and the interaction of soil properties. If the temperature is constant and emitter clogging is negligible, water distribution in drip irrigation units follows a normal distribution. The emitter discharge variability CVq will depend on the hydraulic variation and the manufacturing variation CVm. This hypothesis is better suited when the emitter manufacturing variation is the main cause of the final variation. Normal flow distribution in the unit is characterized by two parameters: the mean and the standard deviation of flow (or its coefficient of variation). Adding these two variations to Eq. (1), results in Ref. [6]:

$$q = [k.h^x.(1 + u.CV_m)] \tag{3}$$

where: u is a normal random variable of mean 0 and standard deviation 1. The coefficient of manufacturing variation (CV_m) is a measure of flow variability of a random emitters' sample of the same brand, model and size, as produced by the manufacturer and before any field operation or aging has taken place [2]. For SDI units, Eq. (3) transforms to:

$$q = [k.(h—h_s)^x.(1 + u.CV_m)] \tag{4}$$

The goal of irrigation systems is to apply water with high uniformity. The system design is frequently based in a given target uniformity that is defined by uniformity indexes. One of the common indexes is the coefficient of variation of emitter flow, CV_q [17], given in Eq. (5):

$$CV_q = [(\sigma_q)/q'] \tag{5}$$

where: σ_q is the flow standard deviation and q' is the mean flow. Other frequently used index is the Christiansen coefficient Cu [8] given below:

$$C_u = 100*[1–(\sigma_{q'})/q'] \tag{6}$$

where: $\sigma_{q'}$ is the mean absolute flow standard deviation. A good irrigation performance requires water distribution uniformities characterized by Cu > 0.9 or $CV_q = 0.1$, and emitters with $CV_m < 0.10$.

Observations on homogenous soils in pots, have shown that the interaction between the effect of soil on emitter discharge in low infiltration soils would work as a self-regulated mechanism [12]. The soil overpressure would act as a regulator, and the emitters with a greater flow rate in surface irrigation would generate a higher overpressure in the soil, which would reduce the subsurface irrigation flow rate to a greater extent than in emitters with a lower flow rate. Consequently, the flow emitter variability would be smaller in buried emitters than in surface ones. Thus, for nonregulated emitters the uniformity of water application in SDI laterals would be greater than the uniformity of surface drip lateral irrigation in homogeneous soils as it is shown in Fig. 8.

For compensating emitters, the flow rate variability in SDI could be similar to the surface drip irrigation in loamy and sandy soils if the head pressure gradient across the emitter and the soil were above the lower limit of the emitter compensation range. Thus, the variability of soil overpressure would be offset by the elastomer regulation.

9.7   FIELD PERFORMANCE OF SDI

Increments between 2.8% and 7.0% on emitter flow-rate have been reported in laterals when nonregulated emitters were excavated [32]. The uniformity of water application in excavated drip tapes, after three years in the field, was lower than the new tapes deployed over the soil. Likewise, field evaluations of SDI laterals resulted in Cu within the interval 75–90 [3, 26].

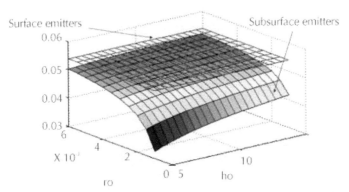

FIGURE 8   Coefficient of variation of flow CVq in laterals with 100 nonregulated emitters of 2 L/h laid on a uniform loamy soil, considering hydraulic variation negligible (Note: r_0, h_0 and hs are expressed in m). (Source: Ref. [12]).

In loamy bare soils, regulated and nonregulated emitters have shown similar behavior in evaluations carried out in two consecutive years (28 to 31). Lateral inflow decreases rapidly within the first 10 to 15 min of irrigation and it approaches a steady value as the time advances (see Fig. 9). Consequently, lateral head losses reduce and the head pressure at the lateral downstream increases. This performance highlights that overpressure hs at the emitter outlet increases at the start of the irrigation until it stabilizes confirming the trend observed in single nonregulated emitters but contradicting the behavior observed in single regulated emitters. Meanwhile, the variation of inlet lateral flow is larger than in nonregulated emitters and differences among laterals are noticeable. Also, the flow in these laterals decreases as the inlet head reduces.

On the one hand, the discharge of regulated emitters decreases over the operating time until it stabilizes. The elastomer material may suffer fatigue when being held under pressure and its structural characteristics may change (28 to 31). When irrigation is shutdown pressure is canceled and the elastomer relaxes and surmounts the deformation caused by pressure. The longer the time elapsed between successive irrigation events the longer the time for the elastomer to return to its initial condition. This behavior would be conditioned by elastomer material and its relative size. On the other hand, soil particles suctioned when irrigation is shutdown, might deposit between the labyrinth and the elastomer and thus, canceled out the self-regulatory effect.

Evaluations performed on a seven years old SDI unit of regulated emitters showed that the hydraulic variability was the major factor affecting emitter discharge, but the effect of the emitter's manufacturing and wear variation was smaller (28 to 31). The uniformity of water application Cu varied between 82 and 92 and the degree of clogging between 25 to 38%. The first improved when reducing pressure variability, thus the emitters behaved as a nonpressure-compensating too. Their performance might be affected by: clogging (deposition of suspended particles, root intrusion) and entrapped air.

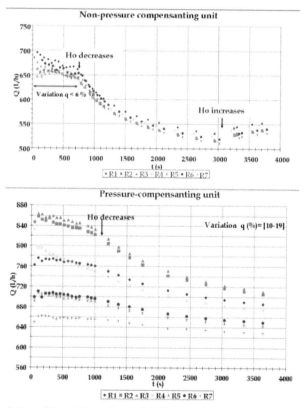

FIGURE 9 Variation of lateral inlet flow of a SDI unit laid on a loamy soil during irrigation (Note: Ri= lateral).

9.7.1   FIELD EVALUATION

Specific methodology for SDI has not been yet developed for field evaluation of SDI units (Fig. 10a). Nevertheless, in spite of the difficulty to measure the emitter discharge in field conditions, its performance may be assessed by a simulation program estimating the distribution of emitter discharges and soil pressures within the unit.

As a practical procedure, the measurement of unit inlet flow and the pressure heads at both ends in the first and last laterals of the unit could suffice to calibrate the simulation program. Both laterals could be easily located in the field and unearthed. Pressures and inlet flow can be measured by digital manometers and by Woltman or portable ultrasonic flow meter, respectively (Fig.10b). Field evaluations can highlight emitter clogging such as the root intrusion observed in Fig. 11.

9.8   SIMULATING SUBSURFACE DRIP IRRIGATION

As it was mentioned above, simulations are a tool for estimating the performance of laterals and units. On the one hand, discharges and pressures of the emitters among the unit (generally looped units) can be computed taking into account the soil properties. On the other hand, the movement of water in the soil and wet bulb geometry can also be estimated.

9.8.1   SIMULATION OF SDI LATERALS AND UNITS

A computer program can be developed to predict water distribution along SDI laterals and units. To do so, we would need to take into account:

- Design variables: length and diameters of laterals; submain and flushing pipe; emitter discharge; emitter's manufacture and wear coefficient and local losses at the insertions of the emitters and the laterals.
- Operation variables: inlet pressure and irrigation time.
- Soil properties: at least texture and saturated hydraulic conductivity and if possible, the other soil parameters of Eq. (2). When dealing with spatial variability, it is required the couple of a specific geo-statistical modeling software and the one predicting water distribution within the unit.

Thus for a target irrigation uniformity, different study cases can be simulated and then, the selection of proper values for the design or operation variables could be achieved. Moreover, the suitability of this method to different soil types and their best management practices can be addressed prior the unit is laid on in the field.

a   b

FIGURE 10   Field evaluation of SDI units: (a) sample with unearthed emitters and (b) manometer and flow-meter for recording head pressure and inlet flow in laterals.

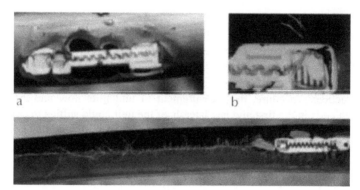

FIGURE 11   Root intrusion in different emitters (36).

9.8.1.1   HYDRAULIC CALCULATION OF LATERALS

In this chapter, we present a simulation program developed by Rodríguez-Sinobas et al. [30], which calculates the flow rates for surface emitters in laterals with Eq. (3) and for buried emitters with Eq. (4). The overpressure hs at the emitter exit was obtained with Eq. (2). Likewise, the head pressure at each emitter insertion h_i was determined by application of the energy equation as follows:

$$h_i = (h_{i+1} + hf_{i-i+1} \pm I_o.s) \tag{7}$$

where: hf_{i-i+1} is the head loss between two consecutive emitters, i–i+1; I_o is the lateral slope and s is the emitter spacing.

Flow regime in laterals is considered smooth turbulent with Reynolds numbers within the range $3000 < Re < 100,000$ since most SDI laterals are made of smooth polyethylene pipe. Thus Blasius head loss equation provides an accurate estimation of the frictional losses inside uniform pipes. Likewise, local head losses at emitter insertions are included as an equivalent length le. Then head loss hf_{i-i+1} between two successive emitters, i–i+1 can be calculated as

$$hf_{i-i+1} = \{0.0246*(v^{0.25})[(Q_{i-i+1})^{1.75} \div D^{4.75}].s.[1 + (l_e \div s)]\} \tag{8}$$

where: Q_{i-i+1} is the flow conveying in the uniform pipe between emitters, i–i+1; D is the lateral internal diameter; and v the kinematic water viscosity.

In most commercial models, local head losses at the emitter insertion are within the interval of 0.2 and 0.5 m (16). Lateral head losses hf can be calculated as

$$hf_{i-i+1} = \{F*0.0246*(v^{0.25})[(Q_o)^{1.75} \div D^{4.75}].L.[1 + (l_e \div s)]\} \tag{9}$$

where: Q_o is the lateral inlet flow; L is the lateral length and F the reduction factor. The reduction factor F (8) is defined below:

$$F = [1 \div (m+1)] + [1 \div (2.N)] + \{[m—1]^{0.5} \div (6.N^2)\} \tag{10}$$

where: N is the number of emitters; and m is the flow exponent of the head loss equation. This formula applies to laterals with the first emitter located at a distant = L/N from the inlet. In general, number of emitters in laterals is large, and water distribution is assumed to be continuous and uniform. Hence, $F = [1/(1+m)]$.

As said, emitter discharge in SDI can be affected by soil properties. It can be influenced by K_s and α, each of them varying throughout the field. Furthermore, the discharge can be also affected by variation in cavity radius r_o. Soil hydraulic properties from close points are expected to be more alike than those far apart, they can be correlated using variograms. Furthermore, these can aim at the estimation of soil properties distribution.

9.8.1.2 EXAMPLES

For the porpoise of illustration, the performance of a lateral in different uniform soils is represented in Fig. 12a, and the effect of soil heterogeneity is shown in Fig. 12b. The comparison of the performance between SDI and surface laterals is addressed in all the examples.

The uniformity index CV_q is higher in SDI than in the surface drip irrigation in all uniform soils except for the sandy one, where uniformities are alike. For a given r_0 and h_0, the possible self-regulation due to the interaction between the emitter discharge and soil pressure is observed in low saturated hydraulic conductivity soils (loamy and clay). The larger the cavity radio the smaller the soil effect, and thus, the SDI lateral behaves as a surface lateral. The irrigation uniformity index in nonuniform soils is less than in uniform soils. It slightly reduces as r_0 and h_0 decrease. For sandy soils the uniformity is very similar in both laterals; even the uniformity in the SDI lateral is higher for certain values of r_0 and h_0.

In loamy soils, for higher r_0 values, the uniformity reduces as emitter flow increases. However, with small r_0 this trend is not shown, and similar uniformities are observed between the smaller emitter flows. The larger the cavity the smaller the soil effect, and thus the SDI lateral would behave as a surface lateral. No difference in water application uniformity is observed between surface and buried laterals in loamy soils for values of $r_0 > 0.01$ m. On the contrary, for small r_0 the self-regulating effect is predominant. Therefore, this is enhanced by higher emitter flows since they develop higher soil overpressures. For heterogeneous loamy soils the tendency is similar to the homogeneous soil: irrigation uniformity decreases as emitter discharge increases (Fig. 12b).

However, in some scenarios, the values of CV_q are higher. The uniformity is similar in all the three discharges for $r_0 > 0.01$; below this value, uniformity rapidly decreases.

Uniformity of water application in all cases reduces as inlet pressure increases. Consequently, it would be advisable to select emitters with small discharge, and to irrigate with inlet pressure not very small (in this case, >10 m).

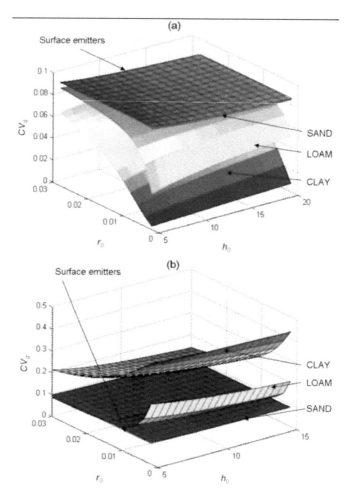

FIGURE 12   Irrigation uniformity index CVq as a function of r_0 and h_0 in different soils for a SDI lateral with noncompensating emitters: (a) uniform soils; (b) nonuniform soils (Note: Units for r_0 and h_0 are expressed in m).

9.8.1.3   HYDRAULIC CALCULATION OF UNITS

Typical SDI units are composed of a looped network (Fig. 2). Water can move either way from head to downstream lateral and reverse. The hydraulic calculation of laterals can be achieved as it was outlined in Section 4.1.1. The inlet pressure h_0 at each lateral would be determined using following equation:

$$H_{oi} = h_{oi+1} + hf_{si-i+1} \pm (I_{os}.s_L) \tag{11}$$

where: i and i+1 are subindices for two successive laterals; hf_{si-i+1} are the head loss in the uniform pipe between laterals, i–i+1; I_{os} is the submain slope and s_L is the lateral spacing. hf_{si-i+1} may be determined with Eq. (8) considering the flow conveying between the laterals i and i+1.

Each lateral inflow can be calculated by adding the flow from all of its emitters. Local head losses at the connection of the lateral with the submain are smaller than those in laterals [28–31], and could be left out in Eq. (8).

Two conditions are met for calculation of looped network. First, the sum of flows conveying along the submain equates the sum of the emitter's flow throughout the unit. Second, the pressure at every location within the network is the same when calculated from head towards the downstream end of the lateral than reversed. The direction of flow in a loop network is not known. A minimum energy line with minimum pressures divides the flow coming from the lateral head and the one coming from its downstream end. Thus from this line, the network calculation is accomplished as two branched networks: one whose length corresponded to the distance from the lateral inlet up to energy line; and the other whose length corresponded to the distance from the downstream end of the lateral to that line

This program predicts discharges and pressures from both directions. The iterative process stops when minimum pressures are met. If the value for minimum pressure from one way is larger than from the other, the energy line moves towards the least pressure. Likewise, the sum of the flows coming and exiting from the flushing pipe is zero.

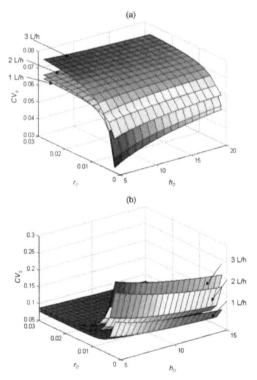

FIGURE 13 Irrigation uniformity coefficient CVq as a function of r_0 and h_0 in a SDI lateral with noncompensating emitters of different discharges: (a) loamy uniform soil; (b) loamy nonuniform soil. Note: Units for r_0 and h_0 are expressed in m.

The program outputs are: the flows and pressures at the inlet and downstream end of the laterals; the minimum energy line; the distribution of emitter discharges and emitter pressures; and the irrigation uniformity coefficient CVq. Simulations can be performed for irrigation units both laid on the surface and buried in the soil. In the last, emitter discharge is calculated with Eq. [4] and soil water pressure at the emitter outlet is calculated with Eq. (2). Soil variability may be determined as detailed for laterals but in this case, the variable will be two-dimensional.

EXAMPLE

A looped unit operating at several pressures in a loamy soil is selected for illustration of the performance of a typical SDI unit in Fig. 13. For all cases, irrigation uniformity improves with small emitter discharges.

In nonuniform soils, the irrigation uniformity indexes are higher than in uniform soils, but their differences were small. Only the clay soil with the highest discharge displayed a significant difference.

In summary, irrigation uniformity is very good for most of the typical selected scenarios. Nevertheless, for clay soils emitter discharge reduced 14% and this should be taken into account in the management of the irrigation system. On the other hand, if higher emitter flows would have been selected, the unit would have performed worse.

9.8.2   HYDRUS 2D/3D

The program HYDRUS-2D/3D [34] includes computational finite element models for simulating the two- and three-dimensional movement of water, heat, and multiple solutes in variably saturated media. Thus it can simulate the transient infiltration process by numerically solving Richards' equation.

The program is a useful tool for research and design of SDI systems. This simulation program may be used for optimal management of SDI systems [9, 22, 23, 35] or for system design purposes [7, 14, 27].

To simulate water movement in SDI emitters, a two-dimensional axis-symmetric water flow around a spherical surface may be simulated reproducing the hydraulic conditions. Simulations can considered a one-half cross section the soil profile. A sphere with a radius r0 can be placed at a given depth below the soil surface, simulating the emitter and the cavity formed around a subsurface source.

A triangular mesh is automatically generated by the program. Dimensions of the finite- elements grid (triangles) may be refined around the emitter and other key points. Boundary conditions can be selected as follows: absence of flux across central axis and the opposite side; the atmospheric condition on the soil surface and free drainage along the bottom of soil profile. For compensating emitters, a constant flux boundary may be assumed around the cavity—it corresponded to the emitter discharge. For noncompensating emitters, a variable flux boundary condition, function of the soil-water pressure, could be assumed calculated from the x and k coefficients of the emitter discharge equation.

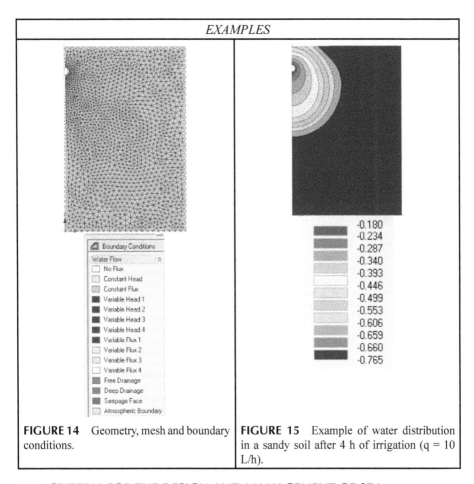

| EXAMPLES |

FIGURE 14   Geometry, mesh and boundary conditions.

FIGURE 15   Example of water distribution in a sandy soil after 4 h of irrigation (q = 10 L/h).

9.9   CRITERIA FOR THE DESIGN AND MANAGEMENT OF SDI

In designing SDI systems: Variables such as emitter discharge, lateral depth and emitter spacing must be selected. Likewise, the management of these systems includes the selection of the operating pressure and the irrigation time. The inlet pressure in the irrigation unit determines the uniformity of water application whereas the irrigation time is a key factor to meet crop water and nutrient requirements.

9.9.1   DETERMINATION OF MAXIMUM EMITTER DISCHARGE

Since soil properties may reduce emitter discharge, the maximum emitter flow rate can be selected by a method called 'soil pits method' [5]. A steady water flow is introduced to the soil through a narrow polyethylene tube that is located underneath the soil, during a given time. The depth at which no surfacing occurs is taken as the minimum depth for SDI laterals. Other method proposed by Gil et al. [14] is based on the following dimensionless emitter discharge equation:

$$q^* = k.(\Delta h^*)^x \tag{12}$$

where: $q^* = (q/h_0)^{x;}$ $\Delta h^* = [(h_0 - h_s)/h_0]$; and h_0 is the emitter pressure. Values of pressure-discharge from several emitter compensating and noncompensating models, arranged according Eq. [12], show the same trend. Each category follows, approximately, a single line that shows a linear relationship in flow rate variation below 50% (see Fig. 16).

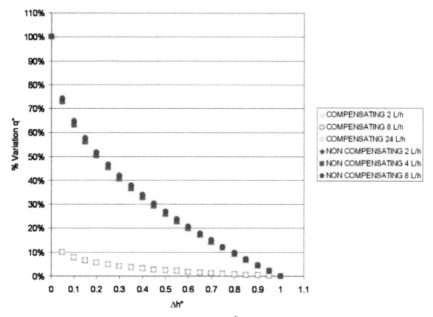

FIGURE 16   Variation of $q^* = [q/(h_0)^x]$ versus $\Delta h^* = [(h_0 - h_s)/h_0]$ for different emitter models [14].

From the above graph, a target uniformity for SDI design is selected and then, Δh^* calculated. Considering a typical variation of $q^* = 10\%$, $\Delta h^* = 0.79$ and 0.05 for noncompensating and compensating emitters, respectively. Then, the overpressure generated in the soil h_s, can be calculated from Eq. (12) for different h_0 values. These would correspond to emitter pressures within the lateral to limit the reduction of emitter discharge to 10%. However, the performance of pressure compensating emitters differs from the noncompensating. In theory, their flow rate keeps constant within their compensating interval but for other values, these would be affected by soil pressure. Thus, considering an emitter head of $h_0 = 10$ m, $\Delta h^* = 0.05$, and $(h_0 - h_s) = 0.5$ m. For antidrain emitters, this value would not be reached since once the pressure gradient is less than 2 m, the emitter would close its discharging orifice.

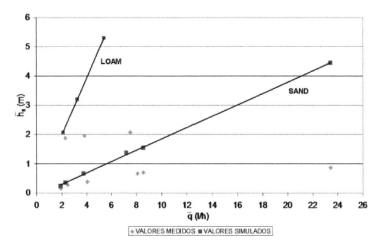

FIGURE 17 Relation between h_s, and q' for different soils considering a constant spherical radius r_0 [14].

Once the value h_s, is known, emitter flow rate could be calculated from its relation with q as the one depicted in Fig. 17. This relation depends upon soil hydraulic properties and it could be determined by numerical models simulating soil water movement. The maximum discharge q_{max} will be obtained by adding to the assigned value the discharge emitter variation desired (10% in the given example).

9.9.2 DESIGN AND MANAGEMENT OF SDI

Computer programs developed for the hydraulic characterization and prediction of water distribution in SDI units have been proved as a useful tool for the design and management of SDI systems under specific scenarios. In addition, these programs might be coupled with geostatistical modeling software for the inclusion of spatial distribution of soil variability. Results show that SDI irrigation is suitable for sandy soils and is also suitable for loamy soils under specific conditions. It is advisable to select emitters with small discharge (1 L/h and 2 L/h), and to irrigate with inlet pressures not very small.

The uniformity of water application index CVq in nonuniform soils is less than in uniform soils (Fig. 13). It is higher in SDI than in the surface drip irrigation in all uniform soils except for the sandy one, where both uniformities are alike.

On the other hand, the wetted bulb dimensions and its water distribution are two main factors determining emitter spacing and lateral depth. Both variables are selected in order to obtain an optimum water distribution within the crop root zone. The depth of wetting bulb coincides with the length of the plant roots.

As an illustration, some guidelines for the project and management of SDI units in a uniform loamy soil are introduced to highlight the applicability of finite element models for simulating water movement in SDI. The example simulates a line source that would correspond to water movement in laterals with small emitter spacing. Thus, the wetted area follows a continuous line with variable width. Two irrigation times and three possible lateral depths are compared considering an emitter pressure of 16 m and a uniform initial water content of 0.1 m³/m³. Figure 18 shows that for this scenario, emitter depths deeper than 10 cm are recommended to prevent soil surface wetting for irrigation times higher than 30 min. Differences observed for 0.2 and 0.3 m depths were negligible.

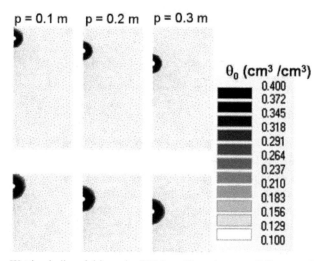

FIGURE 18   Wetting bulbs of drip and a SDI in uniform loamy soil for an emitter pressure of 16 m and initial soil water content $\theta_o = 0.1$.

Wetting bulbs could be simulated for different scenarios by numerical methods such as Hydrus-2D. For each soil, selection of proper design variables (emitter spacing or lateral depth) and/or operation variables (inlet head and irrigation time) could be guided by figures, as those shown above, or by graphs showing wetting bulb dimensions for different conditions.

9.10   PERSPECTIVES FOR SUBSURFACE DRIP IRRIGATION

Irrigation will be one key factor to sustain food supply for the world increasing population. The technical FAO study "World agriculture: towards 2015/2030 [10]" highlights that production of staple crops must follow the same increasing trend than in the last decades. Therefore, their productivity should rise and, thus irrigation will play an important role but it will have to adjust to the conditions of water scarcity and environmental and ecological sustainability.

On the other hand, many areas in the world, as their economy develops, the urban activities expand increasing water demands and develop competing uses of limited water resources. Thus, it raises uncertainty whether the volume of water used in irrigated agriculture can be sustained. Within this framework, water conservation policies are developed by policymakers reinforcing the use of new technologies and more efficient irrigation methods. Farmers are encouraged to use irrigation methods that reduce water use and increase yield in order to allow water to flow to other economic sectors.

In consequence, subsurface drip irrigation perspectives are promising. It shows higher capability for minimizing the loss of water by evaporation, runoff, and deep percolation in comparison to other methods. Thus, the irrigation water saved may become available for other uses. It may also lead to increase crop yields since it reduces fluctuations in soil water content and well-aerated plan root zone. In addition, treated wastewaters could be used with risk reduction in human and animal health in arid areas with water scarcity. Moreover, it may reduce agrochemical application since agrochemicals are precisely distributed within the active roots zone.

Finally, SDI efficiency, in some scenarios, is better than surface trickle irrigation, and it requires smaller inlet heads. Thus energy savings are also enhanced. Furthermore, SDI may eliminate anaerobic decomposition of plant materials and thus substantially reduce methane gas production.

9.11 SUMMARY

In this chapter, subsurface drip irrigation (SDI) is reviewed. The history, perspectives, current status, simulation models, wetting patterns, and uniformity of water application are reviewed in details.

KEYWORDS

- ASAE
- drip irrigation
- dripper
- emitter
- FAO
- Hydrus-2D
- Hydrus-3D
- irrigation
- manometer
- micro irrigation
- simulation models
- Spain
- subsurface drip irrigation
- surface drip irrigation
- uniformity
- water management
- wetting pattern

REFERENCES

1. American Society of Agricultural Engineering (ASAE). ASAE Standards Engineering Practices Data. 43rd edition. August 2005.
2. American Society of Agricultural Engineering (ASAE). Design and Installation of Micro Irrigation Systems. Standard EP-405.1, February 2003.
3. Ayars, J.E., Schoneman, R.A., Dale, F., Meson, B. and Shouse, P. Managing subsurface drip irrigation in the presence of shallow ground water. Agricultural Water Management. April 2001, 47(3):242–264.
4. Bainbridge, A. (2001). Buried clay pot irrigation: a little known but very efficient traditional method of irrigation. Agricultural Water Management. June 2001, 48(2):79–88.
5. Battam, M.A., Sutton, B.G. and Boughton, D.G. Soil pit as a simple design aid for subsurface drip irrigation systems. Irrigation Science. October 2003, 22(3/46):135–141.
6. Bralts, V.F., Wu, I.P. and Gitlin, H.M. Manufacturing variation and drip irrigation uniformity. Transactions of the ASAE, January 1981, 24(1):113–119.
7. Ben-Gal A., Lazorovitch, N. and Shani, U. (2004). Subsurface drip irrigation in gravel filled cavities. Vadose Zone Journal, 3:1407–1413.
8. Christiansen, J.E. (1942). Irrigation by Sprinkling. University of California Agriculture Experiment Station Bulletin, p. 670.
9. Cote, C.M.; Bristow, K.L.; Charlesworth, P.B.; Cook, F.G. and Thorburn, P.J. Analysis of soil wetting and solute transport in subsurface trickle irrigation. Irrigation Science, May 2003, 22(3):143–156.
10. FAO (May 2002). Technical FAO study "World Agriculture: Towards 2015/2030." Food Agriculture Organization of the United Nations, Available from <http://www.fao.org/english/newsroom/news/2002/7828-en.html>
11. Gardner, W.R. Some steady state solutions of unsaturated moisture flow equations with application to evaporation from a water table. Soil Science, May 1958, 85(5):228–232.
12. Gil, M.; Rodríguez-Sinobas, L.; Sánchez, R., Juana, L. and Losada, A. Emitter discharge variability of subsurface drip irrigation in uniform soils: Effect on water application uniformity. Irrigation Science, June 2008, 26(6):451–458.
13. Gil, M.; Rodríguez-Sinobas, L.; Sánchez, R. & Juana, L. Evolution of the spherical cavity radius generated around a subsurface emitter. Biogeoscience (Open Access Journal, European Geosciences Union) Biogeosciences Discuss., 7(January 2010):1–24. Available from <www.biogeosciences-discuss.net/7/1/2010/>
14. Gil, M.; Rodríguez-Sinobas, L.; Sánchez, R. and Juana, L. Determination of maximum emitter discharge in subsurface drip irrigation units. Journal of Irrigation and Drainage Engineering (ASCE), March 2011, 137(3):325–333.
15. Goldberg, D. and Shmuelli, M. Drip irrigation. A method used under arid and desert conditions of high water and soil salinity. Transactions of the ASAE, January 1970, 13(1):38–41.
16. Juana, L.; Rodríguez Sinobas, L. & Losada, A. Determining minor head losses in drip irrigation laterals II: Experimental study and validation. Journal of Irrigation and Drainage Engineering (ASCE), November 2011, 128(6):355–396.
17. Keller, J. and Karmelli, D. (1975). Trickle irrigation design parameters. Transactions of the ASAE, July 1975, 17(4):678–684.
18. Lanting, S. (1975). Subsurface irrigation- Engineering research. Pp. 57–62, In: 34th Report Hawaii Sugar Technology Annual Conference, Honolulu, Hawaii.
19. Lamm, F. R. Advantages and disadvantages of subsurface drip irrigation. Pp. 1–13, In: Proceedings of the Int. Meeting on Advances in Drip/Micro Irrigation, Puerto de La Cruz, Tenerife, Canary Islands, Spain, Dec. 2–5, 2002
20. Lazarovitch N.; Simunek, J.; and Shani, U. System dependent boundary conditions for water flow from a subsurface source. Soil Sci Soc Am J, January 2005, 69(1):46–50.

21. Lazarovitch, N.; Shani, U.; Thompson, T.L. and Warrick, A.W. Soil hydraulic properties affecting discharge uniformity of gravity-fed subsurface drip irrigation. Journal of Irrigation and Drainage Engineering (ASCE), March 2006, 132(2):531–536.

22. Li, J.; Zhang, J. and Rao, M. Modeling of water flow and nitrate transport under surface drip fertigation. Transactions of the ASAE, May 2005, 48(3):627–637.

23. Meshkat, M.; Warner, R.C. and Workman, S.R. Evaporation reduction potential in an undisturbed soil irrigated with surface drip and sand tube irrigation. Transactions of the ASAE, January 2000, 43(1):79–86.

24. Payero, J.O. (2005). Subsurface drip irrigation: Is it a good choice for your operation? Crop Watch news service. University of Nebraska-Lincoln, Institute of Agriculture and Natural Resources, Available from <http://www.ianr.unl.edu/cropwatch/archives/2002/crop02–8.htm>

25. Philip, J.R. (1992). What happens near a quasi-linear point source? Water Resources Research, January 1992, 28(1):47–52.

26. Phene, C.J.; Yue, R.; Wu, I.P.; Ayars, J.E.; Shoneman, R.A. and Meso, B. M. Distribution uniformity of subsurface drip irrigation systems. ASAE Paper No. 92:2569, May 1992, American Society of Agricultural Engineers, St. Joseph, Michigan.

27. Provenzano, G. Using Hydrus-2D simulation model to evaluate wetted soil volume in subsurface drip irrigation systems. Journal of Irrigation and Drainage Engineering (ASCE), July 2007, 133(4):342–349.

28. Rodriguez-Sinobas, L., Juana, L., and Losada, A. Effects of temperature changes on emitter discharge. Journal of Irrigation and Drainage Engineering (ASCE), March 1999, 125(2):64–73.

29. Rodriguez-Sinobas, L.; Juana Sirgado, L.; Sánchez Calvo, R. and Losada Villasante, A.2004. Pérdidas de carga localizadas en inserciones de ramales de goteo. Ingeniería del Agua, May 2004, 11(3):289–296.

30. Rodríguez-Sinobas, L.; Gil, M.; L.; Sánchez, R. and Juana, L. (2009a). Water distribution in subsurface drip irrigation Systems. I: Simulation. Journal of Irrigation and Drainage Engineering (ASCE), December 2009, 135(6):721–728.

31. Rodríguez-Sinobas, L.; Gil, M.; L.; Sánchez, R. and Juana, L. (2009b).Water distribution in subsurface drip irrigation Systems. II: Field evaluation. Journal of Irrigation and Drainage Engineering (ASCE), December 2009, 135(6):729–738.

32. Sadler E.J., Camp, C.R. and Busscher, W.J. Emitter flow rate changes caused by excavating subsurface micro irrigation tubing. Proceedings of the 5th Int. Microirrigation Congress, ISBN 0–929355–62–8, Orlando, Florida, USA, April 2–6, 1995. 763–768.

33. Shani, U.; Xue, S.; Gordin-Katz, R. and Warrick, A.W. Soil-limiting from Subsurface Emitters. I: Pressure Measurements, Journal of Irrigation and Drainage Engineering, (ASCE), April 1996, 122(2):291–295.

34. Simunek, J.; Sejna, M. and van Genuchten, M.T. The Hydrus-2D software package for simulating two-dimensional movement of water, heat and multiple solutes in variably saturated media, version 2.0. Rep. IGCWMC-TPS-53, Int. Ground Water Model. Cent. Colo. Sch. of Mines, Golden-Colorado, p. 251.

35. Schmitz, G.H.; Schutze, N. and Petersohn, U. New strategy for optimizing water application under trickle irrigation. Journal of Irrigation and Drainage Engineering (ASCE), September 2002, 128(5):287–297.

36. Souza Resende, R. Intusão radicular e efeito de vácuo em gotejamento enterrado na irrigação de cana de açúcar. PhD Thesis. Escola Superior de Agricultura 'Luiz de Queiroz" Universidade de São Paulo, Piracicaba, November 2003, p. 143.

37. United State Department of Agriculture USDA-NASS. (2009). Farm and Ranch Irrigation Survey. National Agricultural Statistic Service—USDA. Available from www.nass.usda.gov.

38. Vaziri, C.M. and Gibson, W. Subsurface and drip irrigation for Hawaiian sugarcane. In: 31st Report Hawaii Sugar Technology Annual Conference, Honolulu, 1972 Proceedings by Hawaiian sugar Planters Assoc., 18–22.

PART II
PRINCIPLES OF MICRO IRRIGATION

CHAPTER 10

MODERNIZATION OF FLOATING PUMPING STATION FOR IRRIGATION

RARES HALBAC-COTOARA-ZAMFIR and ALINA GABOR

CONTENTS

10.1   Introduction 188
10.2   Current Irrigation System: Brief Description 189
10.3   General Considerations For Upgrading Existing Irrigation System [3] 192
10.4   The Modernization of Floating Pumping Station 193
10.5   Conclusions 194
10.6   Summary 195
Keywords 195
References 195

*Modified and printed from, "*Rares Halbac-Cotoara-Zamfir and Alina Gabor, 2012. The rehabilitation and modernization of Fantanele-Sagu Arad irrigation system. 12th International Multidisciplinary Scientific GeoConference SGEM 2012, Hydrology and Water Resources, pp. 1–7.*"

10.1 INTRODUCTION

The last years presented in Romania large variations regarding the temperatures and precipitations regimes. The western part of this country (which includes and the Arad County) was affected by floods followed at very short periods of time by other climatic hazards as severe droughts. All of these phenomenons can be understand as results of climatic changes.

Arad County is covered by large areas with irrigation arrangements, systems more than necessary for the agricultural systems from this part of Romania. In 2009, only 8,520 ha in Arad were irrigated. From the analysis results that there is no interest in irrigation for other areas, and the surfaces that must be considered for the rehabilitation of the Şag—Fântânele and Şemlac –Pereg systems represent a maximum of 50% of the facilitated area [1–3].

In 2012, even Romania faced the most severe drought period from the last 25 years, the area which was irrigated in Arad County was reduced to zero.

Fantanele-Sagu irrigation system is located in western part of Romania, Arad County, on Mures River's plain, at south of this river course and is under the administration of National Administration of Land Reclamation and Improvement—West, Arad Territorial Branch (Figs. 1 and 2).

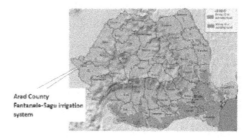

FIGURE 1   Location of Fantanele-Sagu irrigation system [2].

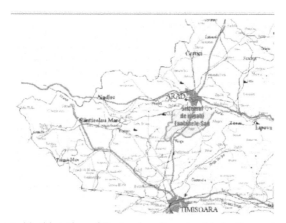

FIGURE 2   Geographical location of Fantanele—Sagu Arad irrigation system [1].

The entire territory of this irrigation system is part of Tisa Plain. Plain district Vinga interposed as a relief step that dominates the low plain of 20–40 m and is dominated by piedmont with 40–60 m and forms the high plains with a general plain relief and with an altitude between 110 m and 145 m, low-sloped. This plain has low ripples, numerous valleys of erosion, frequent micro depressions and light, elongated depressions which accumulate large regular rainfall waters [1, 3].

From geomorphologic point of view, this area is suitable for sprinkler irrigation and other forms of surface irrigation.

Gravitational water is sometimes present in large pores of the soil and is very mobile, poorly retained in this case, so that is rapidly lost through infiltration or leakage. It is in principle accessible to plants but they cannot benefit only slightly from it. Groundwater level was identified in the area at different depths depending on the type of relief and time. In general groundwater is at depths ranging between 0.50 m and 10 m, the highest levels being found in valleys of erosion (0.5–2 m). Most of the groundwater table is at a depth of between 4 m and 10 m [1].

The climate of the studied area is characterized by annual average temperature around 10.8°C and average annual rainfall of 577 mm. Droughts in the region appear with a higher frequency generally from July to August and especially in September. The most common winds are those from southeasterly direction. Wind speed varies from 2.6 to 4.5 m/s. The performed climate studies indicate a less continental climate and more Mediterranean, with less severe winters and summers not excessively warm. Enough precipitation falls during the summer especially, but they are unevenly distributed on months and critical periods of crop plants growing often-causing prejudices. Generally speaking the climate is poor in precipitations as is shown also in the soil balance calculation, resulting in this way a deficit of 206 mm [1, 3].

10.2   CURRENT IRRIGATION SYSTEM: BRIEF DESCRIPTION

Irrigation system Fântânele-Şagu-Arad was designed in 1966, the completing and expansion phase being finalized in 1974 for an area of 9418 ha, 7154 ha being realized and presenting the following scheme of planning hydrotechnical in organizing flow distribution (Fig. 3).

The system has a lifetime of over 40 years, covers an area of 7154 ha, the whole surface being connected to a single hydrotechnique scheme having the supply at Mures River, being served by two adduction channels (I and II) and two pumping stations. The Floating pumping station (Fig. 4) consists of two boats, each supporting two pumps SIRET 900 with a total capacity of 7.4 m^3/s [3]. The fixed pumping station lifts the water with another 18 m from supply channel I to supply channel II and is equipped with three 900 SIRET pump units with an installed capacity of Q = 6.0 m^3/s [3].

FIGURE 3   The scheme of irrigation system and the area proposed for rehabilitation [1].

FIGURE 4   Floating pumping station on Mures River [3].

In the current situation after the studies and technical checks of Floating pumping station, the experts concluded that floating pumping station it is unsafe from operational point of view. Floating pump units and the ships have a very advanced degree of corrosion and require full replacement. Ships are supported with φ24 steel cable of 800 m in length. The cables are worn and corroded and need to be replaced. The φ1000 discharge piping (with a length of 200 m) presents an advanced corrosion and the couplings are not sealed needing to be replaced [3].

FIGURE 5   Ship of floating pumping station on Mures River [3].

FIGURE 6   Repression pipes (Floating pumping station) [3].

Existing pumping aggregates show wear and tear very advanced. Major repairs were made in 2002 and since then have been executed other repairs only for maintenance and operation. Long term disposal is exceeded, from economically point of view being not viable to repair these units, the proposed solution being to replace them with modern and reliable aggregate pumping the same flow. Due to the floods and changes in the level of floating, the pumping units and shore works have suffered damages, the most important being the destruction of bank consolidation, the floating bodies, the base of pumping aggregates and damages to joints and valves [3]. These failures have led to frequent failures of the Floating pumping station which led to a costly operation and failures in providing the required flows for irrigation.

Interruptions were particularly common after 1975. In 1985, floating pumping station was submerged due to a water hammer strike following a power interruption. On this occasion pipes were broken the connections between ship and shore (φ1000 and 20 m length each). Power cable from the transformer station to the ship is damaged due to the age and power interruptions, which may cause accidents or even sinking in case of a short circuit again [3].

Considering the over 40 years of operation it is necessary to change the pumps (due to lower yields of operation), the engines and other components of these stations which don't asnwer to actual requirements. In most pumping stations it is required to change even the electrical panels.

For other parts of the studied irrigation system, which are in the custody of the irrigation water users organizations, their rehabilitation and modernization remains the responsibility of these organizations. Thus were prepared several technical expertise for Fantanele pumping station, Sagu 1 Pumping station, Sagu 2 Pumping station as well as for underground piping network.

The financial part of these problems can be rezolved by applying for European funds through EAFDR.

10.3 GENERAL CONSIDERATIONS FOR UPGRADING EXISTING IRRIGATION SYSTEM [3]

In order to have an irrigation arrangement operating at higher parameters we need to implement retrofitting works (rehabilitation) which must take into account both pumping stations, repumping and pressure stations as well as other components of irrigation systems such as water supply and water distribution channels, underground pipelines, hydraulic structures (bridges, Crossing, weirs, shock devices, etc.). It is also necessary to ensure the mobile devices for the watering application, devices which are currently almost nonexistent. In fact, the whole works remain virtually unusable if not solving the problem of purchasing the equipment necessary to implement watering.

In the situation when is necessary to rehabilitate a floating pumping station, the energy indicators are very important to be considered. These indicators are referring to the energetic efficiency of a pumping stations and aggregate components, specific energy consumption related to energy efficiency of watering plants etc.) The electric power (P, kW) required by a pumping station (power, core booster or pressure release) is determined by the relationship:

$$P = 8,91 \cdot 10^{-3} \cdot \frac{q_0 \cdot S \cdot H}{\eta_H \cdot \eta_P} \cdot \frac{24}{t_f}, \text{kW} \tag{1}$$

$$q_0 = \frac{M}{86,4T} \tag{2}$$

where: M = weighted average irrigation rate (all irrigated crops) in considered month; T = effective irrigation time (days); t_f = daily operating time of the pumping station (hours/day); S = surface covered by pumping station; H = hydraulic head developed by the station (m); η_p = pumping station efficiency; h_H = total efficiency of the water use on the surface serviced by the pumping station. Specific consumption of electricity (kWh/1000 m³ water pumped) at a pumping station is defined in Eq. (3). Specific energy consumption at the plant (kWh/1000 m³ water pump and actually entered into ground) is shown in Eq. (4).

$$e_s = 2,725 \frac{H}{\eta_p} \tag{3}$$

$$e_s = 2,725 \frac{H}{\eta_H \cdot \eta_p} \tag{4}$$

The specific energy consumption for the entire system (kWh/1000 m³ water pumped) is:

$$e_s^s = \frac{t_f \cdot \eta_H}{S_0} \cdot \sum_{i=1}^{N} \left(\frac{S \cdot e_s}{t_f \cdot \eta_H} \right) \tag{5}$$

Pumping aggregates performance depends on the quality of equipment, age and number of hours of use, maintenance standards and specific local factors, such as water quality, operating conditions, etc.

10.4   THE MODERNIZATION OF FLOATING PUMPING STATION

Fantanele floating pumping station will be removed and replaced with a fixed pumping station, equipped with submersible pumps installed in steel pipe of φ1000 mm, located on the left bank of the river Mures, having a slope of 30°. Station base equipment will consists of 6submersible pumps that provide a flow of 950–1000 L/s each at a height of about 26 meters of water column (MWC). Pumping unit is equipped with an engine of 400 kW, wtih direct pumps, the supply power being of 400 V [4, 5].

The suction pipes are practically nonexistent, each electrical pump being mounted below the minimum water level of river Mures, the pumps being equipped with a vacuum cleaner with proper hydraulic form. Discharge pipes, in number of 6, are placed horizontally at 4.5 m apart, having on the first section of about 20 m a diameter of 1000 mm up to the level of 114 MWC. This level is more than up to 1% insurance flood waters of the river Mures. From this point until the station basin outlet, pipes have a diameter of 800 mm.

Construction works referred to Fântânele Floating Pumping Station consist of elastic works at the point of outlet and consolidation of the left bank slope from the share the bed to 114 MWC level, and works to protect and ensure the stability of the 6 discharge pipes on the entire front where they stand. Front that develops these works has a lenght of 56 m [4, 5]. Elastic construction works consist of:

- Earthworks for cleaning and leveling the slope and rough stone filling of eroded areas of the flood waters of the Mures River;
- Geotextile fascine-mattress, ballasted with stone on the bottom of the bed, on about 6 m wide and full length of front pumping station on which are mounted the 6 vacuum;
- Strengthening slope from the bottom bed share (102.5 MWC level) to 114 MWC with a stone layer of 80 cm, 50–150 kg/piece sited on sand layer of 30 cm;
- Hardening each outlet pipe with three pairs of concrete piles, bound together 2 by 2 with metal parts to fix them vertically;
- Filling the spaces between the pipes and covering them with rough stone of 50–150 kg/piece;
- Concrete platform realized at 114 MWC to allow access of equipment and transport equipment, platform size 30 × 10 m [4, 5].

As new fixed pumping station equipped with six submersible pumps will be connected to a voltage of 0.4 kV, the transformer will be changed. It will be installed 2 transformers of 20/0.4 KV, each with a power of 1600 KVA.

The installation of 0.4 KV will be completely new and will provide power, protection and control for the following low voltage electrical consumers:

- 6submersible pumps each with a power of 350 KW;
- Indoor and outdoor lighting installation;
- Installation three phase and single phase outlets;
- Plant automation and dispatching [4, 5].

According to the design objectives, Fantanele Floating Pumping Station rehabilitation solutions will help to improve performance of irrigation system components so that they will can provide: Decreasing of operating and maintenance costs by increasing efficiency and providing modern technology; and reducing power consumption by increase efficiency in water delivery and distribution.

All works proposed for rehabilitation of Fantanele Sagu irrigation system Fântânele Şagu fall in "Land improvement arrangements list or parts of land reclamation arrangements declared as public utility improvements, which is administered by the National Administration of Land Reclamation and Improvements provided in Annex 1 of Government Decision 1582/2006."

The proposed rehabilitation works cover the area currently occupied by existing works, and therefore are not required to be permanently set aside from agricultural circuit. From a legal perspective, the area occupied by objects of infrastructure which follow to be rehabilitated are in administration of National Administration of Land Reclamation and Improvement.

10.5 CONCLUSIONS

The changes in the Romania's climate regime can be integrated within the global context and present pessimistic evolutions for the future of agricultural surfaces from western part of Romania, territory that is partially covered by Fantanele-Şagu Arad irrigation system [3].

In 2005 in Romania irrigation network covered approx. 2.8 million hectares, of which 1.5 million hectares with irrigation infrastructure recently rehabilitated. This large network of irrigation has been exploited in recent years (1998–2003), the percentage of use is between 15.6 to 37.9% of total area recently rehabilitated infrastructure. Rehabilitation and development of irrigation sector is an imperative first order under a semiarid climate with rainfall below 500 mm or less than 250 mm in dry years. The irrigation systems realized before 1990, among which are part and the irrigation system Fantanele—Sag, are mostly damaged, incomplete, obsolete and outdated.

Starting with 2010, there are no more irrigation subsidies, under the current legislation, and therefore is urgently required the rehabilitation and modernization of existing systems. In this situation is an urgent need to be prepared new rehabilitation, modernization and equipping projects of these arrangements, to be made proposals to find opportunities of new resources and solutions to support farmers from the fields arranged. These arrangements must be keep on working and provided with new technical solutions that are economically effective and efficient.

From the presentation of Fanatanele—Sagu Arad irrigation system results the importance of rehabilitation this system using European funds through National Administration of Land Reclamation and Improvement and Organization of Water Users Association, by realizing valuable projects for rehabilitation and provision of equipment [1–3].

10.6   SUMMARY

This chapter presents the rehabilitation and modernization opportunity motivation of a floating pumping station for a large irrigation system, located in western Romania, in the conditions of meeting the private owners requirements related to this irrigation system.

KEYWORDS

- **climatic hazard**
- **drought**
- **effective irrigation time**
- **floating pump**
- **irrigation system**
- **land Reclamation**
- **National Administration of Land Reclamation and Improvement and Organization of Water Users Association [NALRIOWUA]**
- **power, kW**
- **pumping station efficiency**
- **specific energy consumption**
- **station**
- **water delivery**
- **water distribution**
- **western Romania**

REFERENCES

1. Report concerning the environmental impact evaluation of Fantanele-Sagu Arad irrigation system, 2007, Tahal Consulting Engineers, Romania.
2. National Administration of Land Reclamation and Improvement archive.
3. Gabor, A., 2012. The rehabilitation and improvement of Fantanele—Sagu Arad irrigation system, PhD thesis, manuscript, "Politehnica" University of Timisoara, Romania.
4. Technical expertise report of Sagu 2 PPS (proposed for rehabilitation) and of irrigation infrastructure belonging to Organization of Water Users Association Sagu 2, Arad County, Contract no. 1/2009, Politehnica" University of Timisoara, Romania.
5. Technical expertise report of Sagu 1 PPS and Fantanele PPS (proposed for rehabilitation) and of irrigation infrastructure belonging to Organization of Water Users Association Fantanele, Arad County, Contract no. 3/2009, Politehnica" University of Timisoara, Romania.
6. Romanian Ministry of Environment, 2005. Arad County's Environment Report, Arad, Romania.

CHAPTER 11

PRINCIPLES OF MICRO IRRIGATION*

MEGH R. GOYAL

CONTENTS

11.1   Introduction 198
11.2   Advantages 199
11.3   Disadvantages 200
11.4   Micro Irrigation System Layout and its Parts 201
11.5   The Irrigation Duration 208
11.6   The Clogging Agents in the Irrigation Water 208
11.7   The Flushing of Sub Mains and Laterals 208
11.8   The Irrigation Distribution System 209
11.9   Subsurface Drip Irrigation System (SDI) 212
11.10 Summary 219
11.11 Basic Aspects in Micro Irrigation Research 232
Keywords 236
References 236

*This chapter is modified and printed with permission from: "Goyal, Megh R., 2012. Principles of drip/ trickle or micro irrigation. Chapter 5, pp. 103–132, In: *Management of Drip/Trickle or Micro Irrigation*, Editor Megh R. Goyal. Oakville, ON, Canada: Apple Academic Press Inc., ISBN 9781926895123." © Copyright 2013.

11.1   INTRODUCTION

The competitive demand of available water is more and more acute now in most parts of the world. The supplies of good water quality are declining every day. It is, therefore, necessary to find methods to improve water use efficiency in agriculture. The micro irrigation is one of the most efficient irrigation systems that are used in agriculture. This system has been used in arid regions of the world.

Micro irrigation, also known as *trickle irrigation* or *drip irrigation,* applies water slowly to the roots of plants, by depositing the water either on the soil surface or directly to the root zone, through a network of valves, pipes, tubing, and emitters. The goal is to minimize water usage. Micro irrigation may also use devices called microspray heads, which spray water in a small area, instead of emitters. These are generally used on tree and vine crops. Subsurface micro irrigation or *SDI* uses permanently or temporarily buried dripperline or drip tape. It is becoming more widely used for row crop irrigation especially in areas where water supplies are limited. The micro irrigation is a slow and frequent application of water to the soil by means of emitters or drippers located at specific locations/interval throughout the lateral lines. The emitted water moves through soil mainly by unsaturated flow. This allows favorable conditions for soil moisture in the root zone and optimal development of plant.

11.1.1   HISTORY <HTTP://EN.WIKIPEDIA.ORG/WIKI/DRIP_IRRIGATION>

Micro irrigation was used in ancient times by filling buried clay pots with water and allowing the water to gradually seep into the soil. Modern micro irrigation began its development in Germany in 1860 when researchers began experimenting with sub-irrigation using clay pipe to create combination irrigation and drainage systems. In 1913, E.B. House at Colorado State University succeeded in applying water to the root zone of plants without raising the water table. Perforated pipe was introduced in Germany in the 1920s. In 1934, O.E. Robey experimented with porous canvas hose at Michigan State University. With the advent of modern plastics during and after World War II, major improvements in micro irrigation became possible. Plastic microtubing and various types of emitters began to be used in the greenhouses of Europe and the United States.

A new technology of micro irrigation was then introduced in Israel by Simcha Blass and his son Yeshayahu. Instead of releasing water through tiny holes (blocked easily by tiny particles), water was released through larger and longer passageways by using friction to slow water inside a plastic emitter. The first experimental system of this type was established in 1959 in Israel by Blass, where he developed and patented the first practical surface micro irrigation emitter. In 1973, publisher Massada Limited in Israel printed a book in Hebrew by Simcha Blass named "WATER IN STRIFE AND ACTION." Blass was a well known water engineer in Israel in its first years and cheif engineer in Jewish Yishuv. He planned the Israeli main water carrier. It is a common misconception that micro irrigation was invented in Israel. There is no question that much of the product innovation in this field occurred in Israel and that companies in Israel have contributed significantly to the industry, but they can't take all the credit for its development. The facts are that micro irrigation system, with plastic pipes and the new plasic drippers, was invented and first used and developed in Israel by Blass. It does not change the contributions made earlier and later.

This method subsequently spread to Australia, North America, and South America by the late 1960s. In the United States, in the early 1960s, the first drip tape, called *Dew Hose*, was developed by Richard Chapin of Chapin Watermatics (first system established at 1964). In 1969, researchers under Prof. R. K. Sivanappan at Tamil Nadu Agricultural University in Coimbatore, India started research using micro irrigation using the available tubes and microtubes in the market not only in the research station, but also in the farmer's field for banana, grapes, cotton and vegetables. Beginning in 1987–88, Jain irrigation toiled and struggled to pioneer water-management through micro irrigation with the advice of Dr. Sivanappan. Jain irrigation also introduced some hi-tech micro irrigation concepts to Indian agriculture such as 'Integrated System Approach,' One-Stop-Shop for Farmers, 'Infrastructure Status to Micro irrigation and Farm as Industry.' They have established "Research and Demostration Farm" for modern micro irrigation concepts at Hi-tech Agricultural Institute near Jalgaon (Maharashtra, India). The area under micro irrigation is more than 5×10^5 hectares covering about 30 crops including grapes, sugarcane, banana, cotton, horticultural crops, vegetables and fruits.

The first micro irrigation system in Puerto Rico was installed in 1970 for fruit orchard owned by Luciano Fuentes in Coamo municipality, and in mango orchard at Fortuna Agricultural Experimental Substation in Juana Diaz municipality. In 1979, first research experiment on micro irrigation was established at Fortuna Substation by Megh R. Goyal, Father of Irrigation Engineering in Puerto Rico. Today, acreage under micro irrigation for vegetables and fruits has increased to more than 60,000 acres in Puerto Rico. In 2012, *Management of Drip/Trickle or Micro Irrigation* edited by Megh R. Goyal was published by Apple Academic Press Inc.—Only complete bible on micro irrigation.

It is assumed that the micro irrigation will fortify agriculture and increase efficiency of food production. With this system, the plant can efficiently use available natural resources such as: soil, water and air. The micro irrigation is also known as "daily irrigation," "trickle irrigation," "daily flow irrigation" or "drip irrigation." The term "trickle" was originated in England, "drip" in Israel, "daily flow" in Australia and "micro irrigation" in USA. The difference is only in the name, and all these terms have the same meaning. The water in a micro irrigation system flows in three forms:

1. It flows continuously throughout the lateral line.
2. It flows from an emitter or dripper connected to the lateral line.
3. It flows through orifices perforated in the lateral line.

11.2 ADVANTAGES

Well-designed micro irrigation system can increase the crop yield due to following factors:

A. Efficient use of the water:
1. It reduces the direct losses by evaporation.
2. It does not cause wetting of the leaves.
3. It does not cause movement of drops of water due to the effect of wind.

4. It reduces consumption of water by grass and weeds.
5. It eliminates surface drainage.
6. It allows watering the entire field until the edges.
7. It allows applying the irrigation to an exact root depth of crops.
8. It allows watering greater land area with a specific amount of water.

B. Reaction of the plant:
1. It increases the yield per unit (hectare-centimeter) of applied water.
2. It improves the quality of crop and fruit.
3. It allows more uniform crop yield.

C. Environment of the root:
1. It improves ventilation or aeration
2. It increases quantity of available nutrients.
3. The conditions are favorable for retention of water at low tension.

D. Control of pests and diseases:
1. It increases efficiency of sprayings of insecticides and pesticides.
2. It reduces development of insects and diseases.

E. Correction of problem of soil salinity:
1. The increase in salts happens at a distance away from the plant.
2. It reduces salinity problems. A greater reduction is obtained by increasing the water flow. Salts are limited to an outer periphery of a wetted zone.

F. Weed control:
1. It reduces the growth of weeds in the shaded humid space.

G. Agronomic practices:
1. The activities of the irrigation do not interfere with those of the crop, the plant protection and the harvesting.
2. It reduces inter cultivation, since there is less growth of weeds.
3. Helps to control the erosion.
4. It reduces soil compaction.
5. It allows applying fertilizers through the irrigation water- called fertigation.
6. It increases work efficiency in fruit orchards, because the space between the rows is maintained dry.

H. Economic benefits:
1. The cost is lower compared with overhead sprinkler and other permanent irrigation systems.
2. The cost of operation and maintenance is low. The costs are high when the average row spacing is less than 3 meters.
3. It can be used in uneven terrains.
4. The water application efficiency is high. Energy use per acre is reduced due to smaller diameter pipes, and only 50–60% of water is used.
5. The BC ratio is favorable and the pay back period is one to two years.

11.3 DISADVANTAGES

A. The micro irrigation, like other methods of irrigation, cannot adjust to all the specific crops, sites and objectives

B. The system has the following problems and limitations:
1. The drippers are obstructed (or clogged) easily with soil particles, algae or mineral salts.
2. The soil moisture is limited. The soil moisture volume depends on the discharge of drippers, dripper spacing and the soil type.
3. The rodents or insects can damage some components of the micro irrigation system.
4. A more careful high technology management is needed compared to other irrigation systems.
5. The initial investment and annual cost are higher compared to other irrigation methods.

11.4   MICRO IRRIGATION SYSTEM LAYOUT AND ITS PARTS

The main components of the micro irrigation system (Fig. 1) are:
— The water source.
— The pump and the energy unit.
— Main Line (larger diameter Pipe and Pipe Fittings).
— The filtration systems: Sand separator, screen filter, media filters and manifolds, backwash controller.
— Fertigation systems and chemigation equipment.
— The controls in the automation of the system: Plastic Control Valves and Safety Valves
— The water distribution system: Smaller diameter polyethylene tubing (often referred to as "laterals")
— Emitting devices at plants (ex. Drippers, microspray heads, irrigation mats)

To make connections, fittings and accessories are needed for a micro irrigation system, such as: Gate or ball valve, safety or check or one way valve, flushing valve, union, nipple, adapters, reducers, tee, coupling, universal, elbow, double union and cross.

11.4.1   THE WATER SOURCE

The water source can be a treated water, well water, open channel, rivers and lakes. The clean water is essential for the micro irrigation. If water of poor quality is used, physical and chemical or biological polluting agents can obstruct the emitters and drip laterals.

The underground well water is generally of good quality. In some cases, it can contain sand or chemical substances.

Although superficial water from lakes, rivers and open channels can be used, this water source has the disadvantage of possible contamination by bacteria, algae and other organisms.

Almost all water sources contain bacteria and elements that nourish it. A good filtration system is needed to remove all the polluting agents that can obstruct or clog the drippers.

11.4.2 THE PUMP

The pumping and the power units represent a significant part of a initial installation cost of a micro irrigation system. Therefore, for the selection of appropriate equipment, it is convenient to know all the characteristics and operating conditions. It is necessary to acquire effective, reliable and low cost pump and a power unit.

A centrifugal pump (Fig. 2) is suitable for extracting water from superficial sources or shallow wells. It can also be used to increase the pressure in a main or sub main line. The centrifugal pump is relatively cheap and efficient. It is available for wide range of flows and pressures.

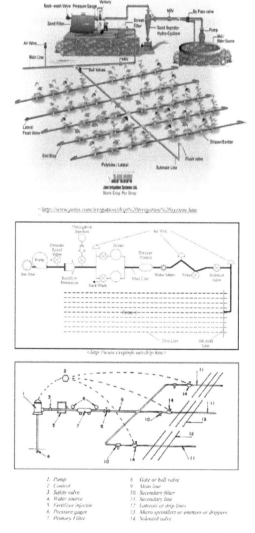

1. Pump	8. Gate or ball valve
2. Control	9. Main line
3. Safety valve	10. Secondary filter
4. Water source	11. Secondary line
5. Fertilizer injector	12. Laterals or drip lines
6. Pressure gages	13. Micro sprinklers or emitters or drippers
7. Primary Filter	14. Solenoid valve

FIGURE 1 Components for a typical drip/micro or trickle irrigation system.

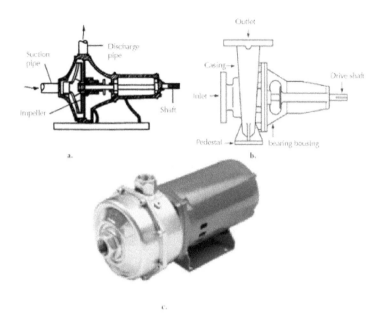

FIGURE 2   Pumps.

While selecting a pump, one should know total pressure in the system, operating pressure, total volume of water for irrigation and horsepower rating. A pump of a slightly greater capacity than necessary can ensure sufficient water at a desired pressure. The pump of a required size guarantees a good uniformity of water application. On the contrary, an undersized pump and oversized drippers can cause non-uniform emitter flow throughout the field. Total pressure for the system generated by a pump depends on the following factors:

1. The highest elevation: Elevation difference between a water source and an emitter located at the highest elevation of the field.
2. The highest friction loss: The pressure required for the most distant emitter. One should also consider discharge flow.

11.4.3   ENERGY OR POWER UNITS

In micro irrigation system, the electrical motors are preferable because of high overall efficiency, ease of automation, quiet operation and necessity of least maintenance.

The gasoline or diesel engines can also be used. However, these may cause variations in the operational pressures and emitter flow. These have high maintenance cost and cause noise during operation.

11.4.4   THE IRRIGATION CONTROLS (FIGS. 3–6)

11.4.4.1   THE VOLUMETRIC VALVE

The volumetric valve measures volume of irrigation application and is necessary for proper management.

11.4.4.2   THE PRESSURE GAGE

The pressure gage measures water pressure in the micro irrigation system. It is especially useful for non-compensating pressure drippers. The pressure gages on the upstream and downstream side of the filtering system indicates pressure loss that helps in programming the flushing process of filters. The intervals and range of pressure gage must match with the operating pressure of the system for good management.

11.4.4.3   THE PRESSURE REGULATOR OR PRESSURE RELIEF VALVE

The pressure regulator or pressure relief valve can be fixed or adjustable. When installed in the principal or secondary lines, it can compensate the changes in elevation or frictional losses of lines. The manual valves, the automatic metering valves, the semiautomatic valves and timers are recommended for the micro irrigation system.

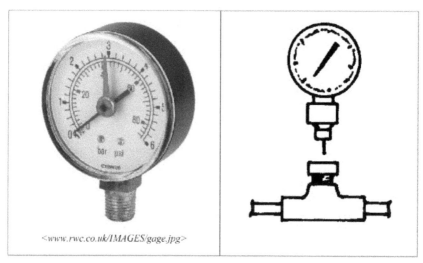

<www.rwc.co.uk/IMAGES/gage.jpg>

FIGURE 3   Pressure gage.

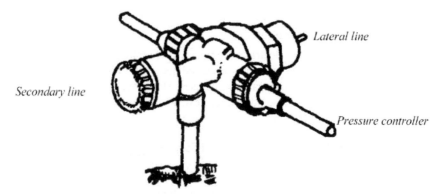

Lateral line

Secondary line

Pressure controller

FIGURE 4   Pressure regulator for the micro irrigation system.

<http://www.alibaba.com/manufacturer/14002360/
Buy_Irrigation_Solenoid_Valve.html>

[Goyal, 1990]

<http://www.duralirrigation.com.au/outlinestore/index.cfm?NavigationID=641>

FIGURE 5   Solenoid valves for micro irrigation system.

<http://www.rainbird.com/about/imagelibrary/drip/
index.htm>

[Goyal, 1990]

<http://www.jains.com/Control%29and%20safety%20valves/jain%20air%20release%20valves>

FIGURE 6   Air relief valve or vacuum breaker.

The automatic flow valves provide desired amount of irrigation in a specified time. These also reduce pressure variations among the lateral lines in an uneven land. The combinations of pressure regulator and flow control valve are also available.

11.4.4.4   AIR RELIEF VALVE OR VACUUM BREAKER

This valve removes air from the micro irrigation system. The suction or vacuum is developed when the irrigation system is shut off. This vacuum can obstruct the drippers if the dirty water or dust is suctioned into the system. Vacuum breaker or air relief valve of "one inch for each 25 gpm of flow" is recommended.

11.4.4.5   FLUSHING VALVE

The flushing valve at the end of each lateral helps in the flushing of the system.

11.4.5   THE DRIPPER OR EMITTER

The water emission devices (drippers or emitters or microsprinklers or sprayers) are unique in the micro irrigation system. The drippers supply water through small orifices in small amounts near the plant. The pressure through the emitters must be sufficiently greater than the pressure difference due to land topography and friction losses in the system. The orifice must be large enough to avoid serious obstructions. These design considerations have resulted in the manufacture of several types of drippers (Fig. 7). There are two categories of drippers for installation in the field:
1.   In-line emitters/drippers.
2.   On-line emitters/drippers.

The in-line emitters are used in the green houses and for row crops such as: vegetables and some fruit orchards. This layout also consists of a series of equally spaced perforations (orifices) throughout the lateral: single or biwall drip laterals. The emitter flow fluctuates from three to four (3 to 4) liters per minute for 30 m of lateral. The operating pressure of this layout fluctuates from 2 to 30 psi. However, most of these systems operate at a pressure less than 15 psi. The lateral or drip lines can be up to 90 m of length depending on the slope. These laterals are limited by irrigation flow rate and by uneven terrain of the land. The on-line drippers are suitable to leveled lands to maintain acceptable uniformity of emitter flow. Because the operating pressure is low, a moderate variation in elevation causes large variation in the emitter flow. The engineers are the right professionals for design of micro irrigation systems.

The flow variations of drippers are due to changes in pressure because of uneven lands. For these situations, compensating pressure drippers provide almost constant flow for wide.

The flow variations fluctuate from moderate to high depending on pressure changes because of non-uniform orifices in a single chamber and biwall drip lines.

Micro tubes or spaghetti connected on the lateral line are used for fruit orchards, ornamentals and flowerpots in the nursery (green house). One or numbers of micro tubes irrigate individual plants or pots, and these are generally connected on the plastic or polyethylene hose. In this system, plants are not near the lateral line.

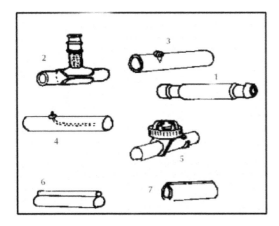

1. In-line emitters.
2. On-line emitters.
3. Micro sprinklers.
4. Micro tube (spaghetti).

5. Compensating pressure emitter.
6. Low pressure emitter lines.
7. Biwall drip lines.

<http://homeharvest.com/dripirrigationemitters.htm>

FIGURE 7   Types of emitters.

The irrigation water enters the dripper at high pressure and leaves the emitter at design pressure. The reduction in pressure is affected by the flow of water through long labyrinths, tortured (zigzag) path, size of orifices and other design variations by the manufacturer. Specially designed emitter can allow free passage of particles of moderate size out of the dripper. These drippers are of self-cleaning type at low operating pressures of 5 to 60 psi with average emitter flow of 2 to 7.6 L per hour.

11.5   THE IRRIGATION DURATION

The irrigation duration depends on the following factors:
1.   The plant water requirements in gallons or liters/ day.
2.   Irrigation interval between each application and irrigation frequency.
3.   Dripper flow rate and volume of water application.
4.   Soil and plant characteristics.
    The irrigation duration can be calculated as follows:
    Hours of irrigation per day = (liters of water required by plant per day)/(Emitter flow rate in liters per hour)(1)

11.6   THE CLOGGING AGENTS IN THE IRRIGATION WATER

Depending on the water resource, the quality of water varies considerably depending on physical/ chemical and biological composition, water demand and rainfall. The clogging agents in the water can be of physical, chemical or biological nature.

The physical clogging agents are sand, silt and clay. The chemical clogging agents include minerals and salts in the water. Many of these clogging agents may stimulate growth of microorganisms. The algae, bacteria, fish, ants, insects and leaves are biological clogging agents. In general, an adequate filtration system, flushing, chemical treatments and good management can prevent obstructions of the drippers and the lateral lines.

11.7   THE FLUSHING OF SUB MAINS AND LATERALS

The maintenance and flushing of the main, secondary and lateral lines are indispensable for good operation of a micro irrigation system. A good filtration system may catch larger size particles. However, smaller size particles may pass through the filter thus arriving at the secondary and lateral lines. These smaller size particles accumulate in the lateral lines and drippers thus causing obstructions. Some particles combined with certain type of slime and bacteria form aggregates that cause emitter clogging. A filter of >100-mesh size is adequate to avoid clogging due to physical agents.

The periodic flushing of lateral lines will reduce these obstructions. The main and secondary lines must have a sufficient flow rate to allow flushing. In the lateral lines, the water flow at 30 meters/second is adequate for a good flushing action. The main line is flushed first, then the secondary line, and finally the lateral line. Several lines are opened at the same time when the water pressure is adequate. If the adequate pressure cannot be maintained with all the lines open, then few lines are opened at a time.

The flushing time must be sufficient to allow all sediments out of the lines. A regular program of inspection, maintenance and flushing helps to reduce clogging of the emitters. The nature of the filtration system, the quality of the water and the experience of the operator will determine when it is necessary to flush the lines. The maintenance and flushing must be a routine procedure. This practice of maintenance reduces obstructions due to sediments in the lines and drippers. It also prevents formation of slime, bacteria and algae. The invasion by ants and insects is also reduced.

11.8   THE IRRIGATION DISTRIBUTION SYSTEM

The irrigation distribution includes pump house, main and sub mains, and drip lateral lines (Fig. 1). The water from the pump is distributed to the field through a main line. The secondary lines of smaller diameters take the water from the main line to the lateral lines. And the lateral lines supply the water to the drippers, which allow slow application to the plant. The main lines can be of high pressure PVC, galvanized iron, and polyethylene or lay flats. The main lines must be buried at least 0.6 meters (2 feet) to avoid mechanical damage during the field operations.

Commonly, polyethylene tubes of 12 to 16 centimeters (approx. ½ to ⅔") of diameter are used for the lateral lines. These lateral lines provide water to the soil surface by means of the drippers. The polyethylene (PE) hoses are available in different sizes, lengths and strength. Some manufacturers produce weather resistant PE hoses. These are suitable for irrigation application.

11.8.1   THE UNEVEN TERRAIN

The unevenness of the agricultural land is an important design criterion. The change in an elevation causes a pressure loss or gain. A change in an elevation of 2.3 feet causes a pressure change of one psi. In a leveled or almost leveled land, the lateral lines with drippers must run throughout crop row. In slopes, drip lines must follow the contour. To install a system in an uneven field, one should consult a specialist. When the lateral lines in uneven field are designed, it is advisable to consider the advantage of the slope. Thus the energy gain with the decrease in elevation is balanced. To maintain the uniform pressure in slopes of five percent, it is recommended to use shorter length of laterals and pressure compensating emitters; and to install pressure regulators.

11.8.2   THE FRUIT ORCHARDS

During dry periods, the fruit trees respond to the micro irrigation with good vegetative and fruit growth. The quantity of fertilizers is reduced to about 30 to 50 percent with fertigation. Generally in fruit orchards, micro irrigation system is permanent with the main and secondary lines buried. The lateral lines with drippers can be buried or left on the soil surface. In fruit orchards, the emitters are located within the shaded area of the tree (Figs. 8 to 11). When a micro irrigation system for fruit trees is designed, more drippers are added as the tree grows. Rule of thumb is to irrigate 60% of the shaded area. When installing the system, it is important to consider the life of the tree (10 to 20 years). This is the principal reason that the submains are buried.

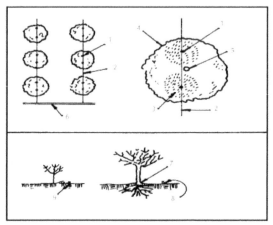

1. Emitter.
2. Lateral line
3. Pattern of wetting zone.
4. Leaf coverage.
5. Trunk.
6. Buried main line.

7. Do not allow flooding of the base of the trunk
   to avoid diseases
8. Move the emitter away from the tree trunk
   depending on the age of a tree.
9. The emitter is near the trunk for the young tree.

FIGURE 8   Micro irrigation system in fruit orchards.

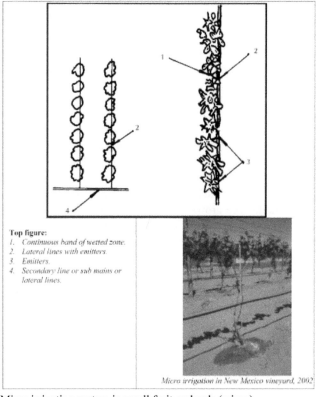

Top figure:
1. Continuous band of wetted zone.
2. Lateral lines with emitters.
3. Emitters.
4. Secondary line or sub mains or
   lateral lines.

Micro irrigation in New Mexico vineyard, 2002

FIGURE 9   Micro irrigation system in small fruit orchards (wines).

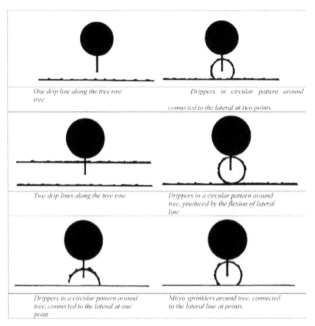

One drip line along the tree row. / Drippers in circular pattern around tree, connected to the lateral at two points.

Two drip lines along the tree row. / Drippers in a circular pattern around tree, produced by the flexion of lateral line.

Drippers in a circular pattern around tree, connected to the lateral at one point. / Micro sprinklers around tree, connected to the lateral line at points.

FIGURE 10 Different arrangements of micro irrigation laterals for fruit orchards.

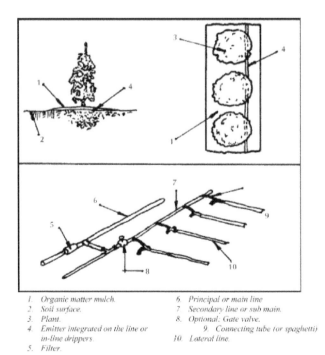

1.	Organic matter mulch.	6.	Principal or main line
2.	Soil surface.	7.	Secondary line or sub main.
3.	Plant.	8.	Optional: Gate valve.
4.	Emitter integrated on the line or in-line drippers.	9.	Connecting tube (or spaghetti)
5.	Filter.	10.	Lateral line.

FIGURE 11 Micro irrigation system in vegetable crops.

11.8.3   THE VEGETABLE CROPS

The micro irrigation system is more beneficial in vegetable crops that are planted in rows (Fig. 11). The emitters apply the desired amount of water throughout the row, and the row spaces are left dry. The drip lines are placed on the soil surface along the rows of plants. It can also be located few inches below the soil surface. Also plant row can have one drip line on either side or two drip lines on both sides.

Commonly, polyethylene tubes of 12 to 16 mm (approx. ½ to ⅔") of diameter are used for the lateral lines. These lateral lines provide water to the soil surface by means of the drippers. The polyethylene (PE) hoses are available in different sizes, lengths and strength. Some manufacturers produce weather resistant PE hoses. These are suitable for irrigation application. Generally, these lines of distribution are located perpendicular to the direction of the secondary main line.

11.9   SUBSURFACE DRIP IRRIGATION SYSTEM (SDI)

Subsurface micro irrigation (SDI) is a variation of traditional micro irrigation where the tubing and emitters are buried beneath the soil surface, rather than laid on the ground or suspended from wires. SDI is the placement of permanent drip tape (trickle) below the soil surface, usually at a depth of between 20 and 40 cm. The products being used today in subsurface micro irrigation come in three basic configurations: hard hose, drip tape and porous tubing (See figure 12). Lateral injector to install subsurface laterals in SDI is shown in Fig. 13. In Western Chestnut Grower Association Inc. 3(2000, 1): 6–7, Dr. D.F. Zoldoske published a simple publication titled: Subsurface Micro irrigation: The Future of irrigation is underground.

"ASAE S526.1 (1999a): Soil and Water Terminology" defines the SDI as "application of water below the soil surface through emitters, with discharge rates generally in the same range as micro irrigation." Earlier, "subirrigation" and "subsurface irrigation" sometimes referred to both SDI and subirrigation (water table management), and "drip/trickle irrigation" can include either surface or subsurface drip/trickle irrigation, or both. Other definitions of SDI require drip lateral placement below specified depths, such as normal tillage depths or a depth that would ensure use for several years. Generally, SDI has been used to describe drip/trickle application equipment installed below the soil surface only for the past 15–20 years.

C. R. Camp, F. R. Lamm, R. G. Evans and C. J. Phene presented a review paper on "Subsurface micro irrigation—past, present, and future" at the 4th Decennial National Irrigation Symposium on November 14–16, 2000 in Phoenix AZ. According to them, subsurface micro irrigation (SDI) has been a part of agricultural irrigation in the USA for about 40 years but interest has increased rapidly during the last 20 years. Early drip emitters and tubing were somewhat primitive in comparison to modern materials, which caused major problems, such as emitter plugging and poor distribution uniformity.

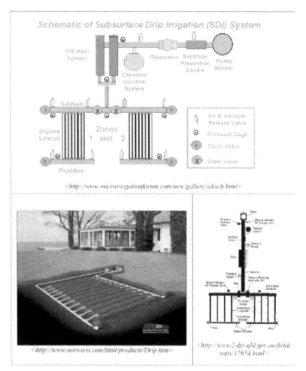

FIGURE 12   Subsurface drip irrigation system.

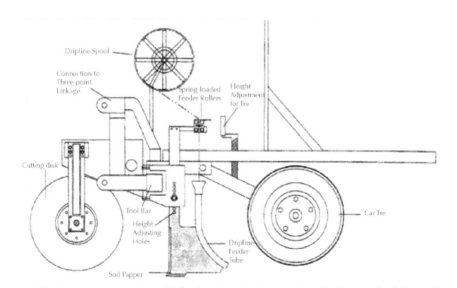

FIGURE 13   Subsurface micro irrigation: Lateral injector (Source: Netafim, Australia) http://www2.dpi.qld.gov.au/fieldcrops/17655.html.

As plastic materials, manufacturing processes, and emitter designs improved, SDI became more popular but emitter plugging caused by root intrusion remained a problem. Initially, SDI was used primarily for high-value crops such as fruits, vegetables, nuts, and sugarcane. As system reliability and longevity improved, SDI was used for lower-valued agronomic crops, primarily because the system could be used for multiple years, reducing the annual cost. Design guidelines have also evolved to include unique design elements for SDI, including air entry ports for vacuum relief and flushing manifolds. Specific installation equipment and guidelines have also been developed, resulting in more consistent system installation, improved performance, and longer life. Crop yields with SDI are equal to or better than yields with other irrigation methods, including surface drip systems. Water requirements are equal to or lower than surface drip. Fertilizer requirements are sometimes lower than for other irrigation methods. Interest in the use of wastewater with SDI has increased during the last decade. The future of SDI is very promizing, including its use in wastewater systems, and especially in areas where water conservation is important or water quality is poor. SDI is a very precise irrigation method, both in the delivery of water and nutrients to desired locations and the timing and frequency of applications for optimal plant growth.

Hard hose products generally have wall thicknesses of 0.75 to 1.25 mm (30 to 50 mils: One mil is a unit of linear measurement equivalent to 0.0254 mm, one thousandth of an inch, often used in measuring the diameter of wires), with nominal inside diameters of 13 to 16 mm. The emitter is either manufactured as an integral part of the tubing or is inserted later, and is typically placed at a repeated spacing interval of 0.5 to 1.5 m. The advantages to hard hose products are: strength and resistance to kinking, punctures and rodent damage. Also, pressure-compensating emitters may be incorporated into hard hose products. A disadvantage of hard hose products can be the initial cost.

Drip tapes, have wall thickness of 0.1 to 0.5 mm (4–20 mils). The thinner materials are more commonly used in single season or throw away applications, typical of strawberry production. The heavier wall materials ranging from 0.35 to 0.50 mm (15–20 mils) are more commonly found in multiyear applications or where the extra strength is required due to stones or other problems identified at the field site. Some tape products have emitter formed by the use of glue or other adhesive material used in the manufacturing process, where the actual water channel is made of the tubing material. Other tape or thin wall products have a premanufactured emitter glued or attached to the tape wall during the manufacturing process. Emission outlets are typically spaced along the tubing from 0.2 to 0.6 m, with wider outlet spacing having the larger flow path emitters.

Tape or thin products are initially the most cost attractive of the three basic configurations on a per unit basis. The disadvantage of tape products is that they are the most prone to mechanical or rodent damage due to reduced wall thickness, and thereby reduced structural strength of the tubing wall. However, the use of new generation polyethylene resins has enhanced mechanical properties of tapes. This has lead to new applications and longer field life.

Porous pipe products emit water all along the length of the tubing. There are literally thousands of orifices (pores) per meter where water weeps out of the tubing. This design has shown resistance to plugging by roots. However, its flow path is by far the smallest of the three configurations. This increases likelihood of plugging by fine particles.

Other design components of subsurface micro irrigation such as filtration and accessories are very similar to those found in conventional drip application. Proper filtration protects the emitter from contamination and clogging. Unfortunately soil particles and other contaminants can be drawn into the emitter from the outside. This generally occurs at system shutdown, when a vacuum is developed in the lines and draws inorganic particles back into the emitter. To keep this from happening, many subsurface irrigation designs incorporate vacuum relief-valves that break the vacuum and keep water and inorganic particles from sucking into the emitter.

Water use requirements of subsurface drip may not differ significantly from conventional uses. Evidence suggests that water applied on top of the soil through conventional drip systems has a cooling affect through evaporation on adjacent plant material. The net difference between surface and subsurface drip systems in total requirements is small.

11.9.1   ADVANTAGES AND LIMITATIONS OF SDI <HTTP://WWW2.DPI. QLD.GOV.AU/FIELDCROPS/17650.HTML>

Graham A. Harris of Australian Cotton Corporative Research Centre at Toowoomba –indicates following advantages and limitations:

1. **Irrigation**: There is a high degree of water application control with the potential for high uniformity of application. For new systems, distribution uniformity (DU) can be 93% or higher compared with 60% to 80% for sprinkler and 50% to 60% for conventional irrigation. The high frequency of irrigation with SDI allows maintenance of optimum soil moisture in the root zone. This is important where salty water is used and with shallow rooted crops, and is an advantage over surface and sprinkler irrigation where the fluctuation in soil water potential is greater, increasing the stress on the crop. Compared with sprinkler irrigation: It is possible to irrigate regardless of wind conditions. Lower pressures are generally needed and lower flow per unit area, requiring smaller diameter mains and laterals. For SDI, land leveling to enable drainage is required. SDI can be installed on a range of paddock sizes and shapes. SDI maintains soil surface structure more effectively than other irrigation types and makes it easier to allow for rainfall events or 'catch up,' provided there is sufficient system capacity. A well maintained SDI system requires less labor to operate than alternative systems.

2. **Agronomical practices**: The partial soil wetting provided by SDI has several benefits:
    — Improved efficiency of nutrient uptake at the fringes of the wetted soil volume
    — Less water lost from soil surface evaporation
    — Less weed germination and growth

— Unrestricted travel for field operations such as spraying and harvesting
— Improved access to rainfall infiltration in some row crop situations.
— Maintains dry crop foliage: The benefits include reduced incidence of foliar disease, reduced loss of applied pesticides, reduced evaporation losses direct from the crop canopy and less leaf burn where saline water is used for irrigation.
— Fertigation can be used with SDI. The possible benefits include savings in labor, more efficient use of nutrients and less risk of nutrient leaching, enhanced timing of nutrient applications to match crop requirements according to development stage and crop condition.

3. **Salinity problems**: Saline irrigation water applied through SDI will have less effect on crops than if applied through surface or sprinkler irrigation. This is due to no foliage absorption of salt. High frequency micro irrigation reduces the effect of increased salt concentration of the soil solution between irrigations compared to the lower-frequency of surface and sprinkler irrigation. There is a leaching of salts from the active root zone to the outer part of the wetted soil volume.

4. **Water use efficiency**: Water use savings range from 0 to 50% when compared with traditional irrigation systems. In situations where water savings are not made there is often a significant yield increase resulting in improved production per unit of irrigation water—improved water use efficiency (WUE). For spray-irrigated Lucerne in the Callide Valley of Australia, the long-term WUE is 1.37 t/ML. The WUE for SDI Lucerne grown by Trevor and Lyn Stringer at Biloela can be seen in table 1, along with a direct comparison between SDI and spray irrigated Lucerne on the Biloela Research Station. Here the improvement in WUE of SDI compared with spray irrigation has been 43% in 1996–97 and 95% in 1997–98. SDI and furrow irrigation comparisons for cotton are given in table 2.

1. **Yield improvement**: Improved yields have been obtained with SDI installations. For Lucerne the yield improvement was 13 and 34% for 1996–97 and 1997–98 respectively. For cotton the yield change ranges from 0 to 21%.

TABLE 1   Comparisons of Lucerne yield between SDI and spray irrigation. "Graham Harris" < http://www2.dpi.qld.gov.au/fieldcrops/17650.html>

Grower	Year	Subsurface drip irrigation				Sprinkler irrigation (Hand shift)			
		Yield	Irrigation	Rain	WUE	Yield	Irrigation	Rain	WUE
		t/ha	ML/ha	mm	t/ML	t/ha	ML/ha	mm	t/ML
Biloela RS	96–97	21.23	5.59	720	1.66	18.88	9.03	720	1.16
Biloela RS	97–98	16.98	6.48	400	1.62	12.68	11.30	400	0.83
Stringer	96–97	22.98	6.91	647	1.72				
Stringer	94–95	24.95	12.19	294	1.65				
Stringer	95–96	26.45	11.19	543	1.59				
Stringer	96–97	22.98	6.91	647	1.72				

TABLE 2 Comparisons of cotton yield between SDI and furrow irrigated cotton at Biloela. "Graham Harris" < http://www2.dpi.qld.gov.au/fieldcrops/17650.html>

		Subsurface drip irrigation				Furrow irrigation			
Grower	Year	Yield	Irrigation	Rain	WUE	Yield	Irrigation	Rain	WUE
		bales/ha	ML/ha	mm	bales/ ML	bales/ha	ML/ha	mm	bales/ ML
B	95–96	10.13	4.69	430	1.13	8.40	5.68	430	0.84
C	95–96	08.65	2.17	430*	1.34	8.65	5.43	4302	0.89
B	96–97	09.26	3.71	364	1.26	8.89	5.19	364	1.01
B	96–97	10.32	3.71	364	1.40	8.89	5.19	364	1.01
Dalby (1 m beds)**									
2000–01		10.0	4.50	396	1.18	7.98	5.30	396	0.86
2001–02		8.78	4.20	440	1.02	8.20	5.60	440	0.82
2002–03		10.1	2.90	608	1.12	9.80	5.60	608	0.84
Moree (2 m beds and 40 cm deep tape)**									
2000–01		7.36	3.73	150	1.41	7.80	6.00	150	1.04
2001–02		7.42	3.29	184	1.45	6.80	5.85	184	0.88
2002–03		8.37	2.60	62	2.60	10.18	7.27	62	1.29

* Accurate rainfall figures not available, therefore rainfall from Grower B was used to estimate WUE for Grower C.

** Source: T-Systems Australia Pvt Ltd).

11.9.2 LIMITATIONS OF SDI <HTTP://WWW2.DPI.QLD.GOV.AU/FIELDCROPS/17650.HTML>

Graham A. Harris of Australian Cotton Corporative Research Centre at Toowoomba indicates following limitations:

1. **Emitter clogging**: The high distribution uniformity inherent in a well-designed SDI system can be readily destroyed through emitter clogging. The orifice diameters of emitters are usually 0.5 to 1 mm^2 and are susceptible to clogging by root penetration, sand, rust, microorganisms, water impurities and chemical precipitates. The clogging is usually the result of insufficient water filtration, lateral flushing and/or chemical injection. Root penetration has generally only been observed in perennial crops like Lucerne where the soil has dried out around an emitter and the roots penetrate the emitter seeking water.

2. **System shutdown**: Water will flow to the lowest point in the field. If air is not allowed to enter the system by means of an air/vacuum release valve, a vacuum will be created and soil is sucked back into the emitter. On undulating fields, it is possible for this to occur throughout the field resulting in emitter

clogging. The soil dries out around these emitters and roots penetrate the emitter seeking out water and further blocking the emitter. This problem can be avoided with an adequate design (including the strategic location of air/vacuum release valves), field preparation (leveling to drain), and high frequency irrigation that produces a permanently saturated soil zone around the emitter.

3. **Chemical treatments**: There are no herbicides registered in Australia to prevent root intrusion. Root hairs that have penetrated an emitter can be burned out with hydrochloric acid. Sand is readily removed from water using centrifugal separators. Suspended organic matter and clay particles can be separated with gravel filters, disk and screen filters. If the water supply has more than 200 mg/L suspended solids then a settling reservoir is recommended before the water enters the filtration unit—above this level the filter system will be overloaded resulting in excessive back flushing. Bacterial slimes and algae growing on the interior walls of the laterals and emitters can combine with clay particles in the water to block the emitters. Bacterial precipitation of sulfur and iron is a further problem. These must be treated with chlorine. Chemical agents can precipitate to cause emitter clogging and these agents are:
    — Iron oxide due to an iron concentration of greater than 0.1 mg/L.
    — Iron bacteria further exacerbate the problem
    — Manganese oxide due to a manganese concentration of greater than 0.1 mg/L.
    — Iron sulfide due to iron and sulfide concentrations of greater than 0.1 mg/L.
    — Carbonates due to bicarbonate levels above 2 meq/L in water with a pH of 7.5 or higher, and Calcium at similar levels or a fertilizer containing calcium.

4. **Salt accumulation**: When saline water is used, salts accumulate at the wetting front. In SDI, this results in an accumulation of salt above and mid-way between the laterals. Where there is insufficient rainfall to move salt below the root zone, this accumulation will affect the growth of the current crop (even a relatively salt tolerant crop like Lucerne) or the establishment and growth of subsequent crops.

5. **Mechanical damage**: SDI laterals must be installed at the required depth below the ground surface along the full length of the field. To achieve this cross-rip, it is ensured that the turnout area at the end of each lateral run is the same level throughout the field; otherwise the lateral will be installed at different depths at the end of the field.

6. **Mice damage** can be significant on cracking soils used to grow grain crops. Do not let these soils crack open between crops by using some irrigation water to keep the soil about the lateral closed up.

7. **Insect damage** has been a significant problem on some SDI sites. There are a number of possible insects that chew through SDI tape. Damage has resulted from crickets, earwigs, false wireworm larvae and white-fringed weevil larvae. Care needs to taken in choosing and preparing the field for SDI, in its installation and maintenance in order to minimize the negative effect of insect

damage. There are no insecticides registered in Australia for the control of insect pests for SDI systems.

8. **Crop establishment**: Soil type and the depth of placement of the SDI laterals will determine the ability of the system to wet the soil surface to aid the crop establishment. In most situations, crops cannot be established using SDI alone. Where it has been installed on farms with existing sprinkler or furrow irrigation systems, these have been used to establish the crop. In some situations, the SDI system has been used to preirrigate the crop and fill the soil profile, with planting following rainfall. Using SDI to fill the profile prior to crop establishment runs the risk of deep drainage as water is being released below the soil surface—this can negate the possible benefits of the system in improving crop WUE.

9. **Soil structural effects**: In certain soils the use of high quality water through SDI has resulted in increasing clay content, exchangeable sodium percentage and calcium—magnesium ratio away from the emitter. The result is a decrease in the lateral spread of water from the emitter during irrigation and a smaller effective root volume.

11.10 SUMMARY

Micro irrigation is an efficient method of water application in agriculture. The micro irrigation system allows a constant application of water by drippers at specific locations on the lateral lines it allows favorable conditions for soil moisture in the root zone and optimal development of plant. Well-designed micro irrigation system can increase the crop yield due to: efficient use of water, improved microclimate around the root zone, pest control, and weed control, agronomic and economic benefits. The main components of the micro irrigation system are water source, pump and the energy unit, filtration system, the chemical injection system, the valves and controls, the water distribution system and the drippers. This complex system requires a more careful high technology management, the high initial investment and adequate maintenance. The subsurface micro irrigation is a placement of drip tape below the soil surface and is compared with other irrigation systems in Tables 3, 4 and 5.

TABLE 3 Summary of information reported for subsurface micro irrigation systems for field, turf, and forest crops (Camp 1998, pp. 1353–67).

Crop	Lateral Depth (m)	Lateral Spacing (m)	Lateral Type*	Sched. Timing	Sched. Amt.	Sched. MSWe	Sched. Other‡	Water Supply§	Other Irrig. Syst.‖	Fert. Mgt.	Irrig. Water Req.	WUE	Plant Meas.	Env. Effects
Alfalfa, 1990	0.35	1.5	FT	ND	ND	ND	PE	ND	S	ND	ND	ND	ND	ND
Alfalfa, 1987	0.12–0.37	1.5	ND	ND	ND	ND	ND	ND	ND	ND	ND	ND	ND	ND
Alfalfa, 1993	0.41	1.2	ND	ND	ND	ND	ND	ND	ND	ND	ND	ND	ND	ND
Bermuda grass, 1988	0.15	0.61, 0.91, 1.22	ND	ND	ND	NP, T	ND	SL	ND	ND	ND	ND	ND	ND
Bermuda grass, 1995	0.15–0.30	Var.	ND	ND	ND	ND	ND	WW	ND	ND	ND	ND	ND	X
Corn, 1992	0.35–0.41	0.91	FD	ND	ND	ND	ND	SL	S	ND	X	ND	ND	ND
Corn, 1989	0.4	1.5	FD	X	X	NP	ND	ND	ND	ND	ND	X	ND	X
Corn, 1989	0.3	0.76, 1.52	RT	X	ND	T	ND	ND	SD	X	ND	ND	ND	ND
Corn, 1996	ND	ND	ND	ND	ND	TDR	ND	ND	SD	ND	ND	ND	ND	ND
Corn, 1997a	0.40–0.45	1.5–3.1	FD	ND	X	T	ND	ND	ND	ND	ND	ND	ND	X
Corn, 1997b	0.40–0.45	1.5	FD	ND	X	T	ND	ND	ND	ND	ND	ND	ND	X
Corn, 1995	0.15, 0.30	1.52	ND	ND	ND	NP	M	ND	F	ND	ND	X	ND	ND
Corn, 1996	0.3	1.52	ND	X	X	NP	IRT	ND	SD	X	ND	X	ND	ND
Corn, 1997	0.3	1.5	FT	X	X	NP	ND	ND	SD	ND	ND	ND	ND	ND
Corn, 1987	0.12–0.37	1.5	ND	ND	ND	ND	ND	ND	ND	ND	ND	ND	ND	ND
Corn, 1991	0.40–0.45	1.5	FD	ND	X	NP	ND	ND	ND	X	ND	X	X	X
Corn, 1995a	0.40–0.45	1.5	FD	ND	X	T	ND	ND	ND	ND	X	X	ND	X
Corn, 1995b	0.45	0.76–3.05	FD	X	X	ND	ND	ND	ND	X	X	ND	ND	ND
Corn, 1997a	0.40–0.45	1.5–3.0	FD	ND	X	NP	ND	ND	ND	ND	ND	X	ND	ND

TABLE 3 (Continued)

Crop	Lateral Depth (m)	Lateral Spacing (m)	Scheduling/Delivery Type*	Timing	Amt.	MSWθ	Other‡	Water Supply§	Other Irrig. Syst.‖	Fert. Mgt.	Irrig. Water Req.	WUE	Plant Meas.	Env. Effects
Corn, 1997b	0.40–0.45	1.5	FD	ND	ND	NP	ND	ND	ND	X	ND	X	X	X
Corn, 1995	0.45	0.76–3.05	ND	ND	ND	ND	ND	ND	ND	ND	ND	ND	ND	ND
Corn, 1982	0.34–0.37	0.76	ND	ND	ND	ND	ND	ND	SD	X	ND	ND	ND	ND
Corn, 1981	0.36	0.9	FD	ND	ND	ND	ND	ND	ND	X	ND	ND	ND	ND
Corn, 1991	0.30	0.95,1.90 0.91,1.82,	ND	ND	ND	ND	ND	WW	SD	ND	ND	ND	ND	ND
Corn, 1993	0.38	2.74	FD	ND	X	ND	ND	ND	ND	ND	ND	ND	ND	ND
Cotton, 1995	0.45	1.7	Var.	ND	ND	ND	ND	ND	ND	ND	ND	ND	ND	ND
Cotton, 1991	0.30,0.45	ND	ND	ND	ND	ND	ND	ND	SD	X	ND	ND	ND	ND
Cotton, 1997a	0.30	1.0,2.0	RT	ND	ND	T	M	ND	ND	X	ND	ND	ND	ND
Cotton, 1994	0.38	ND	RT	ND	ND	ND	ND	ND	F	ND	X	ND	ND	ND
Cotton, 1989	0.20	1	ND	X	X	NP	CWSI	ND	ND	X	ND	X	X	ND
Cotton, 1995	0.20–0.35	1.0–3.1	ND	ND	ND	ND	ND	ND	F	ND	ND	ND	ND	ND
Cotton, 1996	ND	ND	ND	ND	ND	ND	ND	ND	ND	ND	ND	ND	ND	ND
Cotton, 1993, 1995	0.45	1.52	RT	X	X	NP	L	ND	ND	ND	X	ND	X	ND
Cotton, 1991	0.30	0.95,1.9	ND	ND	ND	ND	ND	WW	SD	ND	ND	ND	ND	ND
Cotton, 1992a	ND	ND	ND	ND	ND	NP	PE	ND	F	ND	X	ND	ND	ND
Cotton, 1985	0.40	ND	ND	X	X	NP	PE	ND	SD	ND	ND	ND	ND	X
Cotton, 1985a	0.25	1.0	FD	ND	ND	ND	ND	ND	ND	X	ND	ND	ND	ND
Cotton, 1985b	0.20–0.25	1.9	FD	ND	ND	ND	ND	ND	F	X	X	ND	ND	ND

TABLE 3 *(Continued)*

Crop	Lateral Depth (m)	Lateral Spacing (m)	Type*	Scheduling/Delivery Tim-ing	Amt.	MSWθ	Other‡	Water Sup-ply§	Other Irrig. Syst.‖	Fert. Mgt.	Irrig. Water Req.	WUE	Plant Meas.	Env. Effects
Cotton, 1966	0.40	1.42	R	ND	ND	ND	ND	ND	F	ND	ND	ND	ND	ND
Grain sorghum, 1973	0.20	ND	RI	ND	ND	ND	L	ND	SD	ND	X	X	X	ND
Landscape, 1995	0.15–0.30	Var.	ND	ND	ND	ND	ND	WW	ND	ND	ND	ND	ND	X
Pearl millet, 1995	0.25	0.4	FD	ND	X	ND	ND	ND	ND	ND	ND	X	ND	ND
Sugarcane, 1982	0.30	ND	ND	ND	X	G	PE	ND	F	ND	ND	ND	ND	ND
Sugarcane, 1990	0.10	2.7	FT	X	X	ND	ND	ND	F	ND	ND	ND	ND	ND
Trees, 1994	0.15	ND	ND	ND	X	ND	ND	WW	S, SD	ND	ND	ND	X	X
Turf, 1995	0.15–0.30	Var.	ND	ND	ND	ND	ND	WW	ND	ND	ND	ND	ND	X
Turf, 1992	0.10	Var.	ND	ND	ND	ND	ND	ND	ND	ND	ND	ND	ND	ND
Turf, 1995	0.20	0.46	RT	ND	ND	ND	ND	ND	S	ND	ND	ND	ND	ND
Wheat, 1991	0.30	0.95,1.90	ND	ND	ND	ND	ND	WW	SD	ND	ND	ND	ND	ND
Wheat, 1996	0.18–0.25	0.51	ND	ND	ND	ND	ND	ND	ND	ND	X	ND	X	ND
Wheat, 1985b	0.20–0.25	1.9	FD	ND	ND	ND	ND	ND	F	X	X	ND	ND	ND

* Type code definitions: FD = flexible wall, dual chamber; FT = flexible wall, turbulent flow; R = rigid; RI = rigid wall, insert orifice;
RT = rigid, turbulent flow; Var. = various; ND = no data available.
θ MSW = measured soil water content. Code definition: NP = neutron probe; T = tensiometer.
‡ Other code definitions: CWSI = crop water stress index; IRT = infrared thermometer; L = lysimeter;
M = crop growth model;
PE = pan evaporation.
§ Water supply code definitions: SL = saline/sodic; WW = waste water.
‖ Other irrigation system code definitions: F = furrow; S = sprinkler; SD = surface drip; WT = water table.

TABLE 4 Summary of information reported for subsurface micro irrigation systems on Horticultural (fruits, vegetable, and tree) crops (Camp 1998, pp. 1353–57).

Summ	Lateral Depth (m)	Spacing (m)	Type*	Scheduling/Delivery Tim-ing	Amt.	MSWe	Other‡	Water Supply§	Other Irrig. Syst.||	Fert. Mgt.	Irrig. Water Req.	WUE	Plant Meas.	Env. Effects
Apple, 1995	0.3–0.7	2–5	RS	ND	ND	ND	ND	ND	ND	ND	ND	ND	ND	ND
Asparagus, 1990	0.3	ND	FD	ND	ND	ND	ND	ND	SD, MS	ND	ND	ND	ND	ND
Banana, 1995	0.15–0.3	Var.	ND	ND	ND	ND	ND	WW	ND	ND	ND	ND	ND	X
Bell pepper, 1995	0.15	1	ND	ND	X	ND	PE	ND	ND	X	ND	ND	ND	ND
Bell pepper, 1990	0.1	0.76	ND	ND	ND	ND	ND	ND	ND	X	ND	ND	ND	ND
Broccoli, 1993	0.3	0.76–1.52	RT	X	X	T	ND	ND	SD	ND	ND	ND	ND	ND
Cabbage, 1985b	0.08–0.13	ND	ND	ND	ND	ND	ND	ND	ND	ND	ND	ND	ND	ND
Cabbage, 1989	0.15	ND	ND	ND	ND	ND	ND	ND	F, SD	X	X	ND	ND	ND
Cantaloupe, 1981	0.02–0.15	1.5	FS, FD	X	X	ND	ND	ND	SD, F	ND	ND	ND	ND	ND
Cantaloupe, 1989	0.45	ND	RT	ND	ND	NP	L	ND	SD	ND	ND	X	ND	ND
Carrot, 1981	0.02–0.15	1.5	FS, FD	X	X	ND	ND	ND	SD, F	ND	ND	ND	ND	ND
Carrot, 1996	0.18	1.0	FT	ND	X	NP	TDR	ND	ND	ND	X	ND	ND	ND
Cauliflower, 1996	0.18	1.0	FT	ND	X	NP	TDR	ND	ND	ND	X	ND	ND	ND
Cowpea, 1993	0.30	0.76–1.52	RT	X	ND	T	ND	ND	SD	ND	ND	ND	ND	ND

TABLE 4 (Continued)

Summ	Lateral Depth (m)	Spacing (m)	Type*	Scheduling/Delivery Timing	Amt.	MSWe	Other‡	Water Supply§	Other Ir-rig. Syst.||	Fert. Mgt.	Irrig. Water Req.	WUE	Plant Meas.	Env. Effects
Cucumber, 1996	0.25	0.3	ND	X	ND	ND	ND	ND	SD	ND	ND	ND	X	ND
Grapes, 1990	0.25,0.3	ND	ND	ND	ND	NP	ND	ND	SD	X	ND	ND	ND	ND
Grapes, 1997	0.45	3.67	FT, RT	ND	ND	T	ND	ND	SD	X	X	ND	X	ND
Green bean, 1993	0.3	0.76–1.52	RT	X	X	T	ND	ND	SD	ND	ND	ND	ND	ND
Hops, 1995	0.35–0.45	0.6–1.2	FW	X	ND	ND	ND	ND	ND	ND	ND	ND	ND	ND
Lettuce, 1985a	0.07	0.3	FS	ND	ND	ND	ND	ND	ND	X	ND	ND	ND	ND
Lettuce, 1985b	0.08–0.13	ND	ND	ND	ND	ND	ND	ND	ND	ND	ND	ND	ND	ND
Lettuce, 1996	0.18	1.0	FT	ND	X	NP	TDR	ND	ND	ND	X	ND	ND	ND
Lettuce, 1980	0.10	ND	FD	ND	ND	T	ND	ND	SD, F, S	ND	ND	X	ND	ND
Lettuce, 1995	ND	ND	ND	ND	ND	ND	ND	ND	F	ND	ND	ND	X	ND
Lettuce, 1996a	0.15	1.02	FD	ND	ND	T	ND	ND	ND	X	ND	ND	ND	ND
Lettuce, 1996b	0.15	1.02	FD	ND	ND	T	ND	ND	ND	X	ND	ND	X	X
Muskmelon, 1993	0.3	0.76–1.52	RT	X	X	T	ND	ND	SD	ND	ND	ND	ND	ND
Okra, 1994	0.15	1.0	CP	ND	X	NP	ND	ND	F	ND	X	X	ND	ND
Onion, 1981	0.02 – 0.156	1.5	FS, FD	X	X	ND	ND	ND	SD, F	ND	ND	ND	ND	ND
Onion, 1996	0.18	1.0	FT	ND	X	NP	TDR	ND	ND	ND	X	ND	ND	ND
Papaya, 1995	0.15–0.3	Var.	ND	ND	ND	ND	ND	WW	ND	ND	ND	ND	ND	X
Pea, 1991	0.3	0.95,1.9	ND	ND	ND	ND	ND	WW	SD	ND	ND	ND	ND	ND

TABLE 4 (Continued)

T A B L E 4 Summ	Lateral Depth (m)	Spacing (m)	Type*	Scheduling/Delivery Tim-ing	Amt.	MSWe	Other‡	Water Supply §	Other Ir-rig. Syst.‖	Fert. Mgt.	Irrig. Water Req.	WUE	Plant Meas.	Env. Effects
Pea, 1995	0.3,0.6	5	RS	ND	ND	ND	ND	SL	SD	ND	ND	ND	ND	ND
Potato, 1995	0.3–0.7	2–5	RS	ND	ND	ND	ND	ND	ND	ND	ND	ND	ND	ND
Potato, 1996	0.08–0.58	0.81	F	ND	ND	NP	PE	ND	S	ND	ND	ND	ND	ND
Potato, 1995	0.08	0.9	FD	ND	ND	ND	ND	ND	WL	X	X	ND	X	ND
Potato, 1976	0.05	1.0,1.3	FP	X	X	T	ND	ND	ND	X	ND	ND	X	ND
Potato, 1979	0.02	1.65	FP	X	X	MP	ND	ND	ND	X	ND	X	ND	ND
Potato, 1980	0.1	ND	FD	ND	ND	T	ND	ND	SD, F, S	ND	ND	X	ND	ND
Rape, 1994	0.15	1.0	CP	ND	X	NP	ND	ND	F	ND	ND	ND	ND	ND
Squash, 1993	0.3	0.76–1.52	RT	X	X	T	ND	ND	SD	ND	ND	ND	ND	ND
Squash, 1989	0.15	ND	ND	ND	ND	ND	ND	ND	E, SD	X	X	ND	ND	ND
Sweet corn, 1989	0.3	ND	RT	ND	ND	ND	ND	ND	SD	X	ND	ND	ND	ND
Sweet corn, 1991	0.3,0.45	ND	ND	ND	ND	ND	ND	ND	SD	X	ND	ND	X	X
Sweet corn, 1979	0.45	1.02	RI	X	X	NPT	ND	ND	F, S	X	ND	ND	X	ND
Sweet corn, 1976	0.05	1.0	FP	X	X	MP	ND	ND	E, S	X	ND	ND	X	ND
Sweet corn, 1979	0.05	1.0,1.65	FP	X	X	MP	PE	ND	ND	X	ND	ND	X	ND
Sweet corn, 1977	0.3	1.02	RI	X	ND	T, NPG	ND	ND	S, F	ND	X	ND	ND	ND

TABLE 4 (*Continued*)

Crop, year			Var.											
Tomato, 1995	0.45	1.7	ND	ND	ND	ND	ND	ND	ND	ND	ND	ND	ND	ND
Tomato, 1991	0.3,0.45	ND	ND	ND	ND	ND	ND	ND	SD	X	ND	ND	ND	ND
Tomato, 1994	0.15	1.0	CP	X	ND	NP	ND	ND	F	ND	X	X	ND	X
Tomato, 1989	0.15–0.2	2.0	FD	ND	ND	ND	PE	ND	F	ND	X	X	X	ND
Tomato, 1991	0.25–0.4	1.37	FD	ND	ND	T	ND	ND	WT	X	ND	ND	ND	ND
Tomato, 1993	0.3	1.5	FD	ND	ND	ND	ND	ND	SD	X	ND	ND	ND	ND
Tomato, 1985	0.45	1.63	ND	X	ND	ND	L	ND	S	X	X	ND	X	ND
Tomato, 1996	0.25	0.3	ND	X	ND	ND	ND	ND	SD	ND	ND	X	X	ND
Tomato, 1988	0.25	1.5	ND	ND	ND	ND	ND	ND	S, F	ND	ND	ND	ND	ND
Tomato, 1985	0.45	1.63	RT	X	ND	ND	ND	ND	SD	X	X	X	X	ND
Tomato, 1989	0.08	1.3	FD	ND	ND	NP	ND	ND	ND	ND	ND	X	ND	ND
Tomato, 1996	0.18	1.0	FT	ND	X	NP	TDR	ND	ND	X	ND	ND	ND	ND
Tomato, 1985	0.45	1.7	FT	X	ND	ND	ND	ND	SD	X	ND	X	ND	ND
Tomato, 1987	0.45	1.63	RT	X	ND	ND	L	ND	SD	X	ND	X	ND	ND
Tomato, 1989	0.45	ND	RT	ND	ND	NP	L	ND	SD	X	ND	X	ND	ND
Tomato, 1990, 1992b	0.45	1.63	RT	X	ND	ND	ND	ND	SD	X	ND	X	X	ND
Tomato, 1982	0.46	1.52	RT	ND	ND	MP	PE	ND	F	ND	ND	ND	ND	ND
Tomato, 1990	0.15–0.3	1.5	FT	X	X	ND	ND	ND	ND	ND	ND	ND	ND	ND
Tomato, 1985	0.12	1.32	FD	ND	X	ND	ND	SP	ND	X	X	X	ND	ND

TABLE 4 *(Continued)*

Watermelon, 1995a	0.2	2.0	FD	X	ND	T	ND	ND	ND	X	ND	ND	ND
Watermelon, 1995b	0.2	2.0	FD	X	ND	T	ND	ND	ND	X	ND	ND	X

* Type definition codes: CP = clay pipe; F = furrow; FD = flexible wall, dual chamber;

FP = flexible wall, porous; FS = flexible wall, single chamber; FT = flexible wall,

turbulent flow; FW = flexible wall, orifice in wall; RI = rigid wall, insert orifice;

RS = rigid wall, micro tubing; RT = rigid, turbulent flow; ND = no data available.

ɵ MSW = measured soil water content. Code definition: MP = metric potential sensor;

NP = neutron probe; T = tensiometer.

‡ Other code definitions: L = lysimeter; PE = pan evaporation; TDR = time domain reflectometry.

§ Water supply code definitions: SL = saline/sodic; SP = solar powered pump; WW = waste water.

‖ Other irrigation system code definitions: F = furrow; MS = micro sprinkler S = sprinkler;

SD = surface drip; WL = wheel line; WT = water table; Var. = various

TABLE 5 Design, evaluation, operational, economical and other guidelines (Camp 1998)

Design/Evaluation							Guidelines/Rec.				
	Plugging			Soil							
Uniformity	Deformation	Longevity	Methods	Wetting *	Design	Install	Operation (incl fert)	Filtration	Root Intrusion	Other Effects §	Economics
ND	ND	ND	ND	ND	ND	ND	ND	ND	ND	RL, MD	ND
ND	X	ND	ND	ND	ND	ND	X	ND	ND	ND	ND
ND	ND	ND	ND	ND	ND	ND	ND	ND	ND	PB	ND
ND	ND	ND	ND	M	ND	ND	ND	ND	ND	ND	ND
ND	ND	ND	ND	ND	ND	ND	ND	ND	ND	ND	X
ND	ND	ND	ND	WP	ND	ND	ND	ND	ND	ND	ND
X	X	ND	ND	ND	ND	ND	ND	ND	X	ID, MD	ND
X	X	X	X	ND	ND	X	ND	ND	ND	ND	ND
ND	X	ND	ND	ND	X	X	ND	ND	ND	SD, ID, MD, CT	ND
ND	ND	ND	ND	M	ND	ND	ND	ND	ND	ND	ND
ND	ND	ND	ND	ND	ND	ND	ND	ND	ND	ND	X
ND	ND	ND	ND	M, L	ND	ND	ND	ND	ND	ND	ND
ND	ND	ND	ND	M	X	ND	X	ND	ND	SWM	X
ND	ND	ND	ND	ND	X	ND	X	ND	ND	ND	ND
ND	ND	ND	ND	M	X	ND	ND	ND	ND	SWM	ND
ND	ND	ND	ND	ND	ND	ND	ND	ND	ND	WC	ND
ND	ND	ND	ND	X	ND	ND	ND	ND	ND	SC	ND
ND	ND	ND	ND	X	ND	ND	ND	ND	ND	SWM	ND
ND	ND	ND	ND	ND	X	ND	X	ND	ND	SS	ND

TABLE 5 *(Continued)*

Design/Evaluation					Guidelines/Rec.						
Uniformity	Plugging Deforma-tion	Longevity	Methods	Soil Wetting *	Design	Install	Operation (incl fert)	Filtration	Root Intrusion	Other Effects §	Economics
ND	ND	ND	ND	ND	ND	ND	ND	ND	ND	ND	X
X	X	ND	ND	ND	ND	ND	X	ND	ND	SC	ND
X	X	ND	ND	ND	X	ND	ND	ND	ND	ND	ND
ND	ND	ND	ND	ND	X	ND	ND	ND	ND	MD, RD	ND
ND	ND	ND	ND	ND	X	X	X	ND	ND	ND	ND
ND	ND	ND	ND	ND	X	X	X	X	ND	ND	ND
X	ND	ND	ND	ND	ND	ND	ND	ND	ND	ND	X
ND	ND	ND	ND	ND	X	X	X	X	ND	ND	ND
ND	ND	ND	ND	ND	X	X	ND	X	ND	ND	ND
ND	ND	ND	ND	ND	X	X	X	ND	ND	ND	ND
ND	X	ND	ND	ND	ND	ND	ND	ND	ND	FM	ND
ND	ND	X	ND	ND	ND	ND	ND	ND	ND	SC	ND
X	ND	ND	ND	ND	ND	ND	X	ND	ND	SC, OM, RG	ND
ND	ND	ND	ND	ND	X	X	ND	ND	ND	ND	ND
ND	ND	ND	X	ND	ND	ND	X	ND	ND	ND	ND
X	X	ND	ND	ND	ND	ND	X	ND	ND	SC	ND
ND	ND	ND	ND	ND	ND	ND	ND	ND	ND	ND	X
ND	ND	ND	ND	ND	ND	ND	X	ND	ND	CM, WC	ND

TABLE 5 *(Continued)*

	Design/Evaluation					Guidelines/Rec.						
		Plugging			Soil Wetting							
Uniformity	Deforma-tion	Longevity	Methods	*	Design	Install	Operation (incl fert)	Filtration	Root Intrusion	Other Effects §	Economics
X	ND	X	X	ND	ND	ND	ND	ND	ND	ND	ND
X	ND	ND	ND	ND	ND	X	ND	ND	ND	RG	ND
ND	ND	ND	ND	ND	X	X	X	ND	ND	SC, WU	ND
ND	ND	ND	ND	ND	X	ND	ND	ND	ND	RG	ND
ND	ND	ND	ND	ND	ND	ND	X	ND	ND	CM	ND
ND	ND	ND	ND	ND	ND	ND	X	ND	ND	FM, SC	ND
ND	ND	X	ND	ND	X	ND	X	X	ND	WU, WQ	ND
ND	X	X	ND	ND	X	ND	ND	ND	X	ND	ND
ND	ND	X	ND	X	ND	ND	ND	ND	ND	SWM, EX	ND
X	X	ND	ND	ND	ND	X	ND	ND	ND	SC	ND
ND	ND	ND	ND	ND	ND	X	ND	ND	ND	SE	ND
ND	ND	ND	ND	X	X	ND	X	X	X	SWM	ND
ND	ND	ND	ND	ND	X	ND	ND	ND	ND	EC	ND
ND	ND	ND	ND	M	ND	ND	ND	ND	ND	SWM	ND
ND	ND	ND	ND	M, L	X	ND	ND	ND	ND	ND	ND
ND	ND	ND	ND	ND	X	ND	ND	ND	ND	ND	ND

TABLE 5 *(Continued)*

Design/Evaluation				Guidelines/Rec.						
Plugging			Soil Wetting							
Uniformity	Deformation	Longevity	Methods *	Design	Install	Operation (incl fert)	Filtration	Root Intrusion	Other Effects §	Economics
ND	ND	ND	M	ND	ND	ND	ND	ND	ND	ND
ND	ND	ND	ND	X	ND	ND	ND	ND	ND	ND
ND	ND	ND	M	X	ND	ND	ND	ND	ND	ND
X	X	ND	ND	X	X	X	ND	ND	SWM	ND
ND	ND	ND	ND	ND	X	ND	ND	ND	PB, SWM	ND
ND	ND	ND	M	ND	ND	ND	ND	ND	ND	ND
ND	ND	ND	ND	X	X	X	ND	X	EC	ND
X	X	ND	ND	X	X	X	ND	ND	RG	ND

LEGEND for Table 5:

L = laboratory measurements;	M = model	WP = wetting pattern modification
CM = chemical management	CT = conservation tillage	EC = emitter comparison/evaluation
EX = emitter excavation effect	FM = fertilizer management	ID = insect damage
MD = mechanical damage	OM = organic matter	PB = plastic or foil barrier
RD rodent damage	RG = root growth	RL = row location relative to lateral location
SC = soil chemistry change, e.g. pH	SD = Soil compaction	SE = seeding emergence
SS = soil salinity	SWM = soil water movement	WC = weed control
WU = use of waste water	WQ = water quality guidelines	

11.11   BASIC ASPECTS IN MICRO IRRIGATION RESEARCH

Basic aspects for consideration in micro irrigation research were compiled by the research team of the South-eastern region of the United States Department of Agriculture [S-143], "**Trickle Irrigation in Humid regions**." The following aspects should be considered:

    I. **Atmospheric conditions**
        A.   Precipitation:
            1.   Total amount.
            2.   Effective amount:
                a.   Measurement of soil moisture.
                b.   Estimation of runoff.
            3.   Frequency.
            4.   Rate.
        B.   Evapotranspiration (ET):
            1.   Pan evaporation:
                a.   Class A Pan.
                b.   Pan wire mesh.
            2.   Models to estimate ET:
                a.   Penman (favorite).
                b.   Jensen-Haise.
                c.   Blaney-Criddle.
                d.   Hargreaves-Samani.
                e.   Others.
        C.   Climatic parameters:
            1.   Daily solar radiation.
            2.   Active photosynthetic radiation.
            3.   Maximum and minimum temperatures (daily, monthly or annual).
            4.   Average speed of the daily wind and direction.
            5.   Dew point temperature (a humidity index).
   II. **Soil**
        A.   Classification.
        B.   Physical properties:
            1.   Texture.
            2.   Compaction (apparent density/restrictive layer).
            3.   Hydraulic conductivity curve (soil moisture versus soil tension).
            4.   Infiltration rate.
            5.   Soil temperature at different depths.
            6.   Soil depth:
                a.   Soil profile.
                b.   Water aquifer depth.
                c.   Depth of restrictive (hard) layer.
            7.   Description of the general topography.
        C.   Chemical properties:
            1.   pH.
            2.   Electrical conductivity.

3. Percent of interchangeable sodium.
4. Fertility:
    a. N, P, K, Ca, Mg.
    b. Micro nutrients: Fe, Zn, Mn, Cu.
    c. Cation exchange capacity.
5. Applications of fertilizers:
    a. Amount.
    b. Program (frequency).
    c. Method of application.
    d. Source:
        − Type of fertilizer.
        − Chemical composition.
6. Organic matter (%).
7. Toxic ions:
    a. Boron.
    b. Chloride.
    c. Heavy metals.
    d. Other specific ions.

III. Characteristics of Irrigation System
A. Physical characteristics of the system:
    1. Description of the system:
        a. Type of dripper.
        b. Operating pressure.
        c. Flow rate.
        d. Pattern and number of drippers per plant.
        e. Filtration system.
        f. Method to measure flow.
        g. Description of pumping system.
        h. Water source.
    2. Uniformity of water application and method of evaluation.
    3. Relative volume of the irrigated root zone.
    4. Experimental design:
        a. Treatments.
        b. Replications.
        c. Control.
B. Quality of the water:
    1. Sediments (solid).
    2. Salinity.
    3. Specific ion concentration affecting the soil properties.
        a. Sodium and sodium absorption rate (SAR).
        b. Bicarbonates.
        c. Fe, S.
        d. Ca, Mg.
    4. Toxic ions of the plant:
        a. Boron.

           b.   Chloride.

           c.   Heavy metals.

     5.   pH.

     6.   Potential for organic growth:

           a.   Iron bacteria.

           b.   Sulfur bacteria.

           c.   Algae.

     7.   Clogging or obstruction problems:

           a.   Iron.

           b.   Carbonates.

           c.   Bicarbonates.

C.   Irrigation programming or scheduling:

     1.   Criteria:

           a.   Method/indicator:

           Soil moisture content.

           Drainage capacity of soil.

           Estimated ET.

           Estimated water reserve.

           Simulated model of water reserve.

           Time.

           Time with adjustments for rainfall.

           Plant indicators:

             –   Leaf temperature.

             –   Leaf water potential.

             –   Crop models.

           b.   Location of the measurements:

           Soil measurements in the root zone of the plant.

             i.   Measurements of leaf water potential.

           c.   Frequency of measurements:

            i. Continuous measurements.

            ii. Periodic measurements (specify).

     2.   Necessary measurements:

           a.   Amount of water applied by irrigation.

           b.   Irrigation rate.

           c.   Duration of irrigation.

           d.   Frequency of irrigation application.

IV. Crop Characteristics

A.   Type of crop (species, variety).

B.   Factors that affect the growth:

     1.   Plant density per hectare.

     2.   Plant and row spacing.

     3.   Method of sowing or planting.

    4.   Method of soil management.

    5.   Diseases and location (wet and dry zones).

    6.   Insect/ diseases and location.

    7.   Chemigation:

        a.   Chemical types.

        b.   Method of chemical application.

        a.   Frequency and rate of

    b.   application.

C.   Crop response:

    1.   Growth and development:

        a.   Growth curve.

        b.   Leaf size.

        c.   Plant mass.

        d.   Density.

        e.   Stage of phenological growth.

        f.   Fruit development stage.

        g.   Vegetative development stage.

        h.   Root density and distribution.

    2.   Physiological response:

        a.   Leaf water potential.

        b. Osmotic potential.

        c.   Stomach conductivity.

        d. Chemical analysis of leaf and fruit (N, P, K, Ca, Mg).

D.   Crop yield:

    1.   Amount:

        a.   Total.

        b.   Marketable or commercial (by classification).

    2.   Fruit quality:

        a.   Size.

        b.   Heat.

        c.   Soluble solids.

        d.   Firmness or density.

        e.   Other parameters for fruit quality:

            – Classification.

            – Acidity.

            – Brix°.

            – pH.

KEYWORDS

- Dew Hose
- drip irrigation
- emitters
- green house
- Integrated System Approach
- micro irrigation
- polyethylene
- pressure gage
- solenoid valves
- subsurface irrigation
- trickle irrigation

REFERENCES

1. Goldberg, D., B. Gornat and D. Rimon, 1976. *Micro irrigation: Principles, design and agricultural practices.* Kfar Shmaryahu, Israel: Micro Irrigation Scientific Publications.
2. Goyal, M.R., 2013. *Management of Drip/ Trickle or Micro Irrigation.* Chapters 1 to 16. Apple Academic Press Inc., NJ.
3. Israelson, D.W. and V.E. Hansen, 1965. *Principles and Applications of Irrigation* (Spanish). Editorial Reverte. S. A., Barcelona-España.
4. Jensen, M.E., 1980. *Design and operation of farm irrigation systems.* St. Joseph, MI: ASAE Monograph No. 3, American Society of Agricultural Engineers.
5. Ross, D.S., R.C. Funt, C.W. Reynolds, D.S. Coston, H.H. Fries and N.J. Smith, 1978. *Trickle irrigation and introduction.* The North-east Regional Agricultural Engineering Service (NRAES), NRAE-4, Cornell University, Ithaca, NY-USA.
6. Ross, D.S., R.A. Parsons, W.R. De Tar, H.H. Fries, D. D. Davis, C.W. Reynolds, H.E. Carpenter and E.D. Markwardt, 1980. *Trickle irrigation in the Eastern United States.* Cooperative Extension Service NRAES-4 Cornell University, Ithaca, NY-USA.

CHAPTER 12

CHEMIGATION*

MEGH R. GOYAL

CONTENTS

12.1   Introduction 238
12.2   Chemical Injection Methods 238
12.3   Fertigation 242
12.4   Pestigation 244
12.5   Chloration or Chlorination 244
12.6   Installation, Operation, and Maintenance 255
12.7   Safety Considerations 257
12.8   Examples 258
12.9   Trouble Shooting 259
12.10 Summary 259
Keywords 260
References 262

12.1 INTRODUCTION

Several decades ago, it began the idea of applying fertilizers through irrigation system. During the recent years, the application of chemicals through drip irrigation was adopted. Thus, new technical terms were defined such as: Chemigation, chloration, pestigation, insectigation, fungigation, nemagation, and herbigation [3, 11, 16]. The term chemigation was adopted to include the application of chemicals through the irrigation systems. The chemigation offers following economic advantages compared to other conventional methods [16]:

1. Provides uniformity in the application of chemicals allowing the distribution of these in small quantities during the growing season when and where these are needed.
2. Reduces soil compaction and the chemical damage to the crop.
3. Reduces the quantity of chemicals used in the crop and the health hazards/risks during the application.
4. Reduces the pollution of the environment.
5. Reduces the costs of manual labor, equipment, and energy.

To obtain a better efficiency and to diminish the clogging problems in the lateral lines, filters, drippers or any other part of the system, it is recommended to conduct a chemical and mechanical analysis of the water source [2]. If the chemical analysis shows a high concentration of salts, it can cause clogging problems. Therefore, it is recommended to avoid the use of chemicals that can cause precipitates. For a better efficiency in application, the chemicals should be distributed uniformly around the plants. The uniformity of chemical distribution depends on:

1. The efficiency of the mixture.
2. The uniformity of the water application.
3. The characteristics of the flow.
4. The elements or chemical compounds that are present in the soil.

12.2 CHEMICAL INJECTION METHODS [14, 18, 19]

12.2.1 SELECTION OF A PUMP FOR THE INJECTION OF A CHEMICAL

While selecting an injection pump, one must consider the parts that will be in direct contact with the chemical substances. These parts must be of stainless steel or of material that is corrosion resistant. The injection method consists of basic components such as: Pump, pressure regulator, the gate valve, the pressure gage, the connecting tubes, check valve, and the chemigation tank. The injection pump should be precise, easy to adjust for different degrees of injection, corrosion resistant, durable, rechargeable, with availability of spare parts. The recommended materials are: Stainless steel, resistant plastic, rubber, and aluminum. Bronze, iron, and copper are non acceptable materials. The injection pump must provide a pressure on the discharge line greater than the irrigation pump. Therefore, the operating pressure of the irrigation system exerts a minimum effect in the endurance of the injection pump.

12.2.2 INJECTION METHODS

The efficiency of chemigation depends on the capacity of the injection tank, solubility of a chemical in water, dilution ratio, precision of dilution, the potability, the costs and the capacity of the unit, the method of operation, the experience of the operator, and the needs of the operator. The chemical compounds for the chemigation process must be liquid emulsions or soluble powder. The injector must be appropriate to introduce the substances in the system. In addition, it must be of an adequate size to supply necessary amount of chemical at a desired flow rate. Normally, the injection is carried out with an auxiliary electric pump or by interconnecting the injection pump to the irrigation pump. The most common methods for chemigation are described in Figs. 1 and 2.

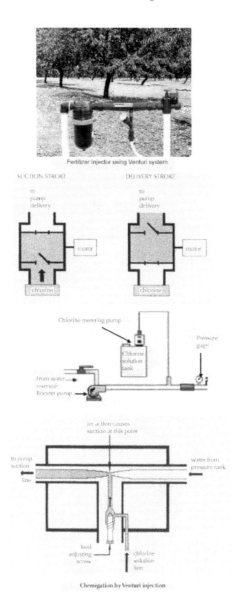

Fertilizer injector using Venturi system.

Chemigation by Venturi injection

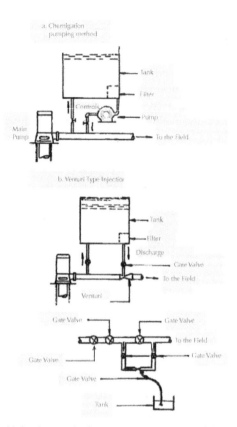

FIGURE 1A   Chemical injection methods: (a) Pressure pump and (b) Venturi type injector.

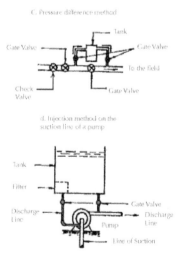

FIGURE 1B   Chemical injection methods: (c) Pressure difference method, and (d) Using the suction line of an irrigation pump.

FIGURE 2 Chemigation process. Top: The installation of a check valve (the nitrogen and pesticide tanks are behind the check valve). Bottom: Bypass Venturi system.

12.2.2.1   INJECTION BY A PRESSURE PUMP

A rotary, diaphragm or piston type pump can be used to inject the chemicals to flow from the chemical tank towards the irrigation line. The chemigation pump must develop a pressure greater than the operating pressure in the irrigation line. The internal parts of the pump must be corrosion resistant. This method is very precise and reliable for injecting the chemicals in the drip irrigation system.

12.2.2.2   INJECTION BY PRESSURE DIFFERENCE

This is one of the easiest methods to operate. In this method a low-pressure tank is used. This tank is connected with the discharge line at two points: one, which serves as a water entrance to the tank and the other is an exit of the mixture of chemicals. A pressure difference is created with a gate valve in the main line. The pressure difference is enough to cause a flow of water through the tank. The chemical mixture flows into the irrigation line. The concentration of chemical in the water is difficult to calculate and control. Therefore, it is recommended to install an accurate metering valve to maintain a precalibrated injection rate.

12.2.2.3   INJECTION BY VENTURI PRINCIPLE

A Venturi system can be used to inject chemicals into the irrigation line. There is a decrease in the pressure accompanied by an increase in the liquid velocity through a Venturi. The pressure difference is created across the Venturi and it is sufficient to cause a flow by suction of the chemical solutions from a tank.

12.2.2.4 INJECTION IN THE SUCTION LINE OF THE IRRIGATION PUMP

A hose or tube can be connected to the suction pipe of the irrigation pump to inject the chemicals. A second hose or tube is connected to the discharge line of a pump to supply water to the tank. This method should not be used with toxic chemical compounds because of possible contamination of the water source. A foot valve or safety valve at the end of a suction line can avoid the contamination.

12.3 FERTIGATION

All fertilizers for the chemigation purpose must be soluble (Table 1). The partially soluble chemical compounds can cause clogging and thus can create operational problems [3, 14].

TABLE 1   Solubility of commercial fertilizers.

Fertilizer	Solubility (grams/liter)
Ammonia	97
Ammonium Nitrate	1185
Ammonium Sulfate	700
Calcium Nitrate	2670
Calcium Sulfate	Insoluble
Di-Ammonium Phosphate	413
Di-Calcium Phosphate	Insoluble
Magnesium Sulfate	700
Manganese Sulfate	517
Mono-Ammonium Phosphate	225
Mono-Calcium Phosphate	Insoluble
Potassium Chloride	277
Potassium Nitrate	135
Potassium Sulfate	67
Urea	1190

12.3.1 NITROGEN

Nitrogen is an element that is most frequently applied in the drip irrigation system. The principal sources of nitrogen for chemigation are: anhydrous ammonia, liquid ammonia, ammonium sulfate, urea, ammonium nitrate, and calcium nitrate. The anhydrous ammonia or the liquid ammonia can increase the pH of the irrigation water, thus a possible precipitation of calcium and magnesium salts. If the irrigation water has high concentration of calcium and magnesium bicarbonates, then these can result in precipitation of chemical compounds. This enhances the clogging problems in the drippers, filters, and laterals. The ammonium salts are very soluble in water and cause less problems of clogging, with the exception of ammonium phosphate. The phosphate salts tend to precipitate in the form of calcium and magnesium phosphates, if there is an abundance of Ca and Mg in the irrigation water. Ammonium sulfate causes little obstruction problems or changes in pH water. The urea is very soluble and it does not react with the irrigation water to form ions, unless the water contains the enzyme urease. This enzyme can be present if the water has large amounts of algae or other biological agents.

The filtration system does not remove urease. This can cause hydrolysis of the urea. Because the concentrations of the enzyme are generally low compared to those in the soil, the urea will not hydrolyze to a significant degree in the irrigation water. The nitrate salts (e.g., calcium nitrate) are relatively soluble in water and do not cause large changes in the pH of the irrigation water. The nitrogen fertigation is more effective than the conventional methods of application, especially in sandy soils. In addition, the nitrogen fertigation is more efficient than the conventional methods in fine textured soils.

12.3.2 PHOSPHORUS

The phosphorus can be applied in the irrigation system, as an organic phosphate compound and glycerophosphates. The organic phosphates (orthophosphates) and urea phosphate are relatively soluble in water and can easily move in the soil. The organic phosphates do not precipitate, and the hydrolysis of an organic phosphate requires large lapse time.

The glycerophosphates react with calcium to form compounds of moderate solubility. The application of phosphorus can enhance obstructions in the drip irrigation system. When phosphoric fertilizers are applied in the irrigation water with high concentrations of Ca and Mg, then the insoluble phosphate compounds are formed that can obstruct the drippers and lateral lines. The phosphorus moves slowly in the soil and the root zone. In addition, the moist soil particles absorb phosphorus to form insoluble compounds. It is not recommended to fertigate phosphorus fertilizers during the growth period of a crop. Instead it should be applied before seeding, during seeding, and during fruit formation. The plant uses phosphorus early in its growth. In the drip irrigated crops, the fertigation of phosphorus can be combined with traditional methods of application. The drip irrigation system is efficient in the application of soluble phosphorus compounds, because the water is applied in the root zone; that facilitates the availability of phosphorus.

12.3.3 POTASSIUM

The potassium can be fertigated in the form of potassium sulfate, potassium chloride, and potassium nitrate. Generally, potassium salts have good solubility in water and cause little problems of precipitation.

12.3.4 MICRO NUTRIENTS

The micronutrients are supplied in the form of chelates. Thus, its solubility in the water is increased and these do not cause any problems of obstruction and precipitation. If the micronutrients are not applied as recommended, then iron, zinc, copper, and magnesium can react with the salts in the soil causing precipitation. This enhances the clogging of the emitters. The chelates should be dissolved before fertigation. The chemigation of micronutrients benefits the plant to accomplish a good development and growth. In addition, the operational cost is lower compared to the foliage application.

12.4   PESTIGATION [2, 11]

Although, sufficient information is available on the application of pesticides through the drip irrigation system in different regions of the world, yet the data does not necessarily adapt to the weather and soil conditions of all regions of the world.

12.4.1   INSECTIGATION

When dispersed emulsions or formulations are used in water, these comprise of the liquid phase and therefore can be distributed uniformly. In order to control foliage insects, the insecticides applied through the drip irrigation system must be systematic and of high solubility.

12.4.2   FUNGIGATION

The fungi dragged by the wind are difficult to control. These fungi produce numerous spores that often are located on the leaves, which are not easy to control. By means of injection of soluble and systematic fungicides through the drip irrigation, the effectiveness of certain fungicides can be increased.

12.4.3   NEMAGATION

The application of fumigants and nematicides through the drip irrigation system is convenient and safe. Using this method, the soil can be fumigated in an efficient way to control nematodes and other detrimental organisms. The success of the operation will depend on many conditions such as: Temperature, soil moisture, soil aeration, content of organic matter, and uniformity of irrigation.

12.4.4   HERBIGATION

In places where the rainfall is limited, the application of herbicides through drip irrigation serves to activate the applied herbicides. This action eliminates the need for mechanical incorporation and reduces the cost of weeding operation. As the drip irrigation system directly takes the water with herbicides to the place where the weeds are to be controlled, a uniform distribution is obtained. There is a significant reduction of losses of nonavailable chemicals as a result of the inactivation by rubbish or organic matter. Thus, the irrigation efficiency can be maximized and the efficiency of the herbicide application can be increased.

12.5   CHLORATION OR CHLORINATION

The obstruction of the filters, the distribution lines, or the emitters is a main problem associated with the operation and management of a drip irrigation system [4, 12, 15]. These obstructions are caused by physical agents (solid particles in suspension), chemicals agents (precipitation of insoluble compounds), or biological agents (macro and microorganisms). The preventive maintenance is the best solution to reduce or to eliminate the obstructions in the emitters or components of the system [1]. The chlorination is an addition of chlorine to water [15]. The chlorine, when it dissolves, acts as an oxidation agent and attacks the microorganisms, such as the algae, fungi, and bacteria. This procedure has been used for many decades to purify the drinking water [5]. Chloration is an injection of chlorine compounds through the irrigation system. The chloration solves effectively and economically the problem of obstruction of the emitters or drippers due to biological agents. **Caution: Do not use any chemical agent through the drip system without consulting a specialist** [12, 13, 15, 17].

The chlorine is the cheapest and effective treatment for the control of bacteria, algae, and the slime in the irrigation water. The chlorine can be introduced at low concentrations (1 ppm) at necessary intervals, or at high concentrations (10–20 ppm) for few minutes. The chlorine can be injected in form of chlorine powder (solid) and chlorine gas. The gas treatment is expensive and dangerous for the operator. The calcium hypochlorite can also be used, but the calcium tends to precipitate. The chlorine also acts as a biocide to iron and sulfur bacteria.

12.5.1   QUALITY OF WATER

12.5.1.1   WATER SOURCE

It is necessary to conduct the physical and chemical analysis of the water before designing a drip irrigation system and choosing an appropriate filtration system [17]. For the chemical analysis, it is important to take a representative sample of the water. If the water source is subsurface (e.g., deep well), the sample must be taken an hour and a half after the pump begins to work. When the water source is from a lake, river, pool, or open channel, the samples must be taken at the surface, at the center, and at the bottom of a water source.

It is important to analyze the sample for suspended solids, dissolved solids and the acidity (pH), macro organisms, and microorganisms [10]. The acidity of the water must be known, since it is a factor that affects the "chemigation directly" and therefore the chloration. For example, the chloration for the control of bacteria is ineffective for a pH > 7.5. Therefore, it is necessary to add acid to lower the pH of irrigation water and to optimize the biocide action of the chlorine compound. If a chemical analysis of the water is known, we can predict the obstruction problems and take suitable measures. In addition, a program of adequate service and maintenance can be developed. The physical, chemical, and biological agents are classified in Table 2. The factors are classified in order of the risk: from low to severe. When the water contains amount of solids, salts, and bacteria within the acceptable limits, then the risk of clogging is reduced.

TABLE 2   Water quality: Criteria that indicates risk of obstruction of the emitters.

Type of Problem	Risk of Obstruction		
	Low	Moderate	Severe
Biological Agents			
Bacteria population**	<10,000	10,000–50,000	>50,000
Physical Agents			
Suspended solids*	<50	50–100	>100
Chemical Agents			
Acidity (pH)	<7.0	7.0–8.0	>8.0
Dissolved solids*	<0.2	0.2–1.5	>2000
Iron*	<0.1	0.1–1.5	>1.5
Hydrogen sulfide	<500	500–2000	>2.0
Manganese*	<0.2	0.2–2.0	>1.5
** Maximum concentration of the representative sample of water. Given in ppm (mg/L). * Maximum number of bacteria per milliliter. Obtained from field samples and laboratory analysis.			

Also the particles of **organic matter** can combine with bacteria and produce a type of obstruction that cannot be controlled with filtration system. The fine particles of organic matter are deposited within the emitters and are cemented with bacteria such as: **Pseudomonas** and **Enterobacter**. This combined mass causes clogging of the emitters. This problem can be controlled with super chloration at the rate of 1000 ppm (mg/L). However, the chloration at these high rates can cause toxicity of a crop. The obstruction caused by the **biological agents** constitutes a serious problem in the drip irrigation system that contains organic sediments with iron or hydrogen sulfide. Generally, the obstruction is not a serious problem if the water does not have organic carbon, which is a power source for the bacteria (Promotes the bacterial growth). There are several organisms that increase the probability of the obstructions when there are ions of iron (Fe^{++}) or sulfur (S–).

Algae in surface water can add carbon to the system. The slime can grow on the inner surface of the pipes. The combination of fertilizer and the heating of the polyethylene pipes (black) due to sunlight can promote the formation and development of these microorganisms. Many of the water sources contain carbonates and bicarbonates that serve like an inorganic power source to promote slime growth; also autotrophic bacteria (that synthesize their own food) are developed. The algae and the fungi are developed in the surface waters. Besides obstructing the emitters, the filamentous algae form a gelatinous substance in the pipes and emitters, which serve as a base for the development of slime. Another type of obstruction can also happen when the filamentous bacteria precipitate the iron into the insoluble iron compounds (Fe^{+++}).

12.5.1.2   GROWTH OF SLIME IN THE DRIP IRRIGATION SYSTEM

The bacteria can grow within the system in absence of light and produce a mass of the slime or cause the precipitation of the iron or sulfur dissolved in water. The slime can act like an adhesive substance that agglutinates fine clay particles sufficiently large enough to cause clogging (4, 13, 15).

12.5.2 GROWTH OF ALGAE IN THE WATER SOURCE OR IN THE IRRIGATION SYSTEM

One of the most frequent problems is the growth of algae and other aquatic plants in the surface water that can be used for drip irrigation. The algae grow well in the surface water. The problem becomes serious if the water source contains nitrogen, phosphorus, or both. In many cases, the algae can cause obstructions in the filtration system. When the screen filters are used, the algae can be entangled in the sieves (screen) of the filter. In high concentration, these aquatic microorganisms can create problems in the sand filters. This requires a frequent flushing and cleaning of the filters.

12.5.3 TYPES OF ALGAE

The main groups of organisms in surface water are classified like protists, plants, and animals. The protists include bacteria, fungi, protozoa, and algae. Algae are unicellular organisms or multicellular autotrophic and photosynthetic, and require organic compounds to reproduce. The major nutriments are carbon dioxide, nitrogen, and phosphorus. The minor elements like iron, copper, and manganese are also important for the development of these organisms.

It is important to note that algae problems usually occur due to ponds being neglected. Often, people think you can simply "dig a hole" and then let the pond take care of itself. Unfortunately, this is not the case. Healthy ponds require proper aeration, bacterial treatments, and adequate pondweed management. The use of copper sulfate has NOT been recommended for algae control as research and field usages have shown a high potential for detrimental environmental effects. In certain waters, copper sulfate is quite toxic to fish and other organisms. Overuse of this product is common due to its short-term effectiveness. This can result in copper build-up in the sediments leading to a sterile bottom, killing important beneficial bacteria, microorganisms, snails, and other beneficial "creatures." Many large pond or lake owners are concerned about "**toxic algae**." Death and sickness to pets, livestock, wildlife, and even man have been attributed to the presence of certain algae, mostly blue–green forming species, in water supplies. Lethal substances produced by these algae are retained within the cells and released after death or are secreted from living cells. Many unattended farm ponds and other waters contain some of these toxic forms, posing a threat to human health and the environment. Medical case histories, biologist reports, and laboratory tests show some of the possible effects of toxic algae. A list complied by the U.S. Department of the Interior Federal Water Pollution Control Administration summarizes medical case histories of algal poisonings for a 120 year period. Exposure to and ingestion of algae caused a variety of "discomforts" including: skin rashes, headaches, nausea, vomiting, diarrhea, fever, muscular pains, and eye, nose, and throat irritation. California State Water Resources Control Board states in the *Water Quality Criteria Handbook* (Second Edition): "There have been reports of rapid deaths of a great variety of animals after drinking water containing high concentrations of blue–green algae such as Microcystis, Aphanizomenon, Nostoc rivulare, Nodularia, Gleotrichia, Gomphosphaeria, and Anabaena. Fatal poisonings have occurred among cattle, pigs, sheep, dogs, horses, turkeys, ducks, geese, and chickens. It is believed that such algae may be toxic to all warm-blooded animals." To find out how one can avoid algae problems in large bodies of water and create a healthier pond, please read an article from: http://www.pondsolutions.com/blue-green-algae.htm

Algae are primitive plants closely related to fungi (Fig. 3). These exhibit no true leaves, stems, or root systems and reproduce by means of spores, cell division, or fragmentation. These "live" from excess nutrients in the water and sunlight for growth. Over 17,400 species of algae have been identified and thousands more probably exist. The simplest algae are single cells (e.g., the diatoms); the more complex forms consist of many cells grouped in a spherical colony (e.g., *Volvox*), in a ribbon like filament (e.g., *Spirogyra*), or in a branching thallus form (e.g., *Fucus*). The cells of the colonies are generally similar, but some are differentiated for reproduction and for other functions. Kelps, the largest algae, may attain a length of more than 200 ft. (61 m). *Euglena* and similar genera are free-swimming one-celled forms that contain chlorophyll but that are also able, under certain conditions, to ingest food in an animal like manner.

Lyngbya colonies (blue–green algae)   Planktonic algae

Filamentous algae   Algae diatoms

Planktonic "pea green soup" algae

Duckweed. Super tiny plant with little leaflets and roots. Can literally cover an entire pond. Commonly transferred from pond to pond by waterfowl.

FIGURE 3   Types of algae.

The green algae include most of the freshwater forms. The pond scum, a green slime found in stagnant water, is a green alga, as is the green film found on the bark of trees. The more complex brown algae and red algae are chiefly saltwater forms; the green color of the chlorophyll is masked by the presence of other pigments. Blue–green algae have been grouped with other prokaryotes in the kingdom *Monera* and renamed *cyanobacteria*. There are four classes of algae for the irrigation system:

12.5.3.1 GREEN ALGAE (CHLOROPHYTA, KLŌROF'UTU)

These **are commonly known as photosynthetic organisms**: Phylum (division) of the kingdom *Protista*. The organisms are largely aquatic or marine. The various species can be unicellular, multicellular, coenocytic (having more than one nucleus in a cell), or colonial. Those that are motile have two apical or subapical flagella. A few types are terrestrial, occurring on moist soil, on the trunks of trees, on moist rocks, and even in snow banks. Cells of the Chlorophyta contain organelles called chloroplasts in which photosynthesis occurs; the photosynthetic pigments chlorophyll *a* and chlorophyll *b,* and various carotenoids, are the same as those found in plants and are found in similar proportions. Chlorophytes store their food in the form of starch in plastids and, in many, the cell walls consist of cellulose. Unlike in plants, there is no differentiation into specialized tissues among members of the division, even though the body, or thallus, may consist of several different kinds of cells. According to *"The Columbia Electronic Encyclopedia,* 6th edition 2006, Columbia University Press." there are four evolutionary lineages of green algae.

1. **Euglenophyta** (yOO"gl*u*nof'*utu*), small phylum (division) of the kingdom *Protista*, consisting of mostly unicellular aquatic algae. Most live in freshwater; many have flagella and are motile. The outer part of the cell consists of a firm but flexible layer called a pellicle, or periplast, which cannot properly be considered a cell wall. Some euglenoids contain chloroplasts with the photosynthetic pigments chlorophyll *a* and *b,* as in the phylum Chlorophyta; others are heterotrophic and can ingest or absorb their food. Food is stored as a polysaccharide, paramylon. Reproduction occurs by longitudinal cell division. The most characteristic genus is *Euglena,* common in ponds and pools, especially when the water has been polluted by runoff from fields or lawns on which fertilizers have been used. There are approximately 1000species of euglenoids.

2. **Dinoflagellata** (dī"nōflăj"*u*lät'*u*, –lā't*u*), phylum (division) of unicellular, mostly marine algae, called dinoflagellates. In some classification systems, this division is called Pyrrhophyta. There are approximately 2000species of dinoflagellates. Most have two flagella that lie perpendicular to one another and cause them to spin as they move through the water. Most have walls, or thecae, that are rigid and armor-like and sometimes take on fantastic shapes. The plates that make up these walls are actually located inside the plasma membrane rather than outside, as cell walls are. Some species are heterotrophic, but many are photosynthetic organisms containing chlorophyll *a* and chlorophyll *c.* The green of these chlorophylls may be masked by various other pigments. Still other species are symbionts, living inside such organisms as jellyfish and corals. Food reserves are largely starch. Reproduction for most dinoflagellates is asexual, through simple division of cells following mitosis. They are unusual in that in each cell, the chromosomes remain compact between divisions, instead of stretching out into slender threads, as in most other organisms. The chromosomes are constricted at regular intervals and do not have centromeres, or fiber-attachment centers. There is no spindle, yet the very numerous chromosomes are divided equally at the time of mitosis.

3. **Chrysophyta** (krusof'*utu*), phylum (division) of unicellular marine or fresh-water organisms of the kingdom *Protista* consisting of the diatoms (class Bacillariophyceae), the golden, or golden–brown, algae (class Chrysophyceae), and the yellow–green algae (class Xanthophyceae). In many chrysophytes the cell walls are composed of cellulose with large quantities of silica. Some have one or two flagella, which can be similar or dissimilar. A few species are ameboid forms with no cell walls. The food storage products of chrysophytes are oils or the polysaccharide laminarin. Formerly classified as plants, the chrysophytes contain the photosynthetic pigments chlorophyll *a* and *c*; all but the yellow–green algae also contain the carotenoid pigment fucoxanthin. Under some circumstances, diatoms will reproduce sexually, but the usual form of reproduction is cell division. The diatoms and golden–brown algae are of great importance as components of the plankton and nanoplankton that form the foundation of the marine food chain.

4. **Phaeophyta** (fēof'*utu*), phylum (division) of the kingdom *Protista* consisting of those organisms commonly called brown algae. Many of the world's familiar seaweeds are members of Phaeophyta. There are approximately 1500 species. Like the chrysophytes, brown algae derive their color from the presence, in the cell chloroplasts, of several brownish carotenoid pigments, including fucoxanthin, in addition to the photosynthetic pigments chlorophyll *a* and *c*. With only a few exceptions, brown algae are marine, growing in the colder oceans of the world, many in the tidal zone, where they are subjected to great stress from wave action; others grow in deep water. Among the brown algae are the largest of all algae, the giant kelps, which may reach a length of over 100 ft. (30 m). *Fucus* (rockweed), *Sargassum* (gulfweed), and the simple filamentous *Ectocarpus* are other examples of brown algae. The cell wall of the brown algae consists of a cellulose differing chemically from that of plants. The outside is covered with a series of gelatinous pectic compounds, generically called algin; this substance, for which the large brown algae, or kelps, of the Pacific coast are harvested commercially, is used industrially as a stabilizer in emulsions and for other purposes. The normal food reserve of the brown algal cell is a soluble polysaccharide called laminarin; mannitol and oil also occur as storage products. The body, or thallus, of the larger brown algae may contain tissues differentiated for different functions, with stemlike, rootlike, and leaflike organs, the most complex structures of all algae.

12.5.3.2 MOTILE

These form colonies when mature. It has flashy green color and it is unicellular and flagellated. The flagella are those that make motile in the water. Filamentous algae, or commonly referred to as "pond scum" or "pond moss" forms greenish mats upon the surface of water. This alga usually begins its growth along the edges or bottom of the pond and "mushrooms" to the surface. Individual filaments are a series of cells joined end to end, which give the thread-like appearance. They also form fur-like growths on bottom logs, rocks, and even on the backs of turtles. Some forms of filamentous algae are commonly referred to as "frog spittle" or "water net."

12.5.3.3   YEL'LOW-GREEN AL'GAE [SYNONYMS FOR YELLOW–GREEN ALGAE]

The diatoms are most important in this group. It is a group of common unicellular and colonial algae of the phylum Chrysophyta, having mostly yellow and green pigments, occurring in soil and on moist rocks and vegetation and also as a slime or scum on ponds and stagnant waters. These are found in fresh water and salt water.

12.5.3.4   BLUE–GREEN ALGAE

Popular name for those microorganisms that are now more properly called cyano-bacteria [sī″unōbăktir′ēu, sī-ăn″ō–] or photosynthetic bacteria that contain chlorophyll. Cyanobacteria are familiar to many as a component of pond scum. Despite their name, different species can be red, brown, or yellow; blooms (dense masses on the surface of a body of water) of a red species are said to have given the Red Sea its name. These are unicellular organisms with flagella. These can form big masses in the surface water. In addition, these can use nitrogen from the atmosphere. Nitrogen-fixing cyanobacteria need only nitrogen and carbon dioxide to live.

12.5.4   PRINCIPLE OF CHLORATION [4–10]

The principle of chloration for treating the water by drip irrigation is similar to the one that is used to purify the water for drinking purpose. Table 3 includes basic reactions of chlorine and its salts. When the chlorine in gaseous state (Cl_2) dissolves in water, the chlorine molecule is combined with water in a reaction called hydrolysis. The hydrolysis produces hypochloric acid (HOCL; Reaction (1)). Following this reaction, the hypochloric acid enters an ionization reaction as shown in Reaction (2). The hypochloric acid (HOCl) and the hypochlorite (OCL) are known as free available compounds and are responsible for controlling the microorganisms in the water. The equilibrium of these depends on the temperature and pH of the irrigation water. When the water is acidic (low pH) the equilibrium moves to the left, resulting in an increase of HOCl. When the water is alkaline (high pH), the chlorine increases in the form of OCl–. The efficiency of HOCl is 40–80 times greater than OCl–. Therefore, the efficiency of the chloration depends greatly on the acidity (pH) of the water source. Reaction (1) produces hydrogen ions (H+) that can increase the acidity. The basicity depends on the amount of added chlorine and the buffer capacity of the water. The sodium hypochlorite [NaOCl] and the calcium hypochlorite [$Ca(OCl)_2$] hydrolyze and produce OH– ions that tend to lower the acidity of the water (Reactions (3) and (4)). If the pH is extremely low, the gaseous chlorine (Cl_2) predominates and can be danger-ous. Therefore, it is recommended to store the sources OCL compounds separate from solids. Also the available free chlorine reacts with oxidizing compounds (like iron, manganese, and hydrogen sulfide) and produces insoluble compounds, which must be removed from the system to avoid clogging.

TABLE 3   Basic forms of chlorine reactions and its salts.

Reactions	Reaction Number
$Cl_2 + H_2O = H^+ + Cl- + Col$	(1)
$HOCl = H^+ + OCl-$	(2)
$NaOCl + H_2O = Na^+ + OH- + Col$	(3)
$Ca(OCl)_2 + 2H_2O = Ca^{2+} + 2OH- + 2HOCl$	(4)
$HOCl + NH_3 = NH_2Cl + H_2O$	(5)
$HOCl + NH_2Cl = NHCl_2 + H_2O$	(6)
$HOCl + NHCl_2 = NCl_3 + H_2O$	(7)
$HOCl + 2Fe^{2+} + H^+ = 2Fe^{3+} + Cl- + H_2O$	(8)
Ferrous to ferric	
$Cl_2 + 2Fe(HCO_3)_2 + Ca(HCO_3)_2 = 2Fe(OH)_3 \text{ (insoluble)} + CaCl_2 + 6CO_2$	(9)
$HOCl + H_2S = S-\text{(insoluble)} + H_2O + H^+ + Cl-$	(10)
$Cl_2 + H_2S = S-\text{(insoluble)} + 2H^+ + 2Cl-$	(11)

The chlorine has two important chemical properties: At a low concentration (1–5 mg/L), it acts as a bactericidal. At a high concentration (100–1000 mg/L), it acts as an oxidizing agent, which can disintegrate particles of organic matter. It is necessary to watch, because the chlorine at these high levels can affect the growth of some plants.

12.5.5   SOURCES OF COMMERCIAL CHLORINE
The most common chorine sources [7] used in a drip irrigation system are sodium hypochlorite, calcium hypochlorite, and gaseous chlorine.

12.5.5.1   SODIUM HYPOCHLORITE, NAOCL
Sodium hypochlorite is liquid and is commonly used as whitener for clothes. It can be easily decomposed at high concentrations, in the presence of light and heat. It must be stored at room temperature in packages resistant to corrosion. This compound is easy to handle. The amounts can be measured precisely and causes few problems.

12.5.5.2   CALCIUM HYPOCHLORITE, CA(OCL)₂
Calcium hypochlorite is available commercially as dust, granulated, or in pellets. It is well soluble in water and is quite stable under appropriate storage conditions. It must be stored at room temperature in a dry place and in packages resistant to corrosion. When this compound is mixed in a concentrated solution, it forms a suspension that contains calcium oxalate, calcium carbonate, and calcium hydroxide. These compounds can obstruct the drip irrigation system.

12.5.5.3   GASEOUS CHLORINE, CL₂ GAS
It is available in liquid form at high pressure in cylinders from 45 kg to 1000 kg. The Cl_2 is very poisonous and corrosive. It must be stored in a well-ventilated place. Table 4 shows equivalent amounts of chlorine for different commercial sources and the required amount to treat 1233 m³ (1 acre-foot) of water to obtain one ppm of chlorine. The NaOCl is safer than Cl₂and avoids calcium precipitates in the emitters, which can

happen when using Ca(OCl)$_2$. It is more economical to use than the Cl$_2$ in large systems. In small systems, it is appropriate to use sodium or calcium hypochlorite. The use of Cl$_2$ is preferred in situations where the addition of sodium and calcium can be detrimental to the crop. It is necessary to observe that Cl$_2$ is dangerous under certain conditions. Thus, the instructions on the label must be followed. It is recommended to install a security valve (one way or check valve) in the tank that is used for injecting the chlorine.

TABLE 4 Equivalent amounts of the commercial sources of chlorine and required amounts to treat one acre-foot of water to obtain one ppm of chlorine.

Commercial source of chlorine	Equivalent Amount to obtain 454 g (1 lb.) of chlorine	Required amount treat one acre-foot (1233 m^3) of water and obtain one ppm of chlorine
Gaseous chlorine (Cl$_2$)	454 g (1.0 lb)	1226 g (2.7 lb.)
Calcium Hypochlorite, Ca(OCl)$_2$		
65–70% of available chlorine	681 g (1.5 lb.)	1816 g (4.0 lb.)
Sodium Hypochlorite, NaOCl		
15% of available chlorine	2.54 L (0.67 gallons)	6.81 L (1.8 gallons)
10% of available chlorine	3.78 L (1.0 gallons)	10.22 L (2.7 gallons)
0.5% of available chlorine	7.57 L (2.0 gallons)	20.44 L (3.4 gallons)

12.5.6 CHLORATION METHOD [8, 10]

The chloration in a drip irrigation system can be continuous or in intervals, depending on the desired results. Application at intervals is appropriate, when the objective is to control the growth of microorganisms in lateral lines, emitters, or in other parts of the system. The continuous treatment is used when we want to precipitate the iron dissolved in the water, to control algae in the system, or where it is not reliable to use the treatment at intervals.

12.5.6.1 GENERAL RECOMMENDATIONS

1. Inject chlorine before the filters. This controls the growth of algae or bacteria in the filters that otherwise would reduce the filtration efficiency. This also allows the filtration of any precipitate caused by the injection of chlorine.
2. Calculate the amount of chlorine to inject. It is necessary to know the volume of water to be treated, the active ingredient of the chemical compound to be used and the desired concentration in the treated water.
3. The chlorine should be injected when the system is in operation.
4. One should take samples from the water at the nearest drippers and most distant drippers to determine the chlorine level at these points. Allow sufficient time so that the lines are filled with the chlorine solution.
5. Adjust the injection ratio. Repeat steps 4 and 5 until the desired concentration in the system is obtained.

12.5.6.2   RECOMMENDED CHLORINE CONCENTRATIONS [7]

1. Continuous treatment (in order to prevent the growth of algae or bacteria): Apply from 1 to 2 mg/L continuously through the system.
2. Treatment at intervals (in order to eliminate the algae or bacteria): Apply from 10 to 20 mg/L for 60 min. The frequency of the treatment depends on the concentration of these microorganisms in the water source.
3. Super chloration (in order to dissolve the organic matter and in many cases the calcium precipitated in the drippers): Inject chlorine at a concentration from 50 to 100 mg/L, depending on the case. After this, close the system and leave it for 24 h, to clean all the secondary and lateral lines. It helps to clean the obstructions in the secondary and lateral lines. We have to be careful while applying these amounts, since these chlorine levels can be toxic to certain crops. Table 5 shows typical dosages of chlorine.

TABLE 5   Typical dosages of chlorine.

Problem due to	Dosage
Algae	1 to 2 ppm continuous, or 10–20 ppm for 30–60 min.
Ferro bacteria	1 + ppm: Varies with the amount of bacteria
Slime	0.5 ppm
Precipitation of iron	$0.64 \times$ [content of Fe^{++}]
Precipitation of Manganese	$1.3 \times$ [content of manganese]
Hydrogen sulfide	3.6 a $8.4 \times$ [the content of H_2S]

12.5.6.3   CHLORINE REQUIREMENTS [5–8, 10]

Chlorine requirements must be known before the chloration. The Cl_2 or NaOCl is a biocide that must be applied in the amounts and at recommended concentrations. The excess of chlorine in the irrigation water can cause damage to the young plants or young trees. On the other hand, the low levels do not solve the problems associated with the growth of microorganisms in the irrigation water. The chlorine is a very active and toxic agent at high concentrations; therefore it must be handled carefully. When it is injected in the irrigation lines, some chlorine reacts with inorganic compounds and organic substances of the water or it adheres to them. In most wells and water sources, from 65 to 81% of the chlorine is lost by this type of reaction. The chlorine (like hypochlorous acid) that adheres to the organic matter or that reacts with other compounds does not destroy microorganisms. For this reason, it does not have value as a biocide agent. The free chlorine (the excess of hypochlorous acid) is the agent that inhibits the growth of bacteria, algae, and other microorganisms in the water. Therefore, it is indispensable to establish the chlorine requirements before the chloration. In this way, we can maintain the desired concentrations of available chlorine.

In order to inhibit the growth of microorganisms, a minimum contact time of 30 min is required (45 min of injection). It also requires a minimum concentration of 0.5–1.0 mg/L of available chlorine measured at the end of the drip line and 2.0–3.0 mg/L of available chlorine at the injection point. The following equations are used to calculate the gallons per hour (gph) of NaOCl that must be injected to obtain the desired concentration of chlorine per minute (gpm):

1. Formula for gpm of 10% NaOCl: (1)
   = [0.0006 × (gpm desirable chlorine) × (Discharge of the pump, gpm)]
2. Formula for gph of 5.25% NaOCl: (2)
   = [0.000114 × (ppm desirable chlorine) × (Discharge of the pump, gpm)]
3. Formula for pounds by hectare of Cl_2 (gas):   (3)
   = [0.000998 × (ppm desirable chlorine) × (Discharge of the pump, gpm)]
4. Gallons of liquid chlorine per hour: (4)
   = [0.06 × ppm of desirable chlorine × discharge of the pump in gpm]/
   [Percentage of chlorine in the material]
5. Dry chlorine in pounds per hour: (5)
   = [0.05 × ppm × gpm]/[percentage of chlorine in the material]
6. Dry chlorine in pounds per 1000 gallons of water: (6)
   = [0.83 × ppm]/[percentage of chlorine in the material]
7. For chlorine gas: (7)
   = Take percentage of chlorine = 100; and calculate as dry chlorine.

12.5.6.4 METHODS TO MEASURE CHLORINE WITH THE D.P.D. METHOD (N, N, DIETIL P-FEMILENEDIAMINA)

It is essential to measure the chlorine when using liquid chlorine as bactericidal and algaecide in irrigation systems of low volume [9, 10]. Most of the methods of measuring chlorine that are used in the swimming pools are not adequate for irrigation systems. This is because many of these equipments measure only total chlorine, but not the residual free chlorine. Equipment "D.P.D." of good quality can measure total chlorine and the free available chlorine. The test equipment D.P.D. is very simple. The directions and procedures come with the equipment. The equipment is used to measure each type of chlorine. When applying these compounds, the water becomes pink in the presence of the chlorine. The more intense is the color, higher is the chlorine concentration. In order to know the chlorine concentration, the color of the water is compared with that of a calibrated chromatic chart. One must remember that the free chlorine is the one that determines the biocide action. If there is not sufficient free chlorine available, the bacteria continue growing even though chlorine has been injected into the system. In other words, if the amount of total chlorine is not sufficient to maintain chlorine free in solution, the treatment gets is of no value. The test equipment D.P.D. can be purchased from the sellers of irrigation equipments or from chemical agents who are specialized in water treatment.

12.6 INSTALLATION, OPERATION, AND MAINTENANCE

12.6.1 INSTALLATION

To be effective, all types of chemigation equipments must be suitable, reliable, and precise. All the electrical devices must resist risks due to bad weather. In addition, all the valves, accessories, fittings, and pipes must resist the operating pressure. Some chemicals can cause corrosion problems. The chemical compounds and the concentrations of chemicals must be compatible with the injection system. The materials of the system must be corrosion resistant. These should be washed with clean water after each fumigation process.

12.6.2   OPERATION

The chemigation procedure must follow a preestablished order. Irrigation system is operated until the soil saturates to a field capacity. Then the chemicals are injected. Once the chemigation has finished, water is allowed to flow free in the system for a sufficient time to remove all the sediments (salts) of chemicals from the laterals and drippers.

12.6.3   MAINTENANCE

The maintenance is a routine procedure. One should inspect all the components of the injection system after each application. It is recommended to replace the defective components before these will stop working altogether. In order to clean the injection system, it is convenient to have the water accessible near the system. After each application, it is recommended to clean and wash exterior of all the parts with water and detergent, and to rinse with clean water. The chemigation system can be cleaned in two ways:
1.   By pressurized air.
2.   By using acids or other chemical agents.

The pressurized air is used to clean the laterals of accumulation of the organic matter. Also, the lines can be cleaned with a commercial grade hydrochloric acid, phosphoric acid, or sulfuric acid at a concentration of 33–38%. When the acid is used, it is convenient to use protective clothing to avoid risks and accidents due to burns. Before using acid, it is recommended to allow the water flow through the system for 15 min. It is safe to fill the tank to 2/3 parts of its capacity and make sure that all components are in good condition. Now add the acid to the tank. The system will operate at a pressure of 0.8–1.0 atmospheres to apply acid. The acid treatment will avoid precipitation of salts and the formation of slime in the system. When the treatment has been completed, allow the water to flow to remove the residues of acid. Other practices for a good operation are:
1.   To lubricate the movable screws and parts, after using the system.
2.   To lubricate the movable screws and parts if the fertigation system was inactive during a prolonged period. It is convenient to activate the system and to make sure that all the parts are in good condition.

Rule of thumb is to chemigate during the middle of the irrigation cycle.

12.6.4   CALIBRATION

The calibration consists of adjustment of the injection equipment to supply a desired amount of chemical. The adjustment is necessary to make sure that the recommended dosage is applied. An excess of chemical is very dangerous and hazardous, whereas a small amount will not give effective results. The use of excess fertilizers is not economical. The amounts less than recommended dosages cause reduction in the crop yield. The precise calibration helps us to obtain accurate, reliable, and desirable results. A simple calibration consists of the collection of a sample of a chemical solution that is being injected to cover the desired area during the irrigation cycle. During the chemigation process, the rate of injection of a chemical compound can be calculated with the following equation:

$$g = (f \times A) / (c \times t_2 \times t_r) \tag{8}$$

where: g = Rate of injection (liters/hour); F = Quantity of chemical compound (Kg/hectare); A = Irrigation area (hectares); C = Concentration of the chemical in the solution (Kg/liter); t_2 = Chemigation time (hours); t_r = Irrigation duration (hours).

12.7 SAFETY CONSIDERATIONS [3, 11, 18, 19]

The safety during the chemigation process is of paramount importance. It is recommended to use specialized equipment to protect the water source, the operators of the system and to avoid risks, health hazards, and accidents. It is necessary to install a safety valve (check valve or one way valve) in the main line between the irrigation pump and the injection point. A manual gate valve is not enough to avoid contamination of the water source. The safety valve allows the water flow in a forward direction. If it is properly installed, it avoids back flow of chemicals towards the water source. It is safer to install an air-relief valve (vacuum breaker) between the irrigation pump and safety valve.

The air-relief valve allows escape of air from the system. It will avoid suction of chemical solution towards the water source. The pump to inject the chemicals and the irrigation pump can be interconnected. The pumps are interconnected in such a way that if one is shut off; the other is shut off automatically. This is convenient when the two pumps are electrical. A safety valve must be installed in the line of chemical injection. This arrangement avoids back flow towards the tank. This back flow can cause dilution of chemical, causing spills and breakage of the system.

The spills of pesticides are extremely dangerous because these can contaminate the water source and can cause health hazards. It is recommended to locate the chemical tank away from the water source. In the case of deep well, the pesticides can wash through the soil and contaminate the well. In addition, the operator and the environment are exposed to the danger of the contamination. The safety valve generally has spring and requires pressure so that the water will flow through those. This valve allows the flow when only an adequate pressure exists in the injection pump. When the injection pump is not in operation, there is no escape of liquid due to small static pressure in the tank. A gate valve at the downstream of a chemigation tank will help to avoid flow of irrigation water towards the tank when the chemigation is not in progress. This valve can be a manual gate valve, ball valve, or solenoid metering valve. The valve should be installed close to the tank. It must be open only during the chemigation. Also it must be corrosion resistant.

The automatic solenoid valve is interconnected electrically to the injection pump. This interconnection allows automatic closing of the valve in the supply line of chemical. Thus, it avoids flow of water in both directions when the chemigation pump is not in operation. The top of a chemical tank should be provided with wide openings that will allow easy filling and cleaning. In addition, the tank must be provided with a sieve or a filter. The tank must be corrosion resistant such as: Stainless steel or reinforced plastic with fiber glass. The flow rate from of the tank should correspond to the capacity of the pump. The tank should be equipped with an indicator to register the level of the liquid. The centrifugal pump provides high volume at low pressure. The pumps of piston and diaphragm provide volumes between moderate and high flows at high pressure. The pumps of roller and gear type provide a moderate volume at low pressure. If a pump is allowed to operate dry, then it can be damaged. It is recommended to follow the instruction manual of the manufacturer for a long life of a pump. Maintain all protectors in place. The injection pump of low volume can inject the concentrated formulation of pesticide. In that way, the problem of constantly mixing the solution in the tank is avoided. In addition, the calibration becomes easier.

One must select the hoses and synthetic or plastic tubes that can resist the operating pressure, climatic conditions, and the solvents in some chemical compounds. Do not allow the bending or kinking of hoses and tubes with another object. Wash the exterior and interior of hoses frequently so that these can last longer. These must be cleaned, washed, and stored well when are not being used. If it is possible, avoid exposure to sun. Due to climatic changes the hoses or tubes show deteriorations on the outer surface. The areas where chemigation is in progress, should display a sign that chemical compounds are being applied through irrigation system. The operator of the chemigation equipment must take all precautions: To use protective clothes, boots, protective goggles, and gloves. The waiting period to enter the field is necessary to avoid health hazards and risks. The precautions are necessary so that persons or animals do not enter the treated area during the application of pesticides and toxic substances.

12.8   EXAMPLES

12.8.1   EXAMPLE #1

A farmer wishes to use a cloth whitener (NaOCl 1–5% active chlorine or available chlorine) to reach a concentration of one ppm of chlorine at the injection point. The flow rate for the system is 100 gpm. In what ratio the chlorine must be injected?

$$IR = [Q \times C \times M]/S \quad (9)$$
$$= (100 \times 1 \times 0.006)/5$$
$$= 0.21 \text{ gph}$$

where: IR = Rate of chlorine injection (gallons/hour); Q = Flow rate of the system (gallons/minute); C = Desired concentration of chlorine (ppm); S = Percent of active ingredient (%); M = 0.006 for the liquid material (NaOCl), or 0.05 for the solid material Ca $(OCl)_2$.

12.8.2   EXAMPLE #2

A farmer wants to inject Cl_2 through the drip irrigation system at a concentration of 10 ppm. What will be the rate of injection of the chlorine gas? The flow rate of the system is 1500 gpm.

$$IR = Q \times C \times 0.012 \quad (10)$$

where: IR = Rate of injection of chlorine (pound/day); Q = Flow Rate of the system (gallons/minute); C = Desired chlorine concentration (ppm);

IR = 1500 × 10 × 0.01.
= 180 pounds per day

12.9 TROUBLE SHOOTING

Cause	Remedy
Uniformity of application is not adequate	
1. Drippers are clogged with precipitates or clay particles.	Replace drippers. Inject HCL according to the instructions. Use dispersing agents such as: Na and Al.
2. Drippers are clogged with microorganisms.	Use biocides, algaecides, and bactericides.
3. Lines are clogged.	Flush the lines by opening the ends.
4. Filters are clogged.	Clean the filters. Open the corresponding gate valves.
Chemical tank is overflowing	
5. Gate valve between two injection points is closed. Gate valve of injection line is closed.	Open the valves.
6. Filters are totally obstructed.	Flush filters.
Signs of chlorosis	
7. Adequate dosage of fertilizers is not used.	Use recommended dosages.
8. Lack of uniformity of application.	Improve the uniformity. Flush and change drippers if necessary.
9. Formulations of nitrogen that precipitate.	Use acid to clean the lines. Do chemical analysis to check the calibration.
Components are leaking	
10. Corrosion of the components.	Use anticorrosive components.
11. Purple color in young leaves: Lost of phosphorus due to precipitation in line.	Use formulations of phosphorus that do not precipitate. Apply the phosphorus in bands not through the irrigation system.

12.10 SUMMARY

The chemigation is an application of chemicals through the irrigation system. The chemigation provides uniformity in the application of chemicals allowing the distribution of these chemicals in small quantities during the growing season when and where these are needed; reduces soil compaction and the chemical damage to the crop; reduces the quantity of chemicals used in the crop and the hazards/risks during the application; reduces the pollution of the environment; and reduces the costs of manual labor, equipment, and energy. The chemical injection unit consists of chemigation pump, pressure regulator, gate valve, pressure gage, connecting tubes, check valve, and chemigation tank. All fertilizers for the chemigation must be soluble. The chlorine can be introduced at low concentrations at necessary intervals, or at high concentrations for few minutes. The chlorine can be injected in form of chlorine powder and chlorine gas. All chemigation equipments must be corrosion resistant, reliable, and precise. The chemigation procedure must follow a preestablished order. Irrigation system is operated until the soil saturates to a field capacity. Then the chemicals are injected. Once the chemigation has finished, water is allowed to flow free in the system for a sufficient time to remove all the sediments from the laterals and drippers. Rule of thumb is to chemigate during the middle of the irrigation cycle. Maintenance is a routine procedure. One should inspect all the components of the injection system after each application. It is recommended to replace the defective components. The calibration consists of adjustment of the injection equipment to supply a desired amount of chemical. It is recommended to use specialized equipment to protect the water source, the operators of the system and to avoid risks, health hazards, and accidents.

Chloration is a process of injection of chlorine compounds through an irrigation system to prevent obstruction. The chloration solves effectively and economically the problem of obstruction of the emitters or drippers due to biological agents. The chloration can be continuous or in intervals, depending on the desired results. The most common chlorine sources are sodium and calcium hypochlorite and chlorine gas. The NaOCl is safer than the Cl_2. The chapter discusses the quality of water, principle of chloration, commercial sources of chlorine, methods of chloration, and examples to calculate the injection rates.

KEYWORDS

- acid treatment
- air relief valve
- atmosphere
- automatic solenoid valve
- back flow
- biological agents
- centrifugal pump
- check valve
- chemigation
- chloration
- chloration
- chlorination or chloration
- chlorophytes
- chloroplast
- chlorosis

- clay
- clogging
- cyanobacteria
- diatoms
- dissolve
- distribution lines
- drip irrigation
- dripper
- emitters
- fertigation
- fertilizer
- field capacity
- filter
- filter, sand
- filtration system
- fungigation
- gate valve
- glycerophosphates
- herbigation
- injection system
- insectigation
- line, distribution
- maintenance
- manual gate valve
- motile
- nemagation
- nitrogen
- nitrogen fertigation
- nutrients
- organic matter
- ph
- phosphorus
- photosynthesis
- polyethylene
- potassium

- precipitation
- pump
- root system
- root zone
- sand
- slime
- soil moisture
- solenoid valve
- solid particles in suspension
- solubility
- term
- urea
- venturi system
- water quality
- water source
- weed

REFERENCES

1. Abbott, J.S., 1985. Emitter clogging: causes and prevention. *ICID Bulletin*, 34(2):17.
2. Apply pesticides correctly—user's guide for commercial application of pesticides [*Aplique los plaguicidas correctamente: Guía para usuarios comerciales de plaguicidas,* Spanish], 1967. US Government—EPA [Environmental Protection Agency], Washington—D.C.
3. BAR-RAM Irrigation Systems: Instructions for the operation and maintenance. Pp. 3–5.
4. Boswell, Michael. J., 1985. *Micro irrigation design manual*. In James *Hardie Irrigation (Now Toro Irrigation)*, Chapter 3, pp. 1–15. Fresno, CA.
5. *Chlorine Kit.* Taylor Chemicals Inc., 7300 York Road, Baltimore-MD- 21204.
6. Ford, H.W., 1979. *A key for determining the use of sodium hyper chlorite to inhibit iron and slime clogging of low pressure irrigation systems in Florida.* Lake Alfred AREC Research Report CS79–3, Florida, FL.
7. Ford, H.W., 1980. *Estimating chlorine requirements.* Lake Alfred AREC Research Florida Report CS80–1, Florida, FL.
8. Ford, H.W., 1980. *The use of chlorine in low volume systems where bacterial slimes are a problem.* Lake Alfred AREC Research Report CS75–5, HWF100, Florida, FL.
9. Ford, H.W., 1980. *Using a DPD test kit for measuring free available chlorine.* Lake Alfred AREC Research Florida Report CS79–1, Florida, FL.
10. Ford, H.W., 1980. *Water quality test for low volume irrigation.* Lake Alfred AREC-Florida Research Report CS79–6 HWF100, Florida, FL.
11. Goyal, M.R, J. Román, F. Gallardo Covas, L.E. Rivera and R. Otero Dávila, 1983. *Chemigation via trickle systems in vegetables and fruit orchards.* Paper presented at 1983 meeting of the American Society of Agricultural Engineers. Montana State University, Montana.
12. Goyal, M.R., L.E. Rivera, 1984. Chemigation through drip irrigation system [*Riego por Goteo: Quimigación*, Spanish]. Bulletin IA61 Series 4, pp. 1–15. Cooperative Agricultural Extension Service, University of Puerto Rico–Mayaguez Campus.

13. Goyal, M.R., L.E. Rivera, 1985. Service and maintenance of drip irrigation system [Riego por Goteo: Servicio *y* Mantenimiento, Spanish]. Bulletin IA64 Series 5, pp. 1–18. Cooperative Agricultural Extension Service, University of Puerto Rico–Mayaguez Campus.

14. Harrison, D.S., 1974. *Injection of liquid fertilizer materials into irrigation systems*. Gainesville, FL: University of Florida, Pp. 3–6, 10–11.

15. Nakayama, F.S. and D.A. Bucks, 1986. *Trickle irrigation for crop production*. Elsevier Science Publisher B.V., The Netherlands. Pp. 142–157 and 179–183.

16. *Proceedings of the National Symposium on Chemigation*. Tifton-CA: Rural Development Center of University of Georgia, August 1982.

17. Rivera, L.E. M.R. Goyal, 1986. Filtration system for drip irrigation [*Riego por Goteo: Sistemas de Filtración*, Spanish]. Bulletin IA66 Series 7 pp.1–42. Cooperative Agricultural Extension Service, University of Puerto Rico—Mayaguez Campus.

18. Rolston, D.E., R.S Rauschkolb, C.J. Phene, R.J. Miller, K. Uriu, R.M. Carlson and D.W. Henderson, 1979. Applying nutrients and other chemicals to trickle irrigated crops. University of California. Pp. 3–12.

19. Smajstrla, A.G., D.S. Harrison, J.C. Good and W.J. Becker, 1982. *Chemigation Safety*. Agricultural Engineering Fact Sheet, Florida Cooperative Extension Service, Institute of Food and Agricultural Sciences, University of Florida.

CHAPTER 13

FERTIGATION: VENTURI INJECTION METHOD

MUKESH KUMAR, T. B. S. RAJPUT and NEELAM PATEL

CONTENTS

13.1   Introduction 266
13.2   Materials and Methods 267
13.3   Results and Discussion 268
13.4   Conclusions 272
13.5   Summary 272
Keywords 273
References 273

13.1 INTRODUCTION

Fertilizer plus irrigation is known as fertigation. Fertigation refers to the application of fertilizers and other chemicals along with irrigation water through drip or sprinkler irrigation to the crops on a continual basis in controlled manner so as to allow for steady uptake of nutrients by plants. Fertigation aims at uniform distribution of applied nutrients in small doses to meet the plants nutrient demand and ensures substantial saving in fertilizer usage [15]. Besides, it is considered ecofriendly as it avoids leaching of fertilizers [13]. Fertigation has the potential to ensure the right combination of water and nutrients to the root zone for meeting the plant's total and temporal requirement of water and nutrients [9]. The application of right combination of water and nutrients in optimal and proper concentration is the key for higher crop yields and the better quality of produce. Fertigation uniformity is an important consideration in design and use of micro irrigation system as nonuniform application of water and fertilizer may result in loss of valuable nutrients, considerable reduction in yield and quality, soil degradation and groundwater contamination. The uniformity of fertilizer concentration depends on injection method [1, 5]. Fertilizer uniformity could be greatly influenced by the injection method and its management during injection process [4]. The application of small amount of fertilizer solution at frequent intervals reduces deep percolation losses [3]. Optimum operating pressure is one of the main factors for uniform distribution of water and fertilizer through micro irrigation system. Fertigation is done using injectors such as fertilizer tank, fertigation pump and venturi. Venturi injector is the most commonly used device for fertilizer application through drip irrigation. The fertilizer application rate of the venturi depends on injection rate besides the concentration of fertilizer solution. The suction rate of venturi mainly depends on pressure gradient (difference in upstream and downstream pressure) and the density of fertilizer solution besides size of venturi. Nakayama and Buck [8] compared all three different methods and rated that fertilizer tank better since it had no moving parts. However, the main disadvantage of fertilizer tank lies in the difficulty to maintain a uniform concentration of fertilizer in the solution during fertigation. Sivanappan and Prasad [14] reported that fertigation pump and venturi can be used to inject fertilizer at constant dilution ratio. Sivanappan et al. [14] suggested five criteria to determine the efficiency of different fertilizer injection devices. They are (a) Dilution ratio, (b) Cost of unit, (c) Portability of the unit, (d) Method of operation and (e) Operating cost. Shani and Sapir [11] discussed the advantages and disadvantages of three fertilizer injection devices and described the venturi as a simple mechanism, capable of giving constant dilution ratio and cheaper in cost. They described the main limitation of venturi as the large pressure drop that occurs during its operation as compared to fertilizer tank and fertigation pump. Kranz et al. [6] found that venturi injectors and diaphragm pump have lower precision and repeatability than fertigation pumps. The venturi is economical and requires no electrical power to operate [8]. Thus, it is important to select an injection device that best suits the irrigation system and the crop to be grown. Rajput et al. [10] reported that the injection rate of venturi directly proportional to the pressure drop across it. Kumar et al. [7] conducted a study and informed that injection rate of venturi varies with fertilizer solution concentration and system operating pressure.

It is clear from the above literature reviewed, a considerable amount of research work has been done on fertigation under drip irrigation in the past few years however, a very few studies have been conducted on evaluation of fertilizer injection devices for fertigation. Therefore, the present study was conducted to the evaluate the performance of a commercially available venturi in terms of its injection rates of water and fertilizer solutions under different system pressures and solute concentrations to generate the information for proper design drip fertigation system.

13.2 MATERIALS AND METHODS

A field study was conducted at the research farm of Water Technology Centre, Indian Agricultural Research Institute, New Delhi, India (Latitude 28°37'30" to 28°30'0" N, Longitude 77°88'45" to 77°81'24" E and AMSL 228.61 m) during October 2010.

TABLE 1   Density of fertilizer solution mixed in different ratios with irrigation water.

Fertilizer: Water	Density (g.cm⁻³)		
			Phosphoric
(kg per liter of water)	Urea	MOP	acid
0.25:1	1.04	1.21	1.09
050:1	1.06	1.23	1.14
0.75:1	1.08	1.26	1.18
100:1	1.11	1.28	1.25

A commercially available venturi injection device (Model: ¾") was procured and installed to evaluate its performance with different fertilizer solution. Pressure gauges were used to monitor the pressures at upstream and downstream of the venturi (Fig. 1). A bypass valve was provided for flow and upstream operating pressure control in the drip irrigation system. Gate valve was used to regulate flow of water and to adjust the operating pressure immediately downstream of the venturi. Upstream pressures of 1.0, 1.2, 1.4, 1.6, 1.8 and 2.0 kg.cm⁻² were maintained and with each upstream pressure, different downstream pressures ranged from 0.1 to 0.6 kg.cm⁻² were fixed for evaluating the injection rate of venturi. A bucket containing water and fertilizer solution was placed near the venturi injector. The fertilizer (urea, muriate of potash (MOP) and phosphoric acid) solutions were prepared by mixing different fertilizers with irrigation water in different ratios of 0.25:1, 0.5:1, 0.75:1 and 1:1 (fertilizer: water). The densities of the different fertilizer solutions were determined (Table 1).

The pressures of upstream and downstream side were adjusted and stabilized at predetermined levels. The injection tube of venturi was placed in the bucket to allow injection of water and fertilizer solution from the bucket. The drip system was operated for 10 min duration and the change in the volume of the water and fertilizer solution in the bucket was recorded. Full experimented was replicated thrice. The volume change in the bucket was divided by operation time to obtain the injection rates of the venturi.

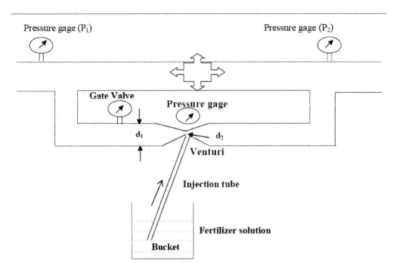

FIGURE 1   Schematic diagram of test setup for venturi.

Injection efficiency of the venturi was calculated using the following relationship suggested by BIS [2]:

$$\eta_{inj} = \frac{SOR}{OP_{min}} x100 \tag{1}$$

where: η_{inj} = Injection efficiency of venture, %; SOR = Suction operating range for venturi injection system (OP_{max}- OP_{min}), kg.cm^{-2}; OP_{max} = Maximum pressure allowed at the inlet of venturi, kg.cm^{-2}; and OP_{min} = Minimum pressure at the outlet of venturi required to operate the venturi, kg.cm^{-2}.

13.3   RESULTS AND DISCUSSION

During the operation of the drip irrigation system, by pass valve was operated to achieve the desired upstream pressure. Different upstream pressures were maintained for testing of different venturi. At each upstream pressure, different downstream pressures were achieved with the help of the gate valve (Fig. 1). Suction rates were determined at all combinations of upstream and downstream pressures.

High upstream pressures resulted in more injection rates of venturi with irrigation water. The injection rate of venturi for irrigation water progressively decreased with increasing downstream pressure at all upstream pressures, that is, with decreasing pressure differential (Fig. 2). Maximum injection rate (106.4 l.h^{-1}) of venturi with irrigation water was at the upstream and downstream pressure of 2.0 kg.cm^{-2} and 0.1 kg.cm^{-2}, respectively. The minimum venturi injection rate of 3.2 l.h^{-1} with irrigation water was at 1.0 kg.cm^{-2} upstream and 0.4 kg.cm^{-2} downstream pressure. These finding are in the line with the findings of Rajput et al. [10].

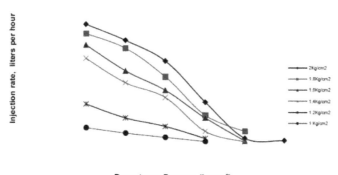

Downstream Pressure (kg cm⁻²)

FIGURE 2   Injection rate at different upstream and downstream pressures for venturi.

Venturi was tested for the purpose of evaluating its performance in the field with different concentrations of fertilizer solutions of different fertilizers as the test fluid. Injection rates of the venturi were determined at different upsteam and downstream pressures with varying concentration of fertilizer solutions.

Injection rates of venturi at different downstream (2.0, 1.6 and 1.2 kg.cm⁻²) and upstream pressures (0.6, 0.5, 0.4, 0.3, 0.2 and 0.1 kg.cm⁻²) for Urea, MoP and Phosphoric Acid fertilizer solutions when dissolved in water in the ratio of 0.25:1, 0.50:1, 0.75:1 and 1:1 (Fertilizer: Water) are presented in Figs. 3 to 5. It is clear from the Figs. 3 to 5 that injection rate decreases as concentration of fertilizer increases, that is, the injection rate is inversely proportional to the concentration of the fertilizer solution.

Maximum injection rate (105.8 l.h⁻¹) for urea fertilizer when dissolved in irrigation water in the ratio of 0.25:1 (Urea: Water) was at the upstream pressure of 2.0 kg.cm⁻² and downstream pressure of 0.1 kg.cm⁻². However, venturi could attain only minimum injection rate 1.78 l.h⁻¹ at upstream and downstream pressure of 1.6 kg.cm⁻² and 0.5 kg.cm⁻² when urea fertilizer was mixed with irrigation water in the ratio of 0.50:1 (Urea: Water).

It is clear from the Fig. 4 that maximum (82.62 l.h⁻¹) and minimum (1.44 l.h⁻¹) injection rate of venturi with MoP solution was recorded when phosphoric acid was mixed with irrigation in the ratio of 1:0.25 and 1:1 respectively, at upstream pressure and (2 and 1.4 kg.cm⁻²) downstream pressure (0.1 and 0.4 kg.cm⁻²), respectively.

Maximum (92.88 l.h⁻¹) and minimum (1.62 l.h⁻¹) injection rate of venturi with phosphoric acid solution was recorded when phosphoric acid was mixed with irrigation in the ratio of 1:0.25 and 1:1 respectively, at upstream pressure and (2 and 1.4 kg.cm⁻²) downstream pressure (0.1 and 0.4 kg.cm⁻²), respectively.

Higher values of injection rates of venturi for Urea, MoP and Phosphoric acid fertilizer solutions when mixed with irrigation in the ratio of 1:0.25 were observed at 2.0 kg.cm⁻² of upstream pressure with downstream pressures of 0.1 kg.cm⁻². However, venturi could attain lower values of inject rates at upstream pressure of 1.2 kg.cm⁻² with downstream pressures of 0.4 kg.cm⁻² for Urea, MoP and Phosphoric acid fertilizer solutions when mixed with irrigation in the ratio of 1:1 as shown (Figs. 3 to 5).

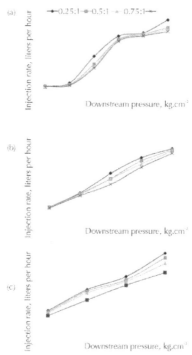

FIGURE 3 Injection rate of enture with different concentration urea fertilizer solutions at (a) 2.0, (b) 1.6 and (c) 1.2 kg.cm⁻² upstream pressure and different downstream pressures.

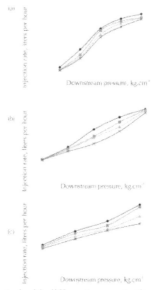

FIGURE 4 Injection rate of venturi with different concentration MoP fertilizer solutions at (a) 2.0, (b) 1.6 and (c) 1.2 kg.cm⁻² upstream pressure and different downstream pressures.

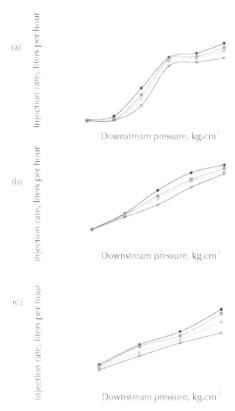

FIGURE 5 Injection rate of venturi with different concentration phosphoric acid fertilizer solutions at (a) 2.0, (b) 1.6 and (c) 1.2 kg.cm^{-2} upstream pressure and different downstream pressures.

   Though the performance of venturi was similar at different upstream and downstream pressures with respect to the maximum injection rate, it was considerably different in terms of fertilizer solutions when mixed with irrigation water. The injection rate of venturi was different for different fertilizer solutions. Higher value of injection rate was for urea fertilizer solution when mixed with irrigation water in the ratio of 0.25:1 at upstream pressure of 2.0 kg.cm^{-2} and downstream pressure of 0.1 kg.cm^{-2} however lower value of injection rates was 1.44 l.h^{-1} at upstream pressure of 1.6 kg.cm^{-2} and downstream pressure of 0.4 kg.cm^{-2} for MOP fertilizer solution when mixed with irrigation water in the ratio of 1:1. Injection rate for urea solution when dissolved in the ratio of 0.25:1 (Fertilizer: Water) is closer to the injection rate of venturi with irrigation water, but the injection rate decreased as the urea concentration increased. Similarly, the injection rate of phosphoric acid and MOP decreased with increase in fertilizer concentrations. It may be due to the variation in density of different fertilizer solutions. Similar results were reported by Rajput et al. [10] while evaluating performance of commercially available ventures under varying inlet and outlet pressure differentials for irrigation water.

Injection efficiency of venture was determined using Eq. (1). Injection efficiency of the venturi was observed to be 95%. This showed that the venturi operates at high efficiency under a wide range of upstream and downstream pressures.

13.4 CONCLUSIONS

The injection rates of venturi with irrigation water and different concentration fertilizer solutions were recorded at different upstream and downstream pressures. The injection rate of venturi was found to be directly proportional to the pressure drop across it for irrigation water. The injection rate of venturi varies with fertilizer solution concentration and system operating pressure. Maximum injection rate of venturi was observed for urea fertilizer solution when dissolved in the ratio of 0.25:1 (Fertilizer: Water), but the injection rate decreased as the urea concentration increased. However minimum value of injection rates was 1.44 l.h^{-1} at upstream pressure of 1.6 kg.cm^{-2} and downstream pressure of 0.4 kg.cm^{-2} for MOP fertilizer solution when mixed with irrigation water in the ratio of 1:1. The injection rate of phosphoric acid and MOP decreased with increase in fertilizer concentrations. The higher value of injection rate was observed for urea fertilizer solution but lower values were observed for MOP fertilizer solutions. Injection efficiency of venturi was observed to be 95% under a wide range of upstream and downstream pressures.

13.5 SUMMARY

Fertigation is an efficient method of fertilizers and chemicals application along with irrigation water to the crop on a continual basis in controlled manner through micro irrigation. Venturi injector is one of the economically available and commonly used devices for fertigation. The injection rate of a venturi mainly depends upon the suction created due to the difference between its upstream and downstream pressures. Performance evaluation of a commercially available venturi was done for injection rate under the range of operating pressures (1.0 to 2.0 kg.cm^{-2}) using normal irrigation water and different concentration fertilizer solutions. Maximum injection rate (106.4 l.h^{-1}) of the venturi with irrigation water was observed at 2.0 kg.cm^{-2} operating pressure. Maximum injection rate (105.8 l.h^{-1}) was recorded at the 2.0 kg.cm^{-2} of upstream and 0.1 kg.cm^{-2} of downstream pressures when urea fertilizer dissolved in irrigation water in the ratio of 0.25:1 (urea: water). However, minimum injection rate of venturi was (1.44 l.h^{-1}) was recorded for muriate of potash (MOP) when dissolved in irrigation water in the ratio of 1:1 (MOP: water), at upstream pressure of 1.6 kg.cm^{-2} and downstream pressure of 0.5 kg.cm^{-2}. Injection rate of venturi was observed to be directly proportional to the pressure gradient, that is, decrease in pressure gradient across the venturi decreased the injection rate. While, it was inversely proportional to the concentration of the fertilizer solution, that is, increase in fertilizer solution concentration decreased the injection rate.

KEYWORDS

- drip irrigation
- fertigation
- fertigation pump
- fertilizer tank
- injection rate
- micro irrigation
- pressure gradient
- solute concentration
- venturi

REFERENCES

1. Bracy, R. P., Parish, R. L., Rosendale, R. M., 2003. Fertigation uniformity affected by injector type. Horttechnology. 13:103–105.
2. BIS. 1997. Indian Standards. Fertilizer and Chemical Injection System-Part I: Venturi injector. IS 14483 (part I).
3. Claudinei, F Souza; Marcos, V Folegatti. 2009. Distribution and storage characterization of soil solution for drip irrigation. Irrig. Sci. 27:277–288.
4. Imas, P., 1999. Recent techniques in fertigation of horticultural crop in Israel. Presented at the IPI–IRII-KKV workshop on Recent Trends in Nutritional Management in Horticultural crops 11–12 February 1999. Dapoli, Maharashtra, India. Cited at website:ipipotash.org/presentn/rti-fohc.html#injection.
5. Jiusheng, Li; Yibin, Meng; Bei, Li. 2007. Field evaluation of fertigation uniformity as affected by injector type and manufacturing variability of emitters. Irrig Sci. 25:117–125.
6. Kranz, W. L., Eisenhauer, D. E., Parkhursrt, A. M., 1996. Calibration accuracy of chemical injection devices. Applied Engineering in Agriculture, 12(2):189–196.
7. Kumar, Mukesh; Rajput, T., B. S., Patel, Neelam. 2012. Effect of system pressure and solute concentration on fertilizer injection rate of a venturi for fertigation. Journal of Agricultural Engineering. 49(4):9–13.
8. Nakayama, F. S., Bucks, D. A., 1986. Trickle irrigation for crop production; Design operation and management Elsevier Science Publishers B.V. Amsterdam, the Netherlands, p 384.
9. Patel, Neelam; Rajput, T., B. S., 2001. Effect of Fertigation on Growth and Yield of Onion. Micro Irrigation, CBIP Publication, no 282, p 451.
10. Prasad, S., 1997. Development of criteria for optimum use of fertilizer through trickle/drip irrigation system. Ph.D. thesis (Unpublished), I.,A.R.I., New Delhi, 91p.
11. Rajput, T., B. S., Patel, Neelam; Kumar, Mukesh. 2004. Performance evaluation of commercially available ventures under varying inlet and outlet pressure differentials. Journal of Agricultural Engineering. 41(2), 38–41.
12. Shani, M., Spair, E., 1994. Advance course on irrigation management held at Indo-Israel training and Education Centre, Jalgaon, India.
13. Shivashankar, K., Khan, M. M., Farooqui, A. A., 1998. Fertigation Studies with Water Soluble NPK Fertilizers in Crop Production. UAS Bangalore, Pp 32.
14. Sivanappan, R. K., Padamakumar, I. O., Kumar, V., 1987. Drip Irrigation. Pub. House (P) Ltd., Coimbatore, India, p. 428.
15. Veeranna, H. K., Abdul, Khalak; Farooqui, A. A., Sujith, G. M., 2001. Effect of Fertigation with Normal and Water-soluble Fertilizers Compared to Drip and Furrow Methods on Yield Fertilizer and Irrigation Water Use Efficiency in Chilli. Micro Irrigation, CBIP Publication, no. 282, p. 461.

CHAPTER 14

CHEMIGATION UNIFORMITY IN SUSTAINABLE MICRO IRRIGATION SYSTEM*

LARRY SCHWANKL

CONTENTS

14.1 Introduction 276
14.2 Injection Equipment 276
14.3 System Management 280
14.4 Summary 285
Keywords 286

*Modified and printed with permission from: "*Schwankl, Larry, 2013. Chemigation uniformity critical through micro irrigation systems. Fluid Fertilizer Foundation Newsletter, Manhattan, Kansas, USA, Spring 2013, 21(2): 8–13.*"

14.1 INTRODUCTION

Chemigation through micro irrigation systems (drip and microsprinklers) is be-coming common-place as micro irrigation systems gain in popularity. The most commonly injected chemicals are fertilizers but other products, such as herbicides, pesticides, nematicides, and even micro irrigation system maintenance chemicals such as chlorine and acids, are injected. Common to all, chemigation practice is the need to apply the chemical as uniformly as possible.

Chemigation uniformity means that all areas of the field serviced by the micro irrigation system receive the same amount of chemical during an injection event. While there may be times when we want to intentionally apply different amounts of chemical to various portions of the field, it takes a very sophisticated micro irrigation injection system to achieve that. In the vast majority of cases, our goal is to apply the injected chemical uniformly as the water is applied. High chemigation uniformity requires:

(1) An irrigation system with high application uniformity; and
(2) An injection system that facilitates high chemigation uniformity. Well-designed and well-maintained micro irrigation systems should have high irrigation application uniformity. Achieving excellent irrigation application uniformity is why growers pay substantial money for micro irrigation systems.

The injection system, especially the injector, is a critical piece of hardware necessary to achieve high chemigation uniformity.

In this chapter, the various injection systems are discussed, including their advantages and disadvantages. The importance of injection system management to maximize the chemigation uniformity is also discussed.

14.2 INJECTION EQUIPMENT

A variety of injection equipment is available in a wide range of cost and capability. The simplest injector systems include batch tanks, while the most sophisticated include electrical, gasoline, or water-driven positive displacement pumps. Which injection system to use depends on injector capabilities, reliability, and cost.

14.2.1 BATCH TANK

Some of the earliest injector systems are a batch tank (Fig. 1). These relatively inex-pensive and simple systems consist of a tank that is plumbed into the irrigation system so that a portion of the irrigation water flows through it. The tank must be able to withstand the operating pressure of the irrigation system. Since there is often pressure loss as water flows through the batch tank system, the tank must be plumbed across a pressure differential so that the batch tank inlet is at a higher pressure than the tank outlet. Examples of micro irrigation system components that can cause a pressure differential are a partially closed valve or a pressure- reducing (pressure regulating) valve. The rate the material is injected is influenced by the flow rate through the tank and the concentration of injection material in the tank at any given time.

The batch tank is filled with the chemical (most frequently, fertilizer) to be applied. During irrigation, water is allowed to flow into the batch tank where it displaces some of the tank's contents, forcing them into the irrigation system.

Initially, the liquid leaving the batch tank is of high chemical concentration, but with time, the concentration in the batch tank becomes more diluted as it mixes with water. The chemical injection starts out at a higher concentration and declines in concentration during the injection period. Batch tanks are appropriate if the objective is to inject a total amount of chemical (e.g., fertilizer) during a long injection period. Batch tanks are not appropriate if a constant injection concentration is required.

14.2.2 VENTURI INJECTION

Injection devices using the venturi principle (Fig. 2) have been used for many years in a wide variety of industrial and agricultural applications. A venturi is a specially shaped constriction in a device's water flow path. As the flow passageway at the venturi section becomes smaller, the velocity of the flowing fluid increases such that a vacuum is formed at the venturi's throat section. An opening located in the venturi's throat allows air or a fluid to be "sucked" in and mixed with the water stream. To create this effect, the inlet pressure to the venture must be at least 15 to 20 percent greater than the outlet pressure. This is achieved by plumbing the venturi device across a system pressure differential (e.g., a partially closed valve, Fig. 3) or using a small pump that draws water from the micro irrigation system and forces it through the venturi injector (Fig. 4). The injection rate of a venturi-type injector depends on the size of the venturi section (1/2-inch to 2-inch sizes are available) and on the pressure difference between the inlet and the outlet of the venturi.

The venturi injector delivers a more constant chemical injection rate than does a batch tank. However, the injection rate of a venturi injector can change (or even stop) if the pressure changes upstream or downstream in the micro irrigation system. This can occur if irrigation sets (applications) with different flow rates are operated from the same water supply pump. A venturi injector installation in which the venturi is plumbed across a pressure differential (Fig. 3) is particularly sensitive to such changes, and it is often inconvenient to readjust the injector. The injector is usually installed parallel to the micro irrigation system pipeline, with valves that can be closed to isolate the injector from the irrigation system when injection is not occurring (see Figs. 3 and 4). A venturi injector installation using a small pump in conjunction with the venturi eliminates the need to install the venturi across a pressure drop, and it also minimizes the venturi's sensitivity to irrigation system pressure fluctuations.

However, this type of installation requires an electrical or gasoline power source.

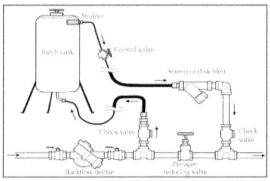

FIGURE 1   Batch tank.

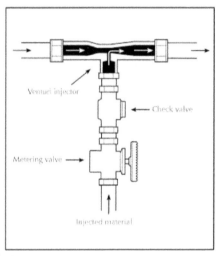

FIGURE 2   Venturi Injector.

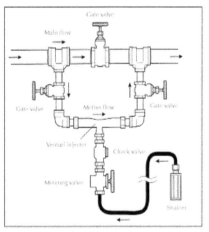

FIGURE 3   Venturi "across a pressure drop" layout.

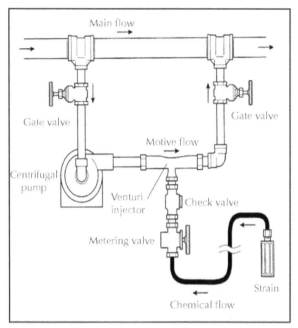

FIGURE 4   Venturi with a small pump.

14.2.3   POSITIVE DISPLACEMENT PUMP

Positive displacement pumps have a relatively small capacity and deliver a very constant rate of injection. The pumps can use a cylinder-piston configuration or a flexible diaphragm to inject a liquid at a pressure higher than that of the irrigation system. Electrically driven, gasoline-engine-driven, and water-driven pump injectors are available. Positive displacement pumps provide the most accurate and constant injection rate, but they are also the most expensive injection devices. They do not need to be installed across a pressure drop since they are externally powered.

Positive displacement pump injectors are available as constant-rate pumps and as proportional pumps. Constant-rate pumps inject at a set rate (often adjustable) no matter what the flow rate is in the irrigation system. Proportional pumps (frequently water-driven) inject at a rate dependent on the flow rate passing through the injector or through the irrigation system. For the electrical or gasoline-driven pumps, the injector is linked to the irrigation system via a flow meter. For example, a proportional positive displacement injector set at 1:250 would inject 1 gallon of material for every 250 gallons of water passing through it. The proportional rate setting can be adjusted, as can the stock tank mixture concentration, to control the injected material's concentration in the irrigation system.

14.2.4   SOLUTIONIZER MACHINE

With the relatively recent advent of solutionizer machines, a wide variety of less-soluble solid materials are now being injected in micro irrigation systems. Foremost of these is gypsum, but other materials such as potassium sulfate and other granular fertilizers are also being injected. The injection rate of the solutionizer machine should be adjusted to ensure that the injected material goes into solution prior to reaching any of the emitters. While the use of solutionizer machines has been an advance in chemigation, some caution must be used. Three issues are of concern: chemical precipitation, injected material solubility, and impurities in the injected material.

As with liquid materials, solid materials containing calcium should be injected with caution if the irrigation water contains substantial bicarbonates (greater than 2 meq/L, or 120 parts per million) and the water pH is greater than 7.5. The jar test should be performed before injecting any new product. Remember that gypsum, which is calcium sulfate, is a ready source of calcium. Some of the materials injected by the solutionizer machine, including gypsum, are not readily soluble in the irrigation water. The concentrations at which these materials are injected should be monitored to ensure that the material is going into solution. In addition, adequate time should be allowed for the materials to go into solution once they are injected into the irrigation system.

Impurities in the injected material are also a major consideration when using a solutionizer machine. Even the "pure," finely ground gypsum materials prepared for use with solutionizer machines contain some impurities. If the material is 95 percent gypsum, 5 percent of it is still "foreign" material. Thus, for every 100 pounds injected, there will be 5 pounds of the foreign material, some of which may not go into solution.

Injecting granular fertilizers using solutionizer machines should be done with even greater caution. Many of these may have been meant to be land-applied, and no particular care was taken in the production to minimize the impurities. Some have oil or wax coatings to prevent water absorption, and introducing insoluble materials such as these into a micro irrigation system may lead to emitter clogging.

One option for dealing with suspended impurities is to inject the slurry from the solutionizer machine upstream of the micro irrigation system's filter(s). A filter cannot be located on the line from the solutionizer machine since the output from the machine is a slurry, and the injected material (e.g., gypsum) does not go into solution until it is injected into the main irrigation line. **Use caution if the micro irrigation system has sand media filters, an automatic back-flush screen, or disk filter system.**

Any material injected while the filters are back-flushing goes out with the flush water—a potential environmental concern when injecting fertilizers or certain irrigation system maintenance products.

14.3   SYSTEM MANAGEMENT

14.3.1   INJECTION POINT

The location in the micro irrigation system at which material is injected should depend on the type of material being injected. Readily soluble materials, such as fluid fertilizers and chlorine, should be injected downstream of the system's main filters. This will prevent the injected material from being part of the filter back flush water if filter cleaning occurs during injection. Also, a small screen or disk filter should be installed in the line from the storage tank to the injector to catch any impurities that may be in the material or tank.

Acidic products for micro irrigation system maintenance should not be injected where low water pH may damage metal components (e.g., some sand media filter tanks). Most plastic components will not be affected by low water pH.

Materials injected by solutionizer machine should be injected upstream of the micro irrigation system's main filters to remove any impurities in the injected material. Ideally, the injection should not occur while the filters are being cleaned.

To ensure that the injected material and the irrigation water are mixed, materials should be injected into the middle of the water stream rather than at the pipe wall. Commercial devices are available that thread into the pipeline through a fitting and extend into the pipe, allowing injection directly into the fast-moving irrigation stream. It is also possible to make such a device using PVC fittings and pipe.

14.3.2 TIMING/DURATION OF CHEMIGATON

When injecting materials through a micro irrigation system, the irrigator should keep in mind two considerations: the irrigation amount applied should be correctly determined so that the applied water; and injected material remain in the plant's root zone.

Applying more irrigation water than the plant's root zone can hold (over irrigation) causes water to percolate below the root zone. Over-irrigation can also cause the injected material to percolate out of the root zone if the injected material travels through the soil easily with the water (e.g., nitrates). Good irrigation scheduling techniques can minimize this hazard and optimize the efficiency of chemigation. A good time to inject material is in the middle of an irrigation set (assuming that the irrigation set is long enough to allow a choice of when to do the injection). This makes it more likely that the injected material will stay in the root zone if over irrigation occurs. Second, the duration of injection should provide a uniform application of injected material throughout the micro irrigation system. It is important to remember that the injected material does not immediately reach all the emitters as soon as injection begins. A period of time is required for the water and injected material to move through the system to the emitters. This travel time depends on the design and layout of the micro irrigation system. In the following discussion of travel times, it should be assumed that injected materials travel through the micro irrigation system at the same speed as the irrigation water.

Water travels through a micro irrigation system's mainline and submain pipelines quite quickly. These pipelines are typically sized so that the flow velocity is less than 5 feet per second (fps) to minimize frictional pressure losses. Flow velocities of 1 to 3 fps are quite common in mainline and submains. Some pipeline systems are long, so movement of water and injected materials through them may take a while: travel times of 20 to 30 min are common, and travel times as long as 65 min have been observed (Table 1).

Irrigation water flows slower in micro irrigation lateral lines than it does through the mainline and submains. The flow velocity is particularly low at the tail end of the lateral lines. Understanding how water flows in drip lateral lines helps explain this.

At the inlet of a drip lateral, the flow rate is that of all the combined downstream emitter discharges. For example, if 60–1-gallon-per-hour (gph) emitters were installed in the lateral line, the flow rate at the head of the drip lateral would be 60 gph or 1 gpm. For typical drip tubing with a nominal inside diameter of 5/8-inch, the resulting flow velocity would be 60 feet per minute (fpm) or 1 foot per second (fps). The flow velocity in the drip line depends on the flow rate and the tubing size. For the same size drip tubing, the higher the flow rate, the higher the flow velocity.

Using the above drip lateral as an example, downstream of the first emitter, the flow rate in the drip tube would be 59 gph; downstream of the second drip emitter, the flow rate would be 58 gph; and so on. Since the flow rate decreases along the drip lateral, so does the flow velocity. The slowest-moving water is between the next-to-last and the last emitter. In our example, the flow rate in this section is only one gph. For the same 5/8-inch drip tubing, the flow velocity would be only about one fpm in this last drip line section.

The total travel time of water along a drip lateral line therefore depends on four factors:

The length of the drip lateral
- The number of emitters installed in the lateral line
- The discharge rate of the emitters
- The inside diameter of the drip tubing.

Knowing these factors, the drip lateral line travel times can be calculated. But the easiest way to determine the travel time is to measure it in the field.

14.3.3   FIELD MEASUREMENT OF DURATION OF CHEMIGATION

Micro irrigation water travel times can be measured by "tracing" the movement of injected chlorine through the system. Injecting chlorine into the irrigation system is a recommended micro irrigation system maintenance procedure. The presence of chlorine in the discharge from emitters can be easily monitored using a **pool or spa chlorine test kit**. The chlorine's passage through the micro irrigation system can be readily traced using this technique, and the water travel time easily determined. The recommended procedure is as follows:

Step 1: Start up the micro irrigation system and allow it to come to full pressure. If the micro irrigation system has not been flushed recently (pipelines and lateral lines), this should be done now. Allow the micro irrigation system to return to full pressure after flushing.

Step 2: Begin injecting chlorine so that the chlorine concentration in the irrigation water is approximately 10 to 20 parts per million (ppm). Note the time when chlorine injection begins.

Step 3: Go to the emitter at the head of the lateral farthest (hydraulically) from the injection point. Using the chlorine test kit, monitor the discharge from that emitter and note the time when the chlorine registers on the test kit. The time from the start of the injection to when the chlorine registers on the test kit is the travel time of water through the mainline—submain system.

Step 4: Go to the last emitter at the tail end of the lateral you just monitored (the lateral farthest hydraulically from the injection point). Monitor discharge from this last emitter until chlorine registers on the test kit and note the time this occurs. The time from the start of the injection to when the chlorine registers on the test kit is the travel time of water through the entire micro irrigation system.

As part of a field study of micro irrigation systems, travel time and chemigation uniformity information on a number of drip systems was collected. Table 1 shows the travel times for these evaluations. This data show that there is no standard water travel time to the far point in a drip irrigation system. Travel time should be measured for each individual micro irrigation system, but it only needs to be measured once.

14.3.4   POST-INJECTION

To ensure chemigation uniformity, it is important that irrigation continue following an injection. This accomplishes two things. **First**, it allows the injected material to be cleared from the micro irrigation system. Second, it maximizes the chemigation uniformity, since all emitters will have discharged nearly the same amount of injected material by the time the irrigation stops.

Clearing the micro irrigation system of the injected material is often important to minimize emitter clogging. For example, leaving fertilizer in the system may encourage biological growth (e.g., biological slimes), which can lead to emitter clogging. Leaving materials containing calcium (e.g., gypsum or calcium nitrate) in the system may lead to chemical precipitation of calcium carbonate (lime), which may also cause emitter clogging. Time and temperature enhance chemical precipitation. The exception to this recommendation may be the injection of system maintenance products such as chlorine or acid. It may be desirable to leave these products in the system at shutdown to maximize their effects and minimize clogging problems.

Just as it takes time for the injected material to travel through the micro irrigation system once injection starts, it takes an equal or greater amount of time for the injected material to clear out of the system. The injected material first clears from the head of the system, and the last point to clear is the emitter hydraulically farthest from the injection point. This is just the opposite of what occurs when injections began, and it balances the amount of injected materials discharged from emitters throughout the micro irrigation system. This gives a uniform chemigation application.

Field evaluations have been done on a single drip lateral line to evaluate the impact on chemical application uniformity of varying the injection and postinjection irrigation times. The results of some of these evaluations are shown in Table 2. The lateral line evaluated was a 500-foot drip lateral (16 mm polyethylene tubing) with 1-gph pressure-compensating drip emitters installed every 5 feet. It was determined through field evaluation that the travel time for water and injected chemicals passing through the lateral line was 25 min.

TABLE 1   Water and chemical travel times through pipelines and drip lateral lines for selected vineyard and orchard field sites.

Site		Mainline and submain pipeline		Lateral line	Total
	Length	Travel time	Length	Travel time	Travel Time
	(ft)	(min)	(ft)	(min)	(min)
1	1,000	22	175	10	32
2	1,500	30	340	10	40
3	5,000	65	340	10	75
4	1,400	15	630	23	38
5	700	8	625	23	31
6	820	17	600	28	45

TABLE 2   Chemigation uniformity in a drip lateral (500 feet long) with 1-gph drip emitters installed at 5-foot intervals) for various injection times and postinjection   clean water irrigations. The water and injected chemical travel time to reach the end of the drip lateral was 25 min.

Injection time	Post-injection Irrigation time	Relative uniformity
(min)	(min)	(%)
50	50	100
50	25	98
25	25	95
25	50	90
13	25	81
50	0	25
25	0	11
13	0	7

Excellent chemical application uniformity was achieved when:
- The injection period was equal to or greater than the water travel time to the end of the drip lateral (25 min in the test case).
- The postinjection irrigation time was equal to or greater than the lateral line's water travel time.

The results in Table 2 also show that there are two injection strategies:

First, avoid injection periods that are less than the micro irrigation system's water travel time to the end (hydraulically) of the system.

Second, an injection should always be followed by a period of "clean" water irrigation. This postinjection irrigation should be at least as long as the water travel time to the end of the system.

The worst chemigation uniformity results from too short of an injection period (less than the end-of-system travel time) followed by immediate micro irrigation system shutdown. In fact, field measurements and laboratory studies have shown that clearing injected material from a micro irrigation system takes even longer than it does to originally move the injected material through the system. Flow velocities at a pipe or tubing wall are very small (theoretically zero), so some of the injected material "hangs up" on the pipe or tubing walls and takes quite a while to clear from the system.

14.4   SUMMARY

While a wide variety of materials are now injected through micro irrigation systems, all of them should be applied with as great application uniformity as possible. This requires a highly uniform micro irrigation system and a high quality injection system that is managed properly. Very important to good management of the injection system is selecting the correct injection time and the appropriate post injection clean water irrigation period. Both the injection time and the postinjection clean water application should be as long as or longer than the travel time of chemical moving from the injector to the farthest point hydraulically in the micro irrigation system. If this is done, the injected chemical application will be as uniform as the water application.

KEYWORDS

- batch tank
- chemigation
- chemigation duration
- chemigation uniformity
- clogging
- emitter
- fertilizer
- injection period
- irrigation uniformity
- micro irrigation
- positive displacement pump
- postinjection
- venturi method

CHAPTER 15

MICRO IRRIGATION: FILTRATION SYSTEMS*

MEGH R. GOYAL

CONTENTS

15.1 Introduction 288
15.2 clogging Agents 288
15.3 Prevention of Obstruction or Clogging 289
15.4 Types of Filters 289
15.5 Principle of Operation of Filters 293
15.6 Selection of filter 296
15.7 Service and Maintenance of the filters 297
15.8 Prevention of clogging 299
15.9 Summary 300
Keywords 300
References 301

*This chapter is modified and printed with permission from: "Goyal, Megh R., 2012. Filtration systems. Chapter 10, pp. 195–212, In: *Management of Drip/Trickle or Micro Irrigation,* Editor Megh R. Goyal. Oakville, ON, Canada: Apple Academic Press Inc., ISBN 9781926895123." © Copyright 2013.

15.1   INTRODUCTION

Drip irrigation is a novel technique, which has extended in different regions of the world. This system provides many benefits compared to other irrigation systems. Also, the drip system involves application of new technology for a guaranteed operation. The success of drip irrigation depends on the filtration of water. The purpose of any filtration system is to reduce any clogging by removing any suspended solids or particles from the irrigation water. Clean water free of contaminants is of vital importance for a better operation of the drip irrigation. An adequate filtration system is an indispensable requirement to eliminate free suspended solids. Partial or total obstructions by different agents may render the drip irrigation system out of order. Thus clogging or obstruction causes economic loss to the farmer. Therefore, the selection of an adequate filtration system is one of the considerations for designing and installing a drip irrigation system (3, 5, 7).

15.2   CLOGGING AGENTS [1, 2, 4, 7]

The presence or development of particles causes the obstruction problem, thus reducing the emitter flow. The problem is progressive, if no corrective measures are taken at a proper time. Once the emitter flow rate decreases, the obstruction is accelerated leading to a complete clogging. The obstruction may occur in any place in the system but predominantly in the emitters/drippers. The problem can be avoided by using clean water and avoid the injection of agents that may precipitate. The clogging agents are grouped in three categories: Physical, chemical or biological. In some cases, the obstruction may be due to two or more of these factors at the same time.

15.2.1   PHYSICAL AGENTS

Different types of soil particles are present in the irrigation water. These particles may consist of soil material of different sizes (sand, silt and clay), sedimentation that builds up in the reservoirs, and small pieces of tubing from pipe lines, etc. These particles can be large enough to pass through the filters or irrigation lines.

Clay particles in combination with salts can accumulate on the internal walls of drippers or the filters, and thus can reduce performance of drippers or the filters. Silt and clay with other chemicals may form aggregates to cause clogging of drippers. During the planning stage of a filtration system, it is important to test the water source for chemical and physical analysis.

15.2.2   CHEMICAL AGENTS

Irrigation water contains many water-soluble salts that can precipitate on the surface of the drippers as the water evaporates between the irrigation intervals. If the salts cannot be dissolved, then mineral deposits can be formed to obstruct the drippers. A high concentration of calcium, magnesium and bicarbonate ions in the irrigation water promotes deposits of calcium and magnesium carbonates. High concentrations of calcium and sulfate ions may cause formation of calcium sulfate on the surface of the dripper. Generally the subsurface water contains iron and manganese salts. On contact with atmospheric air, the insoluble oxides of iron and manganese precipitate. These may cause obstructions of the drippers. Water rich in sulfite ions may produce insoluble sulfur compounds. Besides naturally occurring inorganic compounds, the precipitates can also be formed due to injection of liquid fertilizers or other chemical substances in the irrigation water. Soluble fertilizers may be used with success. There are simple qualitative laboratory tests to identify most of the insoluble compounds. Testing of irrigation water is essential to know the probability of clogging by chemical deposits. It is recommended to identify the insoluble compounds before chemigation; and to use a good filtration system.

15.2.3 BIOLOGICAL AGENTS

The irrigation system can also be obstructed by macro organisms and microorganisms. The irrigation water that comes from the surfaces open to atmospheric air may contain great amounts of organic material. Organic matter content in water may consist of partially decomposed organic matter (mostly of vegetative origin) and microorganisms (algae, bacteria and protozoa). The environmental conditions may favor development and rapid formation of different types of microorganisms. Some species like Sphaertilius and other microbiological cells multiply in darkness and in the presence of iron and sulfur in water. The oxides of iron, manganese and sulfur can be formed in the presence of microorganisms and these oxides may increase the obstruction. Crustaceans, fishes and other microorganisms may cause obstruction problems in the filtration system and irrigation lines. Spiders, ants and other insects may obstruct the drippers. A visual inspection normally helps to identify these organisms. To identify appropriately bacteria, algae and protozoa, a microscopic examination is needed. It is not necessary to obtain an exact identification of the microorganisms, but it is important to identify the presence of nonorganic compounds in the irrigation water.

15.3 PREVENTION OF OBSTRUCTION OR CLOGGING

The need to prevent obstruction by means of an adequate filtration unit should not be underestimated. This importance can be illustrated with the following example: If water is pumped from a pond to a drip irrigation system, this water might contain organic matter, weeds and suspended particles. In the absence of a filtration system, the drippers will be clogged quickly. If the problem is not corrected, the complete system may become useless. Under any circumstances, untreated water should not be used in drip irrigation. The water may look clean. It must be filtered with an adequate filtration system. Depending on the type of impurities and suspended solids, one may use screen filters, disk or ring filters, sand filters and hydrocyclone filters. Other treatments include chloration; acidification and use of compressed air.

15.4 TYPES OF FILTERS (SEE FIGS. 1–5)

15.4.1 GRAVITY FILTERS

Ponds, lakes, open ditches, irrigation channels and water reservoirs are good candidates for the gravity filter. Gravity filters can separate part of the suspended solids from the water. The method is not very reliable if the water has high concentration of microorganisms.

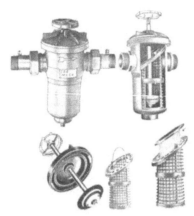

FIGURE 1   Screen filter commonly used in a drip irrigation system.

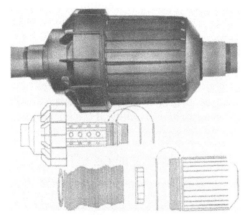

FIGURE 2   The components of disk (ring) filter commonly used in a drip irrigation system.

FIGURE 3   Disk filter: Flushing operation.

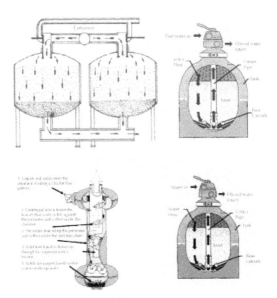

FIGURE 4   Cut section of a typical sand filter.

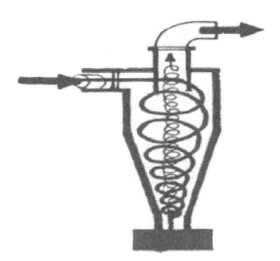

FIGURE 5   Hydrocyclone filter.

15.4.2   PRESSURE FILTRATION SYSTEM

15.4.2.1   SCREEN FILTERS

The screen filter is the simplest of all filtering methods. The principal component is a sieve made of metal, plastic or synthetic fabric enclosed in a special casing. The sieves are classified by number of squares per square inch, with a standard mesh for each size of sieve. The mesh size of a screen increases as the squares become smaller.

For a mesh size of 200, the standard caliber is 75 micrometers (0.0029 inches). Most manufacturers recommend sieves of 75–150 micrometers to avoid clogging in the drippers. However, filters with a mesh size of 30 (600 micrometers) can be used, if water quality is good.

A double screen filter is shown in Fig. 6. It consists of two cylindrical screens (3, 4, 5). The external screen is made of a sieve of 80 mesh size to separate the suspended particles that are larger than 2×10^5 nanometers. The internal screen is made of a sieve of 120 mesh size to separate the particles up to 1.25×10^5 nanometers. Particles smaller than 1.25×10^5 nanometers are not considered harmful to cause clogging of drippers. The double screen filter can be easily dismantled, cleaned and the sieves can be replaced quickly. These filters are available with sieves of different mesh sizes depending on the quality of water. A simple screen filter may consist of one cylindrical sieve. In each type of a screen filter, the sieve separates solid particles. The presence of biological agents in the irrigation system may cause obstructions in the sieve and reduce the filtration capacity. The screen filter can be used in series with other types of filtration devices to make sure that the water is free from sediments and is suitable for the drip irrigation system.

15.4.2.2   DISK OR RING FILTER

The type of filter is similar to a screen filter. It consists of a number of disks encapsulated inside a special cover. The disks have small pathways at the surface by which water is filtered as it passes through these pathways (Figs. 7 and 8). It is important that the disks are tightly fitted together for an efficient filtration and to avoid obstruction.

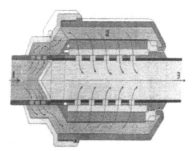

FIGURE 6   Components of the double screen filter.

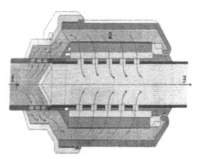

FIGURE 7   The direction of flow of water through a disk filter: 1. Intake of water; 2. Flow through disks or rings; 3. Clean water at the outlet (Exit).

FIGURE 8 Flushing method for the disk filter.

15.4.2.3 SAND FILTER

In the past, sand filters have been used for domestic and industrial purposes. Before the introduction of the trickle irrigation technique, the use of these filters was limited because of high cost. Sand filter (Fig. 4) consists of a fine gravel and sand of selected size. Sand filters are not easily obstructed by algae and can separate high quantities of suspended solids. Sand filters may separate particles from 25 to 100 micrometers. In general, the flow rate through these filters cannot exceed 14 lps (20 gpm) per surface area of filtration. The depth of filtration medium (sand) in the tank should be greater than 500 mm (18 inches). Secondary filters (screen or disk) should follow the sand filters. A flushing valve helps in the cleaning of the sand filters.

15.4.2.4 CENTRIFUGAL FILTER

The centrifugal filter (hydrocyclone or a sand separator) separates the solids from the water. It is a simple device to separate particles of 2×10^3 to 2×10^5 nanometers from water, at a flow rate greater than 600 lph. Basically, the centrifugal filters can separate solid particles that have a specific gravity greater than the water. These are primary filters that are effective to separate particles before the water enters the drip irrigation system. These can also be installed at the intake of the pump (as prefilter) to minimize the wear of the pump. Screen filter (secondary filter) must follow the hydrocyclone filter to avoid clogging. Centrifugal filters are available in different sizes for different discharge rates. These filters are inefficient to separate most of the organic matter and substances.

15.5 PRINCIPLE OF OPERATION OF FILTERS (SEE FIGURES 9–12)

15.5.1 GRAVITY FILTER

The storage of water in a tank for a specified period promotes suspended solids (like gravel, sand, silt and clay) to settle down at the bottom of the tank. Therefore, the solids with a specific gravity greater that water are separated.

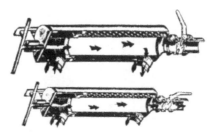

FIGURE 9   Screen filter: Draining operation (top) and the filtration operation (bottom).

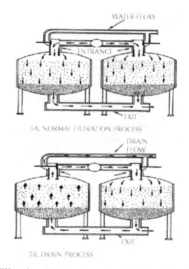

FIGURE 10   Sand Filter: Filtration (5A, top) and principle of flushing (5B, bottom).

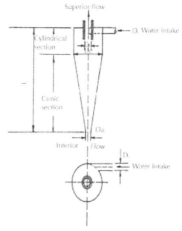

FIGURE 11   Hydrocyclone (centrifugal) filter showing water intake, superior and inferior flow.

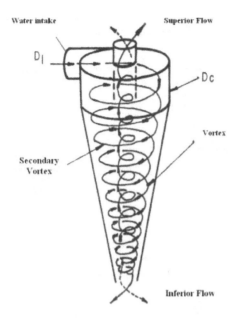

FIGURE 12   Vortex formation with superior and inferior flow in a centrifugal filter and quality of the suspended solids to be removed; the chance of chemical o biological obstructions; and quality of water after filtration.

15.5.2   PRESSURE FILTERS

15.5.2.1   SCREEN AND DISK FILTER

The principle of operation for both types of filters is similar. The water enters a closed chamber and is forced to pass through sieves or pressed disks (or rings) as shown in Fig. 7. In this way, the sediments are trapped in the filtering unit and the clean water is forced out of the filters.

15.5.2.2   SAND FILTERS

Sand filter is a most efficient method to separate organic matter in suspension and organic solids. Contaminated water (without filtration) enters the filter through superior part of the filter and flows (under pressure) through the sand where the suspended particles are trapped. The filtered water flows to the main line.

15.5.2.3   HYDROCYCLONE FILTER

The unfiltered water enters tangentially to the cylindrical section. It creates an angular momentum to cause a rotational movement of the water in the cylindrical section (D_c). A conical form of the filter allows the separation of solid particles from the water. This movement of water is called a vortex.

The solid particles, heavier than water, are thrown at the periphery of a circular path. Due to gravity, the particles go down to a flushing tank that is located at the lowest part of a cylinder. Later the sediments are removed from the flushing tank by opening a gate valve, when it is necessary. Part of the water movement is separated from the principal vortex to form a secondary vortex. The secondary vortex produces an upward movement in the middle and the clean water flows out through the superior exit (D_i).

15.6   SELECTION OF FILTER

An appropriate filtration system is selected to provide the required filtration with minimum cost and at maximum efficiency. The selection of a filter is based on the following factors:

1.   Calculate the size of the irrigation system (the flow rate, operating pressure and volume of water). The capacity of the filter should exceed the total demand of irrigation.
2.   Determine the physical, chemical and biological quality of the irrigation water; the size
3.   Ask the following questions:
     a.   How complex is the filtration system? What are the limitations for flushing?
     b.   Is the manual labor available to perform service and maintenance?
     c.   Is the location of the filter a problem? Where is the location of the water source?
     d.   Is the filtration system flexible? What are possibilities for future modifications?

The filters are available in different sizes on the basis of flow rate (Table 1). The selection of an adequate filtration system is an important consideration in the design of drip irrigation system.

TABLE 1   Guide for the selection of a filter[1].

Flow Rate	Concentrations of solids		**Recommendations
	*Organic	*Inorganic	
Less than 11.4 m³/hr.	L	L	A
(50 gpm)	L	M	A
	L	H	A
	M	L	A
	M	M	B+A
	M	H	B+A
	H	L	B+A
	H	M	B+A
	H	H	B+A
11.4 to 45.6 m³/hr	L	L	A
(50–200 gpm)	L	M	A
	L	H	A
	M	L	A+B
	M	M	A+B or A+C

TABLE 1 *(Continued)*

Flow Rate	Concentrations of solids		**Recommendations
	*Organic	*Inorganic	
	M	H	A+B or A+B+C
	H	L	A+C
	H	M	A+C or A+B+C
	H	H	A+B+C
>45.4 m³/hr	L	L	A
(200 gpm)	L	M	A+C
	L	H	A+C
	M	L	A+C
	M	M	A+B+C
	M	H	A+B+C
	H	L	A+C
	H	M	A+C+C
	H	H	A+B+C

¹The purpose of this guide is to select a correct size of the filter. The specific and individual requirements of the design must be evaluated.
* **Key for concentration of solids:** L = Less than 5 ppm; M = 5–50 ppm; H = Greater than 50 ppm
** **Key for recommendations:** A = Disk or screen filter; B = Hydrocyclone filter; C = Sand filter

15.7 SERVICE AND MAINTENANCE OF THE FILTERS

The filtration system is a pulmonary system of a drip irrigation system. Therefore, we must make sure that the filters are flushed and these perform efficiently. It is necessary to install pressure indicators (gages) at the entrance and at the exit of a filter. When the filters are clean, the gages should indicate the same pressure.

The pressure difference across the filters will increase due to obstruction problems. When this pressure difference is from 10 to 15 psi (depending on the type of filter), then the operator must clean or flush the filters according to the specifications of a manufacturer.

15.7.1 FLUSHING OF FILTERS

Filters should be flushed before each irrigation, or when indicated by the pressure gages. In case of water with large quantities of sediments or contaminants, the filters should be cleaned or flushed frequently. Entrance of dust into the system should be avoided.

15.7.1.1 DISK FILTER

The flushing or the cleaning is achieved by moving the disk (rings) in the direction of the flow. These are then washed with pressurized water (Figs. 3, 7, and 8). After the flushing operation the disks are placed and pressed together. The filter casing is then adjusted and closed tightly. If the disks are not tight enough, the filtration efficiency decreases and the problems of obstruction may occur.

15.7.1.2   SCREEN FILTER

The screen filter is commonly equipped with a cleaning brush or a valve in the inferior part of the filter. If the pressure difference between the gages at the intake and the exit is greater than the allowable difference in pressure, then the flushing is required. Open the drain or flush valve so that the pressurized water escapes through this valve (Fig. 9). Move the cleaning brush up and down, giving it a torsional movement (if it is provided with the unit). This will help to eliminate the sediments through the flushing valve. If the gages still show a need for flushing, shut off the irrigation system and dismantle the filtering assembly. Check if the screen is not obstructed with sediments. Wash the screen and internal parts with pressurized water to make sure that it is in good condition. Broken and defective parts should be repaired or replaced. Assemble the filter to make sure all the parts are installed properly.

15.7.1.3   CENTRIFUGAL FILTER (FIGS. 5, 11, AND 12)

The sediments accumulated in the tank can be removed in two ways:
1. When the system is not in operation, open flushing valve and let the sediments escape out.
2. When the filter is in operation, open the flushing valve and let the sediments escape out.

15.7.1.4   SAND FILTER

Sand filters are flushed by a back flow. For this type of flushing, the two adjacent filters must be interconnected (Fig. 10). The flow is inverted to one of the filters by means of a three-way valve (Fig. 13), while maintaining the second filter in operation under normal conditions. This causes turbulence in the filter, the particles and matter trapped in the sand are escaped out through a flushing pipe (Fig. 10). The water flow for flushing is supplied through a superior part of filter and taken out through a separate flushing line. By using two filters in series, the irrigation operation is not interrupted during the flushing process. This way, two or more filters in series are flushed successively.

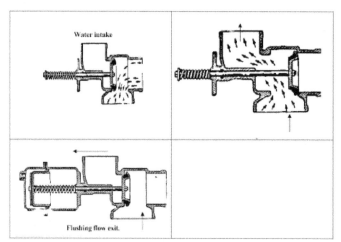

FIGURE 13   Three way valve for flushing of sand filters. (a) Manual valve: Filtration is in process. (b) An automatic three-way valve. (c) Flushing flow exit.

An excessive back flow will expand the sand to a point that the sand itself is expelled out back from the tank during the inverse flow. An insufficient flow will not expand the sand enough to flush the trapped contaminants. To obtain the maximum flushing efficiency, adjust the back flow.

15.8 PREVENTION OF CLOGGING

To avoid obstructions in the drip irrigation system [1–6], one must use an efficient and adequate filtration system. If this is not taken seriously, then it will cost more at later stage and it will be necessary to do modifications in the system. Simple treatments are not always successful to avoid obstructions, because it may be due to a specific condition. Therefore, a particular situation needs a specified solution.

15.8.1 PHYSICAL AGENTS

Besides an adequate filtration system, lines and drippers should be cleaned regularly to prevent obstructions by physical agents. For flushing, open the ends of the principal (main) line and allow expulsion of the accumulated sediments. Then repeat the same process with the secondary (sub main) and lateral (drip) lines. Under extreme conditions of obstructions in the lines and drippers, one can correct these by a pressurized air. Before this process is initiated, start the system and let it run for approximately 15 min. Once the system is charged with (or full of) water, apply pressurized air (approx. 7 bars). The compressed air will force out accumulated material from the lines or drippers.

15.8.2 CHEMICAL AGENTS

Most of the causes of obstruction by chemical agents may be resolved by chloration or treatment with acid. By this process, chemical deposits caused by water and fertilizers are dissolved. In severe cases, the drippers are immersed in a diluted acid solution (approximately one percent). In some cases, these are flushed individually. For less severe cases, it is adequate to inject the acid to lower the pH of water. The process should be repeated till a normal flow rate is obtained through the dripper. The quantity of acid required to lower the pH is determined by testing a small volume of water.

Acids are extremely corrosive, so extreme caution should be taken. Commonly used acids are commercial grade sulfuric, hydrochloric and nitric (muriatic) acids. "**Do not use or apply any chemical agent through the drip system**" without consulting a specialist. The alternative between one method and another will depend on the availability and the cost. The acids may corrode the metal accessories, tubing and casings. Surfaces in contact with acids should be made of stainless steel and plastic. All these parts should be well washed after being in contact with the acid.

15.8.2.1 QUANTITY OF ACID REQUIRED TO LOWER THE PH OF IRRIGATION WATER

The quantity of acid required to lower the pH of irrigation water to a given value can be calculated using Eq. (1) for gallons applied per hour of water flow and Eq. (2) for gallons per each 100 gallons of water. In both equations, the acidity factor and the normality of the acid should be known, for example: Concentrated sulfuric acid (H_2SO_4) =

36N; Concentrated hydrochloric acid (HCl) = 12N. Acidity factor is a milliequivalents of acid per liter of water to lower the pH to a desired level. It is determined by titration of water sample with acid, in a laboratory.

Gallons of acid per hour = [0.06 × acidity factor × gpm]/[Acid normality]   (1)

Gallons of acid per 100 gallons of water = [Acidity factor]/[Acid normality] (2)

15.8.3 BIOLOGICAL AGENTS

When the silt or algae obstruct the emitters, the common treatment is an injection of a biocide (chlorine) followed by complete flushing of the lines to remove any organic matter. The common biocides are chlorine gas (Cl_2) and solutions of hypochlorite ($HOCl^-$). An efficient algae and bacterial control in a drip irrigation system may be maintained by administration of 10–20 ppm during the last 20 min of the irrigation cycle (see Chapters VIII and IX). If the irrigation water has a concentration of one ppm of chlorine, then the obstruction due to biological agents is under control.

15.9 SUMMARY

We must select an adequate filtration system to provide water free of clogging agents. This chapter discusses clogging agents, prevention of clogging, types of filters, and principle of operation of filters, selection of filters, service and maintenance of filters, trouble shooting and procedure to solve problems of clogging.

KEYWORDS

- acidity factor
- back flow
- bar
- chemigation
- chlorination or chloration
- clay
- clogging
- cylindrical section (Dc)
- drip irrigation
- dripper
- emitter
- emitter flow rate
- fertilizer
- fertilizer, liquid
- filter
- filter, double screen
- filter, hydrocyclone

- filter, mesh
- filter, sand
- filter, screen
- filtration system
- flushing valve
- gate valve
- main line
- nipple or union
- organic matter
- pH
- pump
- soluble salts
- water quality
- water, source
- weed

REFERENCES

1. Design, Installation and Performance of Trickle Irrigation Systems. *ASAE Engineering Practice*, ASAE EP. 405, 2005.
2. *Drip Irrigation Management*. Cooperative Extension Service, Division of Agricultural Sciences, University of California, Berkeley-CA. 1981
3. Goldberg D.B., B. Gornat and D. Rimon, 1976. *Drip Irrigation: Principles, Design and Agricultural Practices*. Kfar Shmaryahu, Israel: Drip Irrigation Scientific Publications.
4. Goyal, M.R. and L.E. Rivera, 1985. Service and maintenance of drip irrigation systems [Riego por goteo: servicio *y* mantenimiento, Spanish]. Bulletin IA64, Series 5 by Cooperative Agricultural Extension Service, University of Puerto Rico—Mayaguez Campus.
5. Goyal, M.R., 2013. *Management of Drip/Trickle or Micro Irrigation*. Chapters 1–16. Apple Academic Press Inc., NJ.
6. Jensen E., 1980. *Design and operations of farm irrigation systems*. ASAE Monograph No. 3. American Society of Agricultural Engineers, St. Joseph, MI.
7. *Proceedings of the 3rd International Drip/Trickle Irrigation Congress*. Drip/Trickle Irrigation in Action. Volumes I and II. Fresno, CA, November 18–21, 1985. American Society of Agricultural Engineers, St. Joseph, MI. Publication No. 10–85.

CHAPTER 16

MICRO IRRIGATION DESIGN USING HYDROCALC SOFTWARE*

RARES HALBAC-COTOARA-ZAMFIR and EMANUEL HATEGAN

CONTENTS

16.1   Introduction... 304
16.2   Hydrocalc Software 304
16.3   Case Study 308
16.4   Conclusions... 315
16.5   Summary.. 315
Keywords.. 315
References... 316

*Modified and printed from, "*Rares Halbac-Cotoara-Zamfir and Emanuel Hategan. Designing a drip irrigation system using hydrocalc irrigation planning. Research Journal of Agricultural Science, 2009, 41(1): 419–425.*"

16.1   INTRODUCTION

Drip irrigation also known as *trickle irrigation* or *micro irrigation* is an irrigation method that applies water slowly to the roots of plants, by depositing the water either on the soil surface or directly to the root zone, through a network of valves, pipes, tubing, and emitters. The goal is to minimize water usage. Drip irrigation may also use devices called microspray heads, which spray water in a small area, instead of emitters. These are generally used on tree and vine crops. Subsurface drip irrigation or *SDI* uses permanently or temporarily buried dripper-line or drip tape. It is becoming more widely used for row crop irrigation especially in areas where water supplies are limited. The drip irrigation is a slow and frequent application of water to the soil by means of emitters or drippers located at specific locations/interval throughout the lateral lines. The emitted water moves through soil mainly by unsaturated flow. This allows favorable conditions for soil moisture in the root zone and optimal development of plant. With this system, the plant can efficiently use available natural resources such as: soil, water and air. The drip irrigation is also known as "daily irrigation," "trickle irrigation," "daily flow irrigation" or "micro irrigation." The term "trickle" was originated in England, "drip" in Israel, "daily flow" in Australia and "micro irrigation" in USA. The difference is only in the name, and all these terms have the same meaning. The water in a drip irrigation system flows in three forms: It flows continuously throughout the lateral line; It flows from an emitter or dripper connected to the lateral line; It flows through orifices perforated in the lateral line.

A well designed drip irrigation system can increase the crop yield due to following factors: Efficient use of the water (reduce the direct losses by evaporation, doesn't cause wetting of the leaves, doesn't cause movement of drops of water due to the effect of wind, reduce consumption of water by grass and weeds, eliminates surface drainage, allows watering the entire field until the edges, allows applying the irrigation to an exact root depth of crops, allows watering greater land area with a specific amount of water), reaction of the plant (increases the yield per unit (hectare-centimeter) of applied water, improves the quality of crop and fruit, allows more uniform crop yield), environment of the root (improves ventilation or aeration, increases quantity of available nutrients, conditions are favorable for retention of water at low tension), control of pests and diseases (increases efficiency of sprayings of insecticides and pesticides, reduces development of insects and diseases), correction of problem of soil salinity, weed control (reduces the growth of weeds in the shaded humid space), agronomic practices (the activities of the irrigation do not interfere with those of the crop, the plant protection and the harvesting, reduces inter cultivation, helps to control the erosion, reduces soil compaction, allows applying fertilizers through the irrigation water, increases work efficiency in fruit orchards), economic benefits (cost is lower compared with overhead sprinkler and other permanent irrigation systems, cost of operation and maintenance is low).

16.2   HYDROCALC SOFTWARE

HydroCalc irrigation system planning software is designed to help the user to define the parameters of an irrigation system. The user will be able to run the program with any suitable parameters, review the output, and change input data in order to match it to the appropriate irrigation system set up. Some parameters may be selected from a system list; whereas others are entered by the user according to their own needs so they do not conflict with the program's limitations. The software package includes an opening main window, five calculation programs, one language-setting window and a database that can be modified and updated by the user.

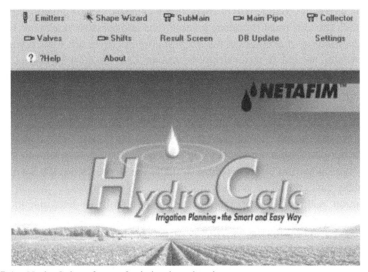

FIGURE 1   HydroCalc software for irrigation planning.

16.2.1   HYDROCALC SOFTWARE: EMITTER SUB-PROGRAM

HydroCalc software includes several subprograms that are listed below:

1. The **Emitters subprogram** calculates the cumulative pressure loss, the average flow rate, the water flow velocity etc. in the selected emitter. It can be changed to suit the desired irrigation system parameters.
2. The **SubMain subprogram** calculates the cumulative pressure loss and the water flow velocity in the submain distributing water pipe (single or telescopic). It changes to suit the required irrigation system parameters.
3. The **MainPipe subprogram** calculates the cumulative pressure loss and the water flow velocity in the main conducting water pipe (single or telescopic). It changes to suit the required irrigation system parameters.
4. The **ShapeWizard subprogram** helps transfer the required system parameters (Inlet Lateral Flow Rate, Minimum Head Pressure) from the Emitters program to the SubMain program.
5. The **Valves subprogram** calculates the valve friction loss according to the given parameters.
6. The **Shifts subprogram** calculates the irrigation rate and number of shifts needed according to the given parameters.

The **Emitters subprogram** is the first application that can be used in the frame of HydroCalc software. There are 4 basic type of emitters, which can be used: Drip Line, online, minisprinklers and sprinklers. According to the previous selection, the user can opt for a specific emitter, which can be a pressure compensated or a non pressure compensated. Each emitter has its own set of nominal flow rate values available.

After the previous mentioned fields are completed, the program automatically fills the following fields: "Inside Diameter," "KD" and "Exponent," values which cannot be changed unless the change will be made in the database. The segment length is next field in which the user must introduce a value. The end pressure represents the actual value for calculation of pressure at the furthest emitter. There are some common values for this field: around 10 m for drippers, around 20 m for mini-sprinklers, between 20–30 m for sprinklers and around 2 m when using the flushing system. There are two more options, which can be filled before starting the computation, options which can also be used with their default values. The Flushing field can be used if the user intends to calculate a system that includes and lateral flushing. Flushing option will work only in subsequently will be used the "Emitter Line Length" calculation method. The second option is about topography. Default value is 0%. Topography field has two subfields: fixed slope and changing slope. Usually the slopes values are not exceeding 10%. In many cases the slope is not uniform.

The option "**Changing**" offers to the user the possibility to determine the altitude along the line, from end line to submain, up to 10 points each consisting of 2 cells: one for distance one for height. The line of distances is used to load the distance from the end of the line to the point of the altitude change. These distances may be unequal. The value of each cell is the net distance from the end of the line point. The value in length in each cell is always greater than the value entered in previous cell. The lower line, line of "heights" is used to enter altitudes of the point whose distance from the end of line is set above them. The values may represent elevation readings from a map or relative elevation to zero level at the end of the line. Positive values mean the point is elevated above the end of the line while negative values mean the point is below the end of line elevation.

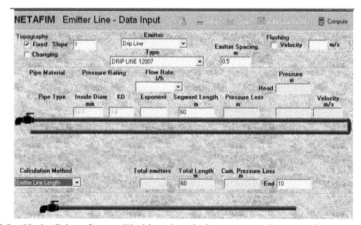

FIGURE 2 HydroCalc software: Working sheet before computation procedure.

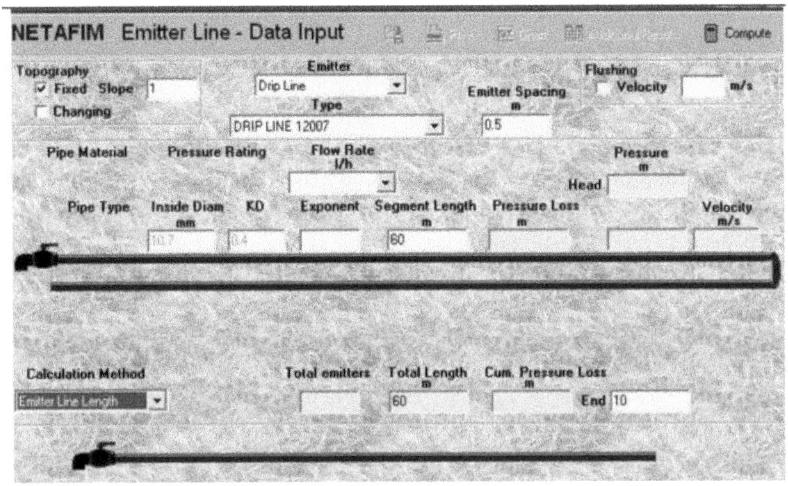

FIGURE 3 HydroCalc software results: working sheet.

HydroCalc uses, for emitters' subprogram, four calculation methods each of them in concordance with the loaded data. The first method is "Emitter Line Length" with which can be realized the computation for the entire designated length. The second method is represented by "Pressure range," a calculation which will be executed in a way that makes sure the maximal pressure variation between maximum emitter's pressure to minimum emitter's pressure does not exceed the pressure range which was introduced by the user. The computation result will also show the maximum lateral length under the designated conditions. "Flow Rate Variation" represents the third computation method, which can be executed to achieve the requested flow variation and will generate the maximum lateral length under these conditions. Flow variation units are in percents. The common values for this field are between 10–15%. The last computation method is "Emission Uniformity" which is similar to "flow rate variation," and will be executed to achieve the maximum lateral length. Emission uniformity units are also in percents but the common value for this field is any value above 85%.

After were entered all the required values can be launched the "Compute" application. The cells which are going to be calculated in the emitter window are: Total Emitters, Total Length, Total Pressure Loss, Pressure Loss, Head Pressure and Velocity per segment. The additional results report can be reached by pressing the "Additional results" button. This report contains the following results: Average Emitter Flow Rate, Inlet Lateral Flow Rate, and Flow Rate Variation (FV, %) that is the difference in percentage between the maximum emitter's discharges to the minimum emitter's discharge using the formula:

$$FV(\%) = \frac{Q_{max} - Q_{min}}{Q_{max}} \cdot 100 \tag{1}$$

Emission uniformity is defined in equation below (ASABE):

$$EU(\%) = \frac{Q_{min}}{Q_{max}} \cdot \left[1 - 1.27 \frac{CV}{\sqrt{n}}\right] \qquad (2)$$

and also Min/Max Emitter Pressure, Inlet Velocity, Total Length, Total Emitters, Emitter Line Pressures with Head Pressure and End Pressure, Emitter Distance and Emitter Line Pressure.

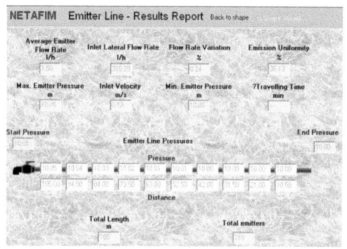

FIGURE 4   Results: Report sheet.

By pressing the "Chart" button it will be displayed the emitter line pressure results chart. The Fig. 4 shows the changes of the pressure and flow rate variation along the line.

16.3   CASE STUDY

The case study is based on designing process of a drip irrigation arrangement in Banloc commune—Timis County—Western Romania. The necessary studies for this project include pedological survey, geotechnical survey, hydrogeological study, general site plan. The project includes the location, the beneficiary, the project and some general information about the conditions of climate, soils, hydrogeological and hydrological conditions and, according to them and the type of arrangement, there is proposed some additional technical measures.

For this case study, it was provided the plan of arrangement, arrangement perimeter, pedological soil mapping, hydrogeological mapping and elements specific to the culture and irrigation. Thus, the area was divided into 20 operational units, each of them having 4 ha with shape square (side 200 m). For this arrangement three crops were selected, 60% for the first culture and the other two crops are grown on about 20% of the arrangement. The water source for the considered irrigation area (S_{TOT} =

80 ha) is the Bârzava River. Irrigation rate for all three crops is 2500 m³/ha, rate which must be given during T = 10 days.

FIGURE 5   The site for the drip irrigation arrangement (Banloc commune—in red, locality in Timis County, western Romania).

It must be stated that the entire arrangements covers a flat surface (slope 0). Units used below are those used by the program. Sizing watering pipes, laterals of main pipes were done by using HydroCalc program, version 2.21 by Netafim. Sizes, diameters, flow rates, pressure pipes that are presented as follows.

Watering pipe sizing for the first crop: On each plot of 4ha will be distributed a watering pipe for two rows of plants, since the diameter of the dispensing wet bulb gives a sufficient volume of water to cover the part of plant surface which must be wetted on both rows. We will start the Emitters subprogram from the main menu of the program. The first step is the introduction of the input data which are the following: drip line, type drainer—Button, emitter spacing—0.8 m, flow rate is 2 L/h, pipe material is PE (polyethylene), pipe operating pressure is 2.5, Type of pipe—32 mm with an inner diameter of 28.8 mm and another 3 features introduced automatically by the program according to the type of pipeline, the segment length is 200 m; downstream pressure have chosen to be 10 m. The next step is to choose the calculation method based on data lengths of drip line.

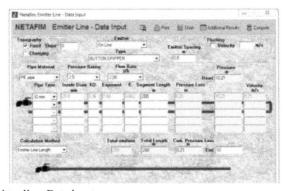

FIGURE 6   Emitter line: Data input.

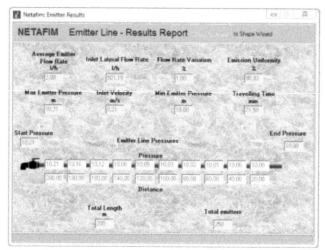

FIGURE 7   Emitter line: Results report.

After running the program we have the following calculations: we have 250 emitters on the drip line, total length is 200 m (we used a single section). Upward pressure is 10.21 m, the pressure loss is 0.21 m and water flow velocity is 0.21 m/s. Additional results of this calculation are shown in Fig. 7. The Graph related to this calculation is shown in Fig. 8.

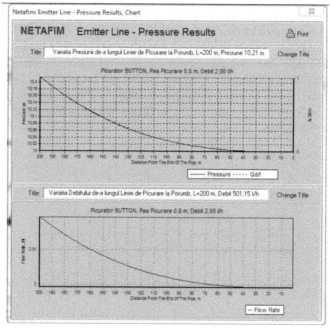

FIGURE 8   Emitter line: Pressure results.

We will proceed in the same way for the watering pipe for the second and the third crop. The input data are: drip line, Button drippers, emitter spacing 0.8 m, flow rate is 1.00 L/h The rest remain the same input data and the size of the pipe from the first crop irrigation. After running the program we have the following calculations: total emitters—500, total length is 200 m (used a single section), upward pressure is 10.29 m, the pressure loss is 0.29 m and the inlet velocity is 0.25 m/s.

Sizing secondary pipe for the first culture: On the area covered by the first crop will be distributed four 600 m long pipelines to distribute the required flow for an optimal functioning of irrigation pipes. Each secondary pipe will be fitted at the downstream end with a valve and a pressure regulator for obtaining the optimum flow rate.

16.3.1 SIZING CALCULATION USING SUBMAIN SUBPROGRAM

Sizing calculation is done in **SubMain** subprogram. Input data (Fig. 9): downstream flow is 501.15 Lph obtained from the calculation of the watering pipe and then we enter the number of lateral irrigation pipes—75. The pipe type is chosen depending on the pressure loss, which should be as low as possible. Pipe material is PVC with a coefficient of friction of 150 (Class 4 pipe). We choose a 180 mm nominal diameter, 170 mm inner diameter, and segment length—600 m. In this subprogram we must introduce the upstream pressure. This is done to achieve the desired downstream pressure that must be inlet pressure of watering pipe (10.21 m).

After running the program we obtain the following results: Total length is 600 m because we used a single section, pressure loss is 0.26 m and the cumulative pressure loss is 0.26 m for the same reason, downstream pressure is 10.24 m, upstream pressure is 10.50 m, water flow speed is 0.47 m/s inflow in secondary pipe is 38.09 m³/h. Additional results of this calculation are shown in Fig. 10. The Graph related to this calculation is shown in Fig. 11.

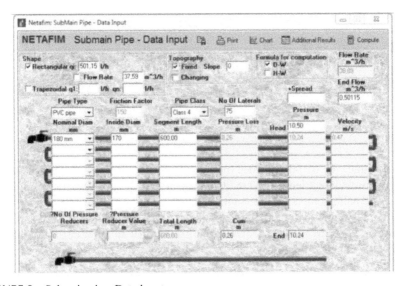

FIGURE 9   Submain pipe: Data input.

| NETAFIM: Submain Pipe - Additional Results | | | | ⊂▢ 🗙 |

NETAFIM   Submain Pipe - Additional Results   🖨 Print

Distance from the beginning m	Delta h m	Height m	Pressure Loss m	Pressure m
0,000	0,000	0,000	0,000	10.500
56,000	0,000	0,000	0,063	10,437
112,000	0,000	0,000	0,052	10,385
168,000	0,000	0,000	0,043	10.342
224,000	0,000	0,000	0.034	10,308
280,000	0,000	0,000	0,026	10,282
336,000	0,000	0,000	0,019	10.263
392,000	0,000	0,000	0,014	10.249
448,000	0,000	0,000	0,007	10.242
504,000	0,000	0,000	0,005	10.237
560,000	0,000	0,000	0,000	10.237
600,000	0,000	0,000	0,000	10.237

Min Pressure	10.24	m	at distance	488.00	m	from the beginning
Max Pressure	10.50	m	at distance	0,00	m	from the beginning
Req. lat. pressure	10.21	m				

FIGURE 10   Submain pipe: Additional results.

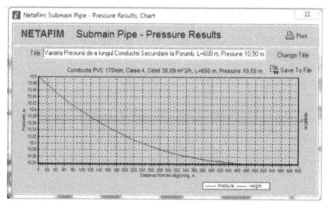

FIGURE 11   Submain pipe: Pressure results.

In the same way we will proceed for the second and third cultures with the related amendments: Segment length 400 m, downstream flow 577.40 Lph, 50 lateral irrigation lines, remaining inputs are the same as the first culture. The results of running the program: total length of 400 m, pressure loss of 0.11 m, the cumulative pressure: 0.11 m, upstream pressure 10.41 m, downstream pressure 10.30 m, Water flow rate 0.36 m/s, flow rate input 29.45 m³/h.

16.3.2 DESIGN OF MAIN PIPE USING MAINPIPE SUB-PROGRAM

The sizing process of the main pipe is done in MainPipe subprogram for four sections of 200 m each. The input data (Fig. 12) for the first 3sections are: Pipe material (type)—PVC, Friction coefficient—150, nominal diameter 280 mm, Class 6, 266.8 mm inner diameter, segment length of 200 m, inflow rates are 59.50 m³/h, 119 m³/h and 186.55 m³/h, pressures in the upstream are 10.6 m, 10.8 m, 11.3 m. As a result of the calculation we obtain the following results for the last segment of 200 m: the inflow rate is 186.55 m³/h, Pressure loss is 0.50 m, downstream pressure is 10.8 m and water flow velocity is 0.93 m/s.

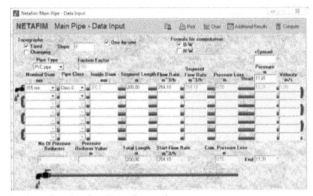

FIGURE 12   Design of main pipe: Data input sheet.

The input data for the last section of the main pipe: Type PVC pipe, Friction coefficient 150, nominal diameter 315 mm, Class 6, 300.2 mm inner diameter, segment length 200 m, 11.81 m upstream pressure resulting a 254.10 m³/h flow. Calculations made by the subprogram resulted in: total length 200 m, pipe entrance flow 254.10 m³/h, Pressure loss 0.50 m, 11.31 m downstream pressure, water flow velocity is 1 m/s. Additional results of this calculation are shown in Fig. 13. The Graph related to this calculation is shown in Fig. 14.

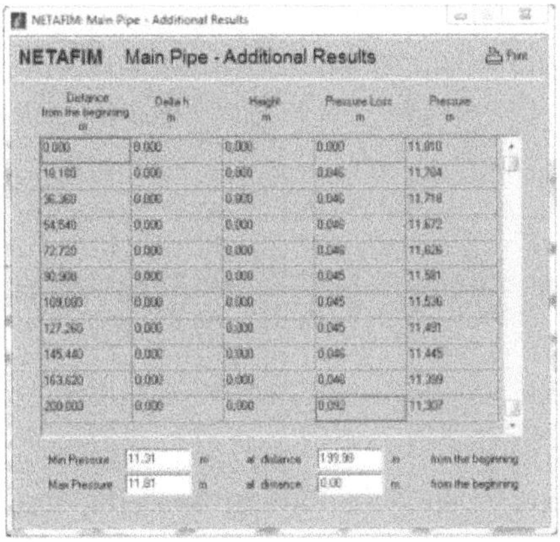

FIGURE 13   Main Pipe—Additional results.

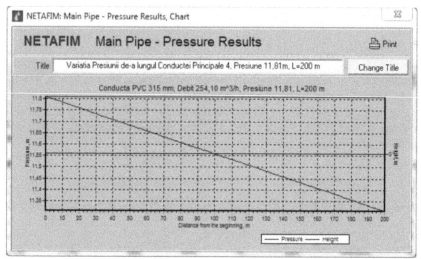

FIGURE 14   Main Pipe—Pressure results.

16.3.3   MAPPING THE PRESSURES LINE FOR THE ENTIRE NETWORK

After the sizing process of the last section of the main line we obtain the necessary data to entry throughout the irrigation system. The calculations are completed with mapping the pressures line for the entire network (Watering lines, laterals, main pipes, pumping stations), with checking the balancing of pressures in all nodes that make up the network and hence the final results. The pumping station flow will be $Q_{SP} = 254.10$ m³/h, pumping station pressure $H_{SP} = 11.81$ m. Depending on these values, we can select the pumps composing the pumping station.

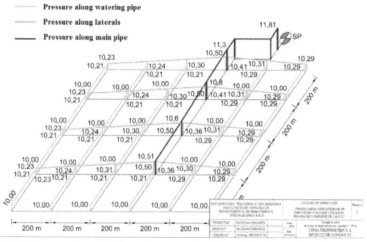

FIGURE 15   Pressure lines for the entire network: Along drip lines (watering), laterals and main pipe.

16.4 CONCLUSIONS

Drip Irrigation is an efficient method of water application in agriculture. The drip irrigation system allows a constant application of water by drippers at specific locations on the lateral lines it allows favorable conditions for soil moisture in the root zone and optimal development of plant. A well designed drip irrigation system can increase the crop yield due to: efficient use of water, improved microclimate around the root zone, pest control, and weed control, agronomic and economic benefits.

HydroCalc, by its characteristics, represents a reliable tool in designing a drip irrigation system. Due to its multiple advantages, this program must be introduced in universities and research centers for facilitating the work of Romanian researches and also for a better understanding by students of the problems regarding drip irrigation.

16.5 SUMMARY

The competitive demand of available water is more and more acute now in most parts of the world. The supplies of good water quality are declining every day. It is, therefore, necessary to find methods to improve water use efficiency in agriculture. The drip irrigation is one of the most efficient irrigation systems that are used in agriculture. Romanian farmers say that irrigation is more expensive than 20 years ago because of the rising temperatures and the land degradation (desertification), processes that are forcing them to irrigate more often. Solution: Drip irrigation systems, which are more economical. Romanian farmers have tried to cut the costs by appealing to new technologies. One such option is to replace sprinklers with drip irrigation systems that consume less water, these systems being especially suited for vegetable growing and horticulture. In dry summers, vegetable growers barely succeed to cover the costs because they have to irrigate the entire growing season (five months length). If in 1990 they had to irrigate once every two weeks, now this time was reduced to 3–4 days. This paper will present Hydrocalc software, an easy calculation tool to perform some basic hydraulic computations, as well as a brief study case for western Romania in which this program is used.

KEYWORDS

- ASABE
- design
- emitter
- emitter flow rate
- emitter flow rate, maximum
- emitter flow rate, minimum
- emitter flow variation
- HydroCalc software, Netafim
- irrigation
- lateral line

- main line
- micro irrigation
- Netafim
- operating pressure
- pressure loss
- Romania
- uniformity coefficient

REFERENCES

1. Burt, C.M., Styles, S.W., 2007. *Drip and Micro Irrigation Design and Management for Trees, Vines and Field Crops*, 3rd Edition, The Irrigation Training and Research Center.
2. Gilary, E., 2008. Hydrocalc Irrigation Planning User Manual, Netafim Corporation, Israel.
3. Goyal, M.R., 2012. Management of Drip/Micro or Trickle Irrigation, Apple Academic Press Inc., New Jersey, USA.
4. Halbac-Cotoara-Zamfir, R., 2011. Hydro-ameliorative arrangements: Designing of irrigation and drainage systems. "Politehnica" University of Timisoara, Romania.

GLOSSARY OF TECHNICAL TERMS: MICRO IRRIGATION MANAGEMENT

Agriculture is a dynamic field that has evolved throughout history and is an integral part of our food chain that sustains the mankind. Agriculture is subdivided into several specialties, such as, plant (crop) science, irrigation, plant physiology, crop protection, crop physiology, soil science, agroclimatology, agronomy, agricultural and biological engineering, economics, climatology, precision farming, water management, soil/water conservation, drainage, agricultural waste management, animal science, and so forth. Each one of these disciplines includes a series of terms and technical words. This chapter defines a series of commonly used terms in irrigation and water management. This will help the farmers, agronomists, soil scientists, engineers, water management scientists, technicians, and other people to broaden their knowledge in water management.

Technical words are often used when talking about water management, irrigation, evapotranspiration, etc. "Glossary of Technical Terms: Micro Irrigation" brings these words together and describes them using everyday language. The reader can use this as a reference tool. It should be useful to researchers, managers, government officials, students and the general public who are involved in the field of water resource management. A common understanding of these terms will make communication easier among the readers and scientists. The ability to communicate effectively and efficiently about water management is extremely important and will become increasingly so, as fresh water supplies for irrigation become scarcer. Use of consistent terminology is critical in presenting clear and meaningful information. Because the purpose of this glossary is to aid communication and understanding within and between water management and evapotranspiration communities, *the definitions contained within are not intended for legal use.* This glossary is provided for *general information only.*

The editor used a variety of other glossaries, dictionaries, atlases, technical papers and several sources that are listed in the bibliography. For the most part, the definitions used by various sources were accepted without modifications. Finally, in the interest of readability, many of the definitions in this chapter have been edited for brevity and clarity, while retaining the original intent as closely as possible.

Abandoned drinking water wells—include those abandoned water wells used for disposal of water.

Abandonment—failure to put a water right to beneficial use for generally five or more years, in which the owner of the water right states that the water right will not be used, or takes such actions that would prevent the water from being beneficially used.

Abandonment of a dam—in a legal sense, abandonment is most precisely described as transfer of all rights, title and interest in a dam to the current property owner. Abandonment may also involve the slow but resolute erosion of rights to a dam by nonuse, physical destruction, lack of maintenance or intent of it. In this latter instance the court holding jurisdiction can only decide the final determination of legal abandonment.

Abiotic—not involving or produced by an organism.

Absolute filter rating—filter rating meaning that 99.9% (or essentially all) of the particles larger than a specific micron rating will be trapped on or within the filter.

Absorption—(1) The entrance of water into the soil or rocks by all natural processes, including the infiltration of precipitation or snowmelt, gravity flow of streams into the valley alluvium into sinkholes or other large openings, and the movement of atmospheric moisture. (2) The uptake of water or dissolved chemicals by a cell or an organism (as tree roots absorb dissolved nutrients in soil). (3) More generally, the process by which substances in gaseous, liquid, or solid form dissolve or mix with other substances. Not to be confused with Adsorption. 4. The process by which one substance is physically taken into and included with another substance. For example, the penetration of water into soil or the entrance of gases, water, nutrients, and other substances into the plant.

Abyssal plains—flat or very gently sloping areas of the ocean floor, at the foot of a continental rise.

AC—abbreviation for alternating current (hertz).

AC pipe—asbestos-cement pipe was commonly used for buried pipelines. It combines strength with light-weight and is immune to rust and corrosion.

Accretionthe gradual build-up of sediments by the settling out of waterborne particles, sometimes resulting in land build-up.

Accumulation tank—a vessel or tank, which receives and stores product water for use on demand.

Acid—a substance, which releases hydrogen ions when dissolved in water. Most acids will dissolve the common metals and will react with a base to form a neutral salt and water. An acid is the opposite of an alkali, has a pH rating lower than 7.0, will turn litmus paper red and has a sour taste.

Acid aerosol—very small liquid or solid particles that are acidic and are small enough to become airborne.

Acid drainage—any drainage from mine workings, waste or tailings, with a low (acidic) pH.

Acid mine—the water discharge from coalmines, usually highly acidic, which can pollute the receiving stream.

Acid neutralizing capacity—measure of the buffering capacity of water; the ability of water to resist changes in pH.

Acid rain—rain with a pH of less than 7.0, due to contact with atmospheric pollutants such as sulfuric oxides.

Acidity—a measure of the capacity of a solution to neutralize bases; the quantitative capacity of water to neutralize a base, expressed in ppm or mg/L calcium carbonate equivalent. The number of hydrogen atoms that are present determines this. It is usually measured by titration with a standard solution of sodium hydroxide.

Acre foot (also acre feet)—a measurement of water quantity most often used in agriculture. The amount of water needed to cover one acre of area with water one foot deep. It is equivalent to 46,560 cubic feet or about 325,851 gallons or 1,233 cubic meters.

Acre inch—see "acre foot" in this chapter and substitute "inch" for "foot."

Activated alumina—a medium made by treating aluminum ore so that it becomes porous and highly adsorptive. Activated alumina will remove several contaminants including fluoride, arsenic and selenium. This medium requires periodic cleaning and appropriate reagent such as alum, acid and or/caustic.

Activated carbon—a water treatment medium, found in block, granulated, or powder form, which is produced by heating carbonaceous substances (bituminous coal or cellulose-based substance such as wood or coconut shell) in the absence of air, creating a highly porous adsorbent material.

Activated carbon filter—a filter made of carbon particles containing numerous pores and channels which trap contaminants as water passes through; not effective for heavy metals, bacteria, nitrates, and dissolved minerals; most effective for organic compounds and general taste and smell problems.

Activated coal—this is the most commonly used adsorption medium, produced by heating carbonaceous substances or cellulose bases in the absence of air. It has a very porous structure and is commonly used to remove organic matter and dissolved gases from water. Its appearance is similar to coal or peat. Available in granular, powder or block form; in powder form it has the highest adsorption capacity.

Activated sludge—oxygen dependent biological process that serves to convert soluble organic matter to solid biomass, that is removable by gravity or filtration; Sludge floc produced in raw or settled wastewater by the growth of zoological bacteria and other organisms in the presence of dissolved oxygen and accumulated in sufficient concentration by returning settled floc previously formed.

Activated sludge process—a biological wastewater treatment technique in which a mixture of wastewater and activated sludge is agitated and aerated. The activated sludge is subsequently separated from the liquid portion (mixed liquor) by sedimentation and digested or returned to the process as needed.

Active groups—really fixed ions bolted on to the matrix of an ion exchanger. Each active group must always have a counterion of opposite charge near itself.

Active root zone—is a root zone where the absorption of nutrients and the water movement is active.

Acute effect (chronic effect)—stimuli severe enough to rapidly induce a response. In aquatic toxicity tests, a response observed in 96-hours or less is typically considered acute. When referring to aquatic toxicology or human health, an acute effect is not always measured in terms of lethality.

Acute toxicity—exposure that will result in significant response shortly after exposure (typically a response is observed within 48 or 96 h).

Adapter—section of a pipe used to change the size of a tube. Using the adapter, we can join pipes with or without threads or inserted threads. There are male, female, and inserted adapters.

Adaptive capacity—the ability of a system (e.g., ecosystem) to adapt to climate change or other environmental disturbances. This may mean moderating potential damages, taking advantage of opportunities or coping with the consequences. In discussions on global warming adaptive capacity often refers to a country. In this case it is currently much lower in developing countries, consequential to poverty.

Adaptive management—a dynamic system or process of task organization and execution that recognizes the future cannot be predicted perfectly. Planning and organizational strategies are reviewed and modified frequently as better information becomes available. Adaptive management applies scientific principles and methods to improve management activities incrementally as decision-makers learn from experience, collect new scientific findings, and adapt to changing social expectations and demands.

Adenosine Triphosphate (ATP)—the primary energy donor in cellular life process. Its central role in living cells makes it an excellent indicator of the presence of living material in water. A measure of ATP therefore provides a sensitive and rapid estimate of biomass. ATP is reported in micrograms per liter of the original water sample.

Adhesion—attraction of water molecules for solid surfaces. Physical attraction of unlike substances to one another. In soils, it is the process that holds water molecules tightly to soil solids at the soil-water interfaces.

Adiabatic expansion—the process that occurs when an air mass rises and expands without exchanging heat with its surroundings.

Adjusted sodium adsorption ratio—index of permeability problems, based upon water quality.

Administrative penalty—a monetary penalty assessed by the regulator. An administrative penalty has a specific and general deterrent effect. The legislation specifies the contraventions for which administrative penalties are available.

Adsorption—Separation of liquids, gases, colloids or suspended matter from a medium by adherence to the surface or pores of a solid; the adhesion of molecules to the surface of solid bodies or liquids with which they are in contact; the tendency exhibited by all solids to exert a molecular attraction for other solids or compounds in wastewater, which are held to their surface until removed and physically or chemically degraded. Activated carbon removes organic matter from wastewater using this property. The physical process occurring when liquids, gases or suspended matter adhere to the surface of, or in the pores of, an adsorbent medium. Adsorption is a physical process, which occurs without chemical reaction.

Advance ratio—ratio of the time for the water to reach the end of the field to the total set time for an irrigation set on a furrow irrigation system. The ratio should be less than 0.5 to have good distribution uniformity.

Advance time—time required for a given stream of irrigation water to move from the upper end of a field to the lower end. Time required for a given surface irrigation stream to move from one point in the field to another.

Advanced waste treatment—wastewater treatment beyond the secondary (biological) stage. It includes the removal of nutrients such as phosphorous and perhaps nitrogen and a high percentage of biochemical oxygen demand and suspended solids. Advanced waste treatment, also known as tertiary treatment, produces a high-quality effluent.

Advection—the process by which solutes are transported by the motion of flowing ground water.

Adverse effect—Impairment of or damage to the environment, human health or safety, or property.

Aerated—supplied with air or oxygen.

Aerated lagoon—a water treatment pond that speeds up biological decomposition of organic waste by stimulating the growth and activity of bacteria, which are responsible for the degradation.

Aeration—technique that is used with water treatment that demands oxygen supply, commonly known as aerobic biological water purification. Either water is brought into contact with water droplets by spraying or air is brought into contact with water by means of aeration facilities. Air is pressed through a body of water by bubbling and the water is supplied with oxygen. Both pressure (closed) aerators and open (gravity) aerators are used. Closed aeration is used chiefly for oxidation; open aeration for degassing; The process of dissolving gases from the air into water by turbulent mixing and molecular diffusion; To supply or impregnate with air; Process of blowing air (or another gas such as carbon dioxide) through a liquid or solid.

aeration (Unsaturated zone)—the zone between the land surface and the water table which characteristically contains liquid water under less than atmospheric pressure and water vapor and air or other gases at atmospheric pressure. The term Unsaturated zone is now generally preferred.

Aeration capacity—Volume fraction of air filled pores in a soil at field capacity.

Aeration tank—a tank that is used to inject air into water.

Aerobic—a process that takes place in the presence of oxygen, such as the digestion of organic matter by bacteria in an oxidation pond (aerobic treatment).

Aerobic digestion—digestion of suspended organic matter by means of aeration.

Aerodynamic transference phenomena—heat transferred by a surface that evaporates and is dragged by means of turbulence convection. The turbulence is created by the wind on the plant cover.

Aerosol—very small liquid or solid particles dispersed in air.

Aerosols—collection of airborne solid or liquid particles that reside in the air for at least several hours. Aerosols have a typical size between 0.01 and 10 nm. Aerosols may influence climate in two ways: directly through scattering and absorbing radiation, and indirectly through acting as condensation nuclei for cloud formation or modifying the optical properties and lifetime of clouds.

Aesthetic contaminants—characteristics of water which affects its taste, odor, color and appearance (and may affect the objects touched by the water) but which do not in themselves have any adverse health effects in otherwise potable water.

Affinity—the keenness with which an ion exchanger takes up and holds on to a counterion. Affinities are very much affected by the concentration of the electrolyte surrounding the ion exchanger.

Affluent (Lake)—a tributary or feeder stream. Streams receiving the run-off from the watershed and flowing into the lake are its affluents; analogous to the affluent of a river. The analogy can be very close where a lake has large inflowing and outflowing streams and is located in a valley or elongated basin. In usage, the term may have the same meaning as influent; although where the reference is to a single inflowing stream, the word influent appears to be the preferred one.

Affluent (Stream)—a stream or river that flows into a larger one; a Tributary.

Afforestation—the artificial establishment of forest crops by planting or sowing on land that has not previously, or recently, grown trees.

Afloat—floating on water.

After bay—the tail race or reservoir of a hydroelectric power plant at the outlet of the turbines used to regulate the flow below the plant; may refer to a short stretch of stream or conduit, or to a pond or reservoir.

Age of groundwater—an approximation of the time between the water's penetration of the land surface at one location and its later presence at another location.

Agglomeration—a process of bringing smaller particles together to form a larger mass.

Aggradation—(1) The build-up of sediments at the headwaters of a lake or reservoir or at a point where stream flow slows to the point that it will drop part or its entire sediment load. (2) Modification of the earth's surface in the direction of uniformity of grade or slope, by deposition, as in a riverbed.

Aggrading—the building up of a stream channel which is flowing too slowly to carry its sediment load.

Aggregate—the arrangement of smaller or bigger size soil particles to form the aggregates; group of primary soil particles that cohere to each other more strongly than to other surrounding particles; Groups of individual soil particles, held together naturally and consisting of particles of sand, silt and clay separated from each other by pores, cracks or planes of weakness; term for the stone or rock gravel needed to fill in an infiltration BMP such as a trench or porous pavement. Clean-washed aggregate is simply aggregate that has been washed clean so that no sediment is associated with. The term, *soil structure*, refers to this arrangement of the soil in natural aggregates. Various types of soil structure are recognized (Massive, platy, prismatic, blocky, granular).

Agribusiness—the sum of all operations involved in the production, storage, processing, and wholesale marketing of agricultural products.

Agricultural—having to do with farming or farms.

Agricultural capability—determines, given the ideal state, what a given area of land is capable of producing in terms of agricultural production and output.

Agricultural drainage—(1) The process of directing excess water away from the root zones of plants by natural or artificial means, such as by using a system of pipes and drains placed below ground surface level. Also referred to as subsurface drainage. (2) The water drained away from irrigated farmland.

Agricultural economics—the application of economic principles to the agribusiness sector of the economy.

Agricultural land—land in farms regularly used for agricultural production; all land devoted to crop or livestock enterprises, for example, farmstead lands, drainage and irrigation ditches, water supply, cropland, and grazing land.

Agricultural levee—a levee that protects agricultural areas where the degree of protection is usually less than that of a flood control levee.

Agricultural pollution—liquid and solid wastes from all types of farming, including runoff from pesticides, fertilizers, and feedlots; erosion and dust from plowing; animal manure and carcasses; and crop residues and debris.

Agricultural restructuring scenario—a term used to describe the sensitivity of agricultural water demand and farm marketing revenues to changes in certain cropping patterns.

Agricultural runoff—the runoff into surface waters of herbicides, fungicides, insecticides, and the nitrate and phosphate components of fertilizers and animal wastes from agricultural land and operations. Considered a Non-Point Source (NPS) of water pollution.

Agricultural suitability—determines how suitable a given area of land is, in it's present state, for agricultural purposes.

Agricultural use—the use of any tract of land for the production of animal or vegetable life; uses include, but are not limited to, the pasturing, grazing, and watering of livestock and the cropping, cultivation, and harvesting of plants.

Agricultural water use—includes water used for irrigation and nonirrigation purposes. Irrigation water use includes the artificial application of water on lands to promote the growth of crops and pasture, or to maintain vegetative growth in recreational lands, parks, and golf courses. Non-irrigation water use includes water used for livestock, which includes water for stock watering, feedlots, and dairy operations, and fish farming and other farm needs.

Agro ecosystem—land used for crops, pasture, and livestock; the adjacent uncultivated land that supports other vegetation and wildlife; and the associated atmosphere, the underlying soils, ground and surface waters, irrigation channels, and drainage networks.

Agro industrial—of or relating to production (as of power for industry and water for irrigation) for both industrial and agricultural purposes.

A-horizon—the uppermost zone in the Soil Profile, from which soluble Salts and Colloids are leached, and in which organic matter has accumulated. Generally this represents the most fertile soil layer. Along with the B-Horizon, this layer constitutes part of the Zone of Eluviation.

Air gap—physical separation of the supply pipe by at least two pipe diameters (never less than one inch) vertically above the overflow rim of the receiving vessel. In this case line pressure is lost. Therefore, a booster pump is usually needed downstream, unless the flow of the water by gravity is sufficient for the water use. With an air gap there is no direct connection between the supply main and the equipment. An air gap may be used to protect against a contaminant or a pollutant, and will protect against both back-siphonage and back-pressure. An air gap is the only acceptable means of protecting against lethal hazards.

Air stripping (Water quality)—a process for the removal of organic contaminants from groundwater. The groundwater flows downward inside a tower filled with materials (the packing) over a large surface area. Air is introduced at the bottom of the tower and is forced upward past the falling water. Individual organic contaminants are transferred from the water to the air, according to the gas and water equilibrium concentration values of each contaminant. Also referred to as Packed Tower Aeration.

Air vent (of a dam)—a pipe designed to provide air to the outlet conduit to reduce turbulence and prevent negative pressures during the release of water. Extra air is usually necessary downstream of constrictions.

Albedo—rate of sun radiations received at the soil surface and the reflected solar radiation; albedo can be expressed as either a percentage or as a fraction of 1. Snow covered areas have a high albedo (up to about 0.9 or 90%) because of their white color, while vegetation has a low albedo (generally about 0.1 or 10%) because of its dark colors and because of the light that is absorbed for photosynthesis. Clouds have an intermediate albedo and are the most important contributor to the earth's albedo. The Earth's aggregate albedo is approximately 0.3.

Alfalfa valve—outlet valve attached to the top of a pipeline riser with an opening equal in diameter to the inside diameter of the riser pipe and an adjustable lid or cover to control water flow.

Algae—any of a group of chiefly aquatic nonvascular plants; most have chlorophyll. Excessive growths are called blooms; Simple single-celled (phytoplankton), colonial, or multicelled, mostly aquatic plants, containing chlorophyll and lacking roots, stems and leaves. Aquatic algae are microscopic plants that grow in sunlit water that contains phosphates, nitrates, and other nutrients. Algae is either suspended in water (plankton) or attached to rocks and other substrates (periphyton). Their abundance, as measured by the amount of chlorophyll *a* (green pigment) in an open water sample, is commonly used to classify the trophic status of a lake. Algae are an essential part of the lake ecosystem and provide the food base for most lake organisms, including fish. The algae population is divided up into green algae and blue algae, of which the blue algae are very damageable to human health. Excessive algae growth may cause the water to have undesirable odors or tastes and decay of algae can deplete the oxygen in the water.

Algaecide—chemical agent to kill or to inhibit the growth of algae.

Algal bloom—a heavy growth of algae in and on a body of water that is often triggered by environmental conditions such as high nitrate and phosphate concentrations. The decay of algal blooms may reduce dissolved oxygen levels.

Algal growth potential—the maximum algal dry weight biomass that can be produced in a natural water sample under standardized laboratory conditions. The growth potential is the algal biomass present at stationary phases and is expressed as milligrams dry weight of algae produced per liter of sample.

Aliquot—a measured portion of a sample taken for analysis. One or more aliquots make up a sample.

Alkali—any strongly basic (high pH) substance capable of neutralizing an acid, such as soda, potash, etc., that is soluble in water and increases the pH of a solution greater than 7.0. It will turn litmus paper blue. Also refers to soluble salts in soil, surface water, or groundwater. Highly alkaline waters tend to cause drying of the skin.

Alkali lake—that containing water very highly impregnated with alkalies. The "alkali" may be sodium carbonate or sodium sulfate and potassium carbonate but includes other alkaline compounds as well. Restricted to arid and semiarid regions.

Alkaline—sometimes water or soils contain an amount of Alkali substances sufficient to raise the pH value above 7.0 and be harmful to the growth of crops. Generally, the term alkaline is applied to water with a pH greater than 7.4.

Alkaline (alkali) soil—soil with pH greater than 7.0; Soil with an exchangeable sodium percentage greater than 15%; Soil that has sufficient exchangeable sodium (alkali) to interfere with plant growth and cause dispersion and swelling of clay minerals.

Alkalinity—a measure of the capacity of a solution to neutralize acids.

Anaerobic—any process that can occur without oxygen; also applies to organisms that can survive without oxygen.

Analysis—a close look at something to find out more about it; can involve looking closely at the individual parts of something and describing them.

Anthropogenic—caused or produced as a result of human activity.

Aquatic—term used to describe any organism growing in, living in, or frequenting water; some plants and animals that live in water are called aquatic species.

Aquifer—an underground layer of rock or soil that contains water and can supply a large quantity of water to wells or springs.

Assessment—a written decision about the importance, size or value of something; for example, an environmental assessment may describe the value of arctic char after a study of the char, the fishermen, the method of fishing and the effect on the environment

Assimilative capacity—the amount of pollutants that a water body may absorb while continuing to meet water quality standards.

Attenuate—reduce in significance or concentration.

Alkalinity—the capacity of water for neutralizing an acid solution. Alkalinity of natural waters is due primarily to the presence of hydroxides, bicarbonates, carbonates and occasionally borates, silicates and phosphates. It is expressed in units of milligrams per liter (mg/l) of $CaCO_3$ (calcium carbonate) or as microequivalents per liter (µeq/l); 20 µeq/l = 1 mg/l of $CaCO_3$. A solution having a pH below 4.5 contains no alkalinity. Low alkalinity is the main indicator of susceptibility to acid rain. Increasing alkalinity is often related to increased algal productivity. Lakes with watersheds that have sedimentary carbonate rocks are high in dissolved carbonates (hard-water lakes). Whereas lakes in granite or igneous rocks are low in dissolved carbonates.

Allocation volume—The specific volume of water allocated to water access entitlements.

Allowable depletion [in, mm]—portion of plant available water that is allowed for plant use prior to irrigation based in plant and management considerations; That part of soil moisture stored in the plant root zone managed for use by plants, usually expressed as equivalent depth of water in acre inches per acre; It is sometimes referred as *allowable soil depletion* or *allowable soil water depletion*.

Allowable stress factor—coefficient used to modify reference evapotranspiration to reflect the water use of a particular plant or group of plants particularly with reference to the water stress.

Allowable voltage loss [volts]—voltage loss in a circuit or portion of a circuit, which if not exceeded, will result in the electrical device working correctly.

Alluvial—an adjective referring to soil or earth material which has been deposited by running water, as in a riverbed, flood plain, or delta.

Alluvial dam lakes—numerous basins which are the sites of both existing and extinct lakes in the arid regions of western U.S. were formed by alluvial dams, especially by the coalescence of fans composed of detritus carried down by streams from opposite sides of valleys. In glaciated regions dams were formed in valleys by glacio-fluvial deposition during the Pleistocene; and barriers of various kinds, which impound water have been created in river flood plains by alluvial deposition.

Alluvial groundwater—ground water that is hydrologically connected to a surface stream that is present in permeable geologic material, usually small rock and gravel.

Alluvial land—areas of unconsolidated alluvium, generally stratified and varying widely in texture, recently deposited by streams, and subject to frequent flooding.

Alluvium—sediments deposited by erosion processes, usually by streams. Depending upon the location in the floodplain of the river, different-sized sediments are deposited.

Alternating current [AC]—current in which the flow of electrons in a circuit flow in one direction and then in the reverse direction.

Alternative sewage system—onsite sewage system other than a conventional gravity system or a conventional pressure distribution system. Properly operated and maintained alternative systems provide equivalent or enhanced treatment performance as compared to conventional gravity systems.

Altimetry—radar technique that measures global elevation of sea, land or ice surfaces compared to the center of the earth.

Alum—a chemical substance that is gelatinous when wet, usually potassium aluminum sulfate, used in water treatment plants for settling out small particles of foreign matter.

Ambient—surrounding or occurring before a location or before an activity occurs. Ambient temperature is the temperature of the surrounding air. Ambient water quality is the water quality in a river, lake, or other water body, as opposed to the quality of water being discharged. In a river or stream, ambient water quality usually refers to the water upstream of a discharge point.

Ambient water quality standards—the allowable amount of materials, as a concentration of pollutants, in water. The standard is set to protect against anticipated adverse effects on human health or welfare, wildlife, or the environment, with a margin of safety in the case of human health.

American Society of Landscape Architects (ASLA)—a professional society that represents landscape architects in the United States and Canada and seeks to better the practice and understanding of landscape architecture through education, research, state registration and other programs.

Ammonification—formation of composites of ammonia.

Amoeba—a single celled protozoan that is widely found in fresh and salt water. Some types of amoebas cause diseases such as amoebic dyzentery.

Ampere (or amp)—a unit of electrical current. The unit is used to specify the movement of electrical charge per unit time through a conductor; one ampere = 1 coulomb per second. One ampere is the movement of 6.28 billion electrons /second.

Amphipods—invertebrate animals of the crustacean order Amphopoda that are generally characterized by laterally flattened bodies; comprising the sand fleas and related forms.

Anadromous—fishes that ascend from their primary habitat in the ocean to fresh waters to spawn.

Anaerobic—a condition in which there is no air or no available free oxygen; such as the digestion of organic matter by bacteria in a UASB-reactor.

Anaerobic digestion— the degradation of organic matter brought about through the action of microorganisms in the absence of elemental oxygen.

Anaerobic organism—that can strive in the absence of oxygen (air), such as bacteria in a septic tank.

Analyte—A specific compound or element of interest undergoing analysis.

Anchor gill net—gill net that has the ends secured by anchors and the top kept at the water surface by floats.

Anion—negatively charged ion, which during electrolysis is attracted towards the anode. The most common anions in soil extracts and waters are bicarbonate, sulfate, carbonate, chloride and nitrate ions.

Anion exchange—a process in which negatively charged ions, or chemicals attach to a filter medium as water passes through.

Anisotrophy—the condition under which one or more of the hydraulic properties of an aquifer vary according to the direction of flow.

Annual flood—the highest peak discharge of a stream in a water year.

Annual flood series—a list of annual floods for a given period of time.

Annual low flow—the lowest flow occurring each year, usually the lowest average flow for periods of perhaps 3, 7, 15, 30, 60, 120, or 180 consecutive days.

Annual plants—plants that grow and complete their life cycle in a perennial year or before.

Annular sea plant—material used to seal the space between the borehole and the casing of the well. Annular sea plants prevent surface contaminants from entering the wall.

Annular space—the open space formed between the borehole and the well casing.

Anode—a site in electrolysis where metal goes into solution as a cation leaving behind an equivalent of electrons to be transferred to an opposite electrode, called a cathode.

Anoxic—denotes the absence of oxygen, as in bodies of water, lake sediments, or sewage. Anoxic conditions generally refer to a body of water sufficiently deprived of oxygen to where *Zooplankton* and fish would not survive.

Anoxic process—the process by which nitrate-nitrogen is converted biologically to nitrogen gas in the absence of oxygen. The process is also known as anoxic denitrification.

Antecedent precipitation—precipitation, which occurred prior to a particular time over a specific area or Drainage Basin. Usually applied as a measure of moisture in the top layer of the soil that would affect runoff from additional rainfall.

Antecedent precipitation index—an index of moisture stored in a basin before a storm, calculated as a weighted summation of past daily precipitation amounts.

Antecedent soil water—degree of wetness of a soil prior to irrigation or at the beginning of a runoff period, typically expressed as an index.

Antecedent streams—antecedent streams are those in place before the rising of mountain chains. As the mountains rise, the streams cut through at the same rate and so maintain their positions.

Anthropogenic emissions—emissions of particles or substances resulting from human activities, such as industry and agriculture.

Anti siphon device—a type of backflow preventer, which seals off the atmospheric vent area when the system is pressurized. Should be installed downstream of a control valve in a location which is at least 12 inches higher than the highest point in the lateral which it serves. When system pressure drops to zero, the float and seal assembly drops, opening the vent to atmosphere and breaking any siphon effect. Synonymous with atmospheric vacuum breaker.

Anti transpirants—composites or materials that are applied to the plant to reduce transpiration. These antitranspirants are not detrimental for the plant physiological processes.

Application efficiency—ratio of the average depth of the irrigation water stored in the root zone to the average depth of irrigation water applied; Ratio of the average depth of irrigation water infiltrated and stored in the root zone to the average depth of irrigation water applied; Amount of water stored in the root zone that is available for plant use divided by the average amount of water applied during an irrigation. Also called water application efficiency (%, decimal).

application efficiency of lower half, water—ratio of the average of the lowest one-fourth of measurements of irrigation water infiltrated to the average depth of water applied.

application efficiency of lower quarter, water—ratio of the average low quarter depth of irrigation water infiltrated and stored in the root zone to the average depth of irrigation water applied; Ratio of the average of the lowest one-fourth of measurements of irrigation water infiltrated to the average depth of water applied.

application efficiency, water—ratio of the average depth of the irrigation water stored in the root zone to the average depth of irrigation water applied; Ratio of the average depth of irrigation water infiltrated and stored in the root zone to the average depth of irrigation water applied; Amount of water stored in the root zone that is available for plant use divided by the average amount of water applied during an irrigation; Ratio of the average depth of irrigation water contributing to target, to the average depth of irrigation water applied.

Application rate—the rate at which water is applied to the turf or ornamental plantings. Also refers to the amount of water applied to a given area in one hour. (In the United States, usually expressed in inches per week. Its metric equivalent is centimeters per week.)

Applied water demand—the quantity of water delivered to the intake of a city's water system or factory, the farm head gate, or a marsh or other wetland, either directly or by incidental drainage. For in-stream use, it is the portion of the stream flow dedicated to in-stream use or reserved under federal or state Wild and Scenic River Acts. Applied water includes the water that returns to groundwater, a stream, canal, or other supply source that can be reused or recycled and thus is not the same as Net Water Demand.

Appurtenant structures (of a dam)—auxiliary features of a dam such as an outlet, spillway, powerhouse, tunnel, etc.

Aquatic—growing in water, living in water, or frequenting water.

Aquatic ecosystem—an aquatic area where living and nonliving elements of the environment interact. This includes the physical, chemical, and biological processes and characteristics of rivers, lakes, and wetlands and the plants and animals associated with them.

Aquatic environment—the components of the Earth related to, living in, or located in or on water or the beds or shores of a water body including (but not limited to) all organic and inorganic matter, living organisms and their habitat (including fish habitat), and their interacting natural systems.

Aquatic fauna—animals growing or living in or frequenting water.

Aquatic Macrophyte—large (in contrast to microscopic) plants that live completely or partially in water.

Aquatic species—the plants and animals living in, or associated with, water bodies, wetlands, and riparian areas.

Aqueous—something made up of water.

Aqueous solubility—the maximum concentration of a chemical that dissolves in a given amount of water.

Aquiclude—a low-permeability unit that forms either the upper or lower boundary of a ground-water flow system.

Aquifer—a geologic formation, group of formations, or part of a formation that contains sufficient saturated permeable material to yield significant quantities of water to wells and springs. If a well taps more than one aquifer, the principal aquifer will be the one that yields the greatest amount of water. Aquifers have specific rates of discharge and recharge. As a result, if groundwater is withdrawn faster than it can be recharged, the underground aquifer cannot sustain itself.

Aquifer adsorptive characteristics—ability of an aquifer to retain atoms, ions, or molecules.

Aquifer degradation characteristics—aquifer contamination can be characterized by parameters such as pH, total organic halogens, total organic carbon, temperature, and specific conductance.

Aquifer recharge wells—wells used to recharge depleted aquifers and may inject fluids from a variety of sources such as lakes, streams, domestic wastewater treatment plants, and other aquifers.

Aquifer remediation related wells—wells used to prevent, control, or remediate aquifer pollution, including but not limited to superfund sites.

Aquifer system—a body of permeable and relatively impermeable materials that functions regionally as a water-yielding unit. It comprises two or more permeable units separated at least locally by confining units (Aquitards) that impede ground-water movement but do not greatly affect the regional hydraulic continuity of the system. The permeable materials can include both saturated and unsaturated sections.

aquifer, Basinfill—an aquifer located in a basin surrounded by mountains and composed of sediments and debris shed from those mountains. Sediments are typically sand and gravel with some clay.

Aquifuge—an absolutely impermeable unit that will neither store nor transmit water.

Aquitard—an impermeable layer of consolidated or unconsolidated material that restricts the movement of water into or out of a confined aquifer; also called confining bed.

Arable land—land capable of being cultivated and suitable for the production of crops.

Arc—portion of a full circle (360°) covered by a part circle sprinkler. The area a part-circle sprinkler irrigates, expressed in degrees of a circle. For example, a 90° arc provides quarter-circle coverage, while an 180° arc provides half-circle coverage.

Area—surface included within a set of lines (Webster). In irrigation, usually used to describe a surface of land or cross section of pipe. [A: in^2, ft^2, acres, ha].

area (sub area), Hydrographic—primarily these are subdrainage systems, typically valleys, within a more comprehensive drainage basin. Hydrographic Areas (Valleys) may be further subdivided into Hydrographic Sub-Areas based on unique hydrologic characteristics (e.g., differences in surface flows) within a given valley or area.

Area actually irrigated—the area, which is actually irrigated at least once in a given year. Often, part of the equipped area is not irrigated for various reasons such as lack of water, absence of farmers, land degradation, damage and organizational problems. It only refers to physical areas, meaning that irrigated land that is cultivated twice a year is counted once.

Area capacity curve—a graph showing the relation between the surface area of the water in a reservoir and the corresponding volume.

Area equipped for irrigation (Equipped lowland areas)—the land equipped for irrigation in lowland areas includes: (1) Cultivated wetland and inland valley bottoms, which have been equipped with water control structures for irrigation and drainage (intake, canals, etc.); (2) Areas along rivers, where cultivation occurs making use of water from receding floods and where structures have been built to retain the receding water; (3) Developed mangroves and equipped delta areas.

Area equipped for irrigation (Full control—total)—the sum of surface irrigation, sprinkler irrigation and localized irrigation. The scientist uses indifferently the expressions "full control" and "full/partial control."

Area equipped for irrigation (Full control localized irrigation)—localized irrigation is a system where the water is distributed under low pressure through a piped network, in a predetermined pattern, and applied as a small discharge to each plant or adjacent to it. There are three main categories: drip irrigation (where drip emitters are used to apply water slowly to the soil surface), spray or microsprinkler irrigation (where water is sprayed onto the soil near individual plants or trees) and bubbler irrigation (where a small stream of water is applied to flood small basins or the soil adjacent to individual trees). To refer to localized irrigation, the following terms are also sometimes used: micro irrigation, trickle irrigation, daily flow irrigation, drip- irrigation, sip irrigation and diurnal irrigation.

Area equipped for irrigation (Full control sprinkler irrigation)—a sprinkler irrigation system consists of a pipe network through which water moves under pressure before being delivered to the crop via sprinkler nozzles. The system basically simulates rainfall in that water is applied through overhead spraying. Therefore, these systems are also known as overhead irrigation systems.

Area equipped for irrigation (Full control surface irrigation)—surface irrigation systems are based on the principle of moving water over the land by simple gravity in order to wet it, either partially or completely, before infiltration. These can be subdivided into furrow, border strip and basin irrigation (including submersion irrigation of rice). Surface irrigation does not refer to the method of transporting water from the source up to the field, which may be done by gravity or by pumping. Manual irrigation using buckets or watering cans should also be included here.

Area equipped for irrigation (Spate irrigation)—spate irrigation can also be referred to as floodwater harvesting. It is a method of random irrigation using the floodwaters of a normally dry watercourse or river bed (wadi). These systems are in general characterized by a very large catchment upstream (200 ha to 50 km²) with a "catchment area: cultivated area" ratio of 100:1 to 10,000:1. There are two types of floodwater harvesting or spate irrigation: (1) Floodwater harvesting within streambeds, where turbulent channel flow is collected and spread through the wadi in which the crops are planted; cross-wadi dams are constructed with stones, earth, or both, often reinforced with gabions; (2) Floodwater diversion, where the floods—or spates—from the seasonal rivers are diverted into adjacent embanked fields for direct application. A stone or concrete structure raises the water level within the wadi to be diverted to the nearby cropping areas.

Area equipped for irrigation (total)—area equipped to provide water to crops. It includes areas equipped for full control irrigation, equipped lowland areas, and areas equipped for spate irrigation. It does not include nonequipped cultivated wetlands and inland valley bottoms or nonequipped flood recession cropping areas.

Area of a lake—the space occupied by the water surface. The area of a lake, generally, is something that cannot be determined with great exactitude; often the figure given is an arbitrary one, and figures from different sources show considerable disagreements. This comes about, because some error is inherent in any of the procedures devised for determining area; because measurements may be made from hydrographic maps, which differ in accuracy and detail, and in time at which the map was made. This latter becomes important where lakes fluctuate greatly in levels. Some differences may arise also where different mathematical procedures are followed in making measurements. Also, often-arbitrary decisions must be made as to location of shoreline, the inclusion or exclusion of islands, and boundaries between a lake and connecting water, all of which consequently affect the computed area. Area is usually expressed square miles and acres; or where the metric system is used in square kilometers and square meters.

area, Cultivated—area under temporary (annual) and permanent crops. It refers to the physical area cultivated meaning that land that is cultivated twice a year is counted only once.

area, Design—where an irrigation system will be installed.

area, Flooded—area of a floodplain that is flooded in a specific stream reach, watershed, or river basin; may be for a single flood event, but is usually expressed as an average, annual value based on the sum of areas from all individual flood events over a long period of time, such as 50 to 100 years, and adjusted to an average value.

area, Surface—area of a soil, usually expressed in square meters per gram.

Arid climate—climate characterized by low rainfall and high evaporation potential. A region is usually considered as arid when precipitation averages less than 10 inches per year.

Aromatic hydrocarbons—class of unsaturated cyclic organic compounds containing one or more ring structures. The name aromatic is derived by the distinctive and often fragrant odors of these compounds.

Aromatics—a type of hydrocarbon that contains a ring structure, such as benzene and toluene. They can be found for instance in gasoline.

Arrhenius, Svante—Swedish scientist that was first to claim in 1896 that fossil fuel combustion may eventually result in enhanced global warming. He proposed a relation between atmospheric carbon dioxide concentrations and temperature. He found that the average surface temperature of the earth is about 15°C because of the infrared absorption capacity of water vapor and carbon dioxide. This is called the natural greenhouse effect. Arrhenius suggested a doubling of the CO_2 concentration would lead to a 5°C temperature rise.

Artesian—a commonly used expression, generally synonymous with Confined and referring to subsurface (ground) bodies of water which, due to underground drainage from higher elevations and confining layers of soil material above and below the water body (referred to as an Artesian Aquifer), result in underground water at pressures greater than atmospheric.

Artesian aquifer—a commonly used expression, generally synonymous with (but a generally less favored term than) confined aquifer. An artesian aquifer is an aquifer which is bounded above and below by formations of impermeable or relatively impermeable material. An aquifer in which ground water is under pressure significantly greater than atmospheric and its upper limit is the bottom of a bed of distinctly lower hydraulic conductivity than that of the aquifer itself.

Artesian pressure—the pressure under which artesian water in an artesian aquifer is subjected, generally significantly greater than atmospheric.

Artesian water—ground water that is under pressure when tapped by a well and is able to rise above the level at which it is first encountered. It may or may not flow out at ground level. The pressure in such an aquifer commonly is called Artesian Pressure, and the formation containing artesian water is an artesian aquifer or confined aquifer.

Artesian well—a well tapping an artesian aquifer. Because the water is under pressure, it rises above the top of the aquifer. The height of the water in the well represents the potentiometric surface of the aquifer.

Artesian zone—a zone where water is confined in an aquifer under pressure so that the water will rise in the well casing or drilled hole above the bottom of the confining layer overlying the aquifer.

Artificial marsh creation—simulation of natural wetland features and functions via topographic and hydraulic modifications on nonwetland landscapes. Typical objectives for artificial marsh creation include ecosystem replacement or storm water management.

Artificial recharge (induced recharge)—the designed (as per man's activities as opposed to the natural or incidental) replenishment of ground water storage from surface water supplies such as irrigation or induced infiltration from streams or wells. There are five common techniques to effect artificial recharge of a groundwater basin: (1) Water spreading consisting of the basin method, stream-channel method, ditch method, and flooding method, all of which tend to divert surface water supplies to effect underground infiltration; (2) Recharge pits designed to take advantage of permeable soil or rock formations; (3) Recharge wells which work directly opposite of pumping wells, although they generally have limited scope and are better used for deep, confined aquifers; (4) Induced recharge, which results from pumping wells near surface, supplies, thereby inducing higher discharge towards the well; and (5) Wastewater Disposal which includes the use of secondary treatment wastewater in combination with spreading techniques, recharge pits, and recharge wells to reintroduce the water into deep aquifers thereby both increasing the available groundwater supply and also further improving the quality of the wastewater.

Artificial wetland—a man-made wetland in an area where a natural wetland did not exist before.

As built plan—a complete plan of an installed irrigation system designating valve, sprinkler and controller locations, routing of pipe and control wire. The plan includes all changes to the original design that were completed during installation.

Assessment monitoring—a program of monitoring ground water under interim status requirements. After a release of contaminants to ground water has been determined, the rate of migration, extent of contamination, and hazardous constituent concentration gradients of the contamination must be identified.

Assessment plan—the written detailed plan drawn up by the owner/operator, which describes and explains the procedures intends to take to perform assessment monitoring.

Assets—in economic terms as work or capital asset. It refers to an irrigation system that offers a productive service to business during a one-year period.

Assimilation—the ability of water to purify itself of pollutants.

Assimilative capacity—the capacity of natural water to receive wastewaters or toxic materials without negative effects and without damage to aquatic life or humans who consume the water.

Atmosphere—a blanket of 560 m of air surrounding the earth that supports life. The atmosphere absorbs energy from the sun, recycles water and other chemicals and works with electrical and magnetic forces to provide a moderate climate. It also protects us from high-energy radiation and the frigid vacuum of space. The atmosphere is primarily composed of nitrogen (78%), oxygen (21%), and argon (1%). Other components include water (0–7%), ozone (0–0.01%) and carbon dioxide (0.01–0.1%). The atmosphere changes from the ground up and consists of four distinct layers: troposphere (8–14.5 km), stratosphere (14.5–50 km), mesosphere (50–85 km) and thermosphere (85–600 km). Temperatures are different for each layer. In the troposphere temperatures drop from about 17 to $-52°C$, in the stratosphere temperatures increase gradually to $-3°C$, in the mesosphere temperatures fall to $-93°C$ as altitude increases, and in the thermosphere temperatures can be as high as $1727°C$. Above the thermosphere, the exosphere starts and continues until it mixes with interplanetary gases or space. This layer primarily consists of low-density particles such as hydrogen and helium; A pressure unit. One atmosphere is equal to 1.013 centimeter of a mercury column, equal to 14.71 psi, equal to 34 feet of water column, equal to 1 bar or equal to 101.3 KPa.

Atmospheric vacuum breaker [AVB]—back flow device configured with a single moving part, a float, which moves up or down to allow atmospheric air into the piping system. The AVB is always placed downstream from all shut-off valves. Its air inlet valve closes when the water flows in the normal direction. But, as water ceases to flow the air inlet valve opens, thus interrupting the possible back-siphonage effect. If piping or a hose is attached to this assembly and run to a point of higher elevation, the backpressure will keep the air inlet valve closed because of the pressure created by the elevation of water. Hence, it would not provide the intended protection. Therefore, this type of assembly must always be installed at least six (6) inches above all downstream piping and outlets. Additionally, this assembly may not have shut-off valves or obstructions downstream. A shut-off valve would keep the assembly under pressure and allow the air inlet valve (or float check) to seal against the air inlet port, thus causing the assembly to act as an elbow, not a backflow preventer. The AVB may not be under continuous pressure for this same reason. An AVB must not be used for more than 12 out of any 24-hour-period. It may be used to protect against either a pollutant or a contaminant, but may only be used to protect against a back-siphonage condition.

Atoll—a coral reef that encloses an area of water. Many Pacific Ocean archipelago are collections of atolls.

Attached growth process—biological treatment process in which the microorganisms responsible for the conversion of the organic matter or other constituents in the wastewater to gases and cell tissue are attached to some inert medium, such as rocks, slag, or specifically designed ceramic or plastic materials. Attached-growth treatment processes are also known as fixed-film processes.

Attenuation—the process of reduction of a compound's concentration over time. This can be through absorption, adsorption, degradation, dilution or transformation; To reduce, weaken, dilute, or lesson in severity, value, or amount such as the attenuation of contaminants as they migrate from a particular source.

Attribution of climate change—the process of establishing the most likely causes for the detected climate change with some defined level of confidence.

Attrition—the action of one particle rubbing against the other in a filter media or ion exchange bed that can in time cause breakdown of the particles.

Audit or irrigation audit—a detailed review of an irrigation system, including tests to determine overall system efficiency, identify problems areas that need correction, and determine an ideal watering schedule.

Aufwuchs—organisms that are attached to or move about on a submerged substrate, but do not penetrate it; also called periphyton.

Automatic system—a sprinklers system, which can be programd to water at regular intervals using either electric or hydraulic controller and valves.

Automation of drip irrigation system—(an automatic or semiautomatic system) provides a number of operational possibilities. This consists of a broad range of controls for a complete automatic system: water meter, pressure gage, solenoid valve, hydraulic valve, timer, and so forth. The system closes as soon as a guaranteed amount of water has been applied.

Automation, non sequential system—includes hydraulic or electrical valves, which independently operate in terms of the amount of water to be replaced and the irrigation frequency. Each unit can supply amount of volume of water and can open at same time due to a predetermined program by use of a solenoid valve. The Control panel contains electrical circuits to operate the pump or the main valve, to add fertilizer and to measure the soil moisture so that the irrigation system can fulfill the needs of a crop.

Automation, periodically electrically operated—it releases a desired amount of water regulated by a volumeter (solenoid valve). The duration of operation is determined by a timer and is activated by solenoid valves. Each unit has its own switch. The periodic irrigation automatically happens with a closed unit and the successive units are operated in sequence. The power unit can be A.C. or D.C.

Autotrophic bacteria—are bacterias that synthesize or produce their own food.

Available capacity of sewer—the difference between the design capacity of a sewer and the average daily flow recorded at a point in time.

Available chlorine—a measure of the amount of chlorine available in chlorinated lime, hypochlorite compounds, and other materials.

Available free chlorine—is an excess of chlorine (hypochloric acid, HOCl) that remains in the irrigation water, after the reaction has been completed with inorganic compounds, organic substances, and the metals.

Available moisture capacity or available water capacity (AWC)—the soil moisture content at field capacity minus the soil moisture content at the wilting point. This is a property that varies among different soil types.

Available moisture content or available water content—the soil moisture content minus the wilting point.

Available soil moisture—difference at any given time between the actual soil moisture content in the root zone soil and the wilting point.

Available water—the portion of water in a soil that can be absorbed by plant roots, usually considered to be that water held in the soil against a tension of up to approximately 15 atmospheres; Water potentially available for consumptive or environmental use; Portion of water in a soil that can be readily absorbed by plant roots. It is the amount of water released between in situ field capacity and the permanent wilting point. See *also available water holding capacity.*

Available water capacity—also known as available water. The portion of water in a soil that can be readily absorbed by plant roots. It is the amount of water between field capacity and permanent wilting point. Usually measured in either %, in/ft, in/in, mm/mm, or m/m.

Available water holding capacity—the capacity of a soil to hold water in a form available to plants. Also, the amount of moisture held in the soil between field capacity (or about one-third atmosphere of tension) and the wilting coefficient (about 15 atmospheres of tension).

Average annual flood damages—the weighted average of all flood damages that would be expected to occur yearly under specified economic conditions and development. Such damages are computed on the basis of the expectancy in any one-year of the amounts of damage that would result from floods throughout the full range of potential magnitude.

Average annual precipitation [in., mm]—long-term historic (generally 30 years or more) arithmetic mean of precipitation (rain, snow, dew) received by an area.

Average annual recharge—the amount of water entering an aquifer on an average annual basis. In many, if not most, hydrologic conditions, "average" has little significance for planning purposes as there may exist so few "average" years in fact.

Average annual runoff (yield)—the average of water-year (October 1 to September 30) runoff or the supply of water produced by a given stream or water development project for a total period of record; measured in cubic feet per second or acre-feet.

Average discharge—in the annual series of the U.S. Geological Survey's (USGS) reports on surface-water supply, the arithmetic average of all complete water years of record whether or not they are consecutive. Average discharge is not published for less than 5 years of record. The term "average" is generally reserved for average of record and "mean" is used for averages of shorter periods, namely daily mean discharge.

Average water year—a term (October 1 to September 30) denoting the average annual hydrologic conditions based upon an extended or existing period of record. Because precipitation, runoff, and other hydrologic variables vary from year to year, planners typically project future scenarios based on hydrologic conditions that generally include average, wet (high-water), and drought (low-water) years.

Average year water demand—the annual demand for water under average hydrologic conditions for a defined level of development.

Average year water supply—the average annual supply of a water development system over a long period. For a dedicated natural flow, it is the long-term average natural flow for wild and scenic rivers or it is environmental flow as required for an average year under specific agreements, water rights, court decisions, and congressional directives.

AWG-UF—the classification of the direct burial wire used for automatic sprinkler systems. Example: #14–1 AWG-UF means a 14 gauge wire, single wire cable, designed for direct burial (no conduit) in the ground. The wire should have this information stamped or printed directly on the wire's plastic insulation. The wire should be buried at least 18″ deep for safety (in most areas this requirement is written into local law).

Axial flow—fluid flow in the same direction as the axis of symmetry of the duct, vessel, or tank.

Axis of a dam—the horizontal centerline of a dam in the longitudinal direction.

Back filling—the return of wastes or other material underground for disposal

Back flow—the flow of water in a medium or pipe or line in a direction opposite to normal flow. Flow is often returned into the system by backflow, if the wastewater in a purification system is severely contaminated; Any unwanted reverse flow of used or nonpotable water or substance from any domestic, industrial or institutional piping system into the pure, potable water distribution system. The direction of flow under these conditions is in the reverse direction from that intended by the system and normally assumed by the owner of the system.

Back flow prevention device—used to prevent the flow of water from the water distribution system back to the water source and may be required by law. Similar to antisiphon valves.

Back ground concentrations—a schedule of sampling and analysis that is compiled during the first year of monitoring. All wells in the monitoring system must be sampled on a quarterly basis to determine drinking water characteristic, ground-water quality, and contamination indicator parameters. For each upgradient well, at least four replicate measurements must be made for the contamination indicator parameters.

Back ground mean—the arithmetic average of a set of data, used as a control value in subsequent statistical tests.

Back ground variance—the variance is the measure of how far and observation value departs from the mean. Background refers to the observations used for control in subsequent statistical tests.

Back pressure—increase of pressure in the downstream piping system above the supply pressure at the point of consideration which would cause, or tend to cause, a reversal of the normal direction of flow; Pressure that can cause water to backflow into the water supply when a user's waste water system is at a higher pressure than the public system.

Back siphonage—reversal of flow (back flow) due to a reduction in system pressure which causes a negative or subatmospheric pressure to exist at a site in the water system; Technically, if one siphons a fluid out of a container or a pipeline, one causes that fluid to flow up over the rim of the container or top of the pipe and then down into a lower elevation through a piece of tubing or, in this case a piece of pipe that is part of the distribution system. In the vernacular, the unwanted fluid is "sucked" into the potable water line. It is important to understand that it is not necessary for the system main to be under a true vacuum (i.e., zero psia) for back siphonage to occur. All that is required is a negative difference in pressure and a piece of tubing or pipe that is completely full of fluid.

Back washing—reversing the flow of water back through the filter media to remove entrapped solids.

Bacteria—single-celled organisms (single form = bacterium) which lack well-defined nuclear membranes and other specialized functional cell parts and reproduce by cell division or spores. Bacteria may be free-living organisms or parasites. Bacteria (along with fungi) are decomposers that break down the wastes and bodies of dead organisms making their components available for reuse. Bacteria cells range from about 1–10 micron in length and from 0.2–1.0 micron in width. They exist almost everywhere on earth. Despite their small size, the total weight of all bacteria in the world likely exceeds that of all other organisms combined. Some bacteria are helpful others are harmful. Some bacteria are capable of causing human, animal or plant diseases; others are essential in pollution control because they break down organic matter.

Bacterial water contamination—the introduction of unwanted bacteria into a water body.

Bactericide—a chemical agent to prevent or kill the bacterial growth.

Bacteriostatic—having the ability to inhibit the growth of bacteria without destroying the bacteria. For example, silver-impregnated activated carbon will reduce bacterial colonization but not eliminate it.

BADCT—Arizona regulatory terminology that means the "best available demonstrated control technology [BADCT]," process, operating method, or other alternative to achieve the greatest degree of discharge reduction determined for a facility by the ADEQ Director under A.R.S. §49–243.

Baffles—deflectors, vanes, guides, grids, gratings, or similar devices constructed or placed in flowing water, wastewater, or slurry systems as a check or to effect a more uniform distribution of velocities; absorb energy; divert, guide, or agitate the liquids; and check eddies.

Ball valve—a simple nonreturn valve consisting of a ball resting on a cylindrical seat within a fluid passageway.

Ballast—the power supply to activate and regulate voltage in an ultraviolet (UV) lamp.

Bank stabilization—methods of securing the structural integrity of earthen stream channel banks with structural supports to prevent bank slumping and undercutting of riparian trees, and overall erosion prevention. To maintain the ecological integrity of the system, for recommended techniques include the use of willow stakes, imbricated riprap or brush bundles.

Bar—is equal to one atmosphere at sea level. It corresponds approximately to 101.3 KPa and 760 mm Hg.

Bar screening—a wastewater treatment process during which the largest particulate matter is separated from the sewage by passing the sewage through coarse screens. This often is the first treatment received by waste.

Barbel—fleshy elongated projection found below the lower jaw, under the snout, or around the mouth in catfishes, cods, sturgeons, and other fishes.

Barnacles—invertebrate animals of the crustacean order Cirripedia with feathery appendages for gathering food; free-swimming as larvae, but fixed to substrate and having a hard outer covering as adult.

Base—an alkaline substance with a pH that exceeds 7.5.

Base flow—the part of stream discharge from ground water seeping into the stream.

Base line data—an initial set of observations or measurements used for comparison; a starting point.

Base line emissions—the emissions that would occur without policy intervention. Baseline estimates are needed to determine the effectiveness of emissions mitigation strategies.

Base plan—In landscape architecture, an essential sheet showing site boundaries and significant site features, used as a basis for subsequent plan development

Basin irrigation—a surface irrigation method where water is applied to a completely level area surrounded by dikes. On high clay content soils and managed correctly basin irrigation can be very efficient.

Bathymetry—measurement on the depth of a body of water, also the mapping of its floor.

Bed load—soil, rock particles, or other debris rolled along the bottom of a stream by the **Base plan**—in landscape architecture, an essential sheet showing site boundaries and significant site features, used as a basis for subsequent plan development.

Bed rock—the consolidated rock underlying unconsolidated surface materials such as soil.

Bench mark—sites measuring natural phenomena in areas where man's activity has essentially no effect. These sites would be expected to continue at least as long as natural conditions exist.

Bench marking—the process of identifying best practices indicating superior performance. Benchmarks are adopted as targets for optimal organizational performance, and may include standards or environmental management processes.

Benthic Invertebrates—animals that live on river and lake bottoms. Many of these inhabitants are immature stages of insects such as mayflies, stoneflies, caddis flies, and midges. Other types of animals include aquatic earthworms or bristle worms, round worms, snails and leeches. The variety and abundance of benthic invertebrates in a river reflects the habitat the river provides.

Benthic organisms (or Benthos)—organisms living in or on bottom substrate in aquatic habitats.

Benthic zone—the lower region of a body of water including the bottom.

Bentonite—a sedimentary rock largely comprised of clay minerals that has a great ability to absorb water and swell in volume.

berm, Earthen—an earthen mound used to direct the flow of runoff around or through a "Best Management Practice."

Best efficiency point for pump—highest efficiency on a pump characteristic curve.

Best management practice (BMP)—an irrigation practice that is both economical and practical that will assist in reducing water use, protect water quality, while maintaining or improving crop production; Structural devices that temporarily store or treat urban storm water runoff to reduce flooding, remove pollutants, and provide other amenities; Techniques and procedures that have been proven through research, testing, and use to be the most effective and appropriate for use. Effectiveness and appropriateness are determined by a combination of: (1) the efficiency of resource use, (2) the availability and evaluation of practical alternatives, (3) the creation of social, economic, and environmental benefits, and (5) the reduction of social, economic, and environmental negative impacts.

Bicarbonates—salts containing the anion HCO_{3-}. When acid is added, this ion breaks into H_2O and CO_2, and acts as a buffer.

Big gun systems—uses a large sprinkler mounted on a wheeled cart or trailer, fed by a flexible rubber hose. The machine may be reeled in by the hose, or moved periodically. Big guns operate at high operating pressures in order to throw water at large distances.

Binder—chemicals that hold short fibers together in a cartridge filter.

Bioaccumulation—occurs when plants or animals collect contaminants in their tissues over time; when low amounts of contaminants are continually absorbed, they build up and can cause illness.

Biochemical oxygen demand (BOD)—a microbiological measurement of the quantity of dissolved oxygen, expressed in milligrams per liter (mg/L), required to stabilize the demand for oxygen in a water sample, usually resulting from the process of microorganisms consuming organic matter (decomposition) and using the available dissolved oxygen in the oxidation process. BOD is expressed as a result of five-day incubation of the sample at 20 degrees centigrade.

Biocide—a chemical that can kill or inhibit the growth of living organisms such as bacteria, fungi, molds and slimes. Biocides can be harmful to humans.

Biocriteria—the biological characteristics that quantitatively describe water body with a healthy community of fish and associated aquatic organisms. Components of biocriteria include the presence and seasonality of key indicator species; the abundance, diversity, and structure of the aquatic community; and the habitat conditions required for these organisms.

Biodegradable—subject to degradation (break down) into simple substances by biological action. For example, the breakdown of detergents, sewage wastes and other organic matter by bacteria.

Biodegradable pollutants—pollutants those are capable of decomposing under natural conditions.

Biodegradation—capable of being broken down especially into innocuous products by the action of living beings (as microorganisms).

Biodeposition—the process by which organisms deposit suspended materials onto the sediments.

Biodisk—a relatively new treatment process, which consists of, a series of flat, parallel disks, which are rotated while partially, immersed in the waste being treated. Biological slime covers the surface of the disks wastewater. This process is quite effective for treating wastes, which are highly concentrated.

Biofilm—population of various microorganisms, trapped in a layer of slime and excretion products, attached to a surface.

Biofiltration—the use of series of vegetated swales to provide filtering treatment for storm water as it is conveyed through the channel. The swales can be grassed, or contain emergent wetlands, or high marsh plants.

Biogeochemical cycle—paths by which elements that are essential to life circulate from the nonliving environment to living organisms and back again to the environment.

Biological agent—growth of slime, algae and, other microorganisms that can obstruct or clog the laterals and drippers (emitters).

Biological contaminants—living organisms such as viruses, bacteria, fungi, and mammal and bird antigens that can cause harmful health effects to humans.

Biological data—the kinds of organisms and how many of each are present in aquatic habitats.

Biological diversity (Biodiversity)—the variability among living organisms and the ecological complexes of which they are a part. This includes the diversity found within and between species and ecosystems.

Biological factors (agents)—growth of microorganisms or macroorganisms that may obstruct the irrigation system.

Biological filter—a bed of sand, gravel, broken stone, or other medium through which wastewater flows or trickles, which depends on biological action for effectiveness.

Biological monitoring—periodic surveys of aquatic biota as an indicator of the general health of a water body. Biological monitoring surveys can span the trophic spectrum, from macroinvertebrates to fish species.

Biological oxidation—decomposition of complex organic materials by microorganisms through oxidation.

Biological wastewater treatment—forms of wastewater treatment in which bacterial or biochemical action is intensified to stabilize, oxidize, and nitrify the unstable organic matter present. Biodisks, contact beds, trickling filters, and activated sludge processes are examples.

Biologically activated carbon—activated carbon that supports active microbial growth, in order to aid in the degradation of organics that have been absorbed on its surface and in its pores.

Biodegradable—material that will decompose under natural, biological conditions and processes.

Biodiversity—the number of different plants and animals that live in a specific area.

Bioindicators—organisms that are used to detect changes in environmental pollutant levels, such organisms are usually sensitive to changes in their surroundings.

Biomagnification—an increase in concentration of a substance at each progressive link in the food chain (e.g., berries birds foxes bears; the concentration of a contaminant such as lead would be highest in a large meat-eater).

Biomass—the quantity of matter present at any given time, expressed as the mass per unit area or volume of habitat; frequently used as a measure of a standing crop. Also: plant material, vegetation, or agricultural waste used as a fuel or energy source.

Biomat—the biological mat (biomat) is a black, jelly like mat about one to two inches thick that forms at the gravel-soil interface at the bottom and sidewalls of the drain field trench. The biomat is composed of microorganisms (and their byproducts) that anchor themselves to soil and rock particles, and whose food is the organic matter in the septic tank effluent. Since the biomat has a low permeability, it serves as a valve to slow down and control the rate of flow out of the trench into the drain field soil, and also serves as a filter to provide effluent treatment. Also known as a clogging mat.

Biomonitoring—the use of living organisms to test the suitability of effluents for discharge into receiving waters and to test the quality of such waters downstream from the discharge.

Bioremediation—the biological treatment of wastewater and sludge, by inducing the breakdown of organics and hydrocarbons to carbon dioxide and water; a process to reduce contaminant levels in soil or water by using microorganisms or vegetation.

Biosolids—treated solid or semisolid residues generated during the treatment of domestic sewage in a wastewater treatment facility. Primarily an organic product produced by wastewater treatment processes that can be beneficially used.

Biosphere—The part of the earth and its atmosphere in which ecosystems and living organisms exist or that is capable of supporting life. The biosphere includes the atmosphere, land (terrestrial biosphere) and water (marine biosphere).

Biostimulatory test—a test that determines the reaction of an organism to a given substance or set of conditions, such as cold, heat, and excessive nutrients. Algae growth potential is an example of this type of test.

Biota—the total plant (flora) and animal (fauna) species living in a specific region or environment.

Biotransformation—conversion of a substance into other compounds by organisms; including biodegradation.

Bioturbation—the process by which bottom sediments are disturbed by infaunal deposit feeders, primarily by their continual ingestion and reingestion of the sediments.

Bivalves—molluscs with paired shells; including clams, oysters, and mussels.

Black water—water containing liquid and solid human body waste generated through toilet usage.

Blaney-Criddle method—it is used to calculate the evapotranspiration in a system based on the correlation between the monthly average of water use and the temperature, the percentage of daily sunshine hours and the crop coefficient.

Bleach—a strong oxidizing agent and disinfectant formulated to break down organic matter and destroy biological organisms.

Blind spots—any place on a filter medium where fluids cannot flow through.

Blinding—a build-up of particles in a filter medium, that prevents fluids from flowing through.

Blood worms—segmented worms of the family Glyceridae that have a dark respiratory fluid, are detritus feeders, and are often used as bait by sport fishermen.

Bloom—high concentrations of algae or phytoplankton that occur when sufficient light and excess nutrients are available.

Blow down—the water discharged from a boiler or cooling tower to dispose of accumulated salts.

Blue green algae—a group of phytoplankton, which often cause nuisance, conditions in water. Some produce chemicals toxic to other organisms, including humans. They often form floating scum as they die. Many can fix nitrogen (N_2) from the air to provide their own nutrient.

BMP fingerprinting—term refers to a series of techniques for locating BMPs (particularly ponds) within a development site so as to minimize their impacts to wetland, forest and sensitive stream reaches.

BOD$_5$—the amount of dissolved oxygen consumed in five days by bacteria that perform biological degradation of organic matter.

Bog (Fen, Marsh, Shallow Open Water, Swamp)—A wetland characterized by peat deposits, acidic water, and extensive surface mats of sphagnum moss. Bogs receive their water from precipitation rather than from runoff, groundwater, or streams, with decreases the availability of nutrients needed for plant growth.

Bog coefficient [BC]—inverse of scheduling coefficient but using the wettest 1% window instead of the driest. Gives no indication of the location of dry areas.

Boiling point—the temperature at which the vapor pressure of a liquid equals the pressure of its surface. The liquid will than vaporize If the pressure of the liquid varies, the actual boiling point varies. For water the boiling point is 100°C.

Border dike—earth ridge or small levee built to guide or hold irrigation or recharge water in a field.

Border irrigation—water is applied at the upper end of a strip having parallel berms or earthen dikes anywhere from 10 to 100 feet apart that serve to confine the water to a strip or rectangular area.

Bore hole—a circular hole drilled or bored into the earth, usually for exploratory or economic purposes, such as water well or oil well.

Bore hole geophysics (Geophysics Borehole)—a general term that encompasses all techniques in which a sensing device is lowered into a borehole for the purpose of characterising the associated geologic formations and their fluids. The results can be interpreted to determine lithology, geometry resistivity, bulk density, porosity, permeability, and moisture content and to define the source, movement, and physical/chemical characteristics of ground water.

Bored wells— a relatively shallow well installed by a mechanically driven auger;

Bottom trawl— any of a variety of cone or wedged-shaped nets towed along the bottom to catch demersal fish.

Boulder placement—placement of large rock material into degraded riffle areas for the purpose of creating small pools and eddies that are preferred habitat for fish and aquatic insects. Use of this technique requires a thorough understanding of existing stream hydraulics and hydrology.

Brackish—refers to water having a salt content between 0.5 and 18 parts per thousand.

Brake horsepower—the power delivered by the motor to drive a pump.

Break through—crack or break in a filter bed that allows the passage of floc or particulate matter through a filter.

Break through—the appearance in the product water of an amount of the contaminant, which exceeds the design performance criteria.

Brine—highly salty and heavily mineralized water, containing heavy metal and organic contaminants.

British thermal units [BTU]—amount of heat required raising the temperature of one pound of water from 63 to 64°F.

Brooding—reproductive mode in which adults retain developing eggs or embryonic stages until they are sufficiently developed to release into the environment.

Brush bundles—three or more fresh-cut cedar trees wired together into a bundle, are placed along an eroding stream bank. The bundles are submerged, thus protecting the bank from erosive forces and at the same time creating cover for aquatic species.

Bubbler—water emission device that tends to bubble water directly to the ground or that throw water a short distance, on the order of one foot (300 mm) before water contacts the ground surface.

Buffer—a substance that reacts with hydrogen or hydroxyl ions in a solution, in order to prevent a change in pH.

Buffering capacity—the ability of a substance to resist an increase or decrease in pH.

Building (construction) permit—an authorization issued by a government agency allowing construction of a project according to approved plans and specifications.

Building codes—regulations specifying the type of construction methods and materials that are allowable on a project.

Built environment—the man-made creation of or alterations to a specific area, including its natural resources. This is in contrast to the "natural environment."

Bulk density of soil—mass of dry soil per unit bulk volume. Generally ranges from 1.3 to 1.6 g/cc.

Bulk water—water supplied by a water provider to another water provider.

Bulk water access entitlement—a water access entitlement allowing an infrastructure operator to access, take and hold water for the supply of water to another person.

Burden—the total mass of a certain gaseous substance in the atmosphere.

Bureau of Land Management (BLM)—A U.S. government agency charged with administering vast areas of public land.

Bushing—a bushing is a small piece used to connect two pipes of different sizes together. A standard reducer bushing has one male end (for the larger pipe) and one female connection (for the smaller pipe).

CAD—acronym for "Computer Aided (i.e., Assisted) Design

CADD—acronym for "Computer Aided (i.e., Assisted) Design and Drafting," a digital design process in which landscape architects use computers to help produce precise drawings and details for the construction of a project.

Cake—solid dewatered residue on a filter media after filtration.

Calcium—one of the principal elements making up the earth's crust. Calcium compounds, when dissolved, make hard water. The presence of calcium in water is a factor contributing to the formation of scale and insoluble soap curds, which are a means of clearly identifying hard water.

Calcium hypochlorite—a chemical that is widely used for water disinfection, for instance in swimming pools or water purification plants. It is especially useful because it is a stable dry powder and can be made into tablets.

Calendar days off—a controller feature that allows you to suspend watering on a specific date.

Calibration—adjustment of a chemical injection equipment so as to apply a desired quantity of chemical compound.

Canal—an artificial waterway designed for navigation or for transporting water for municipal water supply, land irrigation, or drainage.

Capillary action—phenomenon in which water or other liquids will rise above the normal liquid level in a tiny tube or capillary due to the attraction of the molecules in the liquid for each other and for the walls of the tube.

Capillary conductivity—allow the flow of water through soil capillary pores in response to a potential gradient.

Capillary mechanism—the capillary phenomena is due to: the force of attraction of the water by solids in the walls of the pores through which it flows and the surface tension of the water, that resists in any form except a smooth surface in the liquid-air interphase.

Capillary membrane—membrane about the thickness of a human hair, used for Reverse Osmosis, nanofiltration, ultrafiltration and microfiltration.

Capillary pores—are small pores, which maintain the soil moisture at a tension 60 cm or more. It is a fraction of the soil volume that is not occupied by soil particles.

Capillary radius [ft, mm]—additional wetted radius in soil profile beyond surface wetted radius for point source emitters.

Capillary space—is a micropore of the soil, or is a total volume of small pores that retain water in the soil at tension greater than 60 cm of water.

Capillary water—water held in the capillary, or small pores of the soil, usually with soil water pressure (tension) greater than 1/3 bar. Capillary water can move in any direction.

Capillary zone—soil area above the water table where water can rise up slightly through the cohesive force of capillary action.

Carapace—hard protective outer covering such as the shell of a blue crab.

Carbon (C)—an element which is found in almost all living or formerly living matter including plants, proteins, organics and hydrocarbons. Carbon combines readily with oxygen to form carbon dioxide (CO_2). The term "carbon" is sometimes used as a short reference for activated carbon. Diamonds and graphite are pure carbon; carbon is also present, with other substances, on coal, coke, and charcoal.

Carbon cycle—the flow of carbon through the atmosphere, lithosphere, biosphere and hydrosphere.

Carbon dioxide—a colorless, odorless, incombustible gas (CO_2), formed during respiration, combustion, and organic decomposition and used in food refrigeration, carbonated beverages, inert atmospheres, fire extinguishers, and aerosols. It is the principal anthropogenic greenhouse gas that affects the earth's radiative balance. It has a Global Warming Potential (GWP) of 1 and is used as the reference gas for GWPs of other greenhouse gases. It constitutes less than one percent of the Earth's atmosphere

Carbon flux—shifts or flows of carbon over time from one pool to another.

Carbon sequestration—the uptake and storage of carbon, for example by fossil fuels.

Carbonaceous—containing carbon and derived from organic substances such as coal, coconut shells and wood.

Carbonate—chemical compound related to carbon dioxide.

Carbonate hardness—hardness of water caused by carbonate and bicarbonate by-products of calcium and magnesium.

Carbonite environments—refers to sedimentary rock environments composed of calcium or magnesium carbonate.

Carcinogen—any dissolved pollutant that can induce cancer.

Carcinogenic—capable of producing or inducing cancer.

Carnivores—animals that feed on animal tissues.

Carryover soil moisture [in., mm]—moisture stored in the soil within the root zone during the winter, at times when the crop is dormant, or before the crop is planted. This moisture is available to help meet water needs of the next crop to be grown.

Cartilaginous fish—fish of the class Chondrichthyes, having a skeleton that is mostly cartilage rather than bone, such as sharks and rays.

Cartridge—any removable preformed or prepackaged component containing a filtering medium, ion exchanger, membrane or other treatment material which fits inside a housing to make up a cartridge filter.

Cartridge filter—disposable filter device that has a filter range of 0.1 micron to 100 microns; Device often used for single faucet water treatment, made up of a housing and a removable cartridge (element). In residential filtering systems, disposable elements are used.

Case-specific technology limit (End-of-Pipe Limits, Sector-Specific Technology Limit)—a subcategory of technology-based limits. It is a limit based on existing performance, or performance from similar facilities. Unlike a sector-specific technology limit, it is not a published limit. It is derived using best professional judgment.

Casing—the pipe between the intake (screen) section and the surface, serving as a housing for pumping equipment and conduct for the pumped water.

Catadromous—fish that descend from their primary habitat in fresh water to the ocean to spawn.

Catalysis—speeding up (usually) of a chemical reaction by adding a specific substance, the catalyst. Although the catalyst causes the speedup of the reaction, it (the catalyst) is not changed chemically in any way.

Catch basin—a sedimentation area designed to remove pollutants from runoff before being discharged into a stream or pond.

Catch can grid—containers spaced at regular intervals for collecting water for use in a water audit (sprinkler profile test).

Catchment area—A drainage area, especially of a reservoir or river.

Cathode—a site in electrolysis where cations in solution are neutralized by electrons that plate out on the surface or produce a secondary reaction with water.

Cation—positively charged ion, which is attracted towards the cathode during electrolysis. Sodium, potassium, calcium and magnesium are the most common cations in waters and soil extracts.

Cation exchange capacity—the ability of a particular rock or soil to absorb cations. The measure of the positively charged ions in a soil matrix (Fertility and fertilizers); Sum of exchangeable cations (usually Ca, Ma, K, Na, Al, H) that the soil constituent or other material can adsorb at a specific pH. Usually expressed in centimoles of charge per Kg of exchanger (cmol/Kg), or milli equivalents per 100 grams of soil at neutrality (pH = 7.0), meq/100 g.

Caudal—pertaining to the tail, or the posterior end.

Caustic—any substance capable of burning or destroying flesh or tissue. The term usually applies to strong bases.

Cavitation—process when pressure on a liquid falls momentarily below the vapor pressure of that liquid, causing a phase change from liquid to vapor, and then a pressure increase causes another phase change and the gas becomes a liquid again; Process where pressure in the suction line falls below the vapor pressure of the liquid, vapor is formed and moves with the liquid flow. These vapor bubbles or "cavities" collapse when they reach regions of higher pressure on their way through pumps.

CCSM-3—"Community Climate System Model version 3." The world's most advanced climate model used extensively by organizations to predict the consequences of climate change.

Cell volume—refers to the number of cells of any organisms that is counted by using a microscope and grid or counting cell. Many planktonic organisms are multicelled and are counted according to the number of contained cells per sample, usually milliliters (ml) or liters (L).

Cellulose acetate (CA) and cellulose triacetate (CTA)—a cellulose ester obtained by introducing the acetyl radical (CH3CO–) of acetic acid into cellulose (as cotton or wood fibers) to produce a tough plastic material which is used to make the cellulosic type of semipermeable reverse osmosis membranes.

Celsius— the units of temperature in English system of units is commonly used by scientists. Water freezes at 0°C (32°F) and boils at 100°C (212°F). To convert Celsius to Fahrenheit, °F = 9/5°C + 32. Celsius was formerly termed Centigrade.

Center pivot sprinklers—an automated irrigation system composed of a sprinkler lateral rotating around a pivot point and supported by a number of self-propelled towers. Water and power is supplied at the pivot point. They are easily identified from the air as large circular fields. Most center-pivots are one-quarter mile long and irrigate 128- to 132-acre circular fields. Center pivots are the most popular irrigation method in the U.S. because they require low labor input, are fairly reliable, are flexible with crop varieties and soils, can accommodate undulating terrain, are easily adapted for chemigation, and have the best uniformity and efficiency of any.

Centibar—one-hundredth part of a bar. The tensiometer is usually calibrated in centibars.

Centigrade—see Celsius. A temperature scale in which 100 degrees is the boiling point and zero degrees the freezing point for water at sea level.

Centrifugal filter—hydrocyclone or centrifugal filters remove suspended particles of a specific gravity greater than of water. These filters are ineffective to remove most of organic solids.

Centrifugation—a separation process, which uses the action of centrifugal force to promote accelerated settling of particles in a solid-liquid mixture.

Certified Golf Irrigation Auditor (CGIA)—involved in the analysis of turf irrigation water use tailored to the unique conditions found on golf courses. Golf Auditors collect site data, make maintenance recommendations and perform water audits on golf courses. Through their analytical work at the site, these irrigation professionals develop base schedules for greens/tees, fairways and roughs. Prior to certification examination, auditors are required to take an Irrigation Association approved preparatory course.

Certified Irrigation Contractor (CIC)—an irrigation professional whose principle business is the execution of contracts and subcontracts to install, repair and maintain irrigation systems. The CIC must conduct business in such a manner that projects meet the specifications and requirements of the contract.

Certified Irrigation Designer (CID)—engages in the preparation of professional irrigation designs. They evaluate site conditions and determine net irrigation requirements based on the needs of the project. The designer is then responsible for the selection of the most effective irrigation equipment and design methods. The objective of a CID is to establish specifications and design drawings for the construction of an irrigation project.

Certified Landscape Irrigation Auditor (CLIA)—involved in the analysis of landscape irrigation water use. Auditors collect site data, make maintenance recommendations and perform water audits. Through their analytical work at the site, these irrigation professionals develop monthly irrigation base schedules. Prior to certification examination, auditors are required to take an Irrigation Association approved preparatory course. (IA Water Management Committee, 2001)

Certified Landscape Irrigation Manager (CLIM)—an irrigation professional familiar with all areas of turf irrigation design and construction management. CLIMs must be certified as CICs, CID (all Landscape/Turf specialty areas), and either as a CLIA or CGIA. Certified Landscape Irrigation Managers have extensive experience in design, construction, construction management and auditing of turf irrigation systems.

Cesspool—a lined or partially lined underground pit into which raw household wastewater is discharge and from which the liquid seeps into the surrounding soil. Sometimes called leaching cesspool.

Cesspools—include multiple dwelling, community, or regional cesspools or other devices that receive wastes and have an open bottom and sometimes have perforated sides. Must serve more than 20 persons per day if receiving solely sanitary wastes.

CFCs—chlorofluorocarbon: compound containing chlorine and fluorine bonds that have been used as refrigerants before the Montreal Protocol. These compounds have been shown to deplete stratospheric ozone. These compounds can also act as greenhouse gases.

Cfs-day—the volume of water represented by flow of 1 cubic foot per second [cfs] for 24 h. It is equivalent to 86,400 cubic feet, approximately 1.9835 acre feet, about 646,000 gallons or 2,447 cubic meters.

CFU (Colony Forming Units)—a measure that indicates the number of microorganisms in water.

Chain of custody—method for documenting the history and possession of a sample from the time of its collection through its analysis and data reporting to its final disposition.

Channel (Water course)—an open conduit either naturally or artificially created which periodically or continuously contains moving water, or which forms a connecting link between two bodies of water.

Channeling—the flow of water through a limited number of passages in a filter.

Charcoal—an adsorbent carbon product, which has about one-third the surface area of activated carbon.

Check dam—a small dam constructed in a gully or other small watercourse to decrease the stream flow velocity, minimize channel erosion, promote deposition of sediment, and to divert water from a channel; It can be a log or gabion structure placed perpendicular to a stream to enhance aquatic habitat; or an earthen structure, used in grass swales to reduce water velocities, promote sediment deposition, and enhance infiltration.

Check structure—structure to control water depth in a canal, ditch or irrigated field.

Check valve—a valve that allows water to stream in one direction and will then close to prevent development of a back-flow.

Chelating agents—organic compounds that have the ability to draw ion from their water solutions into soluble complexes.

Chemical agent—see chemical factors.

Chemical coagulation—the destabilization and initial aggregation of colloidal and finely divided suspended matter by the addition of a floc-forming chemical.

Chemical factors (Agents)—during the fertigation, pestigation, chloration, chemigation, chemical precipitates may be formed to cause obstruction at different locations of the drip irrigation system.

Chemical oxygen demand (COD)—the amount of oxygen (measured in mg/L) that is consumed in the oxidation of organic and oxidasable inorganic matter, under test conditions. It is used to measure the total amount of organic and inorganic pollution in wastewater. Contrary to BOD, with COD practically all compounds are fully oxidized.

Chemical pollution—introduction of chemical contaminants into a water body.

Chemical spike (Spike)—a sample that contains a measured amount of a known analyte, used for determining matrix interferences.

Chemical standards—materials made from ultra-pure compounds used to calibrate laboratory analytical equipment.

Chemical tissue analysis—measurement of the types and/or amounts of chemical substances present in the tissue of an organism.

Chemical water treatment—chemical treatment of the water to make it acceptable for use in micro irrigation systems. This may include the use of acids, flocculants, fungicides, and bactericides to help prevent emitter plugging or for pH adjustment.

Chemical weathering—dissolving of rock by exposure to rainwater, surface water, oxygen, and other gases in the atmosphere, and compounds secreted by organisms.

Chemigation—process of applying chemicals (fertilizers, insecticides, herbicides, etc.) to crops or soil through an irrigation system with the water.

Chemosynthetic activity—the synthesis of organic matter from mineral substances, with the aid of chemical energy.

Chloramines—a chemical complex that consists of chlorine and ammonia. It serves as a water disinfectant in public water supplies in place of chlorine because chlorine can combine with organics to form dangerous reaction products. In which forms chloramines exist depends on the physical/ chemical properties of the water source. Water containing chloramines may not be used for fish or for kidney dialysis applications.

Chloration or chlorintion—is an application of chlorine through the irrigation system to control the growth of algae, bacteria, and other microorganisms in the irrigation lines and emitters.

Chlorinated hydrocarbons—hydrocarbons that contain chlorine. These include a class of persistent insecticides that accumulate in the aquatic food chain. Among them are DDT, aldrin, dieldrin, heptachlor, chlordane, lindane, endrin, mirex, hexachloride, and toxaphene.

Chlorinated solvent—an organic solvent containing chlorine atoms that is often used as aerosol spray container, in highway paint, and dry cleaning fluids.

Chlorination—the application of chlorine to water or wastewater, generally for the purpose of disinfection, but frequently in the process of photosynthesis, used as an indicator of plants biomass in water.

chlorination, Breakpoint—addition of chlorine to water until there is enough chlorine present for disinfection of water.

Chlorine—a common element, best known as a heavy greenish yellow irritating toxic gas of disagreeable odor used chiefly as a powerful bleaching, oxidizing, and disinfecting agent in water purification and in making pesticides.

Chlorine contact chamber—the part of a water treatment plant where effluent is disinfected by chlorine.

Chlorophyll—a primary plant pigment (green) which functions in the process of photosynthesis, used as an indicator of plant biomass in water.

Chlorophyll-A—a primary group of green pigments that occur in plant cells, chiefly in bodies called chloroplast, that function in the process of photosynthesis, used as a measure of plant (or algal) biomass in the water. Expressed in micrograms per liter (μg/L). The exact levels of chlorophyll A, needed to indicate a healthy stream, river, lake, or estuary have not yet been determined.

Chlorosis—is one of the common symptoms of mineral deficiency of chlorine. It appears as a green color or yellowing of the green parts of a plant, particularly the leaves.

Chronic effect (Acute effect)—A stimulus that lingers or continues for a relatively long period of time, often one-tenth of the life span or more. Chronic effects should be considered a relative term depending on the life span of the organism.

Chronic toxicity—exposure that will result in sub lethal response over a long-term, often one-tenth of the life span or more.

Cilia—hair-like structures of cells, which beat rhythmically and cause locomotion in single cells and multicellular organisms, and which are used to create water currents and/or movement of particles.

Circular mil [CM, circular mils]—unit of measure used to report the cross sectional area of a wire conductor.

Cistern—a tank for storing water or other liquids, usually placed above the ground.

City beautiful movement—a popular social concern of the late 19th and early twentieth centuries aimed at improving the appearance of urban areas through better planning and the addition of formal, romanticized public spaces and gardens.

CL 125, 160, 200, 315—pronounced "class-one-20-five," "class-one-60," "class-two-hundred" and "class-three-15." Pipe classifications (CL = class = classification) based on standard dimension ratios. CL 125 pipe is rated for 125 psi working pressure, CL 200 for 200 psi, etc. Most industry professionals will tell you that it is not wise to use pipe rated at less than twice the actual maximum water pressure level. In other words, for a water pressure of 100 psi, use at least CL 200 pipe.

Cladocerans—invertebrate animals of the crustacean order Cladocera, each having a single sessile compound eye; often called water fleas.

Clam bed—an area of bottom supporting a dense population of clams.

Clarification—any process or combination of processes the primary purpose of which is to reduce the concentration of suspended matter in a liquid.

Clarigester—a treatment process in which clarification and aerobic digestion occur in the same tank.

Class (pipe)—term generally used to describe the pressure rating of SDR-PR (standard dimension ratio-pressure rated). For example, a class 200 pipe has a pressure rating of 200 psi; Term used to identify the physical characteristics of thermoplastic pipe.

class, Soil—group of soils defined as having a specific range in one or more particular properties such as acidity, degree of slope, texture, structure, land- use capability, degree of erosion, or drainage.

Clay—soil separate consisting of particles less than 0.002 mm in equivalent diameter; Soil textural class; Naturally occurring material composed primarily of fine-grained minerals, which is generally plastic at appropriate water contents and will harden when dried or fired.

Clay loam—soil textural class.

Clean development mechanism—investments of developed countries in emissions reducing projects in developing countries to obtain credit to assist in meeting their assigned amounts.

Climate—the average weather for a particular region and time period; the meteorological conditions, including temperature, precipitation, and wind that characteristically prevail in a particular region. Climate in a wider sense is the state, including a statistical description, of the climate system.

Climate (change) scenario—a climate scenario consists of projections of possible climate futures, containing developments of driving forces, greenhouse gas emissions, temperature change and sea level rise and their key relationships. A climate change scenario is the difference between a climate scenario and the current climate.

Climate change—a change in the climate around, persisting for an extended period of time (typically decades or longer). Climate change occurs as a result of natural conditions or anthropogenic sources changing the composition of the atmosphere or the land use type.

Climate change commitment—term introduced by researchers of the National Centre of Atmospheric Research (NCAR) in Boulder, Colorado to clarify the seriousness of climate change to nonclimatologists. It means the amount of climate change that will inevitably occur in the coming century as a result of human behavior in the twentieth century.

Climate feedback—an interaction mechanism between processes in the climate system. A primary process causes changes in a second process, which in turn influences the primary process. A positive feedback intensifies the original process, and a negative feedback reduces it.

Climate forecast—the result of an attempt to produce a most likely description or estimate of the actual evolution of the climate in the long-term.

Climate model—a numerical representation of the climate system that is based on the physical, chemical and biological properties of its components, their interactions and feedback processes and accounts for its known or inferred properties. A climate model is used to study climate characteristics. Climate models used to be fairly simple, but now there is an evolution towards more complex models with active chemistry and biology.

Climate modeling—the simulation of climate using computer-based models.

Climate projection—a description of the response of the climate system to emission or concentration scenarios of green house gases and aerosols, or radiative forcing scenarios, often based upon simulations by climate models. Climate projections are subject to uncertainty, because they are typically based on assumptions concerning future socioeconomic and technological developments that may or may not be realized.

Climate sensitivity—equilibrium change in surface air temperature following a unit change in radiative forcing ($°C/Wm^{-2}$). According to the IPCC climate sensitivity refers to the equilibrium change in global mean surface temperature following a doubling of the atmospheric CO_2 concentration.

Climate system—the highly complex system consisting of the atmosphere, the hydrosphere, the lithosphere and the biosphere and the interactions between them. The evolution of the climate system is a consequence of volcanic eruptions, solar variations and anthropogenic alterations, such as land use change.

Climate variability—mean state and other statistical variations in climate on all temporal scales and spatial scales beyond that of individual weather events. Variations may be caused by natural internal processes within the climate system or by variations in natural or anthropogenic forcing.

Clogging mat—see biomat.

Cloud—collection pf water droplets or ice crystals suspended in the air, which forms when the air is cooled to its dew point and condensation occurs.

Cloud seeding—the use of substances to increase precipitation; silver iodide and dry ice are often used. The process has never been very successful.

Cloudy—the atmosphere loaded with clouds.

CO_2 equivalents—a metric measure used to compare the emissions of various greenhouse gases based upon their Global Warming Potential (GWP): the concentration of CO_2 that would give the same amount of radiative forcing as the given mixture of CO_2 and other greenhouse gases. Essentially, they sum the radiative forcing of all trace gases and treat the total forcing as if it comes from an equivalent CO_2 concentration.

CO_2 fertilization—the enhancement of plant growth as a result of elevated atmospheric CO_2 concentrations.

Coagulation—destabilization of colloid particles by addition of a reactive chemical, called a coagulant. This happens through neutralization of the charges. It will collect dirt and other solids suspended in the water.

Coalescence—liquid particles in suspension that unite to create particles of a greater volume.

Coarse sand—coarse sand is a soil separate. Its physical size varies according to the classification system being used; Soil textural class.

Coarse sandy loam—soil textural class.

Coastal plain—a lowland area extending in a gentle slope inland from the shoreline of an ocean.

Coastal zone—lands and waters near the coast, whose uses and ecology are affected by the sea.

Code of practice—a document governing an activity or a portion of an activity. One example is the *Code of Practice for.*

Coefficient of retardation—values describing the hydraulic frictional characteristics of a pipe material.

Coefficient of runoff [C]—runoff coefficient used in the rational method of predicting a design peak runoff rate. It helps to characterize runoff rate from a watershed area.

Coefficient of uniformity (CU)—a numerical expression, which serves as an index for the uniformity of water applied to a given area within a specific geometric arrangement of sprinklers (e.g., square or triangular).

Coefficient of variation—a statistical measure of the relative dispersion for an independent variable; It compares emitter flow rate of each dripper with the average flow rate of drippers; The standard deviation divided by the mean of a set of data.

Coefficient of variation of the manufacturer—describes the variation in pressure in the pressure controller (regulating) valve in relation to the operating pressure of the drippers.

coefficient, Consumptive use—ratio of volume of irrigation water consumptively used to the total volume of irrigation water that has left the region, both in a specified period of time.

coefficient, Crop—it is a ratio between crop evapotranspiration (ET-cultivation) and the potential evapotranspiration under optimal conditions of growth.

coefficient, Hygroscopic—it is a soil moisture percentage, after it has been dried to the atmospheric conditions or after the soil has reached a balance with a relative humidity of the atmosphere.

coefficient, Pan—ratio between the potential evapotranspiration and class A pan evaporation.

coefficient, Permanent wilting—it is a percentage of soil moisture at which the plants have wilted permanently at a tension of 15 atm. The plants do not recover.

coefficient, Reflection (albedo)—it is a ratio between the amount of solar radiation received by the soil surface and the reflected sun radiation.

coefficient, Wilting—soil moisture content at which the plants have wilted. It corresponds to the upper limit of the usable moisture by plants.

Coefficients of soil water—these are the limits or margins between the three levels of water in the soil: gravitational, capillary, or hygroscopic.

Coelenterates—invertebrate animals of the Phylum Coelenterata (cnidaria) that are basically radially symmetrical; including jellyfishes, sea anemones, and hydroids.

Cold front—the boundary of a cold-air system, usually from the polar regions, that urges into the warmer air masses of the temperate latitudes.

Coliform—rod-shaped bacteria common to the intestinal tract of man and animals. The presence of coliform-type bacteria is an indication of possible pathogenic bacterial contamination. *Fecal coliforms* are those coliforms found in the feces of various warm-blooded animals, whereas the term *coliform* also includes other environmental sources. Expressed as "Most Probable Number (MPN) per 100 milliliters" of water. The presence of coliform does not always mean contamination with human wastes, since coliform can also grow in soil.

Coliform index—a rating of the purity of water based on a count of coliform bacteria.

Collaboration (Partnership)—a process through which parties who see different aspects of a problem can explore their differences and search for (and implement) solutions that go beyond their own limited vision of what is possible. Collaboration is a mechanism for leveraging resources; dealing with scarcities; eliminating duplication; capitalizing on individual strengths; building internal capacities; and increasing participation and ownership strengthened by the potential for synergy and greater impact.

Collector sewers—pipes to collect and carry wastewater from individual sources to an interceptor sewer that will carry it to a treatment facility.

Colloids—matter of very small particle size, in the range of 10.5 to 10.7 in diameter.

Color—color in water can be caused by the presence of such things as plankton, decaying organic matter, industrial wastes, and sewage. A distinction is between "true color" (the color of a water sample after turbidity has been removed by filtration) and "apparent as color" (the color of an untreated water sample). True color is usually measured by comparing the color of a water sample to that of a fixed standard. Color is expressed in terms of "color units" where one color unit is the difference in tint produced by one milligram per liter of the chloroplatinate ion.

Color unit—produced by one milligram per liter of platinum in the form of the chloroplatinate ion. Color is expressed in units of the platinum-cobalt scale.

Comb jellies—invertebrate animals of the phylum Ctenophora that superficially resemble jellyfish (but have few or no tentacles and no stinging nematocysts) and that have cilia, which they use for swimming, arranged in long combs around their bodies; also called ctenophores.

Combined Sewer—Older drainage systems that carry both sanitary waste and storm water runoff. During heavy rain, the capacity of the combined sewer to carry wastewater to the sewage treatment plant may be exceeded, and a discharge of the untreated waste to the river may occur.

Combined sewer overflow (CSO)—excess flow from an overloaded combined sewer.

Command and control approach—a method of environmental management by government that involves specific statutory controls and associated regulatory offenses, which are generally prescriptive in terms of outcomes and behaviors. Examples of this approach include: acts, regulations, approvals, licenses, authorizations, Codes of Practice, and orders.

Comminutor—a device often used with, or in place of, bar screens. Instead of removing large particles, comminutors are designed to grind them into smaller particles, which are then removed by other processes.

Community—a natural assemblage of animals and plants interacting with each other and with abiotic components of the environment.

Compaction—decrease in volume of sediments, as a result of compressive stress, usually resulting from continued deposition above them, but also from drying and other causes.

Competition—interaction between organisms for limited resources needed for survival, including food, shelter, and space.

Complete flow—a term used to indicate that flow (discharge) figures are determined based upon a complete record (full range) of stage, as opposed to partial record flow where figures pertain only to a predetermined limited range in stage.

Compliance assessment—an activity undertaken to determine whether a regulated party's activity/operation complies with a statute, regulation, authorization or Code of Practice. Compliance assessments educate the regulated party on legislative requirements and also identify current or potential noncompliance. Compliance assessments include inspections, reviews, and audits.

Compliance assurance—activities that ensure regulated parties comply with legislation, including the Water Act. These activities include promoting compliance through education and prevention initiatives, and compelling compliance through enforcement responses.

Components of variability—the characteristics that vary from one statistical population to another, such as well locations, and analytical lab errors.

Composite sample—a series of water samples taken over a given period of time and weighted by flow rate.

Compounds—two or more different elements held together in fixed proportions by attractive forces called chemical bonds.

compounds, Insoluble—do not dissolve in the water.

compounds, Soluble—dissolve completely in water.

compounds, Systematic—are composites that can travel through epidermis of the leaves, stems and roots or seeds,

Concentrate—the totality of different substances that are left behind in a filter medium after filtration.

Concentration—the amount of a substance in a given volume of water. For most substances, the concentration is expressed as milligrams per liter (mg/L), which is the same as parts per million (ppm). Technology now exists that can measure substances at the parts per trillion or quadrillion level; the process of separating a mineral from valueless host rock in preparation for further processing; also the amount of a substance in a given weight or volume of another material.

Concentration process—the process of increasing the number of particles per unit volume of a solution, usually by evaporating the liquid.

Concentration profile—graphic representation of the horizontal and vertical locations of contaminant concentration levels on maps and cross-sections.

Concentration scenarios—projections of greenhouse gas concentrations derived from emission scenarios and used as input into a climate model to compute climate projections.

Condensate—water obtained by condensation of water vapor.

Condensation—the process that occurs when an air mass is saturated and water droplets form around nuclei or surfaces.

Conductivity—the quality or power of conducting or transmitting electrical current: in water, a measure of the dissolved ions.

Conductivity of soil—see permeability of the soil.

conductivity, Electrical (conductance)—the property of a substance to transfer an electrical charge. It is reciprocal of an electrical resistance.

conductivity, Electrical, or saturation extract, (ECC)—is an amount of salts in the soil solution when the soil is saturated of water at current or normal conditions.

Conduit—a natural or artificial channel through which fluids may be transported.

Confidence limits—a statistical statement that relates the probability that the true value of a variable falls within the described interval.

Confined aquifer—an aquifer bounded above and below by impermeable beds or beds of distinctly lower permeability than that of the aquifer itself; it contains confined ground water.

Confining bed—an impermeable unit that lies over or under an aquifer and prevents or impedes movement of water into or out of the aquifer.

Confining layer—a geologic stratum exhibiting low permeability and having little or no intrinsic permeability.

Conjunctive use—the use of more than one water source, systematically, to reduce overall environmental impacts. For example, someone might use groundwater instead of surface water during a drought period, and then return to using surface water when runoff became abundant again.

Consensus—when a group of individuals in a decision-making process work towards general agreement by all involved.

Conservation—the planning, management, and implementation of an activity with the objective of protecting the essential physical, chemical, and biological characteristics of the environment against degradation; The process of managing biological resources (e.g., timber, fish) to ensure replacement by regrowth or reproduction of the part harvested before another harvest occurs. A balance between economic growth and environmental and natural resource protection; protection, preservation, management, or restoration of a resource.

Conservation plan—a plan for conserving or protecting various natural or manufactured resources. Such a plan is used as a management tool in making decisions regarding soil, water, vegetation, manufactured objects and other resources at a particular site.

Consolidated—formed into a compact mass; firm and rigid because of the natural interlocking and/or cementation of mineral grain components.

Constituent—individual components, elements, or biological entities such as suspended solids or ammonia nitrogen.

Consumptive pool—The amount of a water resource that can be made available for consumptive use in a particular water resource plan area under the rules of the water resource plan for that water resource plan area.

Consumptive use—The balance of water taken from a source that is not entirely or directly returned to that source. For example, if water is taken from a lake to feed cattle, it is considered a consumptive use of water; Use of water for private benefit consumptive purposes including irrigation, industry, urban and stock and domestic use; Use of water resulting in a large proportion of loss to the atmosphere by evapotranspiration. Irrigation is a consumptive use.

consumptive use, Non-—water drawn for use that is not consumed. For example, water withdrawn for purposes such as hydropower generation. It also includes uses such as boating or fishing where the water is still available for other uses at the same site.

Consumptive water use—water removed from available supplies without return to a water resources system; water used in manufacturing, agriculture, and food preparation; when water is used and not returned to its source, such as through evaporation or by including it in a product.

Contact stabilization process—a modification of the activated sludge process in which raw wastewater is aerated with a high concentration of activated sludge for a short period, usually less than 60 min, to obtain BOD removal by adsorption. The solids are subsequently removed by sedimentation and transferred to a stabilization tank where aeration is continued further to oxidize and condition them before their reintroduction to the raw wastewater flow.

Contact time—the length of time a substance is in contact with a liquid, before it is removed by filtration or the occurrence of a chemical change. Contact time may also be called retention time.

Contaminant—a substance that, in a sufficient concentration, will render water, land, fish, or other things unusable or harmful; Any undesirable physical, chemical or microbiological substance or matter in a given water source or supply. Anything in water, which is not H_2O, may be considered a contaminant.

Contamination—the process of becoming impure; damage to the quality of water sources by sewage, industrial waste, or other matter.

Contents—the volume of water in a reservoir or lake. Unless otherwise indicated, volume is computed on the basis of a level pool and does not include bank storage.

Continental rise—the lower portion of the ocean's continental shelf that leads to the abyssal plain.

Continental shelf—the submerged edge of a continent extending from the lowwater line to a region with a distinct change in slope.

Continental slope—the higher section of the continental shelf; it is the steep section between the continental shelf and rise.

Continuous flow irrigation—system of irrigation water delivery where each irrigator receives the allotted quantity of water continuously.

Contour—the form of the land. Contour lines are map lines connecting points of the same ground elevation and are used to depict and measure slope and drainage. Spot elevations are points of a specific elevation.

Contrails—trails of dangerous chemicals left behind in air by aviation. These were shown in 2003 to contribute to the greenhouse effect and global warming.

Control—designates a feature downstream from the gage that determines the stage- discharge relation at the gage. This feature may be a natural constriction of the channel, an artificial structure, or a uniform cross section over a long reach of the channel.

Control and water treatment station—the control station may include facilities for water measurement, filtration, treatment, addition of amendments, pressure control, and timing of application.

Control dam—a dam or structure with gates to control the discharge from the upstream reservoir or lake.

Control system—group of tubes, accessories, and controls that regulates the water to the lateral lines.

Control valve—a device used to control the flow of water.

Convection—heat transfer from one place to another by the particle movement of a hot liquid gas.

Conventional gravity system—an onsite sewage system consisting of a septic tank and a subsurface soil absorption system with gravity distribution of the effluent.

Conventional pressure distribution system—an onsite sewage system consisting of a septic tank and a subsurface soil absorption system with pressure distribution of the effluent.

Conveyance efficiency—the ratio of the water that is delivered to a field to the total water diverted or pumped into the conveyance system at the upstream end [%, decimal].

conveyance efficiency, Water—ratio of the water delivered, to the total water diverted or pumped into an open channel or pipeline at the upstream end.

Conveyance loss—loss of water from a channel or pipe during transport, including losses due to seepage, leakage, evaporation, and transpiration by plants growing in or near the channel.

Cooling tower—large tower used to transfer the heat in cooling water from a power or industrial plant to the atmosphere either by direct evaporation or by convection and conduction.

Copepoda—minute shrimp-like crustacean that are often the most common zooplankton in estuarine and oceanic waters.

Copepodite—a juvenile stage of copepoda.

Coral reef—a mound or ridge of rock produced by the accumulation of skeletons from living coral.

Core—a continuous columnar sample of the lithologic units extracted from a borehole. Such a sample preserves stratigraphic contacts and structural features.

Coriolis force (Corioois effect)—the force that causes winds to deviate to the right in the (turbulence) of water masses.

Correlation coefficient—an indicator of the degree of interdependence between two variables.

Corrosion—a dissolving and wearing away of metal caused by a chemical reaction (in this case, between water and the piping that the water contacts or between two different metals).

Corrosive environments—subsurface zones containing ground water or soil corrosive to monitoring well construction materials.

Corrosive—substance is disturbed or eroded by the action of chemical agents.

Corrosivity—ability of water to dissolve or break down certain substances, particularly metals.

Corrugation irrigation—surface irrigation method where small channels or corrugations are used to guide water across a field. No attempt is made to confine the water entirely to the corrugations.

Cost benefit analysis—in landscape architecture, a study of the potential cost of site purchase, demolition and improvement in comparison to the income or other benefit to be derived from site development.

Council of Landscape Architecture Registration Boards (CLARB)—a coordinating agency formed in 1961 for state boards that administer licensing exams and maintain records for landscape architects to practice.

Coupler (sprinkler)—device, either self-sealed or mechanically sealed, that connects the ends of two lengths of pipe or pipe to a hose.

Coupling—a section of pipe with female or male threads or without threads. These are used to connect the pipes of different lengths.

Coverage—the area of landscape watered by a sprinkler or grouping of sprinklers.

coverage, Hundred—the design goal of all sprinkler system designers. Sometimes used (incorrectly) in place of the term "head to head coverage." 100% coverage is the objective of head-to-head coverage. But head-to-head coverage does not always result in 100% coverage.

Crab pot—a baited cage with conical-shaped entrance ways, used to capture crabs.

Crumb—a unit or particle of soil composed of many small grains sticking together.

Crustacean—any of a large class Crustacea of mainly aquatic arthropods that

Creeping failure—a condition where the biomat becomes so intensely developed that no water can flow through it eventually causing system malfunction.

Crop coefficient—coefficient used to modify reference evapotranspiration to reflect the water use of a particular plant or group of plants particularly with reference to the plant species; Ratio of the actual crop evapotranspiration to its potential (or reference) evapotranspiration.

Crop growth stages—periods of like plant function during the growing season. Usually four or more periods are identified: *initial*—Between planting or when growth begins and approximately 10 percent ground cover; *crop development*—Between about 10 percent ground cover and 70 or 80 percent ground cover; *mid-season*—From 70 or 80 percent ground cover to beginning of maturity; and *late*—From beginning of maturity to harvest.

Crop reference—is used to develop and to calibrate methods to calculate reference evapotranspiration. The reference crops are alfalfa and grasses.

Crop water requirements—depth of water for the evapotranspiration (ET crop) required by a crop or diversified pattern of crop during a given period (mm/day).

Crop water stress index [CWSI]—index of moisture in a plant compared to a fully watered plant, measured and calculated by a CWSI instrument. Relative humidity, solar radiation, ambient air temperature, and plant canopy temperature are measured.

Crop consumptive use—total amount of water that must be replaced in soil during the crop growth period (see evapotranspiration).

Cropping pattern—sequence of different crops that are seeded in a given order.

Cross connection—any actual or potential connection or structural arrangement between a public or private potable water system and any other source or system through which it is possible to introduce into any part of the potable system any used water, industrial fluids, gas, or substance other than the intended potable water with which the potable system is supplied. By-pass arrangements, jumper connections, removable sections, swivel or changeover arrangements or other "temporary" arrangements through which backflow could occur are considered to be cross-connections.

Cross contamination—the intermixing of two water streams, which results in unacceptable water quality for a given purpose.

Cross—a fitting that joins four sections of pipe at one point forming a "cross." Reducing crosses are available which have different size outlets. Unless you order a custom made cross the outlets opposite each other are always the same size.

Cryptosporidium—a microorganism in water that causes gastrointestinal illness in humans. It is commonly found in untreated surface water and can be removed by filtration. It is resistant to disinfectants such as chlorine.

Crystalline rock—rock made up of minerals in a clearly crystalline state igneous and metamorphic rock, as opposed to sedimentary rock.

Cubic feet—measurement of water quantity, often used by water companies in the USA to measure water use by customers. A cubic foot is one foot in length, one foot in width, and one foot in depth.

Cubic feet per second (cfs, ft³/s)—a rate of water flow at a given point, amounting to a volume of one cubic foot for each second of time. Equal to 7.48 gallons per second or 448.8 gallons per minute or 1.984 acre feet per day.

Cubic feet per second per square mile (cfsm)—the average number of cubic feet of water flowing per second from each square mile of area drained, assuming that the runoff is distributed uniformly in time and area.

Cubic meters—a metric measurement of volume of water quantity, often used by water companies to measure water use by customers. A cubic meter is one meter in length, one meter in width, and one meter deep.

Cultch—shells or other material spread over oyster grounds by man, on which oyster larvae can attach and develop.

Cultural eutrophication—decline of the oxygen rate in water, which has serious consequences for aquatic life, caused by humans.

Cumulative effects—the combined environmental impacts that accumulate over time and space as a result of a series of similar or related individual actions, contaminants, or projects

Cumulative intake [in., mm]—depth of water absorbed by soil from the time of initial water application to the specified elapsed time.

Curie—a measurement of radioactivity equal to $3.7 ´ 10°$ disintegrations per second.

Current—the portion of a stream or body of water, which is moving much faster than the rest of the water. The progress of the water is principally concentrated in the current.

Cycle—the length of time a filter can be used before it needs cleaning, usually including cleaning time.

Cycle + Soak™—Rain Bird 's exclusive feature which allows you to break the total irrigation run time into shorter cycles, segmented by breaks or soaks during which the landscape has time to absorb the water. Optimizes the watering of poor drainage sites, slopes, and heavy soil areas. Helps to prevent run-off.

Cypris—a planktonic larval stage of a barnacle that attaches itself to a hard substrate and then metamorphoses into an adult.

Cyst—a capsule or protective sac produced about themselves by many protozoans (as well as some bacteria and algae) as preparation for entering a resting or a specialized reproductive stage. Similar to spores, cysts tend to be more resistant to destruction by disinfectant. Fortunately, protozoan cysts are typically 2–50 microns in diameter and can be removed from water by fine filtration.

Dam—A barrier constructed on a water body for storage, control, or diversion purposes. A dam may be constructed across a natural watercourse or on the periphery of a reservoir. Natural barriers formed by ice, landslides, or earthquakes are excluded.

Dam height—the difference in elevation between the crest elevation and the lowest point at the downstream toe of a dam.

dam, Arch—curved masonry or concrete dam, convex in shape upstream, that depends on arch action for its stability.

Dam³—a measure of water volume: short for decameters cubed. One dam³ is equal to 1,000 cubic meters.

Darcy's Law—an equation that can be used to compute the quantity of water flowing through an aquifer.

Datum—any level surface, line, or point use as a reference in measuring elevation.

DDT (dichloro-diphenyl-trichloroethane)—a colorless, odorless, water-insoluble crystalline insecticide (C_4HCI_5) that may concentrate in certain organisms and whose effects may be toxic.

Dealkalinization—any process that serves to reduce the alkalinity of water.

Debye—Huckel equation—a means of computing the activity coefficient for an ionic species.

Decant—to draw off the upper layer of liquid after the heaviest material (a solid or another liquid) has settled.

Decapods—invertebrate animals of the crustacean order Decapoda that are characterized by five pairs of legs; including crabs and shrimps.

Decarbonation—the process of removing carbon dioxide from water, using contact towers or air scrubbers.

Decomposers—organisms (chiefly bacteria and fungi) that break down nonliving matter, absorb some of the products of decomposition, and release compounds that become usable by producers.

Decomposition—the breakdown of organic matter by bacteria and fungi, to change the chemical structure and physical appearance of matter.

Deep percolation [DP: in., mm]—movement of water downward through the soil profile below the root zone that cannot be used by plants; Infiltrated water, which moves below the root zone.

Deep percolation percentage [DPP, %]—ratio of the average depth of irrigation water infiltrated and drained out of the root zone to the average depth of irrigation water applied.

Deep well injection—deposition of raw or treated, filtered hazardous waste by pumping it into deep wells, where it is contained in the pores of permeable subsurface rock.

Deficit irrigation—irrigation water management alternative where the soil in the plant root zone is not refilled to field capacity in all or part of the field.

Defluoridation—the removal of fluoride from drinking water to prevent teeth damage.

Defoaming agents—chemicals that are added to wastewater discharges to prevent the water from foaming when it is discharged into a receiving water body.

Deforestation—a land use change induced by removal of forest by humans for various purposes.

Degasification—the process of removing dissolved gasses from water, using vacuum or heat.

Degree of hazard—the type of backflow preventer used to prevent backflow from occurring at the point of a cross-connection depends on the type of substance, which may flow into the potable water supply. A pollutant is considered to be any substance, which would affect the color or odor of the water, but would not pose a health hazard. This is also considered a nonhealth hazard. A substance is considered a health hazard if it causes illness or death if ingested. This health hazard is called a contaminant.

Deionization—process that serves to remove all ionized substances from a solution. Most commonly is the exchange process where cations and anions are removed independently of each other.

Delivery box (irrigation)—structure diverting water from a canal to a farm unit often including measuring devices. Also called "turnout"; Water control structure for diverting water from a canal to a farm unit often including a measuring device. Also called delivery site, delivery facility, and turnout.

Delta—a large collection of sediments deposited at the mouth of a river. When sediment suspended in a river reaches slower-moving waters, it settles and forms a delta.

Demand irrigation—procedure where each irrigator may request irrigation water in the amount needed and at the time desired.

Demersal—living close to or on the bottom.

Demineralization—processes to remove minerals from water, usually the term is restricted to ion exchange processes.

Demiwater—demineralized water. Water that is treated to be contaminant-, mineral- and salt free.

Denitrification—the anaerobic biological reduction of nitrate-nitrogen to nitrogen gas. Also removal of total nitrogen from a system.

Density—mass of a unit volume of a substance. It is usually expressed in kilograms per cubic meter.

density of water, Bulk [lb/ft^3, g/cc]—mass of water per unit bulk volume (approximately 1.0 g/cc or 62.4 lb/ft^3).

Density factor [K_d]—coefficient used to modify reference evapotranspiration to reflect the water use of a particular plant or group of plants particularly with reference to the density of the plant material.

Densogram—pattern of dots that shows the expected coverage from a particular combination of sprinklers, nozzles, pressure and spacing; Graphical representation of precipitation rates.

Deposit feeders—infaunal organisms that meet their nutritional needs by either selectively or indiscriminately ingesting the sediments in which they live.

Deposition environment—a geographically restricted complex where sediment accumulated, described in geomorphic terms and characterized by physical, chemical, and biological conditions (e.g., flood plain, lake, and beach).

Depreciation cost—It is a systematized method to find the value of the irrigation system based on its cost, and is a method as to prorate the cost for the active service life.

Depression spring—a spring formed when the water table reaches a land surface because of a change in topography.

Depression storage—the storage of water from precipitation in low areas, such as ponds, and wetlands.

Depth filtration—a filtration process in which water flows through progressively smaller pore spaces in a filter. Depth filters are designed to entrap particles throughout the mass of the filter media, as opposed to a surface filter where only the surface layer does the actual filtering.

Depth of irrigation—depth of water applied; Depth of soil affected by an irrigation event, [acre inches per acre, in., ft, mm]. Also called irrigation depth.

Desalination—the removal of dissolved inorganic solids (salts) from seawater or brackish water such as water to produce a liquid, which is free of dissolved salts. Desalination is typically accomplished by distillation, reverse osmosis or electro-dialysis.

Desertification—land degradation into dry areas caused by climatic variations and human activities. This often causes a loss of biological and economic productivity of the land. Processes that enhance desertification are for example soil erosion, deterioration of soil properties and long-term loss of natural vegetation.

Design—the creative illustration, planning and specification of space for the greatest possible amount of harmony, utility, value and beauty.

Design area—the specific land area, which is to be irrigated by the micro irrigation system.

Design capacity—measured in GPM, the "design capacity" is a maximum amount of water available for use in an irrigation system. The available design capacity determines how many sprinkler heads may be in operation at the same time.

Design capacity of sewer—the hydraulic capacity of any sewer when installed as indicated by the diameter of the pipe.

Design emission uniformity—an estimate of the uniformity of emitter discharge rates throughout the system.

Design life—the estimated length of time before the system will have to be replaced or rehabilitated.

Design pressure—in the irrigation design tutorials the design pressure is the total pressure available to operate the irrigation system. Other uses of the term vary, but usually refer to the operating pressure at which a specific piece of irrigation equipment is designed to operate.

Designed landscape—a site that might appear to be natural but has elements and features that were planned and specified by a landscape architect.

Desorption—the opposite of adsorption; the release of matter from the adsorption medium, usually to recover material.

Detection of climate change—the process of demonstrating that climate has changed in some defined statistical sense, without providing a reason for that change.

Detention time—the period of time that a water or wastewater flow is retained in a basin, tank, or reservoir for storage or completion of physical, chemical, or biological reaction.

Detergent—a water-soluble cleansing agent, other than soap; A term applied to a wide variety of cleansing agents used to wash clothes, dishes, and other articles. Generally, detergents are organic materials that are surfactant in aqueous solutions.

Deterioration—a reduction in the value of an object due to the elements of nature.

Detritus—particulate matter, especially that of organic origin, floating in the water or settling to the bottom.

Dew—droplets that form when water vapor condenses.

Dew point—the temperature to which air must be cooled to cause condensation of the water vapor it contains.

Dewater—the separation of water from sludge, to produce a solid cake.

Diagenesis—the chemical and physical changes occurring in sediments before consolidation or while in the environment of deposition.

Diameter—dimension or size of a circular pipe, usually but not always the inside diameter [ID]: inside diameter [ID, D]; outside diameter [OD, D]; Nominal diameter [ND].

Diameter of throw (coverage, ft or m)—average diameter of the area wetted by a sprinkler operating in still air.

Diaphragm—rubberized seal which keeps water from flowing through the valve.

Diaphragm emitter—emitter that contain a stretched membrane with a small opening. When particles plug the opening, pressure builds stretching the membrane until the particle is forced out. The membrane and opening then return to normal size.

Diatomaceous earth—soil composed of the siliceous cell walls of dead diatoms that have settled and accumulated on bottom of bodies of water.

Diatoms—unicellular or colonial algae of the division Bacillariophyta that have siliceous cell walls; a major division of primary producers found in aquatic ecosystems.

Dielectric—substance having a very low electrical conductivity.

Dielectric union—pipe connection (union) having an insulator between the two sides of the union for the purpose of reducing corrosion caused by galvanic action.

Diffuser—a component of the ozone contacting system in an ozone generator that allows diffusion of an ozone containing gas.

Diffusion—the process by which both ionic and molecular species dissolved in water move from areas of higher concentration to areas of lower concentration.

Digester—a closed tank for wastewater treatment, in which bacterial action is induced to break down organic matter; called sludge digestion tank.

Digestion—the biological decomposition of organic matter in sludge, resulting in partial gasification, liquefaction, and mineralization.

Digitizing—the process of coding points along various boundaries (watershed, vacant land, gravity spheres) and sewer lines in coordinate system.

Dike (levee)—a long low embankment dam. The term is usually applied to auxiliary dams used to close off areas that would otherwise be flooded by the reservoir.

Diluting water—distilled water that has been stabilized, buffered, and aerated. It is often applied in the BOD tests.

Dilution—to decrease the concentration of a substance by mixing it with another or by adding water

Dilution rate—relation between the chemical agent concentration to be applied and the amount of water that flows through the irrigation system.

Dimension ratio—ratio of the average pipe diameter to the minimum wall thickness. The pipe diameter may be either outside or inside diameter.

Dinoflagellates—aquatic, unicellular algae of the division Pyrrophyta that are major primary producers in temperate estuaries.

Dip net—any of a variety of nets that are held by hand and used to catch fish or aquatic invertebrates; generally used in small streams.

Dipole array—a particular arrangement of electrodes used to measure surface electrical resistivity.

Direct current—current in which electrons flow constantly in one direction.

Direct methods for hydrological investigations—methods (e.g., borehole logging, pump tests), which involve the drilling, collection, observation, and analysis of geologic materials, water samples, and drawdown/ recovery data.

Direct reuse site—Arizona regulatory terminology that means an area where reclaimed water is applied or impounded.

Discharge—flow of surface water in a stream or canal; and is used as a measure of the rate at which a volume of water passes a given point. Therefore, the use of this term is not restricted as to course or location, and it can be used to describe the flow of water from a pipe or a drainage basin.

Discharge rate of drippers—amount of water from drippers at a given pressure that is applied near the plant per unit time.

Discrete sample—a water sample taken at a single instant in time.

Disinfectants—fluids or gasses to disinfect filters, pipelines, systems, etc.

Disinfection—a process of destroying or inactivating pathogenic microorganisms in water and wastewater. To disinfect a fluid or surface a variety of techniques are used, such as ozone disinfection. Often disinfection means eliminating the present microrganisms with a biocide.

Disinfection—treatment of water to inactivate, destroy and /or remove pathogenic (disease producing) bacteria, viruses, cysts,

Dispersivity—ability of a contaminant to disperse within the ground water by molecular diffusion and mechanical mixing.

Disposal—the relocation and/or containment, of unwanted materials.

Dissolution—the process of dissolving a solid in a liquid.

Disposal facility—a facility where hazardous waste is intentionally placed into or on land or water, and at which waste will remain after closure of the facility.

Disposal well—a deep well used for the disposal of liquid wastes,

Dissolve—the process during which solid particles mix molecule by molecule with a liquid and appear to become part of the liquid; formation of a homogenous mixture between a liquid and a solid of which first usually is in greater quantity.

Dissolved—constitutes that material in a representative sample which passes through a 0.45 micron-m membrane filter. This is a convenient operational definition used by federal agencies that collect water data. Determinations of "dissolved" constituents are made on subsamples of the filtrate.

Dissolved air flotation—process of induced flotation with very fine air bubbles [micro bubbles] of 40 to 70 microns. This is accomplished by bubbling air through the liquid, which increases the buoyancy of the solids and lifts them to the surface of the liquid where they can be removed.

Dissolved oxygen—the concentration of oxygen (normally a gas) dissolved in water [ppm or mg/L]. It is a function of temperature and pressure. The colder the water, the more oxygen it will hold. In general, fish require 5.0 mg/L in a stream. It is a measurement of oxygen in water expressed in mg/l or in percent of saturation under existing ambient conditions; the amount present reflects chemicals, physical, and biological activities in the water body. It can only be increased by aeration and the photosynthetic processes of aquatic matter.

Dissolved solids—solid material that totally dissolves in water and can be removed by means of filtration.

dissolved solids, Total—**(TDS)** the weight of matter, including both organic and inorganic matter, in true solution in a stated volume of water. The amount of dissolved solids is usually determined by filtering water through a 0.45 pore-diameter micron filter and weighing the filtration residue left after the evaporation of the water at 180°C; the substance in water such as carbonates, sulfates, chlorides, nitrates, phosphates and metallic ions, iron, calcium, potassium, and others that can pass through a fine, glass fiber filter (filterable residue).

Distillate—the product water, which is mineral-free and potable, from a distiler unit.

Distillation—a process to purify water in which water is heated to steam and the steam is collected and condensed back to water. Some contaminants are left behind with the steam; the process of separating the water from the organic and inorganic contaminants through a combination of evaporation (or vaporization), cooling and condensation.

Distilled water—water which has been cleansed by passing through one or more evaporation-condensation cycles until it contains a very low amount of dissolved solids (usually less than 5.0 ppm TDS).

Distribution system—system of ditches, or conduits and their controls, which conveys water from the supply canal to the farm points of delivery.

Distribution uniformity—test for determining the evenness or uniformity of irrigation water over an area. Numerical indices of uniformity include DU (distribution uniformity of the low quarter) and CU (Christiansen's coefficient of uniformity). It has no units; a calculated value that shows how evenly water is distributed in a sprinkler system to avoid excessively wet or dry areas in the landscape. It depends on the spacing of sprinklers, type of sprinkler used, wind and water pressure among other factors. High distribution uniformity is obtained when an equal amount of water is placed on all areas of the landscape; DU is calculated using a catchment test. Divide the average reading of the lowest one-quarter of catchments by the average reading of all catchments, and multiply the answer by 100. An excellent DU percentage is 75% to 85%, while a good DU is 65% to 70%.

Distribution uniformity of lower quarter [%, decimal]—ratio of the average low quarter depth of irrigation water infiltrated to the average depth of irrigation water infiltrated; Ratio of the average of the lowest one-fourth of measurements of irrigation water infiltrated (or applied) to the average depth of (the total) irrigation water infiltrated (applied).

Disturbed (manipulated) soils—oils with soil profiles that have been altered because of earth-moving activities (or soil amendment).

Diurnal temperature range—The difference between the maximum and minimum temperature during a day.

Diurnal vertical migration—the daily vertical movement of organisms up and down in the water column.

Diversion—the taking of significant quantities of water from a stream or other body of water into a canal, pipe, or other conduit. The term applies to ground water stations when pumping is significant.

Diversion box—structure built into a canal or ditch for dividing the water into predetermined portions and diverting it to other canals or ditches.

Diversion dam—barrier built in a stream for the purpose of diverting part or all of the water from the stream into a canal.

Diversity—a measure of the complexity of an ecological community that is generally a function of both the number of species present and the relative proportions of their numbers.

Diversity Index—a numerical expression of evenness of distribution of aquatic organisms.

Dobson Unit (DU)—this unit measures the total amount of ozone in a vertical column above the earth. One DU corresponds to a column of ozone containing [2.69 ´ 1020] molecules per square meter. The number of Dobson units is the thickness in units of 10^{-5} m, that the ozone column would occupy if compressed into a layer of uniform density at a pressure of 1013 hPa, and a temperature of 0°C.

Dolomite—a carbonate sedimentary rock composed predominantly of $CaMg(CO_3)_2$.

Domestic consumption—water used for household purposes such as washing, food preparation, and showers.

Domestic use—water use in homes and on lawns, including use for laundry, washing cars, cooling, and swimming pools.

Domestic wastewater—a composite of liquid and water-carried wastes associated with the use of water for drinking, cooking, cleaning, washing, hygiene, sanitation or other domestic purposes, together with any infiltration and inflow wastewater, that is released into a wastewater collection system.

Domestic water use—water used for drinking, cooking, washing, and yard use.

Dominant year class—a segment of population composed of a single age group resulting from spawn with a particularly high survival; this segment may make up a significant portion of the total population for a period of several years.

Dosage—amount used of a product.

Double check assembly—two internally loaded, independently operating check valves together with tightly closing resilient seated shut-off valves upstream and downstream of the check valves. Additionally, there are resilient seated test cocks for testing of the assembly. It may be used to protect against a pollutant only. However, this assembly is suitable for protection against either back-siphonage or backpressure.

Down gradient—in the direction of increasing static head.

Down gradient well—well which has been installed hydraulically down gradient of the site, and is capable of detecting the migration of contaminants from a regulated unit. Regulations require the installation of three or more down gradient wells depending upon the site-specific hydrogeological conditions and potential zones of contaminant migration.

Drain—a conduit, channel, or other structure constructed or used to carry water or wastewater by gravity or pumping.

Drain—a small artificial watercourse designed to drain swampy areas or irrigated lands. Theoretically, it is actually a small canal, but it is referred to as a "drain" in many localities.

Drain line—pipeline which is used to carry water from the water treatment system to a waste system.

Drain pipe—a pipe or conduit that carries untreated runoff water to nearby waterways to reduce flooding of pavements. Not to be confused with sewers which carry wastewater.

Drain valve—any valve used to drain water from a line.

Drainage—it is a process by which the superficial or underground water is removed from a soil in an area.

Drainage area—of a stream at a specified location of a site is that area, measured in a horizontal plane, enclosed by a topographic divide from which direct surface runoff from precipitation normally drains by gravity into the stream above the site.

Drainage basin—the total area of land that contributes water and materials to a lake, river, or other water body, either through streams or by localized overland runoff along shorelines.

Drainage district—farmer-led cooperative groups, established under the Drainage Districts Act, that work to improve agricultural water management within a specific area of the province. Districts are formed by Order-in-Council or Ministerial Order at the request of local landowners. Once formed, the district has the power to set and collect taxs, to construct water management works and to enact bylaws.

Drainage divide—a boundary line along a topographically high area that separates two adjacent drainage basins.

Draw—a tributary valley or coulee that usually discharges water only after a rainstorm.

Draw down—A reduction in the water level of a water well when the pump is operating; The lowering of the water surface level of a reservoir due to a release of water from the dam; the depth (from the top of the well) to the water in a well when the pump is operating. The water level typically drops when the pump is running. Artists who paint scenes with geese in them have been known to "draw down."

Dredged spoil—the sediment removed by man from the bottoms of aquatic habitats.

Dredging—cleaning, deepening, or widening of a waterway, using a machine (dredge) that removes materials by means of a scoop or a suction device.

Drift net—gill net that is held upright by floats and counterweights and that drifts freely in the water column.

Drilled well—a well which penetrates fractured bedrock, drilled with a rotating bit; lined with steel or PVC casing.

Driller—a person who is authorized under *The Water Act* to drill or reclaim water well.

Driller's Log—description by the driller of the geologic materials penetrated from the land surface to the greatest depth of the well.

Drilling mud—fluids which are used during the drilling of a borehole or well to wash soil cuttings away from the drill bit and adjust the specific gravity of the liquid in the borehole so that the sides of the hole do not cave in prior to installation of a casing.

Drinking Water—water that has been treated to provincial standards and is fit for human consumption (Potable water); a water treated or untreated which is intended for human use and consumption and considered to be free of harmful chemicals and disease-causing bacteria, cysts, viruses or other microorganisms.

Drip irrigation—type of irrigation system that applies water to the soil very slowly, thus the name "drip" irrigation. Currently it is the most efficient irrigation technology in terms of both water and energy conservation.

Drip line—the circle that can be made on the soil around a tree below the tips of the outermost branches of the tree.

Drip or drop—is formed by water movement through small orifices.

Drip system—an irrigation system that uses drip irrigation.

Dripper, internal spiral type—these are based on the principle of long trajectory. The capillarity effect is produced by means of the union of a pipe extended in an inner cylinder. This is a spiral of a circular or rectangular transverse section. The water enters the terminal and leaves by the side due to small force. The design provides drippers with long trajectory with the integration of components to function uniformly.

Dripper, spiral type—capillary of long trajectory, made of polypropylene, in spirals normally hardened to maintain shape permanent. This dripper is inserted on the lateral and the water leaves the dripper drop by drop.

Drive pipe—casing consisting of the grive shoe and riser. This casing follows the auger bit as it advances.

Drive shoe—steel coupling or band at the bottom edge of the casing reinforced to withstand drive pressures during cable tool and drill through casing driver methods.

Driving forces—climate scenarios contain various *driving forces* of climate change: population growth and socioeconomic and technological development. These drivers encompass various future scenarios that might influence greenhouse gas sources and sinks, such as the energy system and land use change.

Drought—periods of less than average precipitation over a certain period of time. Drought is naturally occurring and can cause imbalances in the hydrologic system; Period of dryness especially when prolonged that causes extensive damage to crops or prevents their successful growth.

Dry pond (wet pond)—relief systems that provide a diversion of excess flow from a storm sewer trunk to an impoundment for temporary storage. A dry pond is often a playground or other open space not normally covered by water. A dry pond reduces flooding downstream.

Dry weight (of soil sample)—oven dry weight of a soil sample.

Dry well—a well which is a bored, drilled, or driven shaft or hole whose depth is greater than its width and is designed and constructed specifically for the disposal of storm water.

Dug well—a large diameter well (30 inches or more) excavated by shovel, usually to a depth just below the water table; commonly lined with brick, stone, or wood to prevent collapse.

Dupuit assumptions—assumptions for flow in an unconfined aquifer: (1) the hydraulic gradient is equal to the slope of the water table (2) the streamlines are horizontal, and (3) the equipotential lines are vertical.

Dupuit Equation—an equation for the volume of water flowing in an unconfined aquifer; based upon the Dupuit Assumptions.

Dynamic pressure—measure of water pressure with the water in motion (also known as working pressure).

Dystrophic lakes—acidic bodies of water that contain many plants but few fish, due to the presence of great amounts of organic matter.

Ecology—a branch of biology dealing with the relationship between living things and their environment.

Ecosystem—an ecological community together with its environment, functioning as a unit. The scale of an ecosystem largely depends on the type of study that is conducted and may range from a small number of populations and their environment to the entire earth; an interactive system which includes the organism of a natural community together with their environment; a community of plants, animals, and nonliving things that exist in the same place. A forest ecosystem has definite boundaries and includes a forest of trees out to the limit of tree growth. Remember that forests are not the only ecosystems. There are hundreds of thousands of defined and undefined ecosystems that can cover the broadest to the tiniest of areas. An ecosystem can be as small as a pond or a dead tree, or as large as the Earth itself.

Ectoparasite—a parasite organism that lives on, rather than in, its host and uses the host for energy and habitat.

Effective precipitation—portion of the total precipitation that is stored in the soil profile for later use by the plants [in, mm].

Effective rainfall—the portion of a rainfall that infiltrates into the soil and is stored for plant use in the crop root zone. During a rainfall event, part of the rain will run off the surface of the soil, get trapped in the foliage, or percolate through the bottom of the root zone. This water is not available for plant growth and, therefore, is not considered to be effective rainfall. Effective rainfall is generally expressed as a depth of water over a unit area (e.g., in/acre).

Efficiency—see application efficiency, irrigation efficiency, conveyance efficiency, pumping plant efficiency, motor efficiency, pump efficiency, or water use efficiency [%, decimal].

Efficiency (media filtration)—the percent of contaminant reduction that occurs with a specified medium volume and specified water contact time.

Efficiency (membrane filtration)—the figure obtained (expressed as a percent) = the volume gallons of water divided by the total volume (gallons).

efficiency, Application—the percentage of the total water applied that is actually stored in the root zone.

efficiency, Irrigation application—the percentage of water applied that can be beneficially used by the crop.

efficiency, Water storage—ratio of the amount of water stored in the root zone during irrigation to the amount of water needed to fill the root zone to field capacity.

efficiency, Water use—ratio of the yield per unit area to the applied irrigation water per unit area.

Effluent—the liquid waste of municipalities, industries, or agricultural operations. Usually the term refers to a treated liquid released from a wastewater treatment process; the discharge from any *on-site sewage* treatment component.

Effluent irrigation—land application of treated wastewater for irrigation and beneficial use of nutrients.

Ejection—the process of forcing something out, expelling it.

Ejector—a device used to inject a chemical solution into wastewater during water treatment.

Electric valve—an automatic valve normally using 24 to 30 volts AC from the controller to activate a solenoid, allowing the valve to open and close.

Electrical charge—the charge on an ion, declared by its number of electrons. A Cl- ion is in fact a Cl atom, which has acquired an electron, and a Ca++ ion is a Ca atom, which has lost two electrons.

Electrical conductivity [EC]—measure of the ability of the soil water to transfer an electrical charge. Used as an indicator for the estimation of salt concentration.

electrical conductivity, Maximum—is a limit of the concentration of salts in the saturated soil (ECC) at which the growth will become paralyzed (null production). Units are mmhos/cm or DS/m.

Electrical resistance block—block made up of various materials containing electrical contact wires that are placed in the soil at selected depths to measure soil moisture content (tension). Electrical resistance, as affected by moisture in the block, is read with a meter.

Electrical resistivity (ER)—a geophysical method where a known current is applied to spaced electrodes in the ground and the resulting electrical resistance is used to detect changes in materials between and below the electrodes. ER is particularly useful for facilities receiving electrically conductive wastes (e.g., inorganic) at sites characterized by settings having minimal quantities of high resistance materials.

Electrodialysis—a process that uses electrical currents, applied to permeable membranes, to remove minerals from water.

Electrolysis—process where electrical energy will change in chemical energy. The process happens in an electrolyte, a watery solution or a salt melting which gives the ions a possibility to transfer between two electrodes. The electrolyte is the connection between the two electrodes, which are also connected to a direct current. If you apply an electrical current, the positive ions migrate to the cathode while the negative ions will migrate to the anode. At the electrodes, the cations will be reduced and the anions will be oxidized.

Electrolyte—substance that dissociates into ions when it dissolves in water.

Electromagnetic conductivity—a geophysical method whereby induced currents are produced and measured in conductive formations from electromagnetic waves generated at the surface. EM is used to define shallow ground water zones characterized by high dissolved solids content.

Electrons—negatively charged building blocks of an atom that circle around the nucleus.

Elevation head—a measurement of pressure.

elevation, Maximum—difference in elevation between the water source and the most distant location in the field.

Ell—fitting used to change the direction of a pipe. Ells look like an "L." PVC ells are available with threads in both ends, threads in one end and a glued socket in the other, or with glue sockets in both ends. Insert ells come with male threads and barbs, female threads and barbs, glue sockets and barbs, glue spigots and barbs, or barbs and barbs. For PVC and poly irrigation uses, these are available in 90 degree and 45 degree bends. Some specialty ells are available in other angles but have limited availability.

Elutriation—a process of sludge conditioning whereby the sludge is washed with either fresh water or plant effluent to reduce the demand for conditioning chemicals and to improve settling or filtering characteristics of the solids. Excessive alkalinity is removed in this process.

Emergent plants—aquatic plants that are rooted in the sediment but whose leaves are at or above the water surface. These wetland plants provide habitat for wildlife and waterfowl in addition to removing urban pollutants.

Emission permit—tradable allocation of entitlements by a government to an individual firm to emit a specific amount of a substance.

Emission point—location where water is discharged from an emitter.

Emission standard—an amount of emission that may not be exceeded legally.

Emission uniformity—index of the uniformity of emitter discharge rates throughout a micro irrigation system. Takes account of variations in emitters and variations in the pressure under which the emitters operate.

Emissions—the release of gaseous substances, such as greenhouse gases, into the atmosphere.

Emissions scenario—Representation of the future development of emissions of greenhouse gases based on a set of assumptions about driving forces and their key relationships.

Emitter—a small watering device, which delivers water at very low rate (measured in gallons per hour) and pressure at the outlet port. Used in drip irrigation to control the flow of water into the soil. Also called a "dripper" or "trickler": *compensating emitter; pressure compensating emitter; continuous flushing emitter*—designed to continuously permit passage of large solid particles while operating at a trickle or drip flow, thus reducing filtration requirements; *flushing emitter*—designed to have flushing flow of water to clear the discharge opening every time the system is turned on; *line source emitter*—water is discharged from closely spaced perforations, emitters, or a porous wall along the tubing; *long-path emitter*- employs a long capillary sized tube or channel to dissipate pressure; *multioutlet emitter*- supplies water to two or more points through small diameter auxiliary tubing; *nonpressure compensating emitter*- designed with a fixed orifice or other components and contains no pressure compensating features; *orifice emitter*—employs a series of orifices to dissipate pressure; *pressure compensating emitter*—designed to discharge water at a near constant rate over a wide range of lateral line pressures; *vortex emitter*—employs a vortex effect to dissipate pressure.

Emitter discharge exponent—the emitter discharge exponent; x, as described by the equation $q = kh^x$, which characterizes the type of emitter. For example, an x value of 0.5 is common for orifice type emitters, whereas the x value for a pressure compensating emitter would range from 0 to 0.5.

Emitter discharge rate—   the instantaneous discharge rate at a given operating pressure from an individual point-source emitter or from a unit length of line-source emitter, expressed as a volume per unit of time.

Emitter manufacturer's coefficient of variation—a statistical term used to describe the variation n discharge rate for a sample of new emitters when operated at a constant temperature and at the emitter design operating pressure.

Emitter performance coefficient of variation—statistical term used to describe the variation in emitter discharge due to the combined effects of emitter manufacturer's variation, emitter wear, and emitter plugging.

Emitter spacing—It is a space between two drippers or emitters. It depends on the type of crop and the emitter flow rates.

Emulsifier—a chemical that helps suspending one liquid in another.

Emulsion—dispersion of one liquid in another liquid, occurs when a liquid in insoluble.

Emulsion formula—is a liquid formulation of a given pesticide that can form emulsion when mixed with water.

End of pipe technique—technique for water purification that serves the reduction of pollutants after they have been formed.

Energy balance—is a physical system that regulates the microclimate in the vicinity of the plant. Incoming solar radiation at the soil surface influences the energy balance.

Enrichment—the addition of nitrogen, phosphorus, carbon compounds or other nutrients into a lake or other waterway, increases the growth potential for algae and other aquatic plants. Most frequently enrichment results from the inflow of sewage effluent or from agricultural runoff.

Enteric bacteria—bacteria, which originate in the intestines of warm-blooded animals.

Entrainment—the process of being drawn in and transported by the flow of water; frequently refers to the process whereby organisms are drawn into the cooling systems of electrical power plants and into water intake systems of other industrial facilities.

Entrance loss—energy lost in eddies and friction at the inlet to a conduit or structure.

Environment—the components of the earth, including air, land, and water, all layers of the atmosphere, organic and inorganic matter, living organisms, and their interacting natural systems.

Environmental design professions—landscape architecture, agricultural engineering, civil engineering, urban planning and architecture. Agronomy is also often included in this group.

Environmental impact—the change in the area of natural resources, including animal and plant life, results from use by man. Some projects may require conducting of an "environmental impact study" before development can proceed.

Environmental impact statement—a report submitted by a company to describe a project or development, the possible positive or negative impacts of its actions, and its plans to reduce, mitigate or avoid these impacts; the information in the report is based on studies that have been carried out; the report is reviewed by the appropriate government agencies and the public.

Environmental inventory—record of an area's natural and man-made resources, including vegetation, animal life, geological characteristics and mankind's presence in such forms as housing, highways and even hazardous wastes.

Environmental Protection Agency—an U.S. government agency responsible for developing and enforcing regulations that guide the use of land and natural resources.

Environmental quality—a measure of the status of the environment, overall or in relation to a media (air, water, land) or the needs of its inhabitants, including humans.

Enzyme—a chemical produced by living cells, which can bring about the digestion (breakdown) of organic molecules into smaller units that can be used by living cell tissues.

Ephemeral wetland—an area that is periodically covered by standing or slow moving water and that has a basin typically dominated by vegetation of the low prairie zone, similar to the surrounding lands. Because of the porous conditions of the soils, the rate of water seepage from these areas is very rapid, and surface water may only be retained for a brief period in early spring.

Ephyrae—flattened juvenile medusoid jellyfish released from the top of polyp form when strobilation is completed.

Epibenthic—term used for organisms that live on the surfaces of bottom substrates.

Epifauna—organisms that live on substrates in aquatic habitats; including crabs, shrimps, snails, oysters, and mussels.

Equalization—the collection of sewage in a storage area to reduce large fluctuations in either its strength or flow.

Equilibrium constant—the number defining the conditions of equilibrium for a particular reversible chemical reaction.

Equipotential line—a line in a two-dimensional ground-water flow field such that the total hydraulic head is the same for all points along the line.

Equipotential surface—a surface in a three-dimensional ground-water flow field such that the total hydraulic head is the same as for all points along the line.

Equipment blank—chemically pure solvent (typically regent grade water) that is passed through an item of field sampling equipment and returned to the laboratory for analysis, to determine the effectiveness of equipment decontamination procedures.

Equivalent weight—the formula weight of a dissolved ionic species divided by the electrical charge. Also known as the combining weight.

Erosion—the natural breakdown and movement of soil and rock by water, wind, or ice. The process may be accelerated by: human activities; eroding down of the soil surface, loosening and movement of the soil or rocks, caused by runoff, wind, ice, or another geologic agent, including processes as the gravitational drags.

Escherichia coli (E. coli)—one of the members of the coliform bacteria group normally found in human and animal intestines and indicative of fecal contamination when found in water. Determination of E. coli in the potable water is often used to measure the microbiological safety of drinking water supplies.

Esters—class of an organic compound derived by the reaction of an organic acid with an alcohol.

Estuary—any area where freshwater meets saltwater, specifically a semi enclosed tidal, coastal body of saline water with a free connection to the sea within which sea water is measurably diluted with fresh water derived from land drainage. Examples are bays, river mouths and salt nurseries, spawning and feeding grounds for a large group of marine life, provide shelter and food for birds and wildlife, and are remarkably able to store and recycle nutrients.

ET or evapotranspiration—the amount of water lost due to evaporation from the soil and transpiration from the plants. ET is used by Smart Controllers to help determine the amount of watering needed by a landscape.

Euglenoids—unicellular algae of the division Euglenophyta, which have one or two flagella.

Euphotic zone—the zone in a water column through which there is sufficient penetration by sunlight for photosynthesis to occur.

Euryecious—an organism adjusted to wide extremes in environmental conditions.

Euryhaline—physiologically adapted for survival in aquatic environments over a broad range of salinities.

Eutrophic—referring to water that is rich in nutrients such as nitrogen and phosphorous.

Eutrophication—the process by which lakes and ponds become enriched with dissolved nutrients, either from natural sources or human activities. Nutrient enrichment may cause an increased growth of algae and other microscopic plants, the decay of which can cause decreased dissolved oxygen levels. Decreased oxygen levels can kill fish and other aquatic life.

Evaporation—the process by which water is changed from the liquid into the vapor state. In hydrology, evaporation is vaporization that takes place at a temperature below the boiling point. It is usually measured with USDA class A evaporation pan.

Evaporation chamber—the part of a distillation system in which water is changed into vapor.

Evaporation pan—standard U.S. Weather Bureau Class A pan (48-inch diameter by 10-inch deep) used to estimate the reference crop evapotranspiration rate. Water levels are measured daily in the pan to determine amount of evaporation; Pan or container placed at or about crop canopy height containing water. Water levels are measured daily in the pan to determine the amount of evaporation.

Evaporation ponds—areas where sewage sludge is dumped and dried.

evaporation, Class a pan—is a common instrument to measure evaporation the water from an open surface. The pan is constructed of galvanized iron sheet with diameter of 125 cm and 25 cm of depth. The pan is painted white.

Evapotranspiration (ET)—combined water lost from both transpiration from plant leaves and evaporation from soil and wet leaves. Also known as crop water use, or consumptive use. The crop ET (ETc) can be estimated by calculating the reference ET for a particular reference crop (ETo for clipped grass, and ETr for alfalfa) from weather data and multiplying this by a crop coefficient. Combination of water transpired from vegetation and evaporated from the soil and plant surfaces: Crop ET is the quantitative amount of ET within the cropped area of a field, and which is associated with growing of a crop; Also called **plant water requirement**—amount of water used by the crop in transpiration and building of plant tissue, and that evaporated from adjacent soil or intercepted by plant foliage. It is sometimes referred to as **consumptive use**; **potential** ET—rate at which water, if available, would be removed from the soil and plant surface expressed as the latent heat transfer per unit area or its equivalent depth of water per unit area; **reference** ET—rate of evapotranspiration from an extensive surface cool-season green grass cover of uniform height of 12 cm, actively growing, completely shading the ground, and not short of water.

Evapotranspiration (ET)—moisture evaporated from the surface and transpired from the plant.

evapotranspiration (ET), Actual—actual consumption of water by a specific crop considering that each crop has different growth characteristics with respect to root and foliage growth, planting dates, and water needs.

evapotranspiration, Potential—the evapotranspiration of a soil area covered with vegetation without limitations of water.

Evapotranspiration rate—the total water used by the plant and the inevitable losses. It includes the amount of water extracted by the plants from the soil, transpired through the leaves, and the water evaporated directly from soil surface.

evapotranspiration, Reference— rate of evapotranspiration from an extensive surface of 8–15 cm of height from the soil with grass or alfalfa, uniform in height, actively growing, completely covering the soil, and without limitations of water.

Exchangeable sodium percentage—fraction of cation exchange capacity of a soil occupied by sodium ions [%]. Exchangeable sodium, (meq/100 gram, soil) divided by CEC (meq/100 gram soil) times 100. It is unreliable in soil containing soluble sodium silicate minerals or large amounts of sodium chloride; Percentage of the cation exchange capacity (meq.) of a soil which is occupied by sodium; Index of the saturation of the soil exchange complex with sodium ions.

Extended detention—a storm water design feature that provides for the detention of a volume of water (1/2 inch of runoff per watershed acre) for longer draw down (2,440 h) times to increase settling of urban pollutants.

External manual bleed—a feature which allows an automatic valve to be opened manually (without controller) by releasing water from above the diaphragm to the outside of the valve. Useful during installation, system start-up and maintenance operations.

External trade—a transaction to transfer a water right from one legal entity to another out of a specified area. The area can be a water system, water management area or a trading zone as defined in respective state and territory legislation.

Externalities—by-products of activities that affect the well being of people or damage the environment, where those impacts are not reflected in market prices.

Extreme weather event—a manifestation of weather, which is rare within its statistical distribution on a particular location. By rare usually means rarer than the 90th percentile. The characteristics of extreme weather vary according to the location.

Eye spot—light-sensitive organelle found on some unicellular organisms, used for orientation to light.

Facultative bacteria—bacteria that can live under aerobic or anaerobic conditions.

Facultative organism—microbes capable of adapting to either aerobic or anaerobic environments.

Facultative process—biological treatment process in which the organisms can function in the presence or absence of molecular oxygen.

Fahrenheit—a temperature scale in which water freezes at 32° and boils at 212° at atmospheric pressure.

Fall Line—a line joining the waterfalls on numerous rivers that marks the point where each river descends from the upland to the lowland. In Virginia, the fall line passes through Fairfax, Fredericksburg, Richmond, Petersburg, and Emporia.

False negative—contamination has occurred, but the results of the t-Test fail to indicate contamination.

False positive—no contamination has occurred, but the results of the t-test indicate contamination.

farm, Adequate size—a farm with resources and productivity sufficient to generate enough income to (a) provide an acceptable level of family living; (b) pay current operating expenses and interest on loans; and (c) allow for capital growth to keep pace with technological growth.

Fecal coliform bacteria—the coliform bacteria of fecal origin from warm-blooded animals expressed as most probable number (MPN) per 100 milliliters of sample. These bacteria produce gas when incubated at 44.5°C in a lactose broth medium. The EPA criteria for primary contact recreational waters in the Potomac River Basin are a maximum log of 200 MPN/100 mL.

Fecal matter—matter (feces) containing or derived from animal or human bodily wastes that are discharged through the anus.

Feed pressure—the pressure at which water is supplied to a water treatment device.

Feed water—the water to be treated that is fed into a given water treatment system.

Female adapter—a fitting used to adapt from solvent welded PVC to a threaded or barbed connection. Never, ever use a plastic female adapter on anything with metal threads. Never tighten a plastic female adapter with a wrench, hand tighten it only! The female adapter will split if you over tighten it. In plumbing, male parts always fit into female parts.

Fermentation—the conversion/breakdown of organic matter by anaerobic bacteria into carbon dioxide, methane and similar compounds of low molecular weight.

Ferric iron—small solid iron particles containing trivalent iron, which are suspended in water and visible as "rusty water." Ferric iron can normally be removed by filtration. Also known as precipitated iron; a divalent iron ion, usually as ferrous bicarbonate which when dissolved in water produces a clear solution. It is usually removed by cation exchange water softening. Also called clear water iron.

Fertigation—application of fertilizers through the irrigation system. A form of chemigation.

Fertilizer—is a chemical compound that is added to soil to increase the crop yield.

fertilizer, Acid—the salts of acidic compounds (as the ammonia and urea) help to acidify the soil; and these have an acidification affect.

fertilizer, Basic—compound that has an alkaline effect on the soil. Some examples are potassium nitrate, sodium nitrate, calcium nitrate, and calcium cynamide.

fertilizer, Neutral—exerts a minimum residual effect on pH of the soil. It is added to resist the tendency of acidification of the nitrogen fertilizers.

Field blank—a water-quality sample where highly purified water is run through the field- sampling procedure and sent to the laboratory to detect if any contamination of the samples is occurring during the sampling process.

Field Capacity (FC)—it is the amount of water left in the soil after all the free water has had a chance to drain out through the force of gravity. [%, in/ft, m/m]; the water content of the soil in the crop root zone after drainage by gravity has occurred, generally 1 to 4 days after an irrigation or flooding event; soil moisture at 0.10 to 0.33 bars of suction. The water content is expressed as the ratio of the volume of water at field capacity to the volume of soil sampled; Moisture remaining in a soil following wetting and natural drainage until free drainage has practically ceased; Amount of water remaining in a soil when the downward water flow due to gravity becomes negligible.

Filter—a device installed as part of the water system through which water flows for the purpose of removing turbidity, taste, color, iron or odor; a canister device with a screen of a specified mesh to catch particles large enough to clog emitters.

Filter feeders—organisms that obtain food particles from the water column by filtering large quantities of water via a wide variety of mechanisms; mostly invertebrates, plus certain fishes.

Filter media—the selected materials in a filter that form the barrier to the passage of filterable suspended solids or dissolved molecules. Filter media are used to remove undesirable materials, tastes and odors from a water supply.

Filter medium—the permeable material that separates solids from liquids passing through it.

Filter pack—sand or glass beads that are placed in the annulus of the wall between the borehole wall and the well screen to prevent formation material from entering through the well screen. Glass beads are smooth, uniform, clean, well rounded, and siliceous. The filter pack typically extends 2 feet above the screen.

filter, Candle—a relatively coarse aperture filter, designed to retain a coat of filter medium on an extended surface.

filter, Disk or ring—it is like a mesh filter. It consists of rings or disks (with grooves) to retain sand particles and suspended solids.

filter, Double mesh—consists of two sieves, which can remove suspended particles of different sizes.

filter, Mesh (screen)—consists of a mesh of metal, plastic, or synthetic fabric. The meshes are classified according to the number of squared holes in one square inch of a screen. The number of mesh increases according to the size of the squared. A sieve of 75 for microns (mesh 200) will have squares with each side of 0.075 mm (0.0029 inches). Drip irrigation we use sieves of 150 or 75 microns (mesh of 100–200).

filter, Sand—consists of fine gravel and sand of selected sizes in a tank. It can eliminate amounts of suspended solids from the irrigation water.

filter, Vortex (primary)—is a hydrocyclone filter to remove suspended solids from the water. The solids heavier than water are thrown outside at the periphery of a filter. Due to the gravitational force, the solid particles descend to a lower portion of the cone. Then, these are expelled out of the filter.

filter, Vortex (secondary)—the secondary vortex elevates the treated water upwards and allows it to escape through the superior flow.

Filtrate—the effluent liquid from a filter system; that part of the feed stream, which has passed through the filter.

Filtration—separation of a solid and a liquid by using a porous substance that only lets the liquid pass through; the process of passing a liquid through a filtering medium (which may consist of granular material, such as sand, magnetite, diatomaceous earth, finely woven cloth, unglazed porcelain, or specially prepared paper) for the removal of suspended or colloidal matter; the assembly of components used to remove suspended solids from irrigation water. This may include both pressure and gravity-type devices and such specific units as settling basins or reservoirs, screens, media beds, and centrifugal units.

filtration, Cross flow—a process that uses opposite flows across a membrane surface to minimize particle build-up.

Flocculent—a chemical added to water that attaches to small particles and helps them to sink; the material that settles on the bottom can be removed to improve the clarity of the water.

Freeboard—the vertical space remaining in a containment structure; the vertical distance between the free water surface and the top of a dam or dyke.

FAC—free available chlorine.

Fines—extremely small particles, which are smaller than the specified size (in millimeters) for the medium.

Finfish—term used to refer to true fishes; excluding shellfish species such as clams and oysters.

First draw—the water that comes out when a tap is first opened. It is likely that is has the highest level of lead contamination from weathering of pipe lines.

Fish habitat—spawning grounds and nursery, rearing, food supply and migration areas on which fish depend directly or indirectly in order to carry out their life processes.

Fish ladder (Fish way)—a series of small pools arranged in an ascending fashion to allow the migration of fish upstream past construction obstacles, such as dams. It may also be an inclined trough which carries water from above to below a dam so that fish can easily swim upstream.

Fishery—an area of water inhabited by fish.

Fishery management—the application of the principles of population dynamics, fish culturing and stocking, marketing, and conservation for the maintenance and sustained use of fish resources.

Fishing effort—a quantifiable measure of the allotment of time, energy, money, and/or equipment for fishing.

Fission—asexual reproduction by cell division.

Fitting—the generic name for the various parts that attach the pipes together. Includes bushings, couplings, crosses, ells, female adapters, male adapters, reducers, and tees. Fittings may be threaded, barbed, soldered, or welded to the pipe. (The glue or cement used on plastic fittings is a solvent, which results in a welded joint). Plastic fittings with threads should never be tightened with wrenches. Hand tighten them only.

Fixed cost—in economic analysis, this is a fixed cost and is not variable during the service life of any equipment. It involves initial capital investment.

Flagella—long hair-like appendages projecting from a cell; primarily used as a mean of locomotion.

Flash flood—a sudden flood usually associated with a sudden heavy rainfall.

Flat worms—unsegmented worms of the phylum Platyhelminthes, which are usually flattened, have soft bodies, and have no appendages.

Flatfish—any of a number of asymmetrical fishes that compose the order Pleuronectiformes and live on the bottom, having a laterally compressed body and both eyes on the same side of the head.

Floaters—light phase organic group water capable of forming an immiscible layer, which can float on the water table.

Floc—A flocculent mass that is formed in the accumulation of suspended particles. It can occur naturally, but is usually induced in order to be able to remove certain particles from wastewater.

Flocculate—to cause (soil) to form small lumps or masses.

Flocculation—in water and wastewater treatment, the agglomeration of colloidal and finely divided suspended matter after coagulation by gentle stirring by either mechanical or hydraulic means. In biological wastewater treatment, where coagulation is not used, agglomeration may be accomplished biologically.

Flood—An overflow of water onto lands that are used or usable by man and not normally covered by water. Floods have two essential characteristics: it is temporary; and the land is adjacent to and inundated by overflow from a river, stream, lake, or ocean.

Flood frequency—a relationship showing the probability that flood of a certain magnitude are equaled or exceeded in any year; or a similar relationship between flood magnitude and frequency of exceedance.

Flood hydrograph—a continuous record of stage and discharge, with respect to time, produced only for the periods when the water elevation exceeds the chosen base.

Flood plain—the lowland that borders a river, usually dry but subject to flooding when the stream overflows its banks.

Flood way—the part of floodplain that, during a flood, has the deepest, fastest, and most destructive flow of water.

Flooded—soil saturated with water.

Flotation—a solids-liquid or liquid-liquid separation procedure, which is applied to particles of which the density is lower than that of the liquid they are in. there are three types: natural, aided and induces flotation.

Flow—the discharge rate of a resource, expressed in volume during a certain period of time. The movement of fluids, through pipe, fittings, valves or other vessels. Generally measured in Gallons Per Minute (GPM), or million gallons per day (mgd) and one cfs = 0.65 mgd, Gallons Per Hour (GPH), Cubic Feet Per Second (ft^3/s), Cubic Meters Per Hour (m^3/h), Liters Per Minute (l/m), or Liters Per Second (L/s).

Friction loss—the amount of pressure lost as water flows through the water meter, pipe, fittings and valves of an irrigation system. As the velocity of water flowing through the system increases, the friction loss will also increase. These losses can be used to calculate the approximate dynamic (working) pressure at any given point of a system.

Flow—the rate or amount of water that moves through pipes in a given period of time. Flow is expressed in gph (gallons per hour), for micro irrigation (drip) devices. It is expressed in gpm (gallons per minute), for high-pressure sprinkler systems.

Flow augmentation—the addition of water to meet flow needs.

Flow control—a valve which modulates in order to maintain a predetermined flow rate without drastically altering the pressure.

Flow controller—an in-line device or orifice fitting which regulates and controls flow of water.

Flow duration—a computed relationship that shows the percentage of time that specified daily discharges were equaled or exceeded in a stated number of complete years.

flow, Inverse (back flow)—a water flow that is in the direction opposite to the direction of a normal flow. The back flow is allowed for flushing of the filters.

Flow net—the set of intersecting equipotential lines and flow lines representing two-dimensional steady flow through porous media.

flow, Normal—amount or volume of water that flows through the lines to maintain uniform application of water.

Flow rate—minimum design pumping rate required to deliver effluent in a timely fashion to a gravity system and to pressurize a pressure manifold of low pressure pipe laterals. The quantity of water which passes a given point in a specified unit of time; Often expressed in U.S. gpm (or L/min), gpm [gal/min], ft^3/s, cfs, Liters/s, Liters/min, m^3/h; Rate of flow or volume per unit period of time.

Flow sensor—a device which actively measures water flow through a piping system and reports its data to the computerized central control system.

flow, Steady—the flow that occurs when, at any point in the flow field, the magnitude and direction of the specific discharge are constant in time.

flow, Unsteady—the flow that occurs when, at any point in the flow field, the magnitude or direction of the specific discharge changes with time. Also called transient flow or nonsteady flow.

Flume—flow measuring device for open channel flow. Water travels through a restriction and the flow rate is determined using the water height on a staff gauge.

Fluoridation—the addition of fluoride compound to a potable water supply to produce the concentration desired (about one PPM) for the purpose of the reduction of dental caries (tooth decay).

Fluoride—a natural occurring constituent of some water supplies, an excess of which (over 2.0 ppm) can cause discolored teeth.

Flush—to open a cold-water tap to clear out all the water, which may have been sitting for a long time in the pipes.

Flushing (cleaning)—to separate suspended solids from the irrigation water to avoid clogging.

Fluvial—of or pertaining to rivers and streams; growing or living in streams ponds; produced the action of a river or stream.

Fluvio-glacial depostitional environment—a complex melange of glacially borns and riverine sediments deposited at the head of a melting glacier. The sediments range in grain size from clays to boulders, and in places that are typically unsorted.

Flux—the rate at which a "Reverse Osmosis Membrane" allows water to pass through it.

Flyway—the principal migratory pathway by which birds travel between their summer breeding grounds and overwintering areas.

Food chain—the dependence of one type of life on another, each in turn eating or absorbing the next organism in the chain. Grass is eaten by cow; cow is eaten by man. This food chain involves grass, cow, and man.

Food web—the complex interaction of food chains in a biological community, including the processes of production, consumption, and decomposition.

Forage species—a species that is actively sought by another as a major food source.

Foraminifera—order of benthic or planktonic protozoans that generally possess shells, which are usually composed of calcium carbonate.

Forest ecosystem—a terrestrial unit of living organisms (plants, animals and microorganisms), all interacting among themselves and with the environment (soil, climate, water and light) in which they live. The environmental "common denominator" of that forest ecological community is a tree, who most faithfully obeys the ecological cycles of energy, water, carbon and nutrients. A forest ecosystem has definite boundaries and includes a forest of trees out to the limit of tree growth. Remember that forests are not the only ecosystems. There are hundreds of thousands of defined and undefined ecosystems that can cover the broadest to the tiniest of areas. An ecosystem can be as small as a pond or a dead tree, or as large as the Earth itself.

Forest Service—an agency of the U.S. Department of Agriculture, primarily responsible for planning and overseeing the use of national forest lands by private, commercial and government users.

Forestry [forestation]—the science and art of cultivating, maintaining, and developing forests.

Fossil fuels—fuels from fossil carbon deposits such as oil, natural gas and coal. These are burned in order to gain energy. During these combustion processes greenhouse gases are released.

Fouling—the deposition of organic matter on the membrane surface, which causes inefficiencies.

Fouling (reverse osmosis)—a phenomenon in which a reverse osmosis membrane adsorbs, interacts with or becomes coated by solutes and or precipitates in the feed stream resulting in a decrease in membrane performance by lowering the flux and /or affecting the rejection solutes.

Fouling in filtration—the accumulation of undesirable foreign matter in a filter causing clogging of pours coating of surfaces and inhibiting or limiting the proper operation of the treatment system.

Fouling organisms— aquatic organisms that encrust submerged surface such as piers, boats, water intake ducts, etc.

FPT—female nominal pipe thread.

Fraction of exhaustion of soil moisture—fraction of the water available in the soil that can be obtained by the crop, allowing an evapotranspiration without limits and the crop growth.

Fragmentation—the subdivision of a solid in fragments. The fragments will then adhere to the nearest surface.

Free available chlorine—the concentration of residual chlorine present as dissolved gas, hypochlorous acid or hypochlorite ion but not including that chlorine combined with ammonia or other less readily available forms of chlorine.

Free drainage—movement of water by gravitational forces through and below the plant root zone. This water is unavailable for plant use except while passing through the soil.

Free energy—a measure of the thermodynamic driving energy of a chemical reaction. Also known as Gibbs free energy or Gibbs function.

Freeboard—the vertical distance between the top of a dam and a particular water level. For example, "freeboard above maximum surface" or "freeboard above normal reservoir level."

Freezing—the change of a liquid into a solid as temperature decreases. The freezing point of water is 0°C.

Frequency analysis—a determination of the distribution of values that a variable has within a variable.

Frequency distribution—values in a sample are grouped into a limited number of classes. A table is made showing the class boundaries and the frequencies (number of members of the sample) in each class. The purpose is to show a compact summary of the data; Measurement and presentation of various fractions of total water applied for selected depth ranges referenced to average depth applied.

Fresh water—water having less than approximately 1,000 mg/L (ppm) of total dissolved solids (TDS).

Freundlich isotherm—an empirical equation that describes the amount of solute adsorbed onto a soil surface.

Friable—soil consistency term referring to the ease with which the soil aggregates may be crumbled (in the hand), i.e., a friable soil is easily crumbled in the hand.

Friction—negative force exerted by the tubes and the components of the system and that prevents the free water movement through system. It is also called drag.

Friction factor (lateral)—factor used to size pipe [psi/100 ft or m /100 m].

friction factor, Christiansen—friction factor or coefficient used in the Christiansen procedure to determine pressure loss in a multiple outlet piping system.

friction factor, Darcy Weisbach—friction factor used in Darcy Weisbach equation.

Friction loss—also, referred to as *pressure loss*. Amount of pressure lost through pipes due to water movement and turbulence; Amount of pressure lost as water flows through an irrigation system (due to friction against the pipe walls); as water moves through an irrigation system, pressure is lost because of turbulence created by the moving water. This turbulence can be created in pipes, valves or fittings.

Friction loss of a pipe—for a pipe of a given length and a particular material, the amount of pressure loss due to the friction depends on the diameter of tube, flow rate, viscosity, and density of water.

Fring marsh creation—planting of emergent aquatic vegetation along the perimeter of open water to enhance pollutant uptake, increase forage and cover for wildlife and aquatic species, and improve the appearance of a pond.

Frost—a layer of ice formed when the temperatures of the Earth's surface (and objects) falls below freezing.

Frost protection—applying irrigation water to affect air temperature, humidity, and dew point to protect plant tissue from freezing. The primary source of heat (called heat of fusion) occurs when water turns to ice, thus protecting sensitive plant tissue.

Fry—the stage in the life of a fish between the hatching of the egg and the absorption of the yolk sac; in a broader sense, all immature stages of fishes.

Full cover—the point in the plant's growth stage where the canopy covers all available ground space and using all available sunlight. Usually refers to row crops.

Full irrigation—management of water applications to fully replace water used by plants over an entire field.

Full supply level—the maximum storage level of a reservoir when it is full. This level usually corresponds to the level of the spillway crest for an un-gated spillway, or to the water level for which the dam is designated.

Fungi—(singular = fungus) plantlike organisms with cells that have distinct nuclei surrounded by nuclear membranes as well as other specialized functional cell parts but that cannot carry photosynthesis. Most fungi are decomposers of wastes and dead bodies from other organisms; a few are parasites. Yeasts, molds, mildew and mushrooms are all fungi.

Fungicide—chemical pesticide that kills fungi or prevents them from causing diseases on plants.

Fungigation—application of fungicides through the irrigation system.

Fungus—talofite plant that does not contain chlorophyll. These are microscopic organisms and some are pathogens.

Furrow—small channel for conveying irrigation water down slope across the field. Sometimes referred to as a rill or corrugation; Trench or channel in the soil made by a tillage tool.

Furrow dike—small earth dike formed in a furrow to prevent water translocation. Typically used with LEPA and LPIC systems. Also used in nonirrigated fields to capture and infiltrate precipitation. Sometimes called reservoir tillage.

Furrow irrigation—Also known as rill irrigation or gravity irrigation. Water is applied to row crops in small ditches or channels between the rows made by tillage implements.

Furrow stream—stream flow in a furrow, corrugation or rill.

Fusiform—tapering toward both ends, as in many diatoms.

Gabion—a large rectangular box of heavy gauge wire mesh, which holds large cobbles and boulders. Used in streams and ponds to change the flow patterns, stabilize banks, or prevent erosion.

gage, Pressure—device that indicates existing pressure in the irrigation system. Absolute pressure = atmospheric pressure + gage pressure.

gage, Vacuum (suction)—important part of a tensiometer that records soil moisture suction values in the soil.

Gallinea ferruginea—one of several types of bacteria that use iron in their metabolism and are capable of depositing gelatinous ferric hydroxide. Also known as iron bacteria.

Gallon—a unit for volume of a liquid. One US gallon is equivalent to 3.785 L.

Gallons per minute—a measurement of water flow primarily used in the U.S.A.; **GPM**—abbreviation for "gallons per minute"; GPH is an abbreviation for gallons per hour.

Gastropods—mollusks of the class Gastropoda, including shell-less as well as frequently coiled, univalve species; usually have distinct heads that bear sensory organs.

gate, slide gate—device used to control the flow of water to, from, or in a pipeline or open channel.

Gate valve—a valve using a rising and descending gate to control the flow of water.

Gated pipe—portable pipe with small gates installed along one side for distributing irrigation water to corrugations or furrows.

Gauge height—the water surface elevation referred to some arbitrary gauge datum. Gauge height is often used interchangeably with the more general term "stage," although gauge height is more appropriate when used with a reading on a gauge.

Gauging station—a particular site on a stream, canal, lake, or reservoir where systematic observations of hydrologic data are obtained.

Geochemical cycle—the process of introduction, storage, transformation, and output of a particular chemical component in the nonliving environment.

Geomorphology—a science that deals with the land and submarine relief features of the surface of the earth or the comparable relief features of a celestial body and seeks a genetic interpretation of them.

Geophysics—the physics of the earth including the fields of meteorology, hydrology, oceanography, seismology, volcanology, magnetism, radioactivity, and geodesy.

Germicidal ultraviolet—ultraviolet light is part of the light spectrum, which is classified into three wavelength ranges: UV-C, from 100 nanometers (nm) to 280 nm; UV-B, from 280 nm to 315 nm; UV-A, from 315 nm to 400 nm. UV-C light is germicidal: it deactivates the DNA of bacteria, viruses and other pathogens and thus destroys their ability to multiply and cause disease. Specifically, UV-C light causes damage to the nucleic acid of microorganisms by forming covalent bonds between certain adjacent bases in the DNA. The formation of such bonds prevents the DNA from being unzipped for replication, and the organism is unable to reproduce. In fact, when the organism tries to replicate, it dies. In this method of disinfection, nothing is added which makes this process simple, inexpensive and requires very low maintenance.

Germination—renewal of the growth of a seed after a period of dormancy in the presence of soil moisture.

Geyzers (or gushers)—natural thermal springs that intermittently eject hot water and steam from crack in the earth. Geyzers are usually associated with volcanically active areas.

Ghyben—Herzberg principle—an equation that relates the depth of a salt-water interface in a costal aquifer to the height of the fresh-water table above the sea level.

Giardia—a common waterborne protozoan that forms cysts and is resistant to disinfectants such as chlorine and ultraviolet light. Giardia can be removed by filters that all particles of four microns and greater in size.

Giardia lamblia—a type of cyst found in the intestines of mammals and in water contaminated by mammal droppings. The giardia lamblia cyst, which is common and is frequently carried by water, is capable of causing a contagious waterborne disease characterized by acute diarrhea. This disease is referred to as beaver fever, because beaver droppings can contain giardia lamblia.

Gigaliter—1 billion liters.

Gill net—a type of gear that captures fish by entangling their gill covers in the meshes of the net.

Glacial till—unsorted and unstratified sediment originating directly from glacial ice (i.e., not rework by glacial melt water).

Glacier—A huge mass of ice slowly flowing over a land mass, formed from compacted snow in an area where snow accumulation exceeds melting and sublimation; flowing bodies of land ice that develop in the colder regions and higher latitudes of the earth.

Global radiation balance—a balance, which implies that globally the amount of incoming solar radiation must on average be equal to the sum of outgoing reflected solar radiation and outgoing infrared radiation emitted by the climate system.

Global surface temperature—the average of the sea temperature in the first few meters of the oceans and the temperature 1.5 m above ground on land surfaces.

Global warming—the warming of the earth's surface, driven by either natural or anthropogenic forces.

Global warming potential—GWP: an index describing radiative characteristics of greenhouse gases relative to that of carbon dioxide. It represents the time greenhouse gases remain in the atmosphere and their potential to absorb infrared radiation.

Goodness of fit—a statistical test to determine the likelihood that sample data have been generated from a population that conforms to a specified type of probability distribution.

Goof plugs—insertable caps to plug holes in mainline and microtubes where drip devices have been removed or are not needed.

GPM (gallons per minute)—a standard measurement of water flow. The available GPM (also known as design capacity) must be known before an irrigation design can be completed. Sprinkler heads have different GPM requirements. The total GPM of all of the sprinkler heads on one zone should not exceed the available GPM.

Grab sample—a single water or wastewater sample taken at a single point in time and location.

Grade—the slope of a plot of land. Grading is the mechanical process of moving earth changing the degree of rise or descent of the land in order to establish good drainage and otherwise suit the intent of a landscape design.

Grain size—the general dimensions of the particles in a sediment or rock, or of the grains of a particular mineral that make up a sediment or rock. It is common for these dimensions to be referred to with broad terms, such as fine, medium, and coarse. A widely used grain size classification is the Udder-Wentworth grade scale.

Granular activated carbon—beds of carbon substances such as coal or coke, through which water is passed; the heating of carbon to encourage active sites to absorb pollutants.

Grass shrimps—shrimps of the genus Palaemonetes, which are most abundant in submerged aquatic vegetation.

Grassed swale—a conventional grass swale is an earthen conveyance system in which the filtering action of grass and soil infiltration are used to remove pollutants from urban storm water. An enhanced grass swale, or biofilter, uses check dams and wide depressions to increase runoff and promote greater settling of pollutants.

Gravitational force—natural force that cause objects to move towards the center of earth.

Gravitational water—soil water that moves into, through, or out of the soil under the influence of gravity.

Gravity flow—a water system that relies on gravity to provide the pressure required to deliver the water. Consists of a water source located at a higher elevation than the water delivery points.

Gravity sphere—aggregation of watersheds and subsheds within which connections to the existing sewer system are possible as defined by the topography.

gravity, Acceleration due to—acceleration caused by the attraction of the mass of earth to bodies at or near its surface (i.e., 32.2 ft/sec^2 or 9.81 m/s^2).

Grease interceptor (trap)—in plumbing, a receptacle designed to collect and retain grease and fatty substances normally found in kitchen or similar wastes. It is installed in the drainage system between the kitchen or other point of production of the waste and the building sewer.

Green belt—a strip of unspoiled, often treed, agricultural or other outlying land used to separate or ring urban areas.

Greenhouse effect—the sun radiates solar energy on earth. The larger part of this energy (45%) is radiated back into space. Greenhouse gases in the atmosphere contribute to global warming by adsorption and reflection of atmospheric and solar energy. This natural phenomenon is what we call the greenhouse effect. It is agreed that the greenhouse effect is correlated with global temperature change. If greenhouse gases would not exist earthly temperatures would be below –18°C. After the industrial revolution of the 1700's, the greenhouse effect was enhanced by greenhouse gas emissions of anthropogenic nature. The main source of anthropogenic greenhouse gas emissions is fossil fuel combustion. The contribution of greenhouse gases to the greenhouse effect varies. The enhanced greenhouse effect is said to cause a cascade of impacts on both environment and society. How and when these effects will take place is still under discussion.

Greenhouse gases—also referred to as GHG. Gases that absorb atmospheric and solar infrared radiation and reflect it back to earth to increase global warming, causing climate change. Natural greenhouse gases include water vapor, carbon dioxide, methane, ozone and nitrous oxide. Past and future anthropogenic emissions of the greenhouse gases carbon dioxide, methane and nitrous oxide enhance global warming.

Grey water—or sullage, is wastewater generated from domestic activities such as laundry, dishwashing, and bathing, which can be recycled on-site for uses such as landscape irrigation and constructed wetlands. Grey water differs from water from the toilets which is designated sewage or black water to indicate it contains human waste; untreated, used water from a household or small commercial establishment (excluding that from toilets or other fixtures and appliances whose wastewater might have come into contact with human waste; the effluent from residential appliances or fixtures except toilets or urinals.

Grit—the heavy suspended mineral matter present in water or wastewater, such as sand, gravel, or cinders.

Grit chamber—a detention chamber or an enlargement of a sewer designed to reduce the velocity of flow of the liquid to permit the separation of mineral from organic solids by differential settling.

Gross domestic product (GDP)—the value of all goods and services produced or consumed within a nation's borders.

Gross primary production (GPP)—the amount of carbon fixed from the atmosphere through photosynthesis.

Gross water price—the transfer price of water as agreed between legal entities inclusive of all applicable transaction costs.

Ground penetrating rader (GPR)—a geophysical method used to identify surface formations, which will reflect electromagnetic radiation. GPR is useful for defining the boundaries of buried trenches and other sub-surface installations on the basis of time-domain reflectrometry.

Ground water—rain and snow water accumulated in the earth's porous rock; Water beneath the earth's surface in a layer of rock or sob called the saturated zone because all openings are filled with water; the water that supplies wells and springs; Water occurring in the zone of saturation in an aquifer or soil; All water under the surface of the ground whether in liquid or solid state. It originates from rainfall or snowmelt that penetrates the layer of soil just below the surface. For ground water to be a recoverable resource, it must exist in an aquifer. Ground water can be found in practically every area of the state, but aquifer depths, yields, and water quality vary.

Ground water aquaculture return flow wells—reinject ground water or geothermal fluids used to support aquaculture. Non-geothermal aquaculture disposal wells are also included in this category.

Ground water discharge—ground water entering coastal waters, which has been contaminated by land-fill leachates, deep well injection of hazardous wastes and septic tanks.

Ground water flow—the movement of water through openings in sediment and rock.

Ground water hydrology—the branch of hydrology that deals with the occurrence, movements, replenishment and depletion, properties and methods of investigation and utilization of ground water.

Ground water injection monitoring program—a monitoring well system capable of yielding ground water samples for analysis. Up-gradient wells must be installed to obtain representative background ground water quality in the uppermost aquifer and be unaffected by the facility. Down-gradient wells must be placed immediately adjacent to the hazardous waste or hazardous waste constituents migrating from the facility.

Ground water mining—the practice of withdrawing ground water at rates in excess of the natural recharge.

Ground water recharge—inflow of water to a ground water reservoir (zone of saturation) from the surface. Infiltration of precipitation and its movement to the water table is one form of natural recharge. Also, the volume of water added by this process.

ground water, Confined—the water contained in a confined aquifer. Pore water pressure is greater than atmospheric at the top of the confined aquifer.

ground water, Perched—the water in an isolated, saturated zone located in the zone of aeration. It is the result of the presence of a layer of material of low hydraulic conductivity, called a perching bed. Perched ground water will have a perched water table.

ground water, Unconfined—the water in an aquifer where there is a water table.

Grout curtain—an underground wall designed to stop ground water flow; can be created by injecting grout into the ground, which subsequently hardens to become impermeable.

Growing season—period, often the frost-free period, during which the climate is such that crops can be produced.

Growth hormone—regulates the plant growth.

Gully—a deeply eroded channel created by the concentrated flow of water.

Gully erosion—large, rapidly eroding water channels, with scoured bottoms and deeply cut sides.

Guttation—it is a water loss in form of drops caused by the root pressure.

Gypsum block—electrical resistance block in which the material used to absorb water is gypsum. It is used to measure soil water content in nonsaline soils.

H_2O—the chemical formula for water.

Habitat—the specific area in which a particular type of plant or animal lives.

Hazardous waste—a waste that contains any substance (solid, liquid, or gaseous) that is harmful or potentially harmful to life or the environment; this type of waste includes toxic flammable, corrosive and oxidizing substances and is subject to special handling, shipping, storage, and disposal requirements.

Hydrocarbons—any substance containing carbon and hydrogen in various combinations (e.g., gasoline and oil).

Hydrogeology—the study of groundwater, with particular emphasis on the chemistry and movement of water.

Hydrologic cycle—the circulation of the Earth's waters from ocean to atmosphere to land and back to ocean.

Hydrology—the science that deals with water, its properties, distribution and circulation over the Earth's surface.

Habitat—the sum total of environmental conditions of a specific place that is occupied by an organism, population, or community. The three components of wildlife habitat are food, water, shelter.

Habitat degradation—the destruction of habitat by physical means: either natural or human activity. The degradation of a barrier island habitat by a hurricane is an example of a natural means; the degradation of wetland habitats by development is an example of human activity.

Half life—the period of time in which a substance loses half of its active characteristics; the time required to reduce the concentration of a material by half. Also referred to as decay constant. The term is used to quantify a first-order exponential decay process.

Halocarbons—Greenhouse gases containing either chlorine, bromine or fluorine and carbon. The chlorine and bromine containing halocarbons are also involved in the depletion of the ozone layer.

Halogenated hydrocarbons—an organic compound containing one or more halogens (e.g., fluorine, chlorine, bromine, and iodine).

Hand move—typically 30 to 40 foot sections of portable aluminum sprinkler pipe with a sprinkler in each section and that is moved manually. Labor requirements are higher than for all other agricultural sprinkler systems.

Hand tongs—a hand-operated, hinged, grasping device used to collect oysters from shallow water.

Hantush–Jacob formula—an equation to describe the change in hydraulic head with time during pumping of a leaky aquifer.

Hard water—water that contains a great number of positive ions. The hardness is determined by the number of calcium and magnesium atoms present. Soap usually dissolves badly in hard water.

Hardness—a property of water causing formation of an insoluble residue when used with soap and causing formation of a scale in vessels in which water has been allowed to evaporate. It is primarily due to the presence of calcium and magnesium ions. It is generally expressed in milligrams per liter; a common quality of water, which contains dissolved compounds of calcium and magnesium and sometimes other divalent and trivalent metallic elements.

Hardscape—elements added to a natural landscape, such as paving stones, gravel, walkways, irrigation systems, roads, retaining.

Harvest intensity—It is a percentage of total area programd for a given period of an irrigated crop.

Haul seine—a net operated in the shallows, usually by hand, which issued to encircle fish and is then hauled to shore.

Hazardous waste management—the collection, source separation, storage, transportation, processing, treatment, recovery, and disposal of hazardous waste.

Hazardous waste—a solid waste which exhibits any hazardous characteristics and has not been specifically excluded as a hazardous waste.

Hazardous waste constituents—a constituent which causes a waste to be classified hazardous.

Hazardous waste management area—the area within a boundary of a property, which encompasses one or more hazardous waste management unit or cell.

Hazen method—an equation that can be used to approximate the hydraulic conductivity of a sediment on the basis of the effective grain size.

Head—the term that irrigation designers use for pressure [ft or m of water].

Head—short for "sprinkler head."

Head ditch—ditch across the upper end of a field used for distributing water in surface irrigation.

Head feet—a measure of pressure expressed in feet of water. Equivalent to 0.433 PSI per foot of water.

Head-to-head—this phrase describes the correct placement of sprinkler heads. One sprinkler must be placed so that it will spray another sprinkler (or 50% of the adjusted diameter). This provides for complete coverage of the irrigated area.

Head gate—water control structure at the entrance to a conduit or canal. (ASAE, 1998)

Head loss—energy loss in fluid flow; Energy loss associated with water flowing through converging or diverging pipe sections.

Head loss in bends—energy loss associated with water flowing through a bend in a pipe.

Head loss, converging and diverging pipes—energy loss associated with water flowing through converging or diverging pipe sections.

head loss, Entrance—energy loss associated with water flowing through the entrance of a pipe.

Head race—the pipe or chute by which water falls downward into the turbine of a power plant.

Head space—the empty volume in simple container between the water level and the cap.

Head spacing—spacing between sprinklers.

Head to head—the movement pattern of a seasick cruise ship passenger. In irrigation "head to head" refers to the situation where sprinklers are spaced so that the water from one sprinkler throws all the way to the next sprinkler. Most sprinklers are designed to give the best performance when head to head spacing is used.

Head to head coverage—the practice of placing sprinklers so that water from one sprinkler overlaps all the way to the next sprinkler head. This helps to increase overall system efficiency and prevents dry spots in the landscape.

Head to head spacing—spacing of sprinkler heads so that each sprinkler throws water to the adjacent sprinkler.

Head works—All structures and associated facilities located at the beginning (upstream end) of a water management project.

head, Discharge—pressure head at the outlet of the pump.

head, Dynamic—specific energy in a flow system.

head, Elevation—head as a result of elevation above a defined datum.

head, Friction—energy head loss caused by the friction of water flowing through a pipe.

head, Pressure—pressure energy in a liquid system expresses as the equivalent height of a water column above a given datum.

head, Static—energy associated with a static liquid system composed of elevation and pressure components.

head, Static discharge—static energy components at the discharge of a pump including elevation and pressure.

head, Static suction—vertical distance from the pump centerline to the surface of the liquid when the liquid supply is above the pump.

Head, Total—the sum of the elevation head, the pressure head, and the velocity head at a given point in an aquifer.

head, Total (dynamic)—Head required to pump water from its source to the point of discharge. Equal to the static lift plus friction head losses in pipes and fittings plus velocity head; Energy in the liquid system expressed as the equivalent height of a water column above a given datum; Sum of static, pressure, friction and velocity head that a pump works against while pumping at a specific flow rate.

head, Total suction—head required to lift water from the water source to the centerline of the pump plus velocity head, entrance losses and friction losses in suction pipeline.

head, Vapor pressure—pressure head at which the liquid (water) will vaporize or boil at a given temperature.

head, Velocity—head or energy caused by the velocity of a moving fluid; Amount of pressure required to generate a specific velocity; the amount of energy or pressure that is used to make the water move at a given velocity.

Headwaters—the source and upper tributaries of a stream or river.

Health contaminant—any substance or condition that may have any adverse effect on human health.

Heat exchanger—a component that is used to remove heat from or ad heat to a liquid.

Heaving sand—unconsolidated sand that cannot maintain the integrity of the borehole wall.

Heavy metals—metals that have a density of 5.0 or higher and a high elemental weight. Most of these are toxic to humans, even in low concentrations.

Heavy water—water in which all the hydrogen atoms have been replaced by deuterium.

Henry's Law—a way of calculating the solubility of a gas in a liquid, based on temperature and partial pressure, by means of constants.

Herbicide—substances to destroy the unwanted weeds that grow along with growth and development of a crop.

herbicide, Selective—herbicide to kill weeds of a certain specie only.

Herbigation—is an application of herbicides through the irrigation system.

Herbivores—organisms that feed on plants (primary producers) and are considered primary consumers.

Heterotrophic plant count (HPC)—a procedure for estimating the total number of live nonphotosynthetic bacteria in water. Colony forming units (CFU) are counted after spreading the sample over a membrane or spread plate and incubating in an amiable growth medium (agar) and at an amiable temperature. These are generally not considered disease-causing bacteria.

Hexametaphosphate—a chemical such as sodium hexametaphosphate, added to water to increase the solubility of certain ions and to deter precipitation of certain chemicals.

High corrosion potential—material with a high propensity for electrochemical degradation.

High density polyethylene [HDPE]—one of several forms of polyethylene used to make pipe and other irrigation components.

High-yield well—a relative term referring to a well capable of quick recovery after it has been purged of at least three casing volume (i.e., samples can be collected immediately after purging).

Historic preservation—this landscape architecture specialization has evolved to encompass maintenance of a site in its present condition; conservation of a site as part of a larger area of historic importance; restoration of a site to a given date or quality; renovation of a site for ongoing use; and interpretation of a vanished landscape.

Holdfasts—structures by which seaweeds attach themselves to a solid substrate.

Holding pond—a pond or reservoir, usually made of earth, built to store polluted runoff.

Holoplankton—group of organisms that are planktonic throughout their lives.

Homeowner water system—a water system that supplies piped water to a single residence.

Hoop net—a tubular net consisting of a series of conical shaped components into which fish enter and become trapped.

Hopkin's Law of bioclimatology—it establishes that there is a delay in the flowering as the number of degrees of latitude or altitude is increased.

horizon, Soil—layer of soil or soil material approximately parallel to the land surface and differing from adjacent genetically related layers in physical, chemical, and biological properties or characteristics such as color, structure, texture, consistency, kinds and number of organisms present, degree of acidity or alkalinity, etc.

Horse power—*water horsepower:* Energy added to water by a pump; *Input horsepower:* Energy added to a motor or engine; *brake horsepower:* Power required to drive a pump; miscellaneous component *energy losses.*

Hose bib—valve configured to be mounted on a wall having threads to accommodate the connection of a water hose.

Household purpose—water used for human consumption, sanitation, fire prevention, and watering animals, gardens, lawns and trees.

Housing and Urban Development, Department of (HUD)—US Federal agency responsible for producing and managing many federally funded public service programs, especially those affecting housing and public spaces.

Humid climate—climate characterized by high rainfall and low evaporation potential. A region generally is considered as humid when precipitation averages more than 40 inches per year.

Humidification—The addition of water vapor to air.

Humidity—water vapor content of the air, usually expressed as relative humidity.

humidity, Absolute—the actual weight of water vapor contained in a unit volume of the atmosphere, usually expressed in grams of water per kilogram of air. Compare to Relative Humidity.

humidity, Absolute—the amount of moisture in the air as expressed by the number of grams of water per cubic meter of air.

humidity, Relative—percent ratio of the absolute humidity to the saturation humidity for an air mass.

humidity, Saturation—the maximum amount of moisture that can be contained by an air mass at a given temperature.

Humus—organic matter in or on a soil; composed of partly or fully decomposed bits of plant tissue derived from plants on or in the soil, or from animal manure.

Hvorslev method—a procedure for performing a slug test in a piezometer that partially penetrates a water-table aquifer.

Hydrant—outlet, usually portable, used for connecting surface irrigation pipe to an alfalfa valve outlet.

Hydraulic—operated, moved, or effected by means of water.

Hydraulic clam dredge—apparatus used in clam harvesting, in which clams are collected on a towed conveyor-dredge.

Hydraulic conductivity—the rate at which water moves through aquifer material under a unit hydraulic gradient, expressed as volume per unit time per unit cross section reduced to feet per day or meters per day.

Hydraulic conductivity of soil, K—ratio of water flow through a cross-sectional area of the soil, for a unit of hydraulic gradient. It is called permeability or transmissibility.

Hydraulic connection—the hydraulic relationship between two different lithologic layers.

Hydraulic contact—that makes one substance with the water or other liquids at rest or movement.

Hydraulic design coefficient of variation—a statistical term used to describe the variation in hydraulic pressure in a sub main unit or throughout a micro irrigation system. Care should be taken not to confuse this term with the emitter discharge coefficient of variation due to hydraulics.

Hydraulic equilibrium—It is a balance reached between the water within stem of a tensiometer and the soil moisture.

Hydraulic gradient—is the difference in the height of the water on and under the soil column. The volume of water

Hydraulic head—water level elevation in a well or piezometer. The evaluation typically referenced to mean sea level to which water rises as a result of hydro statistic pressure.

Hydraulic loading rate—an empirically derived design and operating parameter that relates to ponding, surface shearing rate, and hydraulic detention time. Usually reported in units of volume of wastewater (including recycle) per unit cross-sectional area per day.

Hydraulic system—a sprinkler system that uses water as a replacement for electric current to open and close valves.

Hydraulically overloaded—a condition in which the quantity of flow through a treatment plant is greater than that for which it is designed, which often results in the decrease in operational efficiency of the plant.

Hydraulics—the science of fluids in motion. As it relates to irrigation there are two sub categories for hydraulics. They are open-channel hydraulics (canals, ditches, streams, rivers, etc.) and closed channel (or closed conduit) hydraulics (pipelines, tanks, etc.). It is related to the continuity equation (Q = AV), Bernoulli equation, Darcy-Weisbach equation, etc.

Hydrocarbon—Organic compounds that are built of carbon and hydrogen atoms and are often used in petroleum industries.

Hydrodynamic dispersion—the process by which ground water containing a solute is diluted with uncontaminated ground water as it moves through an aquifer.

Hydroelectric power water use—the use of water in the generation of electricity at plants where the turbine generators are driven by falling water.

Hydrogen ion—the cation H + of acids consisting of a hydrogen atom whose electron has been transferred to the anion of the acid [pH].

Hydrogen sulfide (H_2S)—a gas emitted during organic decomposition by a select group of bacteria, which strongly smells like rotten eggs.

Hydrogeology—the science that deals with subsurface waters and related geologic aspects of surface waters.

Hydrograph—a chart that measures the amount of water flowing past a point as a function of time.

Hydrologic cycle (Water Cycle)—the process by which water evaporates from oceans and other bodies of water, accumulates as water vapor in clouds, and returns to oceans and other bodies of water as rain and snow or as runoff from this precipitation or groundwater.

Hydrologic equation—an expression of the law of mass conservation for purposes of water budgets. It may be stated as inflow equals outflow plus or minus changes in storage.

Hydrologic unit—a geographic area representing part or all of a surface drainage basin or distinct hydrologic feature as delineated by the Office of Water Data Coordination on the State Hydrologic Unit Maps; each hydrologic unit is identified by an 8-digit number.

Hydrology—the science dealing with the properties, distribution, and flow of water on or in the Earth.

Hydrolysis—**the decomposition of organic compounds by interaction with water.**

Hydrophilic—**having an affinity for water.**

Hydrophillic quality—where a substance can be absorbed or be dissolved in water.

Hydrophobic—**having an aversion for water.**

Hydrophyte—a type of plant that grows with root system submerged in standing water.

Hydrophytic—in relation to vegetation, plants that grow in water or in saturated soils that are periodically deficient in oxygen as a result of high water content.

Hydroscopic water—water that clings to the surfaces of mineral particles in the zone of saturation.

Hydrosphere—region that includes all the earth's liquid water, frozen water, floating ice, frozen upper layer of soil, and the small amounts of water vapor in the atmosphere.

Hydrostatic test—**a** pressure test procedure in which a vessel or system is filled with water, purged of air, sealed, subjected to water pressure and then observed and/or tested for leaks, distortion and/or mechanical failure.

Hydro zone—an area of an irrigation system where all the factors that influence the watering schedule are similar. Typical factors to be considered would be the type of plants, the precipitation rate of sprinklers or emitters, solar radiation, wind, soil type, and slope.

Hygroscopic water—water which is bound tightly by the soil solids at potential values lower than 31 bars; Water that is tightly held by soil particles. It does not move with the influence of capillary action or gravity, and it is normally unavailable to plants.

Hypochlorite—an anion that forms products such as calcium and sodium hypochlorite. These products are often used for disinfection and bleaching.

Hypoxic waters—waters with dissolved oxygen concentrations of less than 2 mg/L, the level generally accepted as the minimum required for life and reproduction of aquatic organisms.

Ice—the solid form of water.

ID—abbreviation for the inside diameter of a pipe.

Ideal gas—a gas having a volume that varies inversely with pressure at constant temperature and that also expands by 1/273 of its volume at 0° C for each rise in temperature at constant pressure.

Igneous rock—rock produced through the cooling of molten mineral material.

Illite (Illitic)—a general name for a group of three layers, mica-like clay mineable. These minerals are intermediate in composition and structure (between muscovite and montmorillonite).

Image model—the "Integrated Model for Assessment of the Greenhouse Effect." This is one of the climate models applied for the construction of the IPCC SRES scenarios.

Imhoff cone—a clear, cone shaped container used to measure the volume of settleable solids in a specific volume of water.

Immiscibility—the inability of two or more solids or liquids to readily dissolve into one another.

Impact drive—a sprinkler which rotates using a weighted or spring-loaded arm which is propelled by the water stream and hits the sprinkler body, causing movement around a circle.

Impact power—rate at which drops deliver kinetic energy to the soil; Kinetic energy of impact on soil per unit volume [watts, horsepower].

Impact rate—impact power per unit area; Impact power per square foot [hp/ft^2, kw/m^2].

Impermeable—not easily penetrated by water; Impermeable rocks do not allow water to move through them.

Impervious surface—land where water cannot infiltrate back into the ground such as roofs, driveways, streets, and parking lots.

imperviousness, Total—means the actual amount of land surface taken up with impervious surfaces, often stated as a percentage. Interestingly, a site with a total imperviousness of 60% can act like a site with only 10% imperviousness if strategies such as channeling roof runoff into a garden and using swales to capture rainwater are used.

Impingement—process of being forced and held against structures; frequently refers to the process whereby organisms are forced by water flow against screens in the water intake systems of electrical power plants and industrial facilities.

Impoundment—a structure or location used for confined storage, such as a pond, lake or reservoir for liquid wastes.

Impurity—an unwanted chemical substance that is present within another substance or mixture.

Intermittent stream—a watercourse that does not flow continuously, or flows during spring and summer only.

Inactive water well—a water well that is not currently being used, but is being maintained for future use.

Incineration—consists of burning the sludge to remove the water and reduce the remaining residues to a safe, nonburnable ash. The ash can be disposed of safely on land, in some water, or into caves or other underground locations.

Incorporate—to mix some compound or chemical agent with soil.

Indicator—any biological entity or process, or community whose characteristics show the presence of specific environmental conditions or pollutants.

Indicator organisms—microorganisms, such as coliforms, whose presence is indicative for pollution or for the presence more harmful microorganisms.

Indicator parameters—include: pH, specific conductance, total organic carbon (TOC), total organic halogens (TOX).

Indicator tests—tests for a specific contaminant, group of contaminants, or constituent which signals the presence of something else.

Indirect aerosol effect—aerosols may lead to an indirect radiative forcing of the climate system through acting as condensation nuclei or modifying the optical properties and lifetime of clouds.

Indirect discharge—introduction of pollutants from a nondomestic source into a publicly owned wastewater treatment system. These can be commercial or industrial facilities whose wastes enter local sewers.

Indirect method for hydrogeological investigations—methods which include the measurement or remote sensing of various physical and/or chemical properties of the earth (e.g., electromagnetic conductivity, electrical resistivity, specific conductance, geophysical logging, aerial photography).

Industrial revolution—the complex of radical socioeconomic changes starting around 1760 A.D. Extensive mechanization of production systems resulted in a shift from home-based hand manufacturing to large-scale factory production. The invention of machines like engines were developed. It contributed to the advance of irrigation practices. The industrial revolution marks the beginning of a strong increase in the use of fossil fuels resulting in an increase in greenhouse gases in the atmosphere and therefore contributed to the enhanced greenhouse effect.

Industrial runoff—surface water resulting from precipitation that falls on a plant or facility.

Industrial wastewater—the composite of discarded liquids and unwanted water-carried substances resulting directly from a process carried on at a plant or facility.

Infauna—animals that live or burrow through bottom of sediments.

Infiltration—the flow or movement of water through the interstices of pores of soil or other porous medium; groundwater seeping into a collection system. The act of water entering the soil profile.

Infiltration basin—an impoundment where incoming storm water runoff is stored until it gradually filtrates through the soil of the basin floor.

Infiltration capacity—the maximum rate at which infiltration can occur under specific conditions of soil moisture. For a given soil, the infiltration capacity is a function of the water content.

Infiltration or percolation—instantaneous rate of water that moves downward in a soil.

Infiltration rate—the rate at which water enters the soil. Usually expressed in depth of water per hour. (In the United States, usually expressed in inches per hour. Its metric equivalent is centimeters per hour). Infiltration rate is determined by the type of soil; Downward flow of water into the soil at the air-soil interface; Volume of water infiltrating through a horizontal unit area of soil surface at any instant; How quickly water moves into the soil.

Infiltration trench—a conventional infiltration trench is a shallow, excavated trench that has been backfilled with stone to create an underground reservoir. Storm water runoff diverted into the trench gradually filtrates from the bottom of the trench into the subsoil and eventually into the water table. An enhanced infiltration trench has an extensive pretreatment system to remove sediment and oil. It requires an on-site geotechnical investigation to determine appropriate design and location.

Infiltrometer—device used to measure the infiltration rate / intake rate of water into soil: 1. *Ring infiltrometer* consists of metal rings that are inserted (driven) into the soil surface and filled with water. The rate at which water enters the soil is observed; 2. *Sprinkler infiltrometer* consists of a sprinkler head(s) that applies water to the soil surface at a range of rates of less-than to greater than soil infiltration rates. Maximum infiltration rates are observed and recorded; 3. *Flowing infiltrometer* consists of an inlet device to apply a specific flow rate to a furrow and a collection sump with a pump to return tail water to the inlet device. Water infiltrated by the soil in the test section (typically 10 meters) is replaced with water from a reservoir to keep the flow rate constant. The rate of water infiltrated verses time is also plotted. An equation (typically for a curvilinear line) then represents the intake characteristics for that particular soil condition.

Inflow—direct rain flow, such as roof top drains, into a collection system; Surface water runoff and deep drainage to groundwater (groundwater recharge) and transfers into the water system (both surface and groundwater), for a defined area.

Influent—the stream of water to be treated as it flows into any kind of water treatment unit or device, such as hard water into a water softener or turbid water into a filter; wastewater, partially or completely treated, or in its natural state (raw wastewater) that flows into a reservoir, tank, treatment component, or disposal component.

Infrared radiation—radiation emitted by the surface of earth, atmosphere and clouds. Infrared radiation has a distinctive range of wavelengths called a spectrum. Greenhouse gases strongly absorb this radiation in the atmosphere, and reradiate some back towards the surface, creating the greenhouse effect.

Inhibitor—chemical that interferes with a chemical reaction, such as precipitation.

Initial development stage—is period (for a given crop) during germination or initial growth, when the soil covered by the plant growth is less than 10%.

Injectant—a fluid (water, wastewater, solvent, steam, gas, etc.) for injecting into an enhanced oil recovery project or disposal well.

Injection—the introduction of a chemical or medium into the process water to alter its chemistry or filter specific compounds.

Injection system for chemical products—the system for the application of chemical agents through the irrigation system.

Injection well—a well drilled and constructed in such a manner that water can be pumped into an aquifer in order to recharge it.

Injection, application in the suction line—a hose or tube can be connected on the suction side of a pump to inject chemicals. Another hose or tube is connected on the line to replace water and to mix in a tank. This method should not be used with dangerous substances, since the possibility of contamination of the water source is possible. A foot valve or safety valve at the end of suction line can avoid contamination.

Injection, pressure differential—this method is easy to operate. There is a tank under pressure, it is interconnected to the discharge line at two points, one one serves as water entrance and the second is an exit of the water with chemical agents. A pressure difference is created by use of a ball valve in the main line. The pressure difference across this valve is sufficient to cause a water flow through the tank. The water with chemical agents mixes with the water in the main. The injection rate is recalibrated.

Injection, principle of venture—it injects chemical agents in the irrigation line. The principle of operation includes a pressure reduction at the venture. The difference in pressure created across the venturi is sufficient to cause suction of the chemical solutions from a reservoir.

Injection, use of pressure pump—rotating diaphragm or piston type pump can be used to force the chemical to flow from the tank towards the irrigation line. The internal parts of the pump must be corrosion resistant. This method is precise enough to measure the quantity of chemical agent that is injected into to the irrigation system.

Injector—to inject the gas or liquid into the irrigation system.

Inorganic—the minerals, salts, etc. present in wastewater not attributed to carbon molecules of the organic. Examples include iron, silver, lead, sodium.

Inorganic chemicals—chemical substances of mineral origin, not of basically carbon structure.

inorganic matter—chemical substances which do arise from the process of living growth, are composed of matter other than plant or animal matter. And don't contain hydrocarbons of compounds basically carbon structure. Examples are minerals and metals.

Inoxidable—that cannot be oxidized. The term is related to metals and alloys that are resistant to oxidation.

Insectigation—is an application of insecticides through the irrigation system.

Installation—the connecting or setting up and start up operation of any water treatment system.

Instantaneous discharge—the discharge at a particular instant of time.

Instantaneous precipitation (application) rate—maximum rate [inches/h or mm/h], usually localized, that a sprinkler application device applies water to the soil.

In-stream flow needs (IFN)—the scientifically determined amount of water, flow rate, or water level that is required in a river or other body of water to sustain a healthy aquatic environment or to meet human needs such as recreation, navigation, waste assimilation, or esthetics. An in-stream flow need is not necessarily the same as the natural flow.

Instrumental period—period after 1855 that allowed us to reconstruct temperatures to produce reconstructable data. Before 1855 proxy indicators were used to determine temperatures.

Intake—any structure on the upstream face of a dam or within a reservoir created for directing water into a confined conduit, tunnel, canal, or pipeline.

Intake family—grouping of intake characteristics into families based on field infiltrometer tests on many soils.

Intake rate (Basic)—rate at which water enters the soil after infiltration has decreased to a low and nearly constant value. This value should be used for sprinkler and drip irrigation system designs. [in/h, mm/h]

Intake rate of soil—rate that (irrigation) water enters the soil at the surface.

Intake rate or infiltration rate—the measurement of the speed that water moves into the soil.

intake, Initial—depth (rate) of water absorbed by a soil during the period of rapid or comparatively rapid intake following initial application.

Irrigation central control system—it is a computer based system that enables the programming, monitoring and operation of an irrigation system from a central location.

Integrated environmental assessment—the work field that researches causes and impacts of environmental issues and policy measures (solutions) by combining economic, environmental and social sciences. It is a subdivision of the work field of Environmental Systems Analysis. This is a type of Environmental Science that applies many different tools, including environmental models, environmental impact assessment and environmental indicators, to describe and find solutions for environmental problems.

Intemperization—are edaphic and biotic processes that transform rocks into soil particles.

Interception—part of precipitation or sprinkler irrigation system applied water caught on the vegetation and prevented from reaching the soil surface.

Interceptor—a sewer that receives dry-weather flow from a number of transverse sewers or outlets and frequently additional predetermined quantities of storm water (if from a combined system), and conducts such waters to a point for treatment or disposal.

Interest—is a rate charged by commercial bank on the loan. The interest rate depends on the number of payments, quantity,

Interface region—region in which saltwater and freshwater masses come into contact.

Interflow—the lateral movement of water in the unsaturated zone during and immediately after a precipitation event. The water moving as interflow discharges directly into a stream or lake.

Intermittent Stream—a stream, usually in desert areas, that caries water only part of the time, usually during a flash flood.

Internal manual bleed—feature which allows an automatic valve to be opened manually (without controller) by releasing water from above the diaphragm to the downstream side of the valve. Useful during installation, system start-up and maintenance operations when it is undesirable for water to escape into the valve box.

Internal trade—a transaction to transfer a water right from one legal entity to another within a specified area. The area can be a trading zone, an irrigation district or a water resource plan area as defined in respective state and territory legislation.

International Federation of Landscape Architects (IFLA)—a multinational organization of landscape architects whose purpose is the promotion of landscape design and planning.

Interpolation—obtain a desired value using a reference value but next to it.

Interstate trade—a transaction to transfer a water right from one legal entity to another between different states or territories.

Intertidal habitat—a shoreline area that is alternately covered and uncovered by tidal waters.

Interval (range) of operation of the tensiometer—range of readings of a tensiometer from 2 to 80 centibars when the instrument is operating satisfactorily and under normal wet conditions of a soil.

interval, Confidence—confidence or safety margin for a result that is not exact. This value is used in the uniformity of the water application.

interval, Irrigation—time between the initial or successive applications of irrigation in the same plot (field).

interval, Re-entry—during the pesticide application, an operator must wait to enter into the sprayed field without any risks and without any protective clothing.

Intrastate trade—a transaction to transfer a water right from one legal entity to another within the same state or territory.

Intrinsic permeability—pertaining to the relative ease with which a porous medium can transmit a liquid under a hydraulic or potential gradient. It is a property of the porous medium and is independent of the nature of the liquid or the potential field.

Inverse modeling—modeling technique by which modeling input is estimated from modeling output, rather than vice versa. This is for example used to determine the causes of enhanced greenhouse gas concentrations in the atmosphere and for computing atmospheric transport.

Invertebrate—animal that has no backbone (i.e., lacking a notochord or a dorsal vertebrate column).

Inverted siphon—closed conduit (for conveying water) with end sections above the middle section; used for crossing under a depression, under a highway or other obstruction. Sometimes called a sag pipe.

Iodine (I)—a nonmetallic element which is the heaviest and least reactive of the naturally occurring halogens. It may be used for disinfection. In both its liquid and vapor forms, iodine is readily adsorbed by activated carbon.

Ion—an atom in a solution that is charged, either positively (cations) or negatively (anions).

Ion exchange—the replacement of undesirable ions with a certain charge by desirable ions of the same charge in a solution, by an ion-permeable absorbent.

Ion exchange capacity—measures ability of a formation to adsorb charged atoms or molecules.

Ion exchange—a reversible chemical process in which ions from an insoluble permanent solid medium (the "ion exchanger" usually resin) are exchanged for ions in a solution or fluid mixture surrounding the insoluble medium.

IPCC—"Intergovernmental Panel on Climate Change": the international organization founded by the United Nations that tries to predict the impact of the greenhouse effect according to existing climate models and literature information. The Panel consists of more than 2,500scientific and technical experts from more than 60 countries all over the world. It is therefore referred to as the largest peer-reviewed scientific cooperation project in history.

IPCC SRES scenarios—special reports on emission scenarios by the IPCC, containing information on possible future climate developments and consequences for society and the environment. Emissions scenarios are a central component of any assessment of climate change. See for more information, the reader may go to the weblink of IPCC-SRES scenarios.

Iron (Fe)—a very common element often present in groundwater in amounts ranging from 0.01 to 10.0 ppm (mg/L) Iron can be found in three forms: 1. Soluble form as in ferrous bicarbonate; 2. Bound with a soluble organic compound; 3. As suspended ferric iron particles

Iron bacteria—bacteria which thrives on iron and are able to actually use ferrous iron (as found in water or steel pipes) in their metabolic processes, to incorporate ferric iron in their cell structure, and to deposit gelatinous ferric hydroxide iron compounds in their life processes.

Iron fouling—the accumulation of iron on or within an ion exchange resin bed or filter medium in such amounts that the capacity of the medium is reduced.

Irrecoverable water loss—water loss that becomes unavailable for reuse through evaporation, phreatophytic transpiration, or ground-water recharge that is not economically recoverable.

Irrigable area—the maximum area which can be irrigated in the reference year using the equipment and the quantity of water normally available on the holding. The meaning is therefore similar to the term *area equipped for irrigation*. However, the total irrigable area may differ from the sum of the areas provided with irrigation equipment since the equipment may be mobile and therefore utilizable on several fields in the course of a harvest year; capacity may also be restricted by the quantity of water available or by the period within which mobility is possible; Area capable of being irrigated, principally based on availability of water, suitable soils, and topography of land.

Irrigation—application of water by artificial means. Purposes for irrigating may include, but are not limited to, supplying evapotranspiration needs, leaching of salts, and environment control; Applying water or wastewater to land areas to supply the water and nutrient needs of plants. 1. *Bubbler irrigation* is an application of water to flood the soil surface using a small stream or fountain. The discharge rates for point-source bubbler emitters are greater than for drip or subsurface emitters but generally less than 1 gpm. A small basin is usually required to contain or control the water. 2. *Surface irrigation* is type of irrigation where water is distributed to the plant material by a ground surface distribution network possibly including rows or dikes. Broad class of irrigation methods in which water is distributed over the soil surface by gravity flow. 3. *Basin irrigation* is a flooding of leveled land surrounded by dikes. Used interchangeably with level border irrigation, but usually refers to smaller areas. 4. *Border irrigation* is a flooding of strips of land, rectangular in shape and cross leveled, bordered by dikes. Water is applied at a rate sufficient to move it down the strip in a uniform sheet. Border strips having no down field slope are referred to as level border systems. Border systems constructed on terraced lands are commonly referred to as benched borders. 5. *Cablegation* is a surface irrigation that uses gated pipe to both transmit and distribute water to furrows or border strips. A plug, moving at a controlled rate through the pipe, causes irrigation to progress along the field and causes flow rates from any one gate to decrease continuously from some maximum rate to zero. 6. *Check irrigation* is a modification of a border strip with small earth ridges or checks constructed at intervals to retain water as the water flows down the strip. 7. *Check basin irrigation* is an application of irrigation rapidly to relatively level plots surrounded by levees. The basin is a small check. 8. *Corrugation irrigation* is a method of surface irrigation similar to furrow irrigation, in which small channels, called corrugations, are used to guide water across a field. No attempt is made to confine the water entirely to the corrugations. 9. *Flood irrigation* is a method of irrigation where water is applied to the soil surface without flow controls, such as furrows, borders or corrugations. 10. *Furrow irrigation* is a method of surface irrigation where the water is supplied to small ditches or furrows for guiding across the field. 11. *Alternate set irrigation* is a method of managing irrigation whereby, at every other irrigation, alternate furrows are irrigated, or sprinklers are placed midway between their locations during the previous irrigation. 12. *Alternate side irrigation* is a practice of furrow irrigating one side of a crop row (for row crops or orchards) and then, at about half the irrigation time, irrigating the other side. 13. *Cutback irrigation* is a reduction of the furrow or border inflow stream after water has advanced partially or completely through the field in order to reduce runoff. 14. *Surge irrigation* is a surface irrigation technique wherein flow is applied to furrows (or less commonly, borders) intermittently during a single irrigation set. 15. *Wild flooding* is a surface irrigation where water is applied to the soil surface without flow controls, such as furrows, borders (including dikes), or corrugations. 16. *Sprinkler irrigation* is an irrigation system using mechanical devices with nozzles (sprinklers) to distribute the water by converting water pressure to a high velocity discharge stream or streams.

irrigation, Gross—water actually applied, which may or may not be total irrigation water requirement; i.e. leaving storage in the soil for anticipated rainfall, harvest.

irrigation, Net—actual amount of applied irrigation water stored in the soil for plant use or moved through the soil for leaching salts. Also includes water applied for crop quality and temperature modification; i.e. frost control, cooling plant foliage and fruit. Application losses, such as evaporation, runoff, and deep percolation, are not included.

Irrigation—The controlled application of water for agricultural purposes through man-made systems to supply water requirements not satisfied by rainfall.

Irrigation (water) requirement or net irrigation requirement—depth of water, exclusive of effective precipitation, stored soil moisture, or ground water, that is required for meeting crop evapotranspiration for crop production and other related uses. Such uses may include water required for leaching, frost protection, cooling and chemigation; Difference between evapotranspiration and effective precipitation; Quantity of water needed by the landscape to satisfy the evaporation, transpiration and other uses of the water in the soil. *Gross irrigation requirement* is a total amount of water applied (or desired); Total irrigation requirement including net crop requirement plus any losses incurred in distributing and applying and in operating the system. *Irrigation water requirement* is a calculated amount of water needed to replace soil water used by the crop (soil water deficit), for leaching undesirable elements through and below the plant root zone, plus other needs; after considerations are made for effective precipitation; Plant water requirement adjusted for application uniformity (and efficiency).

Irrigation audit—procedure to collect and present information concerning the uniformity of application, precipitation rate, and general condition of an irrigation system and its components.

Irrigation controller—an electric timing device that operates each (irrigation) zone for a predetermined time and frequency; An automatic timing device that sends an electric signal for automatic valves to open or close according to a set irrigation schedule. *Irrigation CONTROLLER or TIMER or CLOCK* is the brain of the sprinkler system. The controller automatically opens and closes valves according to a preset schedule. Rain Bird controllers have easy-to-set programs to help efficiently manage seasonal adjustments. Rain Bird also has sensors that signal the controller to shut off the system when it rains. An automatic controller is usually more water-efficient that operating sprinklers manually. A feature of Rain Bird controllers that let you easily change the run times of your sprinklers without having to reprogram the controller.

Irrigation controller—is a "timer" to turn on and off an automatic irrigation system. Controllers range from very simple to extremely sophisticated computer systems that use modems, cell-phones, or radios and allow 2-way communication between the controller and the units (valves, meters, weather stations, soil moisture sensors, etc.) being controlled.

irrigation controller, Smart—an irrigation control system that uses weather-based calculations and environmental conditions to determine how much water to apply to a landscape based on the plant water needs.

Irrigation deficit—the percentage of the soil volume in the root zone, which is not wetted to field capacity or above (after irrigation) divided by the total potential root zone of the crop.

Irrigation depth—is an amount of water that is applied to supply the moisture to the roots zone at field capacity.

Irrigation design—plan of an irrigation system with pipe sizing, head layout and valve location.

Irrigation District—an organization that owns and manages a water delivery system for irrigating a given region. The district may also convey water for other purposes, such as municipal use or stock watering; Cooperative, self-governing, semipublic organization set up as a subdivision of a state or local government to provide irrigation water.

Irrigation duration—time interval between the beginning and successive applications of irrigation in a given area.

Irrigation efficiency (water use efficiency)—the percentage of irrigation water which is actually stored in the soil and available for use by landscape as compared to the total amount of water provided to the landscape; Ratio of the average depth of irrigation water that is beneficially used to the average depth of irrigation water applied.

Irrigation frequency—time that passes from irrigation to the other irrigation. It is a number that implies how many times a crop will be irrigated.

Irrigation infrastructure operator—It is an entity that operates water service infrastructure for the purposes of delivering water for the primary purpose of being used for irrigation.

Irrigation interval—average time interval between the commencement of successive irrigations for a given field (or area); Time between irrigation events. Usually considered the maximum allowable time between irrigations during the peak ET period.

Irrigation period—time that it takes to apply one irrigation to a given design area during the peak consumptive-use period of the crop being irrigated.

Irrigation potential—area of land which is potentially irrigable. Country/regional studies assess this value according to different methods, for example some consider only land resources suitable for irrigation, others consider land resources plus water availability, others include in their assessment economic aspects (such as distance and/or difference in elevation between the suitable land and the available water) or environmental aspects, etc. Whatever the case, it includes the area already under agricultural water management.

Irrigation practices—these provide water to agricultural land to compensate for the water shortage. These include a dynamic set of circumstances: micro climatic of the crop, soil characteristics, water resource, and design restrictions.

Irrigation requirement—the quantity of water needed by the landscape to satisfy the evaporation, transpiration and other uses of water in the soil. The Irrigation requirement is usually expressed in depth of water and equals the net irrigation requirement divided by the irrigation efficiency. (In the United States, usually expressed in inches per week or centimeters per week.

Irrigation requirement (water requirement)—the amount of water that must be applied to satisfy the demands of the crop system. It is expressed as a volume or depth. The water requirement includes the soil water deficit, the leaching requirement, and water used by the plants during the irrigation and gravity drainage phases. If other reasons for irrigating are involved, such as frost protection, bed forming, subsidence control, etc., these volumes or depths must be included. Likewise, if the irrigation is not meant to accomplish a task during an irrigation, the component need not be included in the requirement for that particular irrigation. For example, irrigation for frost protection is seasonal and therefore whatever portion of the water above ET needs, being applied strictly to achieve frost protection, need not be included in the water requirement of a summer irrigation; Amount of water necessary to compensate the evapotranspiration needs of a crop; excluding the contributed by the precipitation, the soil water storage, the operational losses, and flushing requirements.

irrigation requirements, Seasonal—total amount of water necessary for the normal plant growth during its growing cycle.

Irrigation sagacity—ratio of volume of irrigation water beneficially or reasonably used to the total volume of irrigation water that has left the region, both in a specified period of time.

Irrigation scheduling—procedure of establishing and implementing the time and amount of irrigation water to apply, determining when to irrigate and how much water to apply based upon measurements or estimates of soil moisture or crop water used by a plant. Set of specifications identifying times to turn on and off water to various zones of an irrigation system. The goal is to schedule irrigation timing and amounts such that the soil water content remains between FC and PWP.

Irrigation set—area irrigated at one time within a field.

Irrigation slope—elevation difference along the direction of irrigation. Sometimes called irrigation grade.

Irrigation system—a set of components which includes the water source (e.g., domestic service or pump), water distribution network (e.g., pipe), control components (e.g., valves and controllers), emission devices (e.g., sprinklers and emitters) and possibly other general irrigation equipment (e.g., quick coupler and back-flow preventer).

irrigation system, Wheel line—also known as side-roll wheel move, or side move. Large-diameter wheels are mounted on a pipeline make it so that the pipeline can be rolled to the next position with less effort than hand-move. These are usually powered by a small gasoline engine.

Irrigation uniformity—a measure of the spatial variability of applied or infiltrated waters over the field. Several indicators can be used to place a numerical value on the irrigation uniformity by statistically representing the deviations in water applied, or stored in the crop root zone, from the field average. The irrigation uniformity is generally expressed as a percent with 100% representing perfect uniformity.

irrigation, Drip—water is applied at very low flow rates (drip) through emitters directly to the soil. It is a highly controlled irrigation method. Here the water is deposited directly in root zone. The water is received slowly at a desired pressure to maintain desired soil moisture at a desired level to allow optimum plant growth. Also it is known as "trickle or micro," "tension," "capillary," "daily," "high frequency," and "continuous moisture." The distribution of the water is by means of the use of secondary lines and emitters. For detailed information, the reader should consult "*Management of Drip/ Trickle or Micro Irrigation* by Megh R. Goyal, Apple Academic Press Inc."

Irrigation duration—is a period of time during which the irrigation system is in operation.

irrigation, Flood—Also known as wild flood irrigation. Water is turned into a field without any flow control such as furrows, boarders or corrugations. This is the least efficient, least uniform and least effective method of irrigation; Water is flooded over a predetermined area.

Irrigation (Infiltration)—the water from the river is taken to the field by leveled irrigation furrows.

irrigation, Overhead—it is a pressure irrigation that uses perforated pipes or the pipes with sprinklers to supply the water similar to a rainfall.

irrigation, Sprinkler—is based on the principle that the water is released at pressure through portable irrigation pipes: Light weight, lines of irrigation of fast connection or united with mounted sprinklers at intervals. An ample range of sprinklers is available to operate under varied pressures, spaces, sizes, and flow. This irrigation system simulates the rainfall, adaptable to several agricultural conditions. Commonly used agricultural sprinkler irrigation systems include: center pivots, wheel lines, hand lines, solid set, and big guns.

irrigation, Subsurface—water is applied to open ditches or to tile lines to bring the water table up to where it is high enough to be used by the plants. Typically this is only practical in locations with high water tables, flat ground, and sandy soils.

irrigation, Surface—irrigation method where the land surface is used to transport the water via gravity flow from the source to the plants. Common surface irrigation methods are: furrow irrigation, corrugation irrigation, border irrigation, basin irrigation, and flood irrigation.

irrigation, Surge—water is sent down a furrow in pulses. After the initial surge, the furrow is allowed to dry. This wetting and drying cycle continues throughout the whole irrigation event. Surge increases uniformity for furrow irrigation and increases application efficiency.

irrigation, Trickle—a low pressure system where water is distributed through closed pipelines. Water is applied directly or very near to the soil surface, either above or below the ground surface, in discrete drops, continuous drops, small streams, or spray. Flows and pressures are typically low. A wide variety of emitters are available to dissipate pressures at points, allowing water applications to the soil in small amounts with little force. The trickle irrigation category includes methods such as drip, subsurface, bubbler, and spray irrigation. See also drip irrigation.

Irrigation—an artificial application of soil moisture that is essential for the plant growth. It supplements the rainfall when it is not sufficient.

Isohalines—lines drawn on a map connecting points of equal salinity.

Isohyetal line—a line, drawn on a map, that passes through points with equal amounts of precipitation (rainfall).

Isotropy—the condition in which hydraulic properties of the aquifer are equal in all directions.

Jackson turbidity unit—formerly used measurement of the turbidity in a water sample. This has been replaced by the nephelometric turbidity unit (NTU).

Jacob straight line method—a graphical method using semilogarithmic paper for evaluating the results of a pumping test.

Jar test—a laboratory test procedure with differing chemical doses, mix speeds, and settling times, to estimate the minimum or ideal coagulant dose required to achieve water quality goals.

Joint implementation—richer countries have the opportunity to achieve their emission reduction goals, formulated in the Kyoto Protocol, by financing energy saving projects for poorer countries that have also signed the treaty.

Juvenile water—water entering the hydrologic cycle for the first time.

Karst topography—a topographic area which has been created by the dissolution of a carbonate rock terrain. This type of topography is characterized by sinkholes, caverns, and lack of surface streams.

Kemmerer sampler—a sampling device that can be lowered either into a deep well or into a lake in order to retrieve a water sample from a particular depth in the well or the lake.

Ketones—class of organic compounds where the carbonyl group is bonded to two alkyl groups.

Kinematic viscosity—the ratio of dynamic viscosity to mass density. It is obtained by dividing viscosity by the fluid density. Measure of the resistance of a liquid to shear forces. Units of kinematic viscosity are square meters per second.

Kinetic energy—energy due to the velocity of moving water.

Kinetic rate coefficient—a number that describes the rate at which a water constituent such as a biochemical oxygen demand or dissolved oxygen rises or falls.

Kinetics—the study of the relationships between temperature and the motion and velocity of very small particles. Kinetic relationships influence the rate of change in a chemical or physical system.

K-Y jelly—clear, water soluble, nontoxic lubricant used for lubricating O-rings in water filters. Also sometimes used to lubricate threads and ease insertion of drip tube into fittings. Because it is water-soluble it doesn't gum up emitters, filters, or sprinkler mechanisms.

Kyoto protocol—protocol adopted in 1998 in Kyoto, Japan. It requires participating countries to reduce their anthropogenic greenhouse gas emissions (CO_2, CH_4, N_2O, HFCs, PFCs, and SF_6) by at least 5% below 1990 levels in the commitment period 2008 to 2012. The Kyoto Protocol was eventually signed in Bonn in 2001 by 186 countries. Several countries such as the United States and Australia have retreated. More information can be found on the web link of Kyoto Protocol.

Laboratory water—purified water used in the laboratory as a basis to create solutions or making dilutions. It contains no interfering substances.

Lag time (flood irrigation)—period between the time that the irrigation stream is turned off at the upper end of an irrigated area and the time that water disappears from the surface at the point or points of application.

Lagoon—a pond containing raw or partially treated wastewater in which aerobic or anaerobic stabilization occurs; The term often refers to a small, artificial body of water usually composed of several cells or compartments used to treat wastewater to a secondary level of treatment

Lake—an inland body of water, usually fresh water, formed by glaciers, river drainage, etc. It is usually larger than a pool or pond.

Laminar flow—that type of flow in which the fluid particles follow paths that are smooth, straight, and parallel to the channel walls. In laminar flow, the viscosity of the fluid damps out turbulent motion.

Land Application—the discharge of raw or treated wastewater onto the ground for treatment or reuse. The wastewater penetrates into the ground where the natural filtering and straining action of the soil removes most of the pollutants. Three techniques are used: spray irrigation, rapid infiltration and overland flow.

Land cover—the vegetation, water, natural surface, and artificial construction of the land surface.

Land fill—a system of trash and garbage disposal in which waste is buried between layers or earth.

Land pan—a device used to measure free-water evaporation.

Land surface datum—a datum plane that is approximately at the land surface at the well.

Land trust—a conservation group that maintains a revolving fund for quickly buying land that is in danger of being developed inappropriately or without regard to proper environmental considerations.

Land use—the management practice of a certain land cover type (a set of human actions). Land use may be forest, arable land, grassland, urban area or other.

Land use change—alteration of the management practice on a certain land cover type. Land use changes may influence the climate system because they impact evapotranspiration and sources and sinks of greenhouse gases.

Landscape—narrowly defined, the amount of countryside and/or city that can be taken in at a glance. Also, an area of land or water taken in the aggregate.

Landscape architect—a professional who designs, plans, and manages outdoor spaces ranging from entire ecosystems to residential sites and whose media include natural and built elements; also referred to as a designer, planner, consultant. Not to be confused with landscapers, landscape contractors or nurserymen.

landscape architecture—the science and art of design, planning, management and stewardship of the land. Landscape architecture involves natural and built elements, cultural and scientific knowledge, and concern for resource conservation to the end that the resulting environment serves a useful and enjoyable purpose. Successful landscape architecture maximizes use of the land, adds value to a project and minimizes costs, all with minimum disruption to nature.

Landscape architecture registration—In the United States, a certification of individuals entitled to use the term "landscape architect" or to practice landscape architecture or both, by means of examination and required degree and experience criteria.

Landscape coefficient—coefficient used to modify reference ET, which includes species factor, density factor and microclimate factor.

Landscape contractor—a trained builder or installer of landscapes, retained to implement the plans of landscape architects.

Langelier index—an index reflecting the equilibrium pH of a water with respect to calcium and alkalinity; used in stabilizing water to control both corrosion and scale deposition.

Langmuir Isotherm—an empirical equation that describes the amount of solute adsorbed onto a soil surface.

Large water system—a water system that services more than 50,000 customers.

Late growth stage—is time between the end of the average stage of vegetative growth and the harvest or maturity.

Lateral—pipe(s) that are located from the control valves to the sprinklers or drip emitter tubes; Secondary or side channel, ditch or conduit. Also call "branch drain" or "spur"; Water delivery pipeline that supplies irrigation water from the main line to sprinklers or emitters; The pipe installed downstream from the control valve on which the (sprinkler heads or) emission devices are located.

Lateral line—nonpressure pipe running from the valve to the sprinklers.

Looped circuit—a piping system, usually a main line, that closes back on itself in a loop, providing more than one path for the water to flow to the valve(s).

Low head drainage—water left in the pipe after a valve is turned off that is gently flowing out of a low elevation sprinkler head.

Lath box—wooden box that is placed in a ditch bank to transfer water from an irrigation ditch to the field to be irrigated (preferred term is *spile*).

Law of mass action—the law stating that for a reversible chemical reaction the rate of reaction is proportional to the concentrations of the reactants.

LC50—the concentration of a substance which is lethal to 50 percent of the test organisms within a specific time period (96 h).

Leach—dissolve by the action of a percolation liquid, such as rainwater seeping through soil and carrying pesticides, fertilizers, or chemical wastes with it.

Leach field—the area where the effluent from a septic tank system is distributed by horizontal underground piping designed to aid in the process of natural leaching and percolation through the soil.

Leachate—a liquid that has been in contact with waste in a landfill or other porous substrate and may have undergone chemical or physical changes as a result, and has subsequently seeped out.

Leachate management system—a method of collecting leachate and directing it to treatment or disposal area.

Leaching—occurs when a liquid (e.g., water) passes through a substance, picking up some of the material and carrying it to other places; this can happen under ground in solid rock, or above ground through piles of material.

Leaching fraction—ratio of the depth of subsurface drainage water (deep percolation) to the depth of infiltrated irrigation water.

Leaching requirement—quantity of irrigation water required for transporting salts through the soil profile to maintain a favorable salt balance in the root zone for plant development.

Leakage—a species of ions in the feed of an ion exchanger present in the effluent.

Legislation—laws such as Acts and Regulations that are established by an elected official.

Length—linear dimension used to describe the quantity/amount/distance of pipe, conductor or similar material in various equations.

Length of run—distance water must flow in furrows or borders over the surface of a field from the head to the end of the field.

LEPA—the use of drop tubes on a center pivot or lateral-move systems to apply water at low pressures very near or directly onto the soil surface. This increases efficiency and uniformity. Furrow dikes are typically required to avoid runoff. Often confused with LESA (low elevation spray application) and MESA (mid elevation spray application); Acronym for *Low Energy Precision Application.*

Level Basin—water is applied at relatively high flow rates to a completely level field. Precision laser leveling is required to achieve the level fields suitable for this method.

Licensee—the individual or organization to whom a license is issued or assigned

Life cycle—the series of like stages in the form and mode of life of an organism. i.e., between successive recurrences of a certain primary stage such as the spore, fertilized egg, seed, or resting cell.

life, Service or useful—period after which the equipment is rendered useless.

Light absorption—the amount of light a certain amount of water can absorb over time.

Lime—common water treatment chemical. Lime can be deposed on walls of showers and bathrooms, after lime has reacted with calcium to form limestone [CaO: Calcium oxide].

Lime scale—hard water scale formed in pipes and vessels (generally more severe on the hot water side) containing a high percentage of calcium carbonate ($CaCO_3$) or magnesium carbonate $MgCO_3$).

Limestone—a general name for rocks composed primarily of the mineral calcite (calcium carbonate).

Limited irrigation—management of irrigation applications to apply less than enough water to satisfy the soil water deficiency in the entire root zone. Sometimes called " deficit" or "stress irrigation."

Limiting factor—the condition that limits treatment capacity usually either groundwater or bedrock.

Limnology—the study of the physical, chemical, hydrological, and biological aspects of fresh water.

Line of distribution—irrigation line that takes the water from the main line to the lateral one (or drip line).

Line source—continuous source of water emitted along a line. Usually associated with drip irrigation where emitters are spaced at regular intervals in poly tubing.

Line spacing—the distance a pipe is moved.

line, Discharge—It is part of an irrigation pipe that takes the water from the pump to the principal line or main.

line, Lateral—It is a part of irrigation piping that takes water from the secondary line/lines to the drippers.

line, Main—it is a part of irrigation pipe that takes the water from the pump house to the secondary line or to distribution.

line, Secondary—it is a part of irrigation piping that takes water from the main line to the lateral lines.

line, Suction—a line on the downstream side of a pump. It takes water from the source to the pump.

Linear—a continuous layer of natural or man-made materials lining the bottom and/or sides of a surface impoundment, landfill, or landfill cell that restricts the downward or lateral escape of hazardous waste, hazardous waste constituents, or leachate.

Liquid—a state of matter, neither gas nor solid, that flows and takes the shape of its container.

Liters per minute—a metric measurement of water flow used worldwide.

Lithology—the systematic description of rocks, in terms of mineral consumption and texture.

Lithophytes—algae (seaweeds) that grow attached to submerged rocks or stones.

Lithosphere—the outer part of the earth, consisting of the crust and upper mantle, approximately 100 km thick.

Lixiviation—washing of salts in the soil by percolation process.

Load discharge—release of contaminants, usually expressed as kg/day, at levels or concentrations above that which can be removed though application of best available technology economically achievable.

load or water pressure is transferred by the arch to the abutments.

Loam—the textural class name of soil having a moderate amount of sand, silt, and clay. Loam soils contain 7–27% clay, 28–50% silt, and <52% sand.

Longitudinal flow—a flow pattern in which water travels from the bottom to top (or vice versa) in either a cartridge or loose media tank-type filtration system. The advantages are greater contact time, higher unit capacity, more complete utilization of medium and more uniform water quality. Also called axial flow.

Loop—a plumbing connection used to bypass water around a location designed for installation of a water treatment system or used when the treatment system is out of service for any reason.

Looped circuit—piping system, usually a main line that closes back on itself in a loop, thus providing water to any location from two routes.

Low brackish—having a salt content between 0.5 and 5 parts per thousand.

Low Energy Precision Application—a water, soil, and plant management regime where precision down-in-crop applications of water are made on the soil surface at the point of use. Application devices are located in the crop canopy on drop tubes mounted on low pressure center pivot and linear move sprinkler irrigation systems.

Low head drainage—condition in which water drains partially or completely out of the lateral line through the sprinkler head after each irrigation cycle is completed.

Low level outlet (Bottom Outlet)—an opening near the bottom of a reservoir, generally used for emptying the reservoir or for scouring sediment.

Low pressure in canopy [LPIC]—low-pressure in-canopy system that may or may not include a complete water, soil and plant management regime as required in LEPA. Application devices are located in the crop canopy with drop tubes mounted on low-pressure center pivot and linear move sprinkler irrigation systems.

Low pressure pipe—lateral 1″–2″ pipe with small orifices (5/32″–1/4″) through which effluent is distributed to trench under low pressure head (2 to 5 feet).

Low yield well—a relative term to a well that cannot recover in sufficient time after well evacuation to permit the immediate collection of water samples.

Lysimeter—device for measuring deep percolation from a soil profile, usually consisting of an enclosed volume of undisturbed soil with some means of collecting drainage water. It may also include some method of measuring changes in the volume of stored soil water; Isolated block of soil, usually undisturbed and in situ, for measuring the quantity, quality, or rate of water movement through or from the soil; Device and monitoring system used to measure the evapotranspiration (ET) rate using a container closed at the bottom and with the top flush with surrounding grade and planted with turf. After rainfall and irrigation application are accounted for, daily change in the weight of the lysimeter is directly related to ET.

Macro invertebrates—invertebrates organisms including larvae of diptera, mayflies, caddis flies and other insects, mollusks, crustaceans and worms that are generally visible to the unaided eye and retained on a screen with 0.595 millimeter (mm) mesh apertures. The community diversity index, D-Bar, is a mathematical expression of the number of individuals within the taxa represented.

Macro phyton—large aquatic plants that can be seen without magnification, including mosses and seed plants.

MAGICC—climate model that calculates average atmospheric temperatures and sea levels. It is used by IPCC for the construction of the SRES scenarios.

Magmatic water—water associated with a magma.

Magnesium (Mg)—one of the elements that make up the earth's crust as part of many rock-forming minerals such as dolomite. Magnesium and calcium dissolved in water constitutes hardness. The presence of magnesium in water contributes to the formation of scale and insoluble soap, which identify hard water.

Main and sub main—the water delivery pipelines that supply water from the control station to the manifolds.

Main line—tubing used in the drip system and is sometimes called lateral line. It is a soft polyethylene material; pipe(s) from the water source to the control valves.

Main or mainline—water delivery pipelines that supply water from the control station to the manifolds; Pipe usually under constant pressure which supplies water from the point of connection to the control valves (or valve-in-head sprinklers).

Main stem—the main body of a river or estuary, excluding its tributaries; the primary channel of a river; the primary river in a drainage basin.

Maintenance and service—revision and periodic care of the irrigation system.

Major ions—major ions include elements, which are (or could be) in fairly high concentration in most natural waters, such as calcium, magnesium, sodium, potassium, bicarbonate, carbonate, sulfate, and chloride.

Male adapter—the fitting used to adapt from solvent weld PVC to a male threaded end. When connecting to metal threads male adapters should be used, so that the plastic male threads screw into the metal female threads. It may seem sexist, but plastic male adapters work better than plastic female adapters.

Management allowable depletion (MAD)—also known as maximum allowable depletion, management allowable deficit, management allowable deficiency, allowable soil depletion or allowable soil water depletion. The portion of plant available water that is allowed for plant use prior to irrigation based in plant and management considerations; Desired soil moisture deficit at the time of irrigation; Portion of *available water* that is scheduled to be used prior to the next irrigation; Planned soil moisture deficit at the time of irrigation. Usually a percentage of the available water capacity [%, decimal].

Manganese (Mn)—an element sometimes found dissolved in groundwater usually in combination with— but in lower concentrations than iron. Manganese is noticeable because in concentrations above 0.05 mg/L it causes black staining of laundry and plumbing fixtures.

Manifold—a group of valves and fittings.

Master valve—a normally closed valve installed at the supply point of the main that opens only when the automatic system is activated. This serves to keep pressure off the mainline except during operation.

Matched precipitation rate—a system or zone in which all the heads have similar precipitation rates is said to have matched precipitation rates.

Manual system—a sprinkler system that is operated by manual valves rather than by automatic controls.

Manifold—the water delivery pipeline that supplies water from the main to the laterals. Pressure regulation is often applied at the inlet to the manifold.

Manipulation of space—in landscape architecture, the organization of areas of land for specific esthetic or functional purposes. This can range from creating small backyard patios to huge urban plazas.

Manning equation—an equation to compute the average velocity of flow in an open channel.

Manometer—measures pressure difference.

Manufacturer's coefficient of variation—measure of the variability of discharge of a random sample of a given make, model, and size of micro irrigation emitter, as produced by the manufacturer and before any field operation or aging has taken place; equal to the ratio of the standard deviation of the discharge of the emitters to the mean discharge of the emitters.

Marsh—a water body covered by water for at least part of the year and characterized by aquatic and grass-like vegetation, especially without peat-like accumulation.

Marsh—soft wetland areas, usually with grasses and other low vegetation, that support a great variety of plant and animal life.

Marsh gas—a combustible gas, mostly methane, that is produced from the decay of vegetation in stagnant water.

Mass erosion—when a hillside becomes saturated with water, large areas of soil can slide or creep downhill, often leading to the formation of gullies.

Mass flow—this theory maintains that a high pressure of turgor in the cells of the leaves soon produces the flow of the sugar solution and other substances through plasmodium to the adjacent cells and through the crib tubes towards the adjacent cells and soon through the crib tubes towards the root cells.

Mass Transport—a volume of water transported per unit time across a given plane in a body of water.

Master plan—A preliminary plan showing proposed ultimate site development. Master plans often comprise site work that must be executed in phases over a long time and are thus subject to drastic modification.

Matched precipitation rate (MPR)—system or zone in which all the heads have similar precipitation rates is said to have matched precipitation rates; Refers to sprinklers that apply water at the same rate per hour no matter the arc of coverage or part of a circle they cover. For instance, a full-circle sprinkler discharges twice the flow of a half-circle sprinkler and a quarter-circle sprinkler discharges half of what the half-circle unit does. MPR allows the same type of sprinklers, no matter what their arc, to be circuited on the same valve and to deliver the same PR rate. Spray heads have fixed arcs and are matched by the manufacturer, while rotors offer a choice of nozzles to match the Designed arc pattern.

Matric potential—the force (measured in units of negative pressure) that the soil exerts (pulls) on water, [bars, kPa]; Dynamic soil property and will be near zero for a saturated soil. Matric potential results from capillary and adsorption forces. This potential was formerly called capillary potential or capillary water.

Mature Karst—Karst environment where the physical features (e.g., sinholes, caves) are well defined.

Maximum allowable deficiency—also called as management allowed depletion. Term used to estimate the amount of water that can be used without adversely affecting the plant and is defined as the ratio of readily available water to available water.

Maximum application rate—maximum discharge at which sprinklers can apply water without causing significant translocation.

Maximum contaminant level (MCL)—the maximum level of a contaminant allowed in water by federal law. Based on health effects and currently available treatment methods.

Maximum number of sun-shine hours—number of hours of sunshine during a period of 24 h.

Maximum value—in a set of data, the measurement having the highest numerical value.

MBAS (Methylene Blue Active Substance)—a measure of apparent detergents. This demonstration depends on the formation of a blue color when methylene blue dye reacts with synthetic detergent compounds.

Mean—the "average" value for a series of counts. The arithmetic mean (x) is found by dividing the sum of all of the counts by the total number of observations. The geometric (log) mean is found by dividing the sum of the logs of the counts by the total number of observations and then transforming that value back into a natural number. Calculation of log as in the case with bacterial counts.

Mean discharge—the arithmetic mean of individual daily mean discharges during a specific period.

Mean low water (MLW)—the average height of all low tides at a given location.

Mean sea level (MSL)—the average relative sea level over a period, such as a month or a year, long enough to average out transients such as waves.

Mechanical aeration—use of mechanical energy to inject air into water to cause a waste stream to absorb oxygen.

Mechanical filter—a pressure or gravity filter designed to physically separate and remove suspended solids from a liquid by mechanical (physical) means rather than by chemical means.

Mechanical flotation—a term used in the mineral industry to describe the use of dispersed air to produce bubbles that measure 0.2 to 2 mm in diameter.

Mechanism of diode vortex—an Israeli designer proposed the unique variation of the orifice concept. This mechanism is a simple device that produces resistance to the flow creating a vortex in a circular chamber. The water enters tangentially at the circumference of the chamber and causes fluid turn around in the chamber at a high speed.

Media—materials that form a barrier to the passage of certain suspended solids or dissolved liquids in filters; A selected group of material used in filters and filter devices to form barriers to the passage of certain solids or molecules which are suspended or dissolved in water.

Median drop size—diameter where half the sprinkler's water volume falls in drops smaller, and half falls in drops larger than the median size; Drop size where 50% of the water volume occurs in drops greater than this size.

Median value—the value of the middle item when items are arrayed according to size.

Medieval warm period—term introduced by the British meteorologist Hubert Lamb in 1965, for a period between the 9th and thirteenth century during which it was extremely warm on many locations in and around Europe. Wine was grown in Scandinavia and agriculture was possible on Greenland. This was determined by studying snow lines in the mountains and temperatures in deep boreholes and has given us the impression that temperature changes may have occurred before. Geochemist Wallace Broecker thinks that cyclic processes in the oceans cause a warmer period once in every 1,500 years.

Medium size water system—a water system that serves 3,300 to 50,000 customers.

Mega liter—1 million liters.

Melting—the change of a solid into a liquid.

Membrane—a thin barrier that allows some compounds or liquids to pass through, and troubles others. It is a semipermeable skin of which the pass-through is determined by size or special nature of the particles. Membranes are commonly used to separate substances; a thin sheet or surface film, either natural or man-made of microporous structure that performs as an effluent filter of particles down to the size range of chemical molecules and ions. Such membranes are termed "semipermeable" because some substances will pass through but others will not.

Mercury—a heavy metal, highly toxic if breathed or ingested. Mercury is residual in the environment, which accumulates in the organs and tissues of aquatic organisms, especially fish or shellfish.

Mesh size—mesh is the number of openings in a square inch of a screen or sieve. It is equal to the square of the number of strains of metal or plastic screening per lineal inch.

Meso haline—having a salt content between 5 and 18 parts per thousand.

Meso trophic—reservoirs and lakes which contain moderate quantities of nutrients and are moderately productive in terms of aquatic animal and plant life.

Meso zooplankton—group of zooplankters that can pass through a plankton net with a mesh size of 202 micrometers.

Metabolic inhibitor—chemical inhibitor of the metabolic reactions of the catalyzed by enzymes.

Metabolize—conversion of food, for instance soluble organic matter, to cellular matter and gaseous byproducts through a biological process.

Metamorphic rock—rock that undergone changes caused by pressure, temperature, sheering stress, or water to result in a more compact and highly crystalline form; gneiss, schist, and marble are examples.

Metamorphic stage—refers to the stage of development that an organism exhibits during its transformation from an immature form to an adult form. This development process exists for most insects, and the degree of difference from the immature stage to an adult form varies from relatively slight to pronounced, with many intermediates. Examples of metamorphic stages of insects are egg-larva-adult or egg-nymph-adult.

Meteorology—study of the Earth's atmosphere, including its movements and other phenomena, especially as they relate to weather forecasting.

Methane—a hydrocarbon that is a greenhouse gas with a global warming potential most recently estimated at 24.5. Methane (CH_4) is produced through anaerobic (without oxygen) decomposition of waste in landfills, animal digestion, decomposition of animal wastes, production and distribution of natural gas and oil, coal production and incomplete fossil fuel combustion; It is colorless, odorless, flammable gas consisting of the hydrocarbons (CH_4).

Methyl bromide—is a soil disinfectant in the form of a gas that is applied through the drip irrigation system.

Micro biocide—a substance that destroys microorganisms.

Micro biota—organisms that are less than 63 micrometers in size, e.g., bacteria.

Micro climate—the unique environmental conditions in a particular area of the landscape. Factors include amount of sunlight or shade, soil type, slope and wind; Atmospheric conditions within or near a crop canopy.

Micro climate factor—factor or coefficient used to adjust reference evapotranspiration to reflect the micro climate of an area.

Micro filtration—the separation or removal from a liquid of particles and microorganisms in the size range of 0.1 to 2.0 microns in diameter.

Micro filtration system—it serves full automatic solid/liquid separation.

Micro grams per grams—a unit expressing the concentration of a chemical element as the mass (micrograms) of the element sorbet per unit mass (gram) of sediment.

Micro grams per liter—a unit expressing the concentration of chemical constituents in solution as mass (micrograms) of solute per unit volume (liter) of water. One thousand micrograms per liter is equivalent to one milligram per liter [ppb, µg/L].

Micro invertebrates—small animals (without backbones) that wilt pass through a U.S. Standard #30sieve (0.595 millimeter mesh opening).

Micro irrigation—the frequent application of small quantities of water directly above or below the soil surface; usually as discrete drops, continuous drops, tiny streams, or microspray; through emitters or applicators placed along a water delivery line. Microirrigation encompasses a number of methods or concepts; such as drip, subsurface, bubbler, and microspray irrigation.

Micro irrigation system—the physical components required to apply water through micro irrigation. System components that may be required include the pumping station, control and water treatment station, main and sub main lines, manifold lines, lateral lines, filtration, emitters, valves, fittings, and other necessary items.

Micro nutrient—a trace element.

Micro organism—a living organism invisible or barely visible to the naked eye and generally observed only through a microscope. Also called a microbe. Microorganisms are generally considered to include algae, bacteria, fungi, protozoa and viruses. Some microorganisms cause acute health problems when consumed in drinking water.

Micro organism—Organism that is so small that they can only be observed through a microscope, for instance bacteria, fungi or yeasts.

Micro pores—are smallest pores of the soil.

Micro spray—a low pressure sprayer device generally placed on a stake that is designed to wet soil with a fan or jet of water.

Micro straining—the filtrating of a fluid through a specially created media designed to remove essentially all of the fluid's suspended solids and colloidal material.

Micro tubes (spaghetti tubes)—the 1/4 inch flexible pipe used to link emitters or sprayers to the mainline. Plastic stakes are frequently used to hold the tubes and the dispensing device attached to the end in place.

Micro watt-seconds per square centimeter—A unit of measurement of intensity and retention or contact time in the operation of ultra-violet systems.

Micro zooplankton—group of zooplankters that can pass through a plankton net with a mesh size of 202 micrometers.

Microbial growth—the multiplication of microorganisms such as bacteria, algae, diatoms, plankton, and fungi.

Micron—a unit to describe a measure of length, equal to 1 millionth of a meter.

Micron rating—a measurement applied to filters or filter media to indicate the particle size at which suspended solids above that size will be removed. As used in the water treatment industry standards, this may be an absolute rating or a nominal rating; A metric unit of length equal to 1 millionth of a meter or one thousandth of a millimeter or about 0.00003937 inches. The symbol for micron is the Greek letter μ.

Mid—depth habitat—term used for a region where waters are approximately 3 to 30 feet in depth; also called shoal habitat.

Mid—ocean Ridge—a ridge of volcanic mountains on the ocean floor, associated with seafloor spreading of crustal plates.

Mid—water trawl—any of a variety of nets towed through the water column to capture pelagic fish.

Milli equivalents per liter—a measure of the concentration of a solute in solution; obtained by dividing the concentration in milligrams per liter by equivalent weight of the ion.

Milli grams per liter (MG/L, mg/L)—a unit expressing the concentration of chemical constituents in solution. Milligrams per liter represent the mass of solute per unit volume (liter) of water. Concentration of suspended sediment also is expressed in mg/L, and is based on the mass of sediment per liter of water-sediment mixture.

Mineral—an inorganic (nonliving) substance which occurs naturally in the earth and has a composition that can be expressed as a chemical formula and a set of characteristics (crystalline structure, hardness etc.) common to all minerals. Examples of minerals are sulfur, salt and stone.

Mineral free water—water produced by either distillation or deionization. This term is sometimes found on labels of bottled water as a substitute term for distiled or deionized water.

Mineral water—contains large amounts of dissolved minerals such as calcium, sodium, magnesium, and iron. Some tap waters contain as many or more minerals than some commercial mineral waters. There is no scientific evidence that either high or low mineral content water is beneficial to humans; Water which is naturally or artificially impregnated with mineral salts or gases (carbon dioxide). The term is also used to designate bottled water that contains no less than 250 ppm total dissolved solids (TDS) and originates from a protected ground water source.

Mineralogy—the study of minerals, including their formation, occurrence, properties, composition, and classification.

Minimum value—in a set of date, the measurement having the lowest numerical value.

MIPT—acronym for male iron pipe thread.

Miscibility—the ability of two liquids to mix.

Mist—liquid particles measuring 40 to 500 micrometers, are formed by condensation of vapor. By comparison, fog particles are smaller than 40 micrometers.

Mist irrigation—method of micro irrigation in which water is applied in very small droplets.

Mister—a device that delivers fog-like droplets of water often for cooling purposes.

Mitigation—a human intervention to reduce the sources or enhance the sinks of substances that pollute the environment, for example greenhouse gases.

Mixed bed—the intermix of two or more filter exchange products in the same vessel during a service run.

Mixed media—the use of two or more media products in a single filtration loose media bed where the products are intermixed rather than in stratified layers. For example, the intermix use of calcite and magnesia in pH modification.

Mixing zone—an area, contiguous to a discharge, in which dilution occurs such that there is a transition between effluent limitations and water quality standards.

Mixture—Various elements, compounds or both, that are mixed.

Moderately brackish— having a salt content between 5 and 18 parts per thousand.

Module—the membrane element and its housing in a reverse osmosis unit.

Moisture deficit, soil moisture depletion—difference between actual soil moisture and soil moisture held in the soil at field capacity.

Moisture equivalent—is a percentage of retained weight of water; saturated previously by a soil mass of 1 cm in thickness after it has been subjected to a centrifugal force of 1,000 times the gravitational force for 30 min.

Moisture meter—device that monitors or measures soil water content or tension.

Moisture percentage—is a ratio of the volume of soil moisture at field capacity in the root zone, to the total the potential root zone.

Moisture sensor—device which monitors the amount of water present in the soil and modifies the watering schedule accordingly.

Molality—a measure of chemical concentration. A one molal solution has one mole of solute dissolved in 1000 grams of water. One mole of a compound is its formula weight in grams.

Mole fraction—the ratio of the moles of one component of a system to the total moles of all components present. Typical values for long-lived greenhouse gases are in the order of mmol/mol (parts per million: ppm) or nmol/mol (parts per billion, ppb). Mole fraction differs from volume mixing ratio, often expressed in ppmv.

Molecule—the smallest particle of an element or compound that retains all of the characteristics of the element or compound. A molecule is made up of one or more atoms.

Monitoring light sensor—an indicator light, electrically or electronically activated, which is positioned in the effluent (product water) stream of a piece of water treatment equipment to detect and signal changes in the water quality which might malfunction of the equipment.

Montreal protocol—protocol on substances that deplete the ozone layer, signed in Montreal in 1987. It controls the consumption and production of chlorine- and bromine-containing chemicals that destroy stratospheric ozone, such as CFCs, methyl chloroform, carbon tetrachloride, and many others.

Morphology—the physical form or structure of plants or animals.

Most probable number (MPN)—a statistical index of the number of coliform bacteria in a given volume of sample as derived by a specific laboratory test. It is used for appraising the sanitary quality of water.

Motor efficiency—ratio of the power delivered to the pump by the motor to the input power supplied to the motor. A three phase electric motor is inherently more efficient, less expensive, and lasts longer than a single phase electric motor [%, decimal].

MPT—male nominal pipe threads

Mud flat—an unvegetated muddy region that is alternately exposed and covered by the tide.

Multiple use—harmonious use of the land for more than one purpose; not necessarily the combination of uses that will yield the highest economic return, e.g., a mix of residential and commercial developments in the same area.

Municipal discharge—discharge of effluent from wastewater treatment plants, which receive wastewater from households, commercial establishments, and industries in the coastal drainage basin.

Municipal sewage—liquid wastes, originating from a community. They may have been composed of domestic wastewaters or industrial discharges.

Municipal sludge—semi liquid residue that remains from the treatment of municipal water and wastewater.

Municipal water—water that has been processed at a central plant to make it potable or "safe to drink" and which is then distributed to homes and businesses via water mains. The term is a general one used to refer to the common source of water in most urban and suburban areas—as opposed to water obtained from separate proprietary sources such as private wells.

Mutagenic—capable of inducing mutation, which can produce permanent change in hereditary material, such as a physical change in the chromosomes.

Municipal water use—purposes usually served by water within a city, town, or village such as household and sanitary purposes, watering of lawns and gardens, and fire protection.

Mysid shrimp—invertebrate animals of the crustacean order Mysidacea that are shrimp- like in appearance and make up a significant component of the diet of many juvenile fish species.

Nano filtration—a membrane treatment process which falls between reverse osmosis and ultrafiltration on the filtration/ separation spectrum.

Nano plankton—group of phytoplankters that are between 5 and 60 micrometers in size.

National Geodetic Vertical Datum of 1929—a geodetic datum derived from a general adjustment of the first order level of both the United States and Canada. It was formerly called "Sea Level Datum of 1929" or "mean sea level" in U.S. Geological Survey reports.

National Park Service—an agency of the U.S. Department of the Interior charged with the planning and administration of all parks and monuments in the federal park system. The NPS is often referred to as the largest single employer of landscape architects in the United States.

National Water Initiative—the Intergovernmental Agreement on a National Water Initiative between the Commonwealth of Australia and the Governments of New South Wales, Victoria, Queensland, South Australia, Western Australia, Tasmania, the Australian Capital Territory and the Northern Territory (as amended from time to time).

Native bacteria—those which are indigenous to a natural body of water.

Natural resources—the elements of supply inherent to an area that can be used to satisfy human needs, including air, soil, water, native vegetation, minerals and wildlife.

Natural sparking water—carbonated water whose carbon dioxide content is from the same source as the water itself.

Natural trout water—water capable of supporting naturally sustaining trout populations.

Naturally soft water—ground surface, or rain water sufficiently free of calcium and magnesium salts so that no curd will form when soap is used and no calcium or magnesium based scale will form when the water is heated.

Nautical river mile—any of various units of distance used for sea and air navigation based on the length of a minute of arc of a great cycle of the earth and differing because the earth is not a perfect sphere.

Navigable water—a body of water that is deep and wide enough for a boat or other floating object to be transported from one place to another. Navigable water includes any body of water capable, in its natural state, of being navigated by floating vessels of any description for the purposes of transportation, recreation, or commerce; as well as any waterway where the public right to navigation exists by dedication of the waterway for public purposes, or by the public having acquired the right to navigate through long use.

Nekton—actively swimming aquatic organisms such as fish and mammals.

Nemagation—is an application of nematicides through the irrigation system.

Nematocysts—stinging cells of many coelenterates.

Nematode—is a multicellular, generally microscopic organism with a main physiological and circulatory systems. In general these are thin worms, are cylindrical, extended, some segmented externally with differentiations in the head and tail.

Nematodes—cylindrical worms of the group Nematoda that have no appendages and are parasitic or free-living, commonly called roundworms.

Net irrigation water requirement—quantity of water that is required for crop production, exclusive of effective precipitation.

Net plankton—group of phytoplankters that are larger than 60 micrometers.

Net positive suction head—head that causes liquid to flow through the suction piping and enter the eye of the pump impeller.

Net positive suction head available—pressure head that is supplied (is available) to the eye of an impeller in a pump based on system characteristics.

Net positive suction head required—minimum pressure head required at the eye of an impeller in a pump to prevent cavitation.

Net precipitation rate—measure of the amount of water that actually reaches the landscape. The net precipitation rate is the gross precipitation rate minus the losses that occur between the sprinkler and the landscape surface.

Net water price—the transfer price of water as agreed between legal entities exclusive of all applicable transaction costs.

Neutral (water chemistry)—the midpoint (neutral) reading of 7.0 on the pH scale, indicating that the solution (water) producing the neutral reading will produce neither an acid nor alkaline reaction. A 7.0 reading on the pH scale means that there are an equal number of free hydrogen (acidic) ions and hydroxide (basic) ions.

Neutralization—the addition of substances to neutralize water, so that it is neither acid, nor basic. Neutralization does not specifically mean a pH of 7.0, it just means the equivalent point of an acid-base reaction.

Neutralization—raising the pH of an acidic material or lowering the pH of an alkaline material to a nearly neutral pH level (7)

National park—a large, public park, often highly scenic and isolated belonging to and operated by the federal government.

Neutron dispersor or scatterer—is an equipment based on the registry of reduction of neutrons installed in an accessible tube in the soil. It consists of a neutron emitter source and a scale. A method to determine soil moisture content and the soil density by means of calibration curve.

Neutrons—uncharged building blocks of an atom that play a part in radio-activity. They can be found in the nucleus.

Nipple—the common plumbing term used in the irrigation trade for a short length of pipe, usually with male threads on both ends.

Nitrate—a natural nitrogen compound sometimes found in well or surface waters. In high concentrations, nitrates can be harmful to young infants.

Nitrification—a biological process, during which nitrifying bacteria convert toxic ammonia to less harmful nitrate. It is commonly used to remove nitrogen substances from wastewater, but in lakes and ponds it occurs naturally; the oxidation of ammonia-nitrogen to nitrate-nitrogen in wastewater by biological or chemical reactions.

Nitrate nitrogen (NO_3N)—the predominant form of inorganic nitrogen that supports algal production. Other inorganic nitrogen forms, such as nitrite nitrogen (NO_2N) and ammonia nitrogen (NH_3N) vary in abundance in streams. Sewage treatment plant effluents may contain high ammonia nitrogen concentration if nitrification is not a part of the treatment process.

Nitrogen—a colorless, tasteless, odorless, gaseous element that constitutes 78 percent of the atmosphere by volume and is a constituent of all living things. It is a main element of the plant growth and development of leaves. Most of nitrogen is absorbed during the vegetative growth period.

Nitrogen oxide—gas consisting of one molecule of nitrogen and varying numbers of oxygen molecules. Nitrogen oxide is produced in the emissions of vehicle exhausts and in power stations. In the atmosphere, nitrogen oxide can contribute to formation of photochemical ozone (smog) and to the greenhouse effect.

Nitrous oxide—a powerful greenhouse gas with a global warming potential of 320. Major sources of nitrous oxide (N_2O) include soil cultivation practices, fossil fuel combustion and biomass burning.

Nominal filter rating—filter rating indicating the approximate size particle, the majority of which will not pass through the filter. It is generally interpreted as meaning that 85% of the particles of the size equal to the nominal filter rating will be retained by the filter.

Nominal size—named size which is usually not the actual dimensions of the product, i.e., a half inch schedule 40 pipe is not 1/2 inch ID or OD.

Non consumptive Use—uses of in which but a small part of the water is lost to the atmosphere by evapotranspiration or by being combined with a manufactured product. Non consumptive uses return to the stream or the ground approximately the same amount as diverted or used. A use of water in which all of the water used is directly returned to the source from which it came. For example, water used in the production of hydroelectricity is a nonconsumptive water use.

Non dedicated sampling equipment—equipment used to sample more than a sampling point.

Non degradable—resistant to decomposition or decay by biological means such as bacteria action or by chemical means such as oxidation, heat, sunlight or solvents.

Non pathogenic—not disease producing.

Non linearity—process without a simple proportional relation between cause and effect. The climate system contains many nonlinear processes, resulting in a system with very complex behavior which may lead to rapid climate change.

Non point source—a discharge which originates over a broad area, such as storm water runoff from forested, agricultural, and urban areas. Distinguished from Point Source.

Non point source pollution—contaminants that enter a water body from diffuse or undefined sources and are usually carried by runoff. Examples of nonpoint sources include agricultural land, coal mines, construction sites, roads, and urban areas. Because nonpoint sources are diffuse, they are often difficult to identify or locate precisely, and are therefore difficult to control.

Non potable—water that is unsafe or unpalatable to drink because it contains pollutants, contaminants, minerals or infective agents.

Non saline sodic soil—soil containing soluble salts that provide an electrical conductivity of saturation extract (ECe) less than 4.0 mmhos/cm and an exchangeable sodium percentage (ESP) greater than 15. Commonly called black alkali or slick spots.

Non volatile memory—a feature in irrigation controllers that will retain the programd information in electronic memory during a power failure, without the need for a battery.

Normal distribution—the character of data that follows the Gaussian distribution (bell) curve.

Normal flow filtration—the flow of the entire feed water stream in one direction directly through the filter media.

Not detectable—a term used in reporting test results to mean that the substance being tested cannot be detected by the equipment or method being used for this particular test. This term implies that it is possible that trace amounts may be present in quantities to small to be detected by the test equipment or method.

Nozzle—the fine orifice through which water passes from the sprinkler or emitter to the atmosphere. Sprinkler flow rate is largely determined by the nozzle size (orifice diameter) and pressure. In most cases the nozzle is removable so that it can be easily cleaned or replaced. With plastic nozzles replacement is generally preferred over cleaning as small scratches in the plastic can cause big problems with water distribution uniformity.

NTU—nephelometric turbidity unit.

Nucleus—the positively charges central part of an atom containing nearly all of the atomic mass a consisting of protons and neutrons (except in hydrogen which consist of one proton only).

Nuisance contaminant—constituents in water, which are not normally harmful to health but may cause offensive taste, odor, color, corrosion, foaming, or staining.

Nuisance species—those species that interfere with normal recreational or occupational uses of a body of water.

Number of Lt detection limit values—the number of times a chemical parameter is not detected by a given analytical procedure over a statistically significant period of time (e.g., one year).

Number of outlets—term used to describe the number of outlets in a lateral.

Nursery ground—area used by juvenile fish or shellfish during their development, such as striped bass in the Potomac estuary.

Nutrient—an element essential for plant or animal growth. Major plant nutrients include nitrogen, phosphorus, carbon, oxygen, sulfur, and potassium.

Nutrient cycle—the movement of a chemical between the living and nonliving worlds, such as carbon migrating from the atmosphere into the human body, then into the atmosphere again.

Nutrient loading—the input of nutrients into a body of water, which leads to increased primary productivity (usually implies excess input of nutrients).

Nutrient pollution—contamination of water resources by excessive inputs of nutrients. In surface waters, excess algal production is a major concern.

Oasis effect—effect of dry zones that surround the microclimate of relatively small soil extensions, where an air mass that move towards an irrigated area can provide sensible heat.

Observation well—a nonpumping well used to observe the elevation of the water table or the potentiometric surface. An observation well is generally of larger diameter than a piezometer and typically is screened or slotted throughout the thickness of the aquifer.

Obstruction—loss in pressure of the system due to particles or suspended solids are accumulated in lines, filters or drippers. The obstructions are a most serious problem in the drip irrigation system. Clogging can cause serious losses.

Ocean—the great body of salt water which occupies two-thirds of the surface of the Earth, or one of its major subdivisions.

Ocean trench—deep, narrow depressions in the Earth's oceans, usually formed by the subduction of one crustal plate under another.

Octanol water partition coefficient—a coefficient representing the ration of solubility of a compound in octanol to its solubility in water. As the octanol water partition coefficient increases, water solubility decreases.

Odor—a sensation resulting from adequate stimulation of the olfactory organ. It can be caused by a variety of materials, both natural and manmade. Odor tests are made by using the human sense of smell, usually by a panel of "testers."

Oligohaline—having a salt content between 0.5 and 5 parts per thousand.

Oligotrophic lake—Deep clear lake with few nutrients, little organic matter and a high dissolved oxygen level.

Omnivores—organisms that consume both animal and plant matter.

Onion shape—pattern of moisture caused by the drip irrigation. It simulates a half onion.

Open space—A relatively clear or forested area left untouched in or near a city.

Operating pressure—the pressure at which a system of sprinklers operates. Static pressure less pressure losses. Usually as measured at the base or nozzle of a sprinkler; the pressure at which a device or irrigation system is designed to operate. Can mean just about anything depending on usage. There can be "optimum operating pressure" "minimum operating pressure," "maximum operating pressure" and "operating pressure range"; The manufacture's specific range of pressure expressed in pounds per square inch (psi) or kPa within which a water processing device or water system is designed to function. A range of 30 to 100 psi is often indicated. Also called working pressure.

Operating temperature—the manufacturer's recommended feed water or inlet water temperature for a water treatment system.

Operational life—the designed or planned useful period during which a facility remains operational while continuing to be subject to permit conditions, including closure requirements. Operational life does not include postclosure activities.

Opportunity time—time that water inundates the soil surface with opportunity to infiltrate.

Organic—having the characteristics of or being derived from a living organism, plant and animal. Containing carbon (although a few very simple carbon compounds such as carbon oxides, the carbides, carbon disulfides and metallic carbonyls and carbonates are considered inorganic).

Organic—the molecules, cells, etc. in wastewater from living organisms based on elemental carbon.

organic carbon, Total (TOC)—the total amount of organic material in water; determined by combustion method and expressed in mg/L. Natural waters usually contain TOC less than 5 mg/L and greater values indicate organic contamination.

Organic compounds—chemical compounds containing carbon, usually of living origin.

Organic groups—this component refers to the reporting of the presence of organic groups such as phenols or the menthols, rather than of specific organic molecules, such as chloroform or DDT. Results are obtained from the application of analytic techniques such a mass spectrometry, NMR (Nuclear Magnetic Resonance) and IR (Infrared Spectroscopy).

Organic matter—substances consisting of, or derived from plant or animal matter, as opposed to inorganic matter which is derived from rocks, ore and minerals. Organic matter is characterized by its carbon hydrogen structure; The residual of plants and animals in several states of decomposition, alive organisms of the soil and substances synthesized by these organisms.

Organic polymers—drilling fluid additives comprised of long-chained heavy organic molecules. Drilling fluid additives are used to increase drilling rates and drilling fluid yields, thereby decreasing operational costs.

Organic vapor analyzer—a field monitoring device used to determine the concentrations of organic compound in air using flame ionization or photoionization detection systems.

Organically overloaded—a condition in which the poundage of organic wastes entering a sewage treatment plant is greater than that for which it is designed. As with hydraulic overloading, this often results in decreasing the operational efficiency of the plant.

Organism—any living entity, such as an insect, phytoplankter, or zooplankter.

Organism count area—refers to the number of organisms collected and enumerated in a sample and adjusted to the number per area habitat, usually square meters (m), acres, or hectares. Periphyton benthic organisms and macrophytes are expressed in these terms.

Organism count volume—refers to the number of organisms collected and enumerated in a sample and adjusted to the number per sample volume, usually milliliters (ml) or liters (I). Numbers of planktonic organisms can be expressed in these terms.

Orifice—opening with a closed perimeter through which water flows. Certain shapes of orifices are calibrated for use in measuring flow rates; Opening in system component such as pipe, tubing, or nozzle.

Ortho phosphate (O-PO$_4$)—the soluble form of phosphorus that is used by algae for growth and generally is associated with domestic sewage discharge, industrial waste and agricultural runoff. Ortho phosphate greater than.03 mg/L in lakes can support algal blooms.

Osmatic mechanism—is a pressure exerted on the living bodies due to unequal concentration of salts on both sides of a cellular wall or a membrane. Water will move from a low concentration of salts through the membrane towards the high concentration of salts, and therefore, exerts additional pressure on the side of membranes.

Osmosis—the natural tendency for water to spontaneously pass through a semipermeable membrane separating two solutions of different concentrations (strength). The water will naturally pass from the weaker (less concentrated) solution containing fewer particles of dissolved substance to the stronger (more concentrated) solution containing more particles of a dissolved substance. Thus natural osmosis causes the stronger solution to become more diluted and tends to equalize the strength of the solution on both sides of the membrane.

Osmotic potential—The equivalent force that a salinity differential can cause across a semipermeable membrane (plant root) [bars, kPa].

Osmotic pressure—the pressure and potential energy difference which exists between two solutions on either side of a semipermeable membrane because of the tendency of water to flow in osmosis.

Outfall—the place where a wastewater treatment plant discharges treated water into the environment.

Outlet—A discharge opening lower than the spillway crest designed to release reservoir water through or around a dam.

Outwash sand—stratified sediment (usually sand and gravel) removed from a glacier by melt water streams deposited beyond the active margin of a glacier.

Oven dry—drying of soil samples in an over for a sufficient period of time to reach a constant weight; Refers to soil samples that have been dried in an oven at 105°C for 24 h

Over burden—material that must be removed to gain access to an ore, particularly at a surface (open pit) mine.

Oxidation—occurs when a substance is exposed to air.

Overflow rate—one of the guidelines for design of the settling tanks and clarifiers in a treatment plant to determine if tanks and clarifiers are used enough.

Overland Flow—the flow of water over a land surface due to direct precipitation. Overland flow generally occurs when the precipitation rate exceeds the infiltration capacity of the soil and depression storage is full. Also called Horton overland flow.

Overlap—area which is watered by two or more sprinklers.

Oxidation—a chemical reaction in which ions are transferring electrons, to increase positive valence.

Oxidation pond—a man made body of water in which waste is consumed by bacteria.

oxidation process, Advanced—one of several combination oxidation processes. Advanced chemical oxidation processes use (chemical) oxidants to reduce COD/BOD levels, and to remove both organic and oxidizable inorganic components. The processes can completely oxidize organic materials to carbon dioxide and water, although it is often not necessary to operate the processes to this level of treatment.

Oxidation reduction potential—the electric potential required to transfer electrons from the oxidant to the reductant, used as a qualitative measure of the state of oxidation in water treatment systems.

Oxidizing acids—an acid (e.g., HNO_3) which tends to lose electrons in a reaction.

Oxidizing agent—a chemical substance that gains electron (is reduced) and brings about the oxidation of other substances in chemical oxidation and reduction (redox) reactions.

Oxygen depletion—the reduction of the dissolved oxygen level in a water body.

Ozonation—the process of feeding ozone into a water supply for the purpose of decolorization, deodorization, disinfectant or oxidation.

Ozone (O_3)—An unstable, poisonous allotrope of oxygen; O_3, that is formed naturally in the ozone layer from atmospheric oxygen by electric discharge or exposure to ultraviolet radiation, also produced in the lower atmosphere by the photochemical reaction of certain pollutants. It is a highly reactive oxidizing agent used to deodorize air, purify water, and treat industrial wastes. It also is a greenhouse gas.

Ozone generator—a device that generates ozone by passing a voltage through a chamber that contains oxygen. It is often used as a disinfection system.

Pan coefficient—factor to relate actual evapotranspiration of a crop to the rate water evaporates from a free water surface in a shallow pan. The coefficient usually changes by crop growth stage.

panel, Central control—the central panel allows a complete control of all the field operations, sending instructions to the field panels, and obtaining continuous data on the operation of the irrigation system. Irrigation programr, a unit for the transmission information consists of a unit, units standard for the control of water in lateral and a unit for danger signs.

panel, Field—in computerized irrigation system with a series of field controls the system manually or by remote control. In the completely automated system, the signals are transmitted to individual points in the field; the field panel also collects information (value of water pressure, etc.) and transmits to the principal panel.

Parameter—a variable, measurable property whose value is a determinant of the characteristics of a system such as water. Temperature, pressure, and density are examples of parameters.

Parameterization—a rough approximation of current climate processes on average effects of the larger-scale variables in climate models. This is done when processes cannot be modeled explicitly, for example when they are much smaller than the computer grid.

Parasite—it is an organism that obtains its food while living in or on the body of other animals or plants.

Park way—a road laid through a garden or park-like landscape, usually with median and roadside plantings.

Partial pressure—that pressure of a gas in a liquid, which is in equilibrium with the solution. In a mixture of gases, the partial pressure of any one gas is the total pressure times the fraction of the gas in the mixture (by volume or number of molecules).

Partial record station—a particular site where limited streamflow and/or water-quality data are collected systematically over a period of years for use in hydrologic analyzes.

Particle—a very tiny, separate subdivision matter.

Particle filtration—filtration of particles in the size range of 2 microns or larger in diameter. Particle filtration is typically handled by cartridge filters and media filters.

Particle size—the diameter, in millimeters (mm), of suspended sediment or bed material determined by either sieve or sedimentation methods. Sedimentation methods (pipet, bottom- withdrawal tube, visual-accumulation tube) determine fall diameter of particles in either distiled water (chemically dispersed) or in native water (the river water at the time and point of sampling); As used in water industry standards, this term refers to the size expressed in microns, of a particle suspended in water as determined by the smallest dimension.

particles, Suspended—particles found suspended in water. These consist of soil particles of several sizes (sand, silt, and clay), soil materials washed in open channels and taken through the pumping units, eroded accumulated material in the deposits, and particles eroded from the surface of pipes and fittings.

Particulate loading—the mass of particulates per unit volume of water.

Parts per billion—expressed as ppb; a unit of concentration equivalent to the µg/l.

Parts per million (ppm)—a measure of proportion by weight which is equivalent to one unit of weight of solute (dissolved substance) per million weights of solution. Since one liter of water weighs 1 million milligrams, one ppm is equal to one milligram per liter (mg/L). PPM is the preferred unit of measure in water or wastewater analysis.

Parts per thousand (ppt)—parts per thousand parts of water (by weight), term generally used for salinity measurements as parts salt per thousand part water.

Pasteurization—the elimination of microorganisms by heat applies for a certain period of time.

Patent tongs—hydraulically operated grab-like apparatus operated from a boom on a boat to harvest oysters.

Pathogen—A disease-causing biological agent such as a bacterium, parasite, virus or fungus. Examples in wastewater include Salmonella, Vibro Cholera, and Entamoeba histolytica.

PCB (Polychlorinated Biphenyls)—a class of industrial compounds that are toxic environmental pollutants and tend to accumulate in animal tissue.

Peak available capacity of sewer— the difference between the design capacity of a sewer and the peak flow which is calculated based on average daily flow.

Peak evapotranspiration—the maximum rate of daily evapotranspiration as determined from the average of the seven highest consecutive days expected to occur.

Peak irrigation water requirement—the water quantity needed to meet the peak evapotranspiration.

Peak use rate—maximum rate at which a crop uses water.

Peat land—permanent wetlands characterized by a bed made of highly organic soil (>50% combustible) composed of partially decayed plant material.

Pelagic—pertaining to organisms living in open waters.

Percent composition—a unit for expressing the ration of a particular part of a sample or population to the total sample or population, in terms of types, numbers, mass, or volume.

Percent recovery—the percentage of the feed water, which becomes product water. Determined by the number of gallons (or liters) of product water divided by the total gallons (or liters) of feed water and multiplied by100. The percent recovery is called recovery rate in reverse osmosis and ultra filtration.

Percent rejection—(reverse osmosis/ultra filtration)—the percentage of TDS in the feed water that is prevented from passing the membrane with the permeate. The formula used is: the difference obtained from the TDS in permeate divided by TDS in feed water; then multiply the answer obtained by 100 to obtain a percentage.

Percent saturation—the amount of a substance that is dissolved in a solution compared to the amount that could be dissolved in it.

Percentage fines—percentage of water volume falling in fine (< 1 mm in diameter) drops. Term also used relative to soil particle size.

percentage, Wetting—the percentage of the soil volume in the root zone which is wetted to field capacity or above (after irrigation) divided by the total potential root zone of the crop.

Perched water—ground water that occurs above the water table "perched" above a layer of unsaturated rock or soil.

Percolating water—Water that passes through rocks or soil under the force of gravity.

Percolation—beneficial deep percolation-leaching. It is a beneficial use when it leaches salts from the root zone to a level required for acceptable crop production; non beneficial (excess) deep percolation—If the actual depth of deep percolation at a given location is more than the required beneficial leaching depth, which is in excess of the requirement is nonbeneficial.

Percolation rate—rate at which water moves through porous media, such as soil.

Performance curve—graph showing the capability of a product with varying inputs, i.e., the dynamic head of a pump as it varies with discharge.

period, Growth—the time between the day of sowing, transplanting, and harvesting for a given crop.

Periodic chart—arrangement of elements in order of increasing atomic numbers, created by a scientist called Mendelejef.

Periphyton—the community of microorganisms that are attached to or live upon submerge surfaces.

Perma frost—perennially frozen ground, occurring wherever the temperature remains at or below 0°C for two or more years in a row.

Permanent irrigation—irrigation having underground piping with risers and sprinklers. Preferred term is stationary sprinklers.

Permanent wilting percentage—soil moisture at 15 atmosphere of tension. At this limit, water is not available to plants causing wilting.

Permanent wilting point—moisture content, on a dry weight basis, at which plants can no longer obtain sufficient moisture from the soil to satisfy water requirements. Plants will not fully recover when water is added to the crop root zone once permanent wilting point has been experienced. Classically at 15 atmospheres (15 bars), soil moisture tension is used to estimate PWP; Moisture content of the soil after the plant can no longer extract moisture at a sufficient rate for wilted leaves to recover overnight or when placed in a saturated environment. Also known as wilting percentage, wilting coefficient or wilting point.

Permeability—qualitatively, the ease with which gases, liquids, or plant roots penetrate or pass through a layer of soil; Quantitatively, the specific soil property designating the rate at which gases and liquids can flow through the soil or porous media.

Permeable—having pores or openings that permit liquids or gases to pass through.

Permeameter—a laboratory device used to measure the intrinsic permeability and hydraulic conductivity of a soil or rock sample.

Permeate—that portion of the feed water, which passes through the membrane to become product water.

Persistence—refers to the length of time a compound stays in the environment, once introduced.

Pesticide—any chemical compound used to control unwanted species that attack crops, animals, or people. This diverse group of chemicals includes herbicides, fungicides, and insecticides.

Petrographic analysis—systematic description and classification of rocks.

pF—logarithm of the soil moisture tension in centimeters of a column.

pH (potential of hydrogen)—a measure of the acid or base quality of water that is the negative log of the hydrogen ion concentration. A measure of the acidity or alkalinity of a solution; the pH scale ranges from 0–14, with 7 representing neutral solutions; a solution with a pH greater than seven is described as alkaline, and one with a pH below seven is called acidic; vinegar is an example of an acid, while household bleach is an alkaline solution

Phosphate—a salt of phosphoric acid. In the water treatment industry, poly phosphates are used a sequestering agents to control iron and hardness, and as a coating agent to control corrosion by formation of a thin passivating film on metal surfaces.

phosphorus, Total (TP)—all the phosphorus forms in the sample (mg/l as P) including ortho-phosphate, poly phosphate and organic phosphorus. Total phosphorus concentrations greater than.1 mg/l in streams can support algal blooms.

Photosynthesis—the process in green plants and certain other organisms by which carbohydrates are synthesized from carbon dioxide and water using light as an energy source. Most forms of photosynthesis release oxygen as a byproduct.

Photosynthesis—The process of conversion of water and carbon dioxide to carbohydrates. It takes place in the presence of chlorophyll and is activated by sunlight. During the process oxygen is released. Only plants and a limited number of microorganisms can perform photosynthesis.

Phylogenetic group—a group of organisms defined by evolutionary origin; also called taxonomic category or taxon; categories from least to most specific include phylum (for animals) or division (for plants), class, order, family, genus, species, subspecies, variety, and form.

Physical agent—accumulation of sand, clay, and organic matter that can obstruct lines and drippers.

Physical and chemical treatment—Processes generally used in wastewater treatment facilities. Physical processes are for instance filtration. Chemical treatment can be coagulation, chlorination, or ozone treatment.

Physical weathering—Breaking down of rock into bits and pieces by exposure to temperature and changes and the physical action of moving ice and water, growing roots, and human activities such as farming and construction.

Phytoplankton—free-floating, mostly microscopic aquatic plants.

Piezometer—a non pumping well, generally of small diameter, that is used to measure the elevation of the water table or potentiometric surface. A piezometer generally has short well screen through which water can enter.

Pilot tests—the testing of a cleanup technology under actual site conditions in a laboratory in order to identify potential problems before implementation.

Pipe dope—common name for commercial products used to apply to pipe fittings to assist in the appropriate fit of the threaded joints.

Pitot tube—a small ell shaped tube, which can be attached to a pressure gauge or other measuring device to measure the velocity head of water discharging from a nozzle.

Planned unit development (PUD)—in zoning, a housing or commercial development composed of individual units that are regulated as a whole.

Planning—the illustration and description of problem-statements and large-scale design solutions that affect extensive areas of land; the anticipation of problems that will be encountered as human use and development of land continues.

Plant available water—The available water located in the root zone. Calculated as the root zone depth multiplied by the available water capacity [%, decimal, in/ft, m/m].

Plant population—number of plants per unit of area of a crop.

Plasmolysis—is a separation of the cytoplasm of the cellular wall as a result of the osmotic water movement. The plasmolysis increases the concentration of salts. The osmotic movement of the water occurs from a cell towards a concentrated soil solution.

Plastic resin—is an adapted material to mold emitters (emitting or jets). It must be strong, resistant to the solar radiation and black color to avoid the growth of algae. The finished product must maintain its original dimensions under climatic variations, and must not alter the flow characteristics of an emitter.

Pleistocene glacial age—a geological epoch of the Quaternary period of the Cenozoic Era, lasting from 2.5 million years to about 10,000 years ago, during which there were four glacial and three interglacial periods.

Plow layer (compact soil layer)—is a plow layer or hard soil that does not permit penetration and growth of the roots and affects water infiltration through this layer.

Plug—an accessory that is threaded or is without threads. It is placed at the end of a pipe to stop flow of water.

Pollutant—a contaminant that negatively impacts the physical, chemical, or biological properties of the environment

Poly pipe—Polyethylene pipe is a black, flexible pipe popular for lateral line use in areas susceptible to long freezes in the winter. Soil moves as it freezes, and because poly pipe is flexible, it will hold up to this movement while the more rigid PVC pipe may break. An insert fitting with a hose clamp or a compression fitting is used with poly pipe.

Point of connection [POC]—location where irrigation system is connected to a (potable) water supply.

Point of entry (POE) treatment—full service water treatment at the inlet to an entire building or facility (outside faucets may be excepted from treatment).

Point of use (POU) treatment—water treatment at a single outlet or limited number of water outlets in a building, but for less than the whole building or facility. POU treatment is often used to treat water for drinking and cooking only.

Point source—a discharge with a definite outlet such as a pipe, tunnel on channel. Usually industrial and/or municipal discharges. Distinguished from NON-POINT SOURCE; A stationary location from which pollutants are discharged. It is a single identifiable source of pollution, such as a pipeline or a factory.

Point source pollution—pollution that originates from one, easily identifiable cause or location, such as a sewage treatment plant or feedlot.

Polar coordinates—the means by which the position of a point in a two dimensional plane is described; based upon the radial distance from the origin to the given point and an angle between a horizontal line passing through the origin and a line extending from the origin to the given point.

Polar substance—a substance that carries a positive or negative charge, for instance water.

Polishing—the final treatment process which involves the conversion of ammonia nitrogen into nitrate and nitrite nitrogen (referring to wastewater treatment).

Polishing filter—a filter installed for use after the primary water treatment stage to remove any trace of undesirable matter or to polish the water.

Pollutant—a contaminant in a concentration or amount that adversely alters the physical, chemical, or biological properties of the natural environment.

Pollutant load—the amount of pollutant entering a water body. Loads are usually expressed in terms of a weight and a time frame, such as kilograms per day (kg/d).

Pollution—the destruction or impairment of natural environment's purity by contaminants.

Poly ethylene—is an ethylene of the cured thermoplastic resin group (–CH2 CH2–). It is very resistant to chemical agents and any degree of flexibility or rigidity can be given during the manufacturing process. Each manufacturing step adds chemical agent to create desired properties. A plastic used for manufacturing irrigation tubing. "Poly" for short. Poly pipe is almost always black in color, sometimes with a strip of a different color for identification. It is very flexible, and is usually sold in coils of 100 feet or more of tube. Poly pipe is often used in areas where the ground freezes 12″ deep or more, and also in mountainous areas that are extremely rocky. Poly pipe uses insert type fittings where a barbed shank is shoved into the end of the tubing. These fittings must be clamped, the barbs alone will not hole the tube on the fitting. Exception-special barbed fittings made for sprinkler risers do not need to be clamped. Also see "PVC" for the most commonly used pipe material.

Poly ethylene (PE) pipe—flexible black pipe used in irrigation systems.

Poly haline—having a high salt content, between 18 and 30 parts per thousand.

Poly phosphate: A form of phosphate polymer consisting of a series of condensed phosphoric acids containing more than one atom of phosphorous. Polyphosphate is used as a sequestering agent to control iron and hardness, and as a coating agent that forms a thin passivating film on metal surfaces to control scale.

Poly propylene—it is plastic resin. It is used frequently to manufacture drippers. This plastic is a propylene polymer, a substance that is similar to polyethylene.

Poly vinyl chloride, PVC—is a vinyl chloride polymer. It is a thermoplastic resin ($-CH_2CH_1-$) resistant to ambient conditions. If it is exposed to sunlight for longer periods, it will lose the plastic properties gradually. Therefore, the PVC tubes should not be left on the soil surface but should be buried or installed below the soil surface.

Polymer—an organic molecule composed of many similar molecules, for example, the starch is composed of glucose units.

POP—[Persistent Organic Pollutants]: complex compounds that are very persistent and difficultly biologically degradable.

Pop up sprinkler head—a sprinkler head that retracts below ground level when it is not operating. Pop-up sprinklers which stick in the raised position are known as "lawn mower food."

Population—in general, a group of organisms of the same species occupying a particular habitat.

Pore—an opening in a membrane or medium that allows water to pass through.

Pore space—empty space between soil particles that is normally filled with either air and/or water.

Porosity—the quality of being porous, or full of pores or openings; a measure of the amount of space in a material or of the water storage capacity of a substance; A measure of the volume of pores in a material. Porosity is calculated as a ration of the interstices of material (e.g., the volume of spaces between the media particles in a filter bed) to the volume of its mass, and is expressed as a percentage.

Porous—full of pores through which water, light, etc. may pass.

Porous ceramic tip—consists of porous membrane cup with a vast number of small pores. It is closed at one end and other end is connected to tensiometer. It is installed in the root zone at the desired depth, to measure tension.

Porous pavement—an alternative to conventional pavement whereby runoff is diverted through a porous asphalt layer and into an underground stone reservoir. The stored runoff then gradually infiltrates into the subsoil.

Porous space—is the total space that is not occupied by soil particles in the total volume of soil.

Positive displacement pump—pump that moves a fixed quantity of fluid with each stroke or rotation, such as a piston or gear pump.

Potability—the quality of being suitable for drinking.

Potable—suitable for drinking.

Potable (drinking) water—a water supply which meets U.S. EPA and/or state water quality standards and is considered safe and fit for human consumption.

Potable Water—water meant for human consumption. Domestic or drinking water. Water that is provided by a waterworks system (private or municipal) and is used for drinking, cooking, dishwashing, or other domestic purposes requiring water that is suitable for human consumption. Water that is safe for drinking and cooking.

Potassium—the plant need high quantities of potassium that is supplied by the natural reserves in the soil or chemical or chemical fertilizers.

Potassium chloride (KCl)—a colorless potassium salt which can be used as a regenerate in cation exchange water softeners.

Potential—is an energy to take a unit weight of the object from a datum line to a desired location. Units are force x distance. It can also be measured in terms of potential head ($P=\gamma h$).

Potential—soil water potential: amount of work that must be done per unit quantity of pure water in order to transport reversibly and isothermally an infinitesimal quantity of water from a pool of pure water at a specified elevation at atmospheric pressure to the soil water at the point under consideration; *total potential:*

Sum of matric, pressure, solute and gravitational potentials; *matric potential:* attraction of the solid soil matrix for water; *pressure potential:* potential caused by water pressure; *solute or osmotic potential:* Potential caused by salinity; *gravitational potential:* Relative height of a point above or below a reference elevation.

potential application efficiency of Low Quarter, Water—low quarter application efficiency obtainable with a given irrigation system when the depth of irrigation water infiltrated in the quarter of the area receiving the least water equals some predetermined value of the soil moisture deficit.

potential of the soil moisture, Total—is a sum of all the contributing forces that act on the soil moisture. $Pt = Pg + Po$.

potential, Capillary—is a work that is required to move a unit mass of water against the capillary forces from the surface of the water to a specified location.

potential, Gravitational—is due to a gravity force on the free soil water.

potential, Osmatic (soil)—it is attributed to the salts dissolved in the soil solutions. The dissolved substances may be or may not be ionic, but the net effect to reduce the free energy of the water. The osmotic potential has little effect the movement of the soils water. It may affect the water absorbed by the plant roots. The membrane of the roots, which transmits water more water vapor movement because the water vapor pressure reduces the presence of un-dissolved salts.

potential, Pressure—includes the effect that may increase or decrease the pressure in the free energy of the soil water.

potential, Soil water—is a result of absorption and capillary forces. The net effect of these two forces is to reduce the free energy of the soil water. It affects the soil moisture retention during the soil water movement.

Potentiation—the ability of one chemical to increase the effect of another chemical.

Potentiometric surface (in reference to artisan or confined aquifers, in which water is under pressure)—an imaginary surface defined by the level to which ground water will rise if released from its confining bed by a well or conduit. This surface, signifying the height of water in confined aquifers, is similar to the water table that denotes the height of water in unconfined aquifers.

Potomac estuary.

POU—Point of use.

Pound net—a type of stationary fishing gear with a long net attached to stakes (a wall), which directs the fish through a maze and into a trap or pocket.

Pounds per square inch (psi)—unit of measure for expressing pressure.

Power—rate of doing work.

ppb—parts per billion.

ppm—parts per million.

ppt—parts per trillion.

Precipitate—an insoluble reaction product in an aqueous chemical reaction.

Precipitation—total amount of precipitation (rainfall, drizzle, snow, ice, hail, fog, and condensation), expressed in depth of water that could cover a horizontal plane in absence of runoffs infiltration or evapotranspiration (mm/day).

Precipitation (For agricultural water purposes)—the natural deposition on the earth's surface of water in the forms of rain, sleet, hail, snow, or mist (fog). In humid areas where freezing temperatures do not generally occur, precipitation is used synonymously with rainfall. The noun form of the term precipitation is generally associated with the depth or volume of water that fell over a unit area.

Precipitation of chemical compounds—deposit of solid compounds in the lines and emitters as a result of chemical reactions or chemigation.

Precipitation process—the altering of dissolved compounds to insoluble or badly soluble compounds, in order to be able to remove the compounds by means of filtration.

Precipitation rate—the speed at which a sprinkler or an irrigation system applies water. The common units of measure for precipitation rates are inches per hour or millimeters per hour.

Precursors—atmospheric compounds that are no greenhouse gases but can enhance greenhouse gas in-duced processes by contributing to physical or chemical processes regulating greenhouse gas production or destruction rates.

Predation—process of feeding by which an animal preys on another animal.

Preemergent—a chemical agent (fertilizers or herbicides or pesticides) that is applied before the emer-gency of the weeds or crops.

Pressure—the force of water, measure in PSI or foot head.

Program—the watering schedule of an automatic sprinkler.

precipitation rate, Lowest—lowest precipitation rate in a defined contiguous area.

Pressure—the force per unit area measured. (In the United States, usually expressed in pounds per square inch. Its metric equivalent is Bars.) Insufficient water pressure can result in poor sprinkler coverage, while excessively high water pressure may cause misting and fogging leading to water waste. This is the pres-sure that moves water through pipes, sprinklers and emitters. Static pressure is measured when no water is flowing and dynamic pressure is measured when water is flowing. Pressure and flow are affected by each other [bars, psi, kPa].

Pressure at the head—when manufacturers discuss pressure for sprinkler use or nozzle performance, they are referring to the dynamic pressure as measured at the base of the sprinkler head.

Pressure rating—the estimated maximum internal pressure that can be continuously exerted in a pipe or container with a high degree of certainty that it will not fail.

Pressure compensating emitter—an emitter designed to maintain a constant output (flow) over a wide range of operating pressures and elevations.

Pressure controller—is used in pressure irrigation systems such as sprinkler or trickle. It regulates pressure in the system. It is also called pressure regulator valve or pressure-relief valve.

Pressure differential the difference in the pressure between two points is a water system. The difference may be due to the difference in elevation and/or to pressure drop resulting from water flow.

Pressure distribution—a system of small diameter pipes equally distributing effluent throughout a trench or bed.

Pressure drop—a decrease in the water pressure (in psi) which occurs as the water flows. The difference between the inlet and outlet water pressure during water flow through a water treatment device.

Pressure due to surge—water pressure caused due to changes in water velocity in a pipe system. Also referred to as **surge pressure**.

Pressure gauge—a device used to measure water pressure. The best pressure gauges are "liquid filled," however, most cheap gauges work good enough for irrigation use. If you do use a cheap gauge, don't leave it connected to the water pipe. The constant pressure will ruin it.

pressure, Gage—measures the pressure across the filters and the pressure in the secondary lines or at any location.

Pressure head—measurement of water pressure based on the water depth. Measurement is stated as "feet of head" or "meters of head." One foot of head is the pressure at the bottom of a 1 foot high column of water, which is also equal to 0.433 PSI. So it's really a measure of the weight of water of a given depth. It doesn't matter how much water is present, the pressure head is only determined by the depth of the water. The water pressure at the bottom of a 2″ diameter, 20 foot tall water filled pipe is the same as the water pressure at the bottom of a 20 foot deep lake.

Pressure loss—the amount of pressure lost as water flows through a system. Synonymous with Friction Loss. The term given for the loss of energy, in the form of pressure, that occurs whenever water moves through a pipe or any other piece of irrigation equipment. Pressure loss also occurs when water moves uphill against the force of gravity. If the total pressure loss in a piping system exceeds the available static water pressure the water will not flow. In landscape irrigation systems no flow means no grow.

Pressure manifold—three inch to eight inch pipe with larger orifices (1/2″–1″) through which effluent is distributed to gravity supply lines that in turn feed conventional gravity trenches.

Pressure membrane—is a permeable membrane to water and less permeable to the wet gasses: through which the water can escape from the soil sample in response to a pressure gradient.

Pressure rating—estimated maximum internal pressure that can be continuously exerted in a pipe or container with a high degree of certainty that it will not fail.

Pressure regulator—a device which maintains constant downstream operating pressure which is lower than the upstream operating pressure. Typical household water pressure is 50–60 psi while drip systems are designed to operate at around 20–30 psi depending on the manufacturer.

pressure, Root—pressure developed in the roots as a result of the osmosis. It is a mechanism to fill the guttation and exudation from the cut branches.

pressure, Saturated vapor—is a vapor pressure when the air is saturated at given temperature (millibar).

pressure sensitive emitter—an emitter that releases more water at the higher pressures and less at lower pressures found with long mainlines or terrain changes.

Pressure sewers—a system of pipes in which water, wastewater, or other liquid is pumped to a higher elevation.

Pressure tank—enclosed container attached to a water system usually containing an air pocket so that it behaves as a temporary water supply.

pressure, Turgor—pressure exerted in the liquid by the walls of the plant cells.

Pressure vacuum breaker [PVB]—backflow device configured with a spring loaded float and an independent spring loaded check valve. Check valve which is designed to close with the aid of a spring when flow stops. It also has an air inlet valve, which is designed to open when the internal pressure is one psi above atmospheric pressure so that no nonpotable liquid may be siphoned back into the potable water system. Being spring-loaded it does not rely upon gravity as does the atmospheric vacuum breaker. This assembly includes resilient seated shut-off valves and test cocks. The PVB must be installed at least 12 inches above all downstream piping and outlets. The PVB may be used to protect against a pollutant or contaminant, however, it may only be used to protect against back siphonage. It is not acceptable protection against backpressure.

pressure, Vapor—pressure exerted by the water vapor contained in the air (millibar).

Pre-treatment—processes used to reduce or eliminate wastewater pollutants from before they are discharged.

Primary consumers—plant-eating organisms that make up the second trophic level.

Primary producers—those organisms capable of synthesizing complex organic compounds from simple inorganic substances by photosynthesis (green plants) or by chemosynthesis, first trophic level.

Primary productivity—a measure of the rate at which radiant energy is converted into new organic matter and accumulated through photosynthetic and chemosynthetic activity of producer organisms (chiefly green plants). The rate of primary production is estimated by measuring the amount of oxygen released (oxygen method) or the amount of carbon assimilated by the plants (carbon method).

Primary treatment—the first major (sometimes only) treatment in a wastewater treatment works, usually sedimentation; removes a substantial amount of suspended matter, but little or no colloidal and dissolved matter.

Primary wastewater treatment—the removal of suspended, floating and precipitated solids from untreated wastewater.

Principal amount—total sum of money that is borrowed for the purchase of the irrigation system.

Process water—water that serves in any level of the manufacturing process of certain products.

Produced Water—water that is released with hydrocarbons (oil, gas, and crude bitumen) from an oil or gas well. Produced water is separated from the oil and gas and is measured and reported to the Energy Resources Conservation Board.

Product water—water that has been through the total treatment process and meets the quality standards required for the use to which the water will be used.

Production rate—the amount (gallons or liters) of product water the system produces per minute or (especially for reverse osmosis) per 24-hour period.

Profile (soil)—vertical section of the soil through all its horizons and extending into the C horizon.

Profile (sprinkler)—chart showing the application rates vs. distance of throw for a sprinkler head.

Program—the watering plan or schedule that tells the controller exactly when and how long to run each set of sprinklers. Many Rain Bird controllers offer multiple programs, which can be useful on sites where different plant groups have different irrigation needs.

Project efficiency—overall efficiency of irrigation water use in a project setting that accounts for all water uses and losses, such as crop ET, environmental control, salinity control, deep percolation, runoff, ditch and canal leakage, phreatophyte use, wetlands use, operational spills, and open water evaporation.

Propeller pump—pump which develops most of its head by the lifting action of vanes on the water. properties affect these factors.

Propylene—is a flammable gaseous hydrocarbon, usually obtained from petroleum and propane.

Protons—positively charged building blocks of an atom that are centered in the nucleus.

Prototype—an original water treatment equipment unit on which a specific equipment line is modeled.

Protozoa—large microorganisms, which consume bacteria.

Proxy climate indicator—variables that are an indirect measure of some combination of climate-related variations back in time. These are used to determine temperature in a time when the thermometer was not yet invented. Examples include tree ring records, characteristics of corals, fraction of melted ice, concentration of salts and acids and the load of pollen trapped in air bubbles.

psi—abbreviation for pounds per square inch.

Psycrometer—an instrument to measure the air humidity.

Public water system—a system that provides piped water for human consumption to at least 15 service connections or regularly serves 25 individuals.

Puddling—when water gathers in one location, such as at the base of a sprinkler or at a low spot on the site. Can be caused by low-head drainage, overirrigation, or slow soil infiltration.

Pump—mechanical device that converts mechanical forms of energy into hydraulic energy: *centrifugal pump* consisting of rotating vanes (impeller) enclosed in a housing and used to impart energy to a fluid through centrifugal force; *jockey pump*—Usually a small pump used to provide pressure and flow in a multipump system; *mixed flow pump* is a centrifugal pump in which the pressure is developed partly by centrifugal force and partly by the lifting action of the impellers on the water; *multistage* is a pump having more than one impeller mounted on a single shaft; *radial flow pump* is a centrifugal pump that uses diffuser vanes to transform the velocity head into pressure head. Commonly called a "turbine pump"; *submersible pump* is a pump where the motor and pump are submersed below the water surface; *trash pump* is a pump designed to pump large sized particulate matter in addition to liquid; *pto pump* is a pump driven by a separate power supply connected to the pump by a power takeoff (pto) drive; *vertical turbine* is a pump having one or more stages, each consisting of an impeller on a vertical shaft, surrounded by stationary and usually symmetrical guide vanes. Combines the energy-imparting characteristics of axial-flow and propeller pumps.

Pump chamber—a tank or compartment following the septic tank, which contains a pump, floats, and volume for storage of effluent. Effluent is pumped from the pump chamber to another pretreatment process or to the disposal component. In certain types of pressure distribution systems, this may also be called a "surge tank." If a siphon is used in lieu of a pump, this will be called a "siphon chamber."

Pump column—pipe through which water from well pumps (vertical turbine impellers) is conveyed to the ground surface (pump discharge head).

Pump efficiency—Ratio of the waterpower produced by the pump to the power delivered to the pump by the motor [%, decimal].

Pump house—is a small house that has permanent installation of pump, filtration system, controls, and other regulators.

Pump start circuit—the feature on automatic controllers which supplies 24 VAC, which can be used to activate a pump through an external pump start relay.

Pump start relay—low-amperage or electric switch designed for use with pump start circuits.

pump, Booster—a device to increase the water pressure is a system where some pressure already exists. For example, if water comes from a water company at 40 PSI of pressure but you need 80 PSI of pressure for the irrigation system, you would use a booster pump to increase the pressure.

pump, Centrifugal—consists of impellers. It can be vertical or horizontal.

pump, Fertigation—is used to inject the fertilizer solution into the irrigation system. These can be of two types: (a) The pump that is operated by an external power and (b) The proportional pump that is operated by the pressure of the irrigation water.

pump, Submersible turbine—is connected directly to an electrical motor. The pumping unit has analogous characteristics to the classic pumps for deep wells with the same pressures and capacities. The high flow rates at desired pressures are possible. These types of pumps are used for wells whose depth can exceed 400 meters. Pumps with are submersible motor are available.

pump, Vacuum (manually operated)—is used to remove the air from the stem of a tensiometer.

Pumping—the artificial removal of water from an aquifer.

Pumping cone—the area around a discharging well where the hydraulic head in the aquifer has been lowered by pumping. Also called cone of depression.

Pumping station—the pump or pumps that provides water and pressure to the system, together with all necessary equipments such as base, pump, screens, valves, motor controls, motor protection devices, fences, shelters, and backflow prevention device.

Pumping test—a test made by pumping a well for a period of time and observing the change in hydraulic head in the aquifer. A pumping test may be used to determine the capacity of the well and the hydraulic characteristics of the aquifer. Also called aquifer test.

Pure water—this term has no real meaning unless the word "pure" is defined by some standard such as pharmaceutical grade water.

Purified water—a USP grade water produced from water meeting U.S. EPA standards for potable drinking water which has microbiological content under control and is free from foreign substances.

Putrefaction—biological decomposition of organic matter; associated with anaerobic conditions.

PVC pipe—generally white in color, PVC (poly vinyl chloride) pipe is more rigid than black poly pipe, and requires the use of PVC solvents (glue) for connections. The pipe manufacturers also recommend the use of primer just prior to the application of the solvent to ensure a strong, watertight connection. PVC pipe can be used both as a mainline and as a lateral line.

Pycnocline—a layer of water that exhibits rapid change in density, analogous to thermocline.

Pyrogen—substance that is produces by bacteria and it fairly stable. It causes fever in mammals.

Qualification test—test and verifications performed to validate water treatment equipment conformance to a specific standard.

Qualified professional in Geology—a professional, by degree, experience, or certification, specializing in the study of the earth material science.

Qualitative water assessment—analysis of water used to describe the visible or esthetic characteristics of water.

Quality of irrigation water—refers to the purity of the water to the obstruction problems in the lateral and emitter.

Quantification limit—the lower limit to the range in which the concentration of a solute can be determined by a particular analytical method.

Quartz sleeve or quartz jacket—clear, pure fused quartz tube used to protect the high intensity ultraviolet lamps in ultraviolet systems. It usually retards less than 10% of the ultraviolet radiation dose.

Quick coupling valve—permanently installed valve which allows direct access to the irrigation mainline. A quick coupling key is used to open the valve.

Quicksilver water—a solution of mercury nitrate used in gilding.

Radial flow—the flow pattern in which water flows from the outside of a filter element to the center core.

Radiation (solar)—the process in which energy (as waves or particles) is emitted from the sun, transmitted through space, and absorbed by the Earth.

radiation, Extraterrestrial—amount of received radiation horizontally at the top of atmosphere.

radiation, Net—balance between all the radiation of long wave that enters and leaves (Rn = Rns + Rnt).

radiation, Net (long wave)—balance between the radiation of long wave that leaves and enters. (Negative)

radiation, Net solar—difference between the radiation of short wave received at the soil surface and the reflected radiation by the soil (surface of water or crop).

radiation, Solar—amount of radiation of short wave received in a horizontal plane at the soil surface.

Radiative forcing—A perturbation of the global radiation balance. This may be human-induced or natural.

Radical—a group of atoms acting as a single atom, which go through chemical reactions without being changed.

Radio active source—is a compound or an element that emits radioactive rays.

Radio chemical species—refers to the individual radioactive elements that produce radioactivity, such as radium 226, cobalt 60, strontium 90, and tritium.

Radio isotope—an isotope (or element) exhibiting radioactivity, an unstable condition in which an element is transformed into another element and radiation is given off.

Radio active—having the property of releasing radiation.

Radius of throw—the distance from the sprinkler head to the furthest point of water application.

Rain delay—lets the irrigation system off for a specific number of days without having to remember to turn it back on.

Rain gage—any instrument used for recording and/or measuring the amount of rainfall.

Rain shutoff device or rain sensor—a device which prevents the controller from activating the valves when a preset amount of rainfall is detected.

Rain Switch—a feature on many controllers, which allows the owner to stop watering during rainy periods without interrupting the program.

Rainfall—the quantity of water that falls as rain, only in a given period of time.

Rainfall intensity—rate of rainfall for the design storm frequency and for the time of concentration of the drainage area. (Used in rational method to compute runoff.)

rainfall, Effective—usable rainfall in agreement with the water requirements of the crop. This excludes deep percolation, run-off, and interception.

Range—the highest and lowest values in a set of data.

Rate of migration—the time a contaminant takes to travel from one stationary point to another. Generally expressed in units of time/distance.

Rated capacity (filtration or adsorption)—the manufacturer's statement regarding the expected number of days the equipment will be in service or the expected number of gallons of product water is delivered before backwash, rinse or replacement is needed.

Rated flow rate—the specified maximum and minimum flow rate at which a particular piece of water treatment equipment will continuously produce the desired quality of water.

Rated pressure drop—the expected pressure drop in psi as stated by the equipment manufacturer or obtained under test conditions.

Rational equation—equation used to predict the runoff from a watershed.

Ratoons—are plants that bloom repeatedly conserving all and freshness and greenish. All the meristem does not enter the reproductive phase at same time.

Raw sewage—untreated wastewater and its contents.

Raw water—water, usually from wells or surface sources, which has had no previous treatment and is entering the water processing system or device. The water at the inlet side of any water treatment device.

Reach—a group of river segments with similar biophysical characteristics. Most river reaches represent simple streams and rivers, while some reaches represent the shorelines of wide rivers, lakes and coastlines.

Reaction turbine—a type of water wheel in which water turns the blades of a motor, which then drives an electrical generator or other machine.

Readily available water—portion of available water that is more readily available for plant usage. It varies with plant type.

Reaeration—renewing air supplies in the lower layers of a reservoir in order to raise oxygen levels.

Recarbonization—process in which carbon dioxide is bubbled into treatment water in order to lower the pH.

Receiving waters—a river, lake, ocean, stream or other watercourse into which wastewater or treated effluent is discharged.

Recharge—the flow of water into the saturated zone; the return of water to an aquifer.

Recharge area—an area in which there are downward components of hydraulic head in the aquifer. Infiltration moves downward into the deeper parts as a function of the elevation of the water surface.

Recharge area—an area where rainwater soaks through the ground to reach an aquifer.

Recharge basin—a basin or pit excavated to provide a means of allowing water to soak into the ground at rates exceeding those that would occur naturally.

Recharge boundary—an aquifer system boundary that adds water to the aquifer. Streams and lakes are typically recharge boundaries.

Recharge well—a well specifically designed so that water can be pumped into an aquifer in order to recharge the ground-water reservoir.

Recirculation—recycling water after it is used. Often it has to pass a wastewater purification system before it can be reused.

Reclamation—any attempt to restore to beneficial use land that has lost its fertility and stability; most often applies to mining reclamation, such as the restoration of strip mines and quarries.

Reclaimed water—water that has been collected after an original use for reuse.

Recycled water—water that is used more than one time before it passes back into the natural hydrologic system; A type of reuse water typically run repeatedly through a closed system; Produced water that is reused for conventional water flooding or enhanced recovery steam injection, after its recovery (with hydrocarbons) from production wells. It is the total quantity of water injected at a project, minus the make-up water.

Redox—short-term for reduction/ oxidation reactions. Redox reactions are a series of reactions of substances in which electron transfer takes place. The substance that gains electrons is called oxidizing agent.

Reduced pressure principle assembly—consists of two internally loaded independently operating check valves and a mechanically independent, hydraulically dependent relief valve located between the check valves. This relief valve is designed to maintain a zone of reduced pressure between the two check valves at all times. The RP also contains tightly closing, resilient seated shut-off valves upstream and downstream of the check valves along with resilient seated test cocks. This assembly is used for the protection of the potable water supply from either pollutants or contaminants and may be used to protect against either back siphonage or backpressure.

Reducer—a fitting used to change from one size pipe to another. Two types are generally available. The first, and most common is the reducer bushing (abbreviated "red bush" or RB). The reducer bushing fits inside a coupling or other fitting on the large end. The pipe fits into the *red bush* on the small end. The other common reducer is a "bell reducer." The pipe fits inside the reducer on both ends of the bell reducer. It is also called a "bell" reducer.

Reduction—a chemical reaction in which ions gain electrons to reduce their positive valence.

Regeneration—putting the desired counterion back on the ion exchanger, by displacing an ion of higher affinity with one of lower affinity.

Regeneration (ion exchange, softening)—the use of a chemical solution (regenerant) to displace the contaminant ions deposited on the ion exchange resin during the service run and replace them with the kind of ions necessary to restore the capacity of the exchange medium for reuse.

Regional Administrator—the Regional Administrator of the appropriate regional Office of the Environmental Protection Agency, or the authorized representative.

Regulated unit—hazardous waste management unit. The number of regulated units will define the extent of the hazardous waste management area.

Regulated water resource—a water resource which has its flows controlled through the use of infrastructure to store and release water.

Regulation—the artificial manipulation of stream flow.

Reject water—a term used in distillation, reverse osmosis and ultrafiltration to describe that portion of the incoming feed water that has passed across the membrane but has not been converted to product water and is being sent to drain.

Rejection rate—in a reverse osmosis or ultrafiltration system, rejection rate is the quantity of feed water that does not pass through the membrane expressed as a percent of the total quantity of incoming fed water.

Relative humidity—ratio of the amount of water vapor present in the atmosphere to the amount required for saturation at the same dry bulb temperature.

Relative Sea Level (RSL)—sea level measured by a tide gauge with respect to the land upon which it is situated.

Relay—electrical or electronic device which uses a signal current to actuate a separate electrical circuit.

Reliability—the frequency with which water allocated under a water access entitlement is able to be supplied in full.

Relief storm water trunk—a pipeline designed to receive an excess volume of water, storm water, or wastewater. Its purpose is flood protection.

Remote control—device that can be used to activate electronic irrigation valves from a given distance away from the irrigation controller.

Remote control valve—a valve which is actuated by an automatic controller by electric or hydraulic means. Synonymous with automatic control valve.

Removable—capable of being taken away from a water treatment equipment unit using only simple tools such as a screw driver, pliers, or open ended wrench. Readily removable indicates capable of being taken away from a water treatment unit without the use of tools.

Reserve area—an area of land approved for the installation of a conforming system and dedicated for replacement of the onsite sewage system upon its failure.

Reserve capacity—extra treatment capacity built into wastewater treatment plants and sewers to be able to catch up with future flow increases due to population growth.

Reservoir—a man-made lake that collects and stores water for future use. During periods of low river flow, reservoirs can release additional flow if water is available.

Reservoir tank—same as storage tank on a reverse osmosis system.

Residential equipment—the term sometimes used to denote smaller sized water processing equipment which has been designed primarily for home use.

Residual—the amount of a specific material which remains in the water after the water has been through water treatment step.

Residual chlorine—chlorine allowed to remain in a treated water after a specified period of contact time and allowed to provide disinfection protection throughout the distribution system. The amount of residual chlorine is the difference between the total chlorine added and that consumed by the oxidizable matter.

Residual value—is a junk value after which the equipment is not useful.

Residue—the dry solids remaining after the evaporation of a sample of water or sludge.

Resin—used in the water processing industry, this term refers to ion exchange resin products which are usually specifically manufactured organic polymer beads used in softening and other ion exchange processes to remove dissolved salts from water.

Resistance—in electrical systems, it is the resistance to the flow of current. It can be compared to friction loss in an irrigation piping system. Resistance causes a drop in voltage along the length of a wire and is measured in ohms.

Resolution—the breaking of an emulsion into its individual components.

Respiration—the process whereby living organisms convert organic matter to CO_2, releasing energy and consuming oxygen.

Response time—The time needed for the climate system or its components to reequilibrate to a new state.

Restrictive layer—a stratum impeding the vertical movement of water, air, and growth of plant roots, such as hardpan, clay pan, fragipan, caliche, some compacted soils, bedrock, and unstructured clay soils.

Restoration—the renewing or repairing of a natural system so that its functions and qualities are comparable to those of its original, unaltered state.

Retardation—preferential retention of contaminant movement in the subsurface zone. Retention may be a result of adsorption processes or solubility differences.

Retention—contact time.

Retraction—operation when the pop-up riser of a sprinkler such as a spray head or rotor returns to the case in the ground. Also called pop-down.

Return flow—Water that has been diverted under the terms of a *Water Act* license for a specific purpose.

Reverse osmosis—a water treatment process that removes undesirable materials from water by using pressure to force the water molecules through a semipermeable membrane. This process is called "reverse" osmosis because the pressure forces the water to flow in the reverse direction (from the concentrated solution to the dilute solution) to the flow direction (from the dilute to the concentrated) in the process of natural osmosis. RO removes ionized salts, colloids and organic molecules down to a molecular weight of 100.

Reverse osmosis process—process uses a semipermeable membrane to separate and remove dissolved solids, organics, pyrogens, submicron colloidal matter, viruses, and bacteria from water. The process is called 'reverse' osmosis since it requires pressure to force pure water across a membrane, leaving the impurities behind.

Reynolds number—a number, defined by an equation that can be used to determine whether flow will be laminar or turbulent.

Riffle—a stretch of choppy water caused by a shallow extending across a stream bed.

Rill erosion—surface flow forms small, eroding channels called rills, detaching increasing amounts of soil from the sides and bottoms, and enlarging and joining with other rills as they move downslope.

Riparian—area of flowing streams that lies between the normal water line and some defined high water line; Pertaining to the banks of a body of water; riparian owner is the one who owns the banks; Riparian water right is the right to use and control water by virtue of ownership of the banks.

Riparian restoration—the replanting of the banks and floodplains of a stream with native forest and shrub species to stabilize eroding soils, improve both surface and ground water quality, increase stream shading, and enhance wildlife habitat.

Riparian right—a right (as access to or use of the shore, bed, and water) of one owning riparian land.

Riprap—a layer of stone, precast blocks, bags of concrete, or other suitable materials, generally placed on the upstream slopes of an embankment or along a watercourse as protection against wave action, erosion, or scour. Riprap is usually placed by dumping or other mechanical methods, but is occasionally hand placed.

Rip-Rapping—covering stream banks and dam faces with rock or other material to prevent erosion from water contact.

Riser—length of pipe which has male nominal pipe threads on each end and is usually affixed to a lateral or submain to support a sprinkler or antisiphon valve.

Risk—the uncertainty that surrounds future events and outcomes; the expression of the likelihood and impact of an event with the potential to influence the achievement of an organization's objectives.

Risk management—the process of identifying, analyzing, assessing, and evaluating risks; assigning ownership; taking actions to mitigate or anticipate them; and monitoring and reviewing progress.

Risk range—one talks about the variable amounts of water provided to the soil: the discharge of the drippers, the amount of water applied during an irrigation interval. It is possible to predict the effect of each one of these factors, because the soil

River basin—the total area drained by a river and its tributaries.

Root depth—preferred term *root zone*. *Effective Root Depth:* Depth from which roots extract water. The effective rooting depth is generally the depth from which the crop is currently capable of extracting soil water. However, it may also be expressed as the depth from which the crop can extract water when mature or the depth from which a future crop can extract soil water. Maximum effective root depth depends on the rooting capability of the plant, soil profile characteristics, and moisture levels in the soil profile.

Root zone—Also referred to as the root depth or effective rooting depth. The depth or volume of soil from which plants effectively extract water from. Typically this is defined as the depth of soil containing 80 percent of the plant roots. [ft, m]

Run time—length of time to operate an irrigation system or an individual zone.

Run time multiplier—multiplier used to compensate for the lack of perfect uniformity in a sprinkler system.

Rotors—gear-driven sprinklers that spray a solid stream of water and rotate slowly in a circular pattern.

R.P.A.—abbreviation for *reduced pressure assembly*.

Runoff—portion of precipitation, snow melt or irrigation, that flows over the soil, eventually making its way to surface water supplies; Surface water that leaves the subject region in liquid form.

Runoff in inches— shows the depth to which the drainage area would be covered if all the runoff for a given time period were uniformly distributed on it.

Runoff rate—rate at which water flows above ground from a watershed or field; Water, which is not absorbed by the soil and drains to another location. Runoff occurs when water is applied in excessive amounts or too quickly.

runoff, Direct—water that flows from the ground surface directly into streams, rivers, and lakes.

Rural water utility—an organization that supplies water for irrigation, town and community use in a rural area.

Safe water—water that does not contain harmful bacteria, toxic materials, or chemicals, and is considered safe for drinking.

Safe yield—the amount of naturally occurring ground-water that can be economically and legally withdrawn from an aquifer on a sustained basis without impairing the native ground- water quality or creating an undesirable effect such as environmental damage. It cannot exceed the increase in recharge or leakage from adjacent strata plus the reduction in discharge, which is due to the decline in head caused by pumping.

Sail index—cross-sectional area per unit volume of water of the drop comprising the sprinkler spray; Cross sectional area of spray (drops) per unit volume of water.

Saline sodic soil—soil containing both sufficient soluble salts and exchangeable sodium to interfere with the growth of most crops. The exchangeable sodium percentage (ESP) is greater than or equal to 15, and electrical conductivity of the saturation extract is greater than 4 mmhos/cm. It is difficult to leach because the clay colloids are dispersed.

Saline soil—nonsodic soil containing soluble salts in such quantities that they interfere with the growth of most crops; Soil that has sufficient soluble salts to interfere with crop growth; Non-sodic soil containing sufficient soluble salts to impair its productivity for growing most crops. The electrical conductivity of the saturation extract is greater than 4 mmhos/cm, and exchangeable sodium percentage (ESP) is less than 15; i.e., nonsodic. The principal ions are chloride, sulfate, small amounts of bicarbonate, and occasionally some nitrate. Sensitive plants are affected at half this salinity, and highly tolerant ones at about twice this salinity.

Saline water intrusion barrier well—used to inject water into freshwater, aquifers to prevent intrusion of salt water.

Salinity—refers to the amount of salts dissolved in soil water; Excess of soluble salts in the soil solution with sufficient concentration to harm the plants and to reduce the crop yield.

Salinity (saline)—concentration of salts in irrigation water. High concentrations can limit the crops ability to take up and use the water in the soil [dS/m, mg/l, ppm].

Salt balance—phenomenon of salt discharge equaling salt input in a given body of water.

Salt index—is defined as the degree of increase in the osmatic pressure of the soil solution due to a fertilizer, in comparison with the osmatic pressure created by an equivalent amount of $NaNO_3$. With an increase in the osmatic pressure of the soil solution, the ability of the plant to absorb reduces and the rate or plant growth and crop yield are reduced.

Salt water—the general term for all water over 1,000 ppm (mg/L) total dissolved solids.

Salts—amount of soluble salts in the soil, expressed in terms of percentage or parts by million, and so forth.

Sampling and analysis plan—a detailed document describing the procedure used to collect, handle and analyze ground-water samples for all quality control measures which will be implemented to ensure that sample collection, analysis, and data presentation activities meet the prescribed requirements.

Sand—soil particles ranging from 0.05 to 0.2 mm in diameter; Soil material containing 85% or more in this size range.

Sand filter—the oldest and most basic filtration process, which generally uses two grades of sand (coarse and fine) for turbidity removal or as a first stage-roughing filter or prefilter in more complex processing systems.

Sand filtration—sand filtration is a frequently used and very robust method to remove suspended solids from water. The filtration medium consists of a multiple layer of sand with a variety in size and specific gravity. Sand filters can be supplied in different sizes and materials both hand operated and fully automatically.

Sanitary sewer—a pipe that carries liquid and water-carried wastes from residences, commercial buildings, industrial plants, and institutions, together with varying quantities of ground water and storm water that are not admitted intentionally.

Saturated zone—the part of water-bearing material where voids are filled with water; the zone in which ground water occurs.

Saturation—the condition of a liquid when it has taken into solution the maximum possible quantity of a given substance.

Saturation (of soil)—condition where all soil pores/voids are filled with water.

Scale—a coating or precipitate deposited on surfaces such as kettles, water pipes or steam boilers that are in contact with hard water. Waters that contain carbonates or bicarbonates of calcium or magnesium are especially likely to cause scale when heated.

Scavenger—(water treatment) a polymer matrix or ion exchanger that is used specifically to remove organic species from the feed water before the water is to pass through the deionization.

Scenic easement—a legal means of protecting beautiful views and associated esthetic quality along a site by restricting change in existing features without government approval.

Sch 40, Sch 80—means "schedule 40, 80." Sometimes abbreviated "SCHED." A set of standards for pipe diameter and wall thickness used for both plastic and steel pipe. SCH 40 pipe is the standard type used for water pipes sized 1–1/2 and smaller. SCH 80 is a similar standard having thicker walls than SCH 40.

Schedule—method of specifying the dimensions and thus the allowable operating pressure of pipe. For a specific schedule rating, the wall thickness remains relatively constant for different pipe diameters, resulting in lower allowable operating pressures for larger diameter pipe.

Scheduling—procedure of establishing and implementing the time and amount of irrigation water to apply; Determining when to irrigate and how much water to apply, based upon measurements or estimates of soil moisture or crop water used by a plant.

Scheduling coefficient—the SC is a measure of uniformity of water distribution in any specific irrigated area as it relates to the precipitation rate of the entire area. Stated another way, the SC is an indication of the additional sprinkler run time necessary to compensate for dry areas. In a catchment test, the average precipitation rate for all catchments is divided by the one lowest precipitation rate to establish the SC of the system. A perfect SC of 1.0 states that all catchments in a zone fill to the same level.

Scheduling program—set of specifications identifying times to turn on and off water to various zones of an irrigation system.

Scintillometer—is a scientific device used to measure small fluctuations of the refractive index of air caused by variations in temperature, humidity, and pressure. It consists of an optical or radio wave transmitter and a receiver at both ends of an atmospheric propagation path. The receiver detects and evaluates the intensity fluctuations of the transmitted signal, called scintillation. The magnitude of the refractive index fluctuations is usually measured in terms of Cn^2, the structure constant of refractive index fluctuations, which is the spectral amplitude of refractive index fluctuations in the inertial subrange of turbulence. Some types of scintillometers, such as displaced-beam scintillometers, can also measure the inner scale of refractive index fluctuations, which is the smallest size of eddies in the inertial subrange. Scintillometers also allow measurements of the transfer of heat between the Earth's surface and the air above, called the sensible heat flux. Inner-scale scintillometers can also measure the dissipation rate of turbulent kinetic energy and the momentum flux.

Screen record—information about openings that permit water to enter a well, including perforations and incased sections of the aquifer.

Screening—use of screens to remove coarse floating and suspended solids from sewage.

SDI—Silt density index.

Sea (salt) water—the average salinity of seawater is 35 parts of salt by weight per 1,000 parts of water (equals 35 ppt). Brackish water has salinity of over 0.5 ppt.

Sea level alteration—a change in global average sea level brought about by volume changes in the world ocean. This may be caused by changes in water density or changes in the total mass of water.

Sea weed—macroscopic, multicellular species of benthic algae.

Secondary treatment—removal of biodegradable organics and suspended solids. Disinfection is also typically included in the definition of conventional secondary treatment.

Sediment—particulate organic and inorganic matter that originates mostly from disintegrated rocks and is transported by, suspended in, or deposited from water; it includes chemical and biochemical precipitates and decomposed organic material, such as humus. The quantity, characteristics, cause of the occurrence of sediment in streams are influenced by environmental factors. Some major factors are degree of slope, length of slope, soil characteristics, land usage, and quantity and intensity of precipitation.

Settling pond—a natural or artificial water body used to contain wastewater in order to enable solids to be removed from it before it is released to the natural environment.

Sewage—toilet wastes and gray water. *Grey water* gets its name from its cloudy appearance and from its status as being between fresh, potable water (known as "white water") and sewage water ("black water"). In a household context, gray water is the leftover water from baths, showers, hand basins and washing machines only. Some definitions of gray water include water from the kitchen sink. Any water containing human waste is considered black water.

Sedimentary rock—the part of water-bearing material where voids are filled with water;

Sedimentation—the process of subsidence and deposition of suspended matter carried by water, wastewater, or other liquids, by gravity. It is usually accomplished by reducing the velocity of the liquid below the point at which it can transport the suspended material. Also called settling.

Sediments—soil, sand, and minerals washed from land into water, usually after rain.

Seepage—the flow or movement of water through a dam, its foundation, or abutments.

Seepage velocity—the actual rate of movement of fluid particles through porous media.

Seismic prospecting—any of the various geophysical methods for characterising subsurface properties based on the analysis of elastic waves artificially generated at the surface (e.g., seismic reflection, seismic refraction).

Seismic refraction—a method of determining subsurface geophysical properties by measuring the length of time it takes for artificially generated seismic wave to pass through the ground.

Selectivity—the tendency of an ion exchanger to "prefer" (have more attraction for) certain kinds of ions over others, as if the resin were ranking the types of ions in order to be removed; most preferred ion, second most preferred, etc.

Self supplied industrial use—water supply developed by an individual industry or factory for its own use.

Semi confined aquifer—an aquifer partially confined by soil layers of low permeability through which recharge and discharge can still occur.

Semi permeable—a medium that allows water to pass through, but rejects dissolved solids, so that it can be used to separate solids from water.

Semiarid climate—climate characterized as neither entirely arid nor humid, but intermediate between the two conditions. A region is usually considered as semiarid when precipitation averages between 10 and 20 inches per year.

Separate sewer—a sewer system that carries only sanitary sewage; no storm-water runoff. When a sewer is constructed this way, wastewater treatment plants can be sized to treat sanitary wastes only and all of the water entering the plant receives complete treatment at all times.

Separation—the isolation of the various compounds in a mixture.

Septage—septic tank sludge that is a combination of raw primary sludge and an anaerobically produced raw sludge; Wastewater removed and hauled from a septic tank, holding tank, pit toilet, or similar system that receives only domestic wastewater. This does not include wastes from grease traps, industrial processes, commercial processes or agricultural processes.

Septic—producing or characterized by bacterial decomposition.

Septic system—a wastewater and sewage disposal system in which solids settle to the bottom of a tank and liquids are discharged to the soil through a series of lines in a drain field.

Septic tank—an on-site system used for domestic wastes when a sewer line is not available to carry them to a treatment plant. The wastes are piped to underground tanks directly from the home or homes. Bacteria in the waste decompose the organic matter and the sludge settles on the bottom of the tank. The effluent flows out of the tank into the ground through a system of drainage pipes. The sludge should be pumped out of the tanks at regular intervals.

Serial filtration—the arrangement of two or more filtering steps, one following the other, in order to remove increasingly finer particles at each stage and provide for filtration of all sizes of suspended solids.

Service connector—the pipe that carries tap water from the public water main to a building. In the past, these were often made of lead.

Service flow—the rate in U.S. gallons per minute (gpm) or liters per minute (L/min) at which a given water processing system can deliver product water. The rating may be intermittent peak flow or constant flow.

Service line—section of piping connecting larger municipal supply line with water meter.

Set time—elapsed time between the beginning and end of water application to an irrigation set; Amount of time required to apply a specific amount of water during one irrigation to a given area, typically refilling the plant root zone to field capacity minus expected rainfall.

Sets per line—the number of times each line must be moved [Sprinkler irrigation].

Setting—process by which pelagic larvae attach to substrates.

settle out and plug the emitter.

Settleable solids—suspended solids, in mL/L, that will settle out of suspension within a specified period of time.

Settling—the process of sinking of a substance sinking in water. This occurs when the substance does not dissolve in water and its density is larger than that of water.

Settling pond—an open *lagoon* into which wastewater contaminated with solid pollutants is placed and allowed to stand. The solid pollutants suspended in the water sink to the bottom of the lagoon and the liquid is allowed to overflow out of the enclosure.

Sewage—untreated wastes from toilets, baths, sinks, lavatories, laundries, and other plumbing fixtures in places of human habitation, employment, or recreation.

Sewage contamination—the introduction of untreated sewage into a water body.

Sewage disposal—the area and structures designed to contain Facilities and treat sewage.

Sewage sludge—sludge produced in a public sewer.

Sewage system (Onsite wastewater treatment facility)—an integrated arrangement of components for a residence, building, industrial establishment, or other places not connected to a public sewer system which: conveys, stores, treats, and/or provides subsurface soil treatment and disposal on the property where it originates or upon adjacent or nearby property; and includes piping, treatment devices, other accessories, and soil underlying the disposal component of the initial and reserve areas.

Sewer—a pipe or conduit that carries wastewater. Not to be confused with drain pipes that carry untreated runoff water to nearby waterways to reduce flooding of pavements.

sewer system, Conventional—system that was traditionally used to collect municipal wastewater in gravity sewers and convey it to a central primary or secondary treatment plant, before discharge on receiving surface waters.

Sewerage—The entire system of sewage collection, treatment, and disposal.

Shallow habitat—term used for a region where waters are less than three feet in depth.

shallow well—a well sunk in easily penetrated ground to a point which is below the water table but usually less than about 30 feet in depth.

Sheet erosion—when rain falls faster than the ground can absorb it, water collects and flows over the surface, carrying with it particles detached by splash erosion. The same occurs during periods of heavy snowmelt and irrigation.

Shrub sprinkler head—a sprinkler head mounted above ground level on a pipe, usually used for watering shrubs.

Shutoff head—pressure head on the outlet side of a pump at which the discharge drops to zero. Maximum pressure a pump will develop at a given speed.

Significant digits—the number of digits reported as the result of a calculation or measurement (exclusive of following zeroes).

Silt—mineral portion of soil have particle sizes ranging from 0.05 to 0.002 mm in diameter.

Silt density index—a test used to measure the level of suspended solids in feed water for membrane filtration systems.

Siltation—the deposition, in a water body, of sediments (e.g., sand and clay) that appear as tiny suspended particles

Single leg profile—precipitation rate profile of an individual sprinkler head operating at a known, constant pressure.

Single stage system—a filtering system that uses only a single filtering cartridge.

Sink—any process, activity or mechanism that removes a greenhouse gas, an aerosol or a precursor from the atmosphere. It is now possible to achieve Kyoto emission reduction goals by planting forests that take up a sufficient amount of carbon dioxide emissions in a country.

Sink hole—a depression formed by the natural collapse of the surface, usually in association with karst (limestone) topography.

Sinker—dense phase organic liquid which coalesces in an immiscible layer at the bottom of the saturated zone.

Siphon—closed conduit used to convey water across localized minor elevation raises in grade. It generally has end sections below the middle section. A vacuum pump is commonly used to remove air and keep the siphon primed. The upstream end must be under the water surface. Both ends must be under water, or the lower end must be closed to prime the siphon.

Siphon tube—relatively short, light-weight, curved tube used to convey water over ditch banks to irrigate furrows or borders.

Site plan—a dimensioned drawing indicating the form of an existing area and the physical objects existing in it and those to be built or installed upon it.

Slide gate—head control valve, which slides on rails, used to control drainage or irrigation water.

Slip configuration—a thread less connection which is solvent-welded.

Slip configuration or slip joint—connection without threads (of PVC pipe or fittings) which is solvent welded.

Slip—the term used to describe a solvent welded connection on a fitting. *Slip-thread ell* that has threads on one side and a glued socket on the other.

Slope effect—is a variation in the hydrostatic pressure in the main, submains, and laterals of irrigation system to a change in elevation in ground surface.

Slough (Marsh)—a marshy or reedy pool, pond, inlet, or backwater.

Sludge—the accumulated solids separated from liquids, such as water or wastewater, during processing, or deposits on bottoms of streams or other bodies of water. The precipitate resulting from chemical treatment, coagulation, or sedimentation of water or wastewater that settles to the bottom, or is skimmed from the surface in the sedimentation tanks and must be disposed of by appropriate means.

Slug—a temporary abnormally high concentration of an undesirable substance that shows up in the product water.

Slug test—a single well test to determine the in-situ hydraulic conductivity of an aquifer by the instantaneous addition or removal of a known quantity (slug) of water into or from a well, and the subsequent measurement of the resulting well recovery time.

Slurry—a thin watery mixture of a very fine insoluble substance.

Socket—a socket is a female connection on a fitting. It can be threaded, or glued, but most of the time the term is used for glued fittings.

Sodic soil—non saline soil containing sufficient exchangeable sodium to adversely affect crop production and soil structure.

Sodium (Na+)—a metallic element found abundantly in compounds in nature, but never existing alone.

Sodium adsorption ratio—portion of soluble sodium ions in relation to the soluble calcium and magnesium ions in the soil water extract.

sodium adsorption ratio, Adjusted—sodium adsorption ratio of a water adjusted for the precipitation or dissolution of CA^{2+} and Mg^{2+} that is expected to occur where a water reacts with alkaline earth carbonates with a soil.

Sodium chloride (NaCl)—the chemical name for common table salt.

Sodium percentage—percentage of total cations that is sodium in water or soil solution.

Soft scape—the natural elements with which landscape architects work, such as plant materials and the soil itself.

Soft water—any water that is not "hard." Water is considered to be hard when it contains a large amount of dissolved minerals, such as salts containing calcium or magnesium. You may be familiar with hard water that interferes with the lathering action of soap.

softened water—any water which has been processed in some manner to reduce the total hardness to 17.1 mg/L or ppm (1.0 grain per gallon) or less expressed as calcium carbonate equivalent.

Softening—the removal of calcium and magnesium from water to reduce hardness.

Soil—a natural substance of the intemperization of minerals and the decomposition of organic matter that covers the soil surface, in which plants grow.

Soil absorption capacity—in subsurface effluent disposal, the ability of the soil to absorb water.

soil absorption system, Subsurface—a system of trenches of three feet or less in width, or beds between three and ten feet in width, containing distribution pipe within a layer of clean gravel designed and installed in original, undisturbed soil for the purpose of receiving effluent and transmitting it into the soil. Also known as leach fields, drain fields, and absorption fields.

soil coverage, Complete—when the soil is 100% covered by crop or vegetative growth.

soil coverage, Effective—is a percentage of covered soil offered by the crop when the evapotranspiration is near its maximum value.

Soil flushing requirements—fraction of the irrigation water that enters the soil and that flows indeed to traverse and beyond of the roots zone by salinity. It is the minimum amount of water necessary for the control of soil salinity.

Soil horizon—layer of soil differing from adjacent genetically related layers in physical, chemical, and biological properties or characteristics; Layer of soil, usually approximately parallel to the soil surface, with distinct characteristics produced by soil forming processes.

Soil moisture (Soil Water)—water diffused in the soil immediately below the land surface (zone of aeration), from which water is discharged by transpiration in plants or by evaporation.

Soil moisture (water) depletion (deficit)—difference between field capacity and the actual soil moisture in the root zone soil at any given time. It is the amount of water required to bring the soil in the root zone to field capacity; Amount of water required to fill the plant root zone to field capacity.

Soil moisture content (as a depth)—for a specific layer of soil of a known depth (usually the root zone), the soil moisture can be expressed as a depth (soil moisture content as a proportion times the root depth). For example, if the soil moisture content is 0.35 and the *root depth* is 300 mm, then the soil moisture content as a depth is $0.35 \times 300 = 105$ mm.

Soil moisture content (as a proportion)—the volume of water per volume of soil. Maximum soil moisture occurs when the soil is saturated. When fully saturated the soil moisture is equal to the **porosity** of the soil.

Soil moisture deficit—also known as soil moisture depletion. The difference between field capacity and the actual soil water content in the root zone [%, decimal, in/ft, m/m].

Soil moisture depletion—the actual depletion of water at a particular time that reflects the amount of water to fill the plant root zone to field capacity.

Soil particles—soil particles of diameter between from 0.20 to 0.005 microns.

Soil permeability—a process with which the gases, liquids or plant roots penetrate or pass through a soil.

soil profile—vertical section of the soil from the surface through all its horizons into the parent material; A vertical cross section through a soil which shows any layers or horizons of which the soil is composed.

Soil reservoir capacity—preferred term plant available water.

soil sampler (auger)—is used for taking soil samples from a desired depth. It is necessary for the installation of tensiometer.

Soil series—lowest category of U.S. system of soil taxonomy. A conceptualized class of soil bodies having similar characteristics and arrangement in the soil profile.

soil structure—is an arrangement of soil particles to form aggregates, which are in a variety of recognized forms: large, small, and so forth.

soil texture—is a percentage of sand, silt, clay in a given soil class: (a) Sand: mineral particles with diameter of 0.05–1.0 millimeter; (b) The silt: mineral particles with diameter of 0.002–0.05 millimeter; (c) Clay: mineral particles smaller of diameter of <0.002 millimeter.

Soil type—the texture and structure of the soil particles which affects its ability to take in and store water for use by plants. Soils range from clay to loam to sand. Clay soils take in water more slowly than loam or sandy soils (lower infiltration rate).

Soil water (moisture)—all water stored in the soil.

soil water characteristic curve—soil-specific relationship between the soil-water matric potential and soil-water content.

Soil water content—amount of water in a given volume (or weight) of soil; Amount of water in a soil sample based on dry weight of the soil sample.

Soil water content (SWC)—the volume of water present in a unit volume of soil. SWC can be expressed as the ratio of the volume of water in a soil to the volume of the soil sampled. SWC is also referred to as soil moisture content.

Soil water deficit (SWD)—the amount of water that is depleted by a crop between irrigations that needs to be replenished during the forthcoming irrigation. SWD corresponds to the difference between field capacity (FC) and the soil water content (SWC) at the time of interest, or the amount of water that must be added to the root zone to return it to field capacity. SWD is generally expressed as a volume of water over an irrigated area (e.g., acre–inches/acre or inches).

Soil water tension—measure of the tenacity with which water is retained in the soil. It is the force per unit area that must be exerted to remove water from the soil.

soil, Alkaline—soil class that has high percentage of clay an excessive degree of saturation with free sodium, and with an non appreciable amount of soluble salts.

soil, Apparent density—is a ratio mass of soil to volume of soil particles including the pore space of the sample. It is expressed in grams per cubic centimeters.

soil, Cation exchange capacity—is a total amount of cations that a soil can adhere by cation exchange, usually expressed in milliequivalents in 100 grams of soil. The measurement of the cationic capacity depends on the method employed.

soil, Clayey—is applied to any class of ground that contains a high percent of clay. Soil with high fine particle content, particularly clay, is difficult to plow.

soil, Field capacity—amount of water retained by the soil after abundant irrigation or heavy rainfall; and when the rate of downward water movement has ceased; and the gravitational water has been eliminated. It is water content in the soil at a tension of 0.33 bar.

soil, Infiltration capacity—is a speed at which the water moves through the soil.

soil, Low water tension of the—in drip irrigation, the soil moisture content does not exceed the field capacity. Deficit implies that the matrix tension does not exceed 30–50 bars. Low moisture tension is difficult to maintain with any other irrigation system.

soil, Sandy—soil class with high percentage of sand and is easy to cultivate.

soil, Silty—soil class with 1–21% clay, 28–50% silt, and <52% of sand.

Solar effect—is determined by calculating the theoretical length of the day during the vegetative period of different crops.

Solder—a metallic compound used to seal the joints between pipes. Until recently, most solders contained about 50 percent lead.

Solenoid—is device on an automatic electric valve that regulates the flow of water into the upper chamber of the valve, permitting it to either open or close. It is a simple electromagnet.

Solenoid valve—an automatic valve operating under low voltage (24 V-AC), which may be remotely located in the landscape and controlled via a signal cable from the central controller (timer).

Solid set—refers to a stationary sprinkler system. Water-supply pipelines are generally fixed—usually below the soil surface—and sprinkler nozzles are elevated above the surface. In some cases, hand-move systems may be installed prior to the crop season and removed after harvest, effectively serving as solid set. Solid-set systems are commonly used in orchards and vineyards for frost protection and crop cooling. Solid-set systems are also widely used on turf and in landscaping.

Solidification—removal of wastewater from a waste or changing it chemically to make it less permeable and susceptible to transport by water.

Solubility—property with which a solid is dissolve in a liquid. The solubility can depend on the dissolvent, the dissolved substance and temperature.

Soluble—the relative capacity of a solid compound to dissolve in a given liquid.

Solute—any substance derived from the atmosphere, vegetation, soil, or rocks that is dissolved in water.

Solvent—Substance (usually liquid) capable of dissolving one or more other substances.

Solid waste disposal facilities—the area and associated structures designed to contain solid wastes.

Solubility—the quantity of material that dissolves in a given volume of water.

Source water—raw/untreated water received for treatment to provide potable water to municipal, industrial or private users. Sources may include high quality groundwater, groundwater under the influence of surface water and surface water from lake, stream, river or watercourse.

Spacing—*between sprinklers:* Distance from one sprinkler to the next one along a row; *between rows (of sprinklers):* Distance between adjacent rows of sprinklers; *between drains:* Distance between parallel subsurface drain lines; *equilateral triangular spacing:* Sprinklers are spaced in an equilateral triangular pattern; *rectangular spacing:* Sprinklers are spaced in a rectangular pattern; *row spacing:* Distance between the rows of sprinklers; *head to head spacing:* Sprinklers are spaced so that each head throws water to the adjacent heads; *triangular spacing:* Sprinklers are spaced in a triangular pattern.

Sparger—a device that introduces compressed air into a liquid.

Sparging—injection of air below the water table to strip dissolved volatile organic compounds and to facilitate aerobic biodegradation of organic compounds.

Spatial scale—climate may vary on a large range of spatial scales. Spatial scales can be local (less than 100,000 km^2), regional (100,000–10,000,000 km^2), or even continental (10–100,000,000 km^2).

specialist, Irrigation—the Certified Agricultural Irrigation Specialist is involved in the management and operation of on-farm irrigation systems. These systems include surface irrigation methods, as well as pressurized systems like micro irrigation and sprinklers. Prior to certification examination, specialists are required to take an Irrigation Association approved preparatory course.

Species factor—factor or coefficient used to adjust reference evapotranspiration to reflect plant species.

Specific conductance—method to estimate the dissolved solid content of a water supply by testing its conductivity.

Specific gravity—ratio of a substance's density (or specific weight) to that of some standard substance such as water. For liquids, the standard is water at sea level and 60°F with a specific gravity of one.

Specific gravity of the soil—is a ratio of the density of the soil divided by the density of water; ratio of weight of dry soil to soil volume.

Specific retention—the amount of water tightly bound by physical forces in an aquifer; the amount of water unavailable for use.

Specific source—an artificial conduit or other conveyance where pollutants are discharged

Specific speed—index of pump type related to impeller speed, discharge and total head.

Specific volume—volume occupied by a unit mass of fluid (Reference Manual).

Specific weight—weight per unit volume of a substance.

Specific yield—the amount of water in an aquifer available to wells; the quantity of water, which a unit volume of aquifer, after being saturated, will yield by gravity.

Spigot—a spigot is a male connection on a fitting. A spigot fits inside a socket. It can be threaded, or glued, but most of the time the term is used for glued fittings.

Spiked sample—a water sample to which a known quantity of a solute has been added so that the accuracy of the laboratory in analyzing the sample can be determined.

Spile—conduit, made of lath, pipe or hose, placed through ditch banks to transfer water from an irrigation ditch to a field.

Spill way capacity—the maximum flow a spillway is capable of discharging when the reservoir is at its highest water surface elevation.

Spiral wound—a very common construction configuration for one style of reverse osmosis membrane and cartridge filter element. In RO membranes, the membrane sheets are assembled in layers around a perforated mandrel product water tube with coarse mesh spacers screens between the layers, to form a complete module element. In cartridge filter elements, the filtration material such as fiber cord, is continually would around a perforated mandrel core tube.

Splash erosion—rain drops (or sprinkle irrigation) break to natural physical an chemical bonds between soil particles, moving them a short distance and making them more vulnerable to surface water flow.

Spore—a small reproductive body, often single celled, capable of reproducing the organism under favorable conditions. The spore is sometimes considered the resting stage of the organism. Among the organisms that may produce spores are algae, bacteria and certain protozoan. In water, most spores resist adverse conditions, which would readily destroy the parent organism.

Spray head—a type of fixed spray sprinkler that pops up from underground and waters a set pattern, usually from 4 to 15 feet in range. Used for lawns and shrubbery areas.

Spray heads—sprinklers that emit a fan-type spray of small droplets of water. These heads generally have a radius of 19′ and shorter.

Spray irrigation—application of water by a small spray or mist to the soil surface, where travel through the air becomes instrumental in the distribution of water. In some countries other than the US, sprinkler irrigation is called spray irrigation.

Spring—1. A place where the water flows from the ground. In most cases springs are the points at which the underground water table intercepts the earth's surface, and thus, water flows from the ground. Some springs are the outlets of underground streams or rivers that flow through channels in the ground. 2. The coiled metal device that retracts a sprinkler head or helps a hydraulic valve close.

Sprinkler—a hydraulically operated mechanical device which discharges water through a nozzle or nozzles.

Sprinkler distribution pattern—two dimensional water depth-distance relationship measured from a single (or multiple) sprinkler head(s).

Sprinkler (head)—device for distributing water under pressure; Nozzle or device, which may or may not rotate, for distributing water under pressure through the air; Hydraulically operated mechanical device which discharges water through a nozzle or nozzles. *Gear drive sprinkler:* Sprinkler containing gears as part of its rotational drive mechanism. *Impact drive:* Sprinkler which rotates using a weighted or spring-loaded arm which is propelled by the water stream and hits the sprinkler body, causing movement. *Rotor:* Sprinkler that rotates, but may more specifically refer to a gear driven sprinkler. *Spray head:* Sprinkler head that does not rotate. *Valve-in-head:* Sprinkler head having an integrally mounted valve.

Sprinkler irrigation—method of irrigation in which the water is sprayed, or sprinkled, through the air to the ground surface.

Sprinkler precipitation rate—precipitation rate of a group of heads used together and all having the same arc, spacing and flow.

Sprinkler spacing—an individual pipe section length.

Square spacing—the term given to a sprinkler head layout pattern where the sprinklers, when viewed from above, appear as more or less a square with one sprinkler in each corner. See also the more efficient "triangle spacing." Square spacing is the preferred sprinkler layout of nonhip people.

Stabilization triangle—space in de graph depicting greenhouse gas emissions (tones per year) over the past century, between the trend line and the line representing stability at current levels.

Stage (Complete)—the stage of a stream or lake is the height of the water surface above an established datum plane.

Stage discharge relation—the relation between gage height (stage) and volume of water per unit of time flowing in a channel.

Stake gill net—gill net held in position by long stakes or poles spaced along the length of the net.

Standard deviation—statistical term used to describe the distribution of values.

Standard deviation—the positive square root of the variance. The variance is the average of the squares of the differences between the actual measurements and the mean.

Standard dimension ratio—dimension Ratio is defined as the ratio of the diameter of a pipe to its wall thickness. Outside diameter is used for OD rated pipe while ID is used for ID rated pipe. Certain dimension ratios have been selected by convention and standards to be used for construction of pipe. These dimension ratios are referred to as standard dimension ratios.

Standing crop—the amount of organic material in a trophic level; usually expressed in terms of the number per unit area or in terms of biomass (i.e., the quantity of living matter); also called standing stock.

Start times—when you program a controller, you schedule the precise time you want to begin watering on water days. The start time is the time the first station in a program begins to water. All other stations in the program follow in sequence. Remember, start times usually apply to the entire program, not to the individual stations.

Static lift—vertical distance between water source and discharge water levels in a pump installation. Same as *total static head.*

Static pressure—the measurement of water pressure when the water is at rest in a closed system.

Static water pressure—the water pressure as measured when the water is not moving. The "not moving" part is critical, if the water is moving it isn't "static." When measuring static water pressure all the water outlets on the pipe must be closed. So if you're measuring the static pressure at a house you connect the pressure gauge, then take the reading while all the faucets, the ice maker, etc., are turned off. Static water pressure is a measure of the water's energy potential.

Station—circuit on a controller which has the ability to be programd with a run time unique and separate from other circuits and provides power to one or more remote control valves (or valve-in-head sprinklers).

station, Control —can include facilities to measure and to regulate the water flow, operation, filtration, chemigation, pressure, and other irrigation instruments.

station, Pump—a permanent small house that has a pump, engine or motor, filtration system, chemical injection equipment, main irrigation controller and all necessary fittings, valves, and accessories.

station, Selector—an indicating point that is used for two purposes: 1. A visual reference that demonstrates to the operator which field station is being irrigated during an automatic cycle of irrigation; 2. Manual selector that is manually operated to irrigate a desire field.

station, Tensiometer—a representative location in the field where the tensiometer is installed to obtain suction readings.

Stationary sprinklers—irrigation having underground piping with risers and sprinklers.

Statistical uniformity—an estimate of the uniformity of emitter discharge rates throughout an existing micro irrigation system.

Stilling basin—a pond or reservoir, riprapped or in a natural state, formed downstream of a dam, usually by means of a small auxiliary dam or weir. Its purpose is to protect the streambed from scouring caused by spillway and outlet discharges. The basin serves to dissipate energy.

Stoke's Law—a method to calculate the rate of fall of particles through a fluid, based on density, viscosity and particle size.

Stomate—is a small opening between the epidermic cells specialized in epidermis of the leaves and herbaceous stems, through which the exchange of gasses occurs.

Stop-a-matic® valve—a spring-loaded check valve used beneath a sprinkler to prevent low-head drainage. The check valve feature may also be built into the sprinkler.

Storage capacity—the maximum volume of water available for use from the water storage tank.

Storage coefficient—the volume of water released from storage in a vertical column of 1.0square foot, when the water table or piezometric surface declines 1 foot.

Storm sewer—a conduit for carrying off surplus precipitation from the surface of the land

Storm water—water discharged from a surface as a result of rainfall or snowfall.

Storm water drainage system—any structure for collecting, storing, or disposing of stormwater and the connections between them as outlined in the *Environmental Protection and Enhancement Act.* The system includes storm water sewers, pumping stations, storage areas, management facilities, treatment facilities, and outfall structures.

Stratification—the phenomenon occurring when a body of water becomes divide into distinguishable layers.

Stratigraphy—the science (study) of original succession and age of rock strata, also dealing with their form, distribution, lithologic composition, fossil content, and geophysical and geochemical properties. Stratigraphy encompasses unconsolidated materials (i.e., soils).

Stream—a body of water flowing in a natural channel as distinct from a canal.

Stream buffer—a variable width strip of vegetated land adjacent to a stream that is preserved from development activity to protect water quality, aquatic and terrestrial habitats.

Stream flow—the discharge that occurs in a natural channel. Although the term "discharge" can be applied to the flow of the canal, the word "stream flow" uniquely describes the discharge in a surface stream course. The term "stream flow" is more general than "runoff" as stream flow may be applied to discharge whether or not it is affected by diversion or regulation.

Stream gaging station—a gaging station where a record of discharge of a stream is obtained. Within the U.S. Geological Survey this term is used only for those gaging stations where a continuous record of discharge is obtained.

Stress irrigation—management of irrigation water to apply less than enough water to satisfy the soil water deficiency in the entire root zone. Preferred term is *limited irrigation* or *deficit irrigation*.

String wound element—a cartridge style element constructed by continuous spiral winding of natural or synthetic yarn around a preformed product water tube core and then building it up in layers to form a depth type filter.

Structural anomaly—an ecologic feature, especially in the subsurface, distinguished by geophysical, geological, or geochemical means, which is different from the general surroundings.

Structure (soil)—combination or arrangement of primary soil particles into secondary units or peds. The secondary units are characterized on the basis of size, shape, and grade (degree of distinctness); Combination of various soil particle types into a uniform mixture that behaves as a single unit. Structural classes include granular, blocky, and columnar massive, platy, prismatic, among others. Aggregation of primary soil particles, into units, which are separated from each other by surfaces of weakness. An individual natural soil aggregate is called a ped in contrast to a clod caused by disturbance, or a concretion caused by cementation.

Sub humid climate—climate characterized by moderate rainfall and moderate to high evaporation potential. A region is usually considered subhumid when precipitation averages more than 20 inches per year, but less than 40 inches per year.

Sub irrigation—application of irrigation water below the ground surface by raising the water table to within or near the root zone; Applying irrigation water below the ground surface either by raising the water table or by using a buried perforated or porous pipe system that discharges water directly into the plant root zone.

Sub main unit—independently controlled irrigation unit usually covering from 1 to 5 ha and including a submain manifold lateral lines and emitters.

Sub micron filter—a cartridge type membrane filter used in fine particle separation applications to remove particles of less than one micron in size.

Sub surface drip irrigation—application of water below the soil surface through emitters, with discharge rates generally in the same range as drip irrigation. The method of water application is different from and not to be confused with subirrigation where the root zone is irrigated by water table control.

Sublimation—The transitions of water directly from the solid state to the gaseous state, without passing through the liquid state.

Subsidence—the lowering of the elevation of the land surface due to withdrawal of subsurface fluids, resulting from the compaction of sediments composing the aquifer system.

Substrate—the physical surface upon which an organism lives.

Suction (static) lift—vertical distance between the elevation of the surface of the water source and the center of the pump impeller.

Sulfur (S)—a yellowish solid chemical element. "Sulfur" is also often used to refer to sulfur water.

Sulfur bacteria—Thio-Bacillus.

Sulfur water—water containing objectionable amounts of hydrogen sulfide gas, which causes an offensive "rotten egg" odor.

Sump—an excavation for the purpose of catching or storing liquids such as gray water; the water drains into the soil

Suspended solids—organic and inorganic particles, such as solids from wastewater, sand, and clay that are suspended and carried in water.

Sump pit—a single-chamber oil/grit separator used to pretreat runoff before it enters an infiltration trench.

Sunshine day light hours—number of daily sunshine hours. It is also defined as the duration or burns on a hydrograph of "Campbell Stokes" recorder.

Supply demand—water replaced independently to satisfy the necessity of irrigation needs.

Supply factor—the rate between the possible maximum supply and the real supply in cubic meters per second.

Supply requirement—the rate of the maximum requirement by daily provision during the period of maximum water use.

supply, Maximum irrigation—average water requirement during the highest crop water use, based on cropping patter, crop rotation, and climate.

Surface area—of a lake is the boundary of the lake and measured by a planimeter in acres. In localities not covered by topographic maps, the areas are computed from the best maps available at the time planimetered.

Surface filtration—filtration that occurs at the surface layer (as opposed to within the body depth) of the filter and is accomplished by passing the material to be filtered over a grating screen, sieve or membrane fabric with micro sized holes. The size of the holes in the filter determines what materials will pass through and what will filter out (held back).

Surface impoundment—a facility or part of a facility which is a natural topographic depression, man-made excavation, or diked area formed primarily of earthen materials (although it may be lined with man-made materials), which is designed to hold an accumulation of liquid wastes or wastes containing free liquids, and which is not an injection well. Examples of surface impoundments are holding, storage, settling, and aeration pits, ponds, and lagoons.

Surface tension—the elastic-like force in a body, especially a liquid, tending to minimize, or constrict, the area of the surface.

Surface water—water bodies such as lakes, ponds, wetlands, rivers, and streams, as well as groundwater with a direct and immediate hydrological connection to surface water (for example, water in a well beside a river).

Surface water telemetry—the surface water telemetry component identifies the type of system or equipment being used to transmit surface water information (primary stage data) from the data collection site to a central receiving site.

Surge pressure—water pressure caused by changes in water velocity in a pipe system.

Suspended—(as used in tables of chemical analyzes) refers to the amount (concentration) of the total concentration in a water-sediment mixture. The water-sediment mixture is associated with (or sobbed on) that material retained on a 0.45 micrometer filter.

Suspended growth process—biological treatment process in which the microorganisms responsible for the conversion of the organic matter or other constituents in the wastewater to gases and cell tissue are maintained in suspension within the liquid.

Suspended sediment—the sediment that at any given time is maintained in suspension by the upward components of turbulent currents or that exists in suspension as a colloid: *Suspended sediment concentration:* the velocity-weighted concentration of suspended sediment in the sampled zone (from the water surface to a point approximately 0.3 ft. above the bed) expressed as milligrams of dry sediment per liter water-sediment mixture (mg/l); *Suspended sediment discharge (tons/day):* the rate at which dry weight of sediment passes a section of a stream or is the quantity of sediment, as measured by dry weight or volume, that passes a section in a given time; *Suspended sediment load:* quantity of suspended sediment passing a section in a specified period; *Total sediment discharge (tons/day)*—the sum of the suspended sediment discharge and the bed-load discharge. It is the total quantity of sediment, as measured by dry weight or volume, that passes a section during a given time; *Mean concentration:* the time-weighted concentration of suspended-sediment passing a stream during a 24-hour day.

Suspended solids (SS)—the term "suspended solids" (colloidal and particulate matter such as clay, sand, and finely divided organic material, etc.) is synonymous with "suspended- sediment"; however, it is generally used by sanitary engineers in connection with water treatment facilities when referring to small particles of solid pollutants present in wastewater which resist separation from the water by treatment, while "suspended-sediment" is generally used by civil or hydraulic engineers in connection with sediment transport studies. Most of these materials settle out of the body of water and become bottom sediment as the water flow decreases. In sufficient quantities, these materials can clog the gills of fish and choke out plants, shellfish, and other life forms on the river bottom.

suspended, Recoverable—the amount of a given constituent that is in solution after the part of a representative water-suspended sediment sample that is retained on a 0.45 um membrane filter has been digested by a method (usually using a dilute acid solution) that results in dissolution of only readily soluble substances. Complete dissolution of all the particulate matter is not achieved by the digestion treatment and this determination represents something less that the "total" amount (that is, less than 95 percent) of the constituent present in the sample. To achieve comparability of analytical data, equivalent digestion procedures would be required of all laboratories performing such; analyzes because different digestion procedures are unlikely to produce different analytical results.

suspended, Total—the total amount of a given constituent in the part of representative water-suspended sediment sample that is retained on a 0.45 um membrane filter. This term is used only when the analytical procedure assures measurement of at least 95 percent of the constituent determined. Acknowledge of the expected form of the constituent in the sample, as well as the analytical methodology used, is required to determine when the results should be reported as "suspended, total."

Swale—a natural depression or wide shallow ditch used to temporarily store route, or filter runoff.

Swamp—wet land saturated with water, sometimes inundated with water; vegetation usually dominated by shrubs and trees.

Swing assembly—an assembly of flexible swing pipe and fittings that are used to connect a sprinkler to the lateral pipe. Allows you to easily adjust the sprinklers to grade level and also helps to prevent breakage due to force on the sprinkler.

Swing joint—a threaded connection of pipe and fittings between the pipe and sprinkler, which allows movement to be taken up in the threads rather than as a sheer force on the pipe. Also used to raise or lower sprinklers to a final grade without plumbing changes.

Synergism—The combined action of several chemicals, which produces a total effect greater than the effects of the chemicals separately.

Synthetic organic compounds—manufactured compounds of carbon chains or rings.

System—a complete integrated series consisting of various components and perhaps multiple water treatment processes, which can be tested, installed and operated as a single unit of equipment. For example, a single RO treatment system generally consists of two or more stages of media filtration plus cross flow membrane filtration and water storage.

Systemic—are compounds with a property to pass through epidermis of the leaves, stems, roots or seeds; and can go up it selves to the sap, which them transports it to all the parts of plant.

System capacity—ability of an irrigation system to deliver the net required rate and volume of water necessary to meet crop water needs plus any losses during the application process. Crop water needs can include soil moisture storage for later plant use, leaching of toxic elements from the soil, air temperature modification, crop quality, and other plant needs.

System precipitation rate—precipitation rate for a system is the average precipitation rate of all sprinklers in a given area regardless of the arc, spacing, or flow rate of each head.

System vulnerability—the degree to which a system is susceptible to, or unable to cope with, adverse effects of climate change, including climate variability and extremes. This is a function of the character, magnitude, and rate of climate change and variation to which a system is exposed, its sensitivity, and its adaptive capacity.

T test—a statistical method used to determine the significance of difference or change between sets of initial background and subsequent parameter values.

Tail water—water in a stream or canal, immediately downstream from a structure; Excess irrigation water which reaches the lower end of a field.

Taste threshold—the minimum concentration of a chemical or biological substance, which can just be tasted.

TCE—trichloroethylene.

TDS—Total dissolved solids. The weight per unit volume of water of suspended solids in a filter media after filtration or evaporation.

Tee—a tee is a fitting used to branch a side pipe off of a pipeline. A related fitting is the "Y" which is used primarily for sewer pipelines not sprinklers.

Temperature—a measure of the intensity aspect of heat energy present in a body of water.

temperature, Dew point—temperature at which the air is cooled off and is saturated of water and at which the water vapor begins to condense.

temperature, Wet bulb—temperature given by a thermometer whose bulb is surrounded by a fine cotton piece. This allows reducing the temperature by means of loss of latent heat due to evaporation.

Temporal scale—climate may vary on a large range of temporal scales. Temporal scales may be seasonal to even geological, which goes up to hundreds of millions of years.

Tensiometer—instrument, consisting of a porous cup filled with water and connected to a manometer or vacuum gauge, used for measuring the soil-water matric potential. It measures the soil suction caused by soil capillary pores. It consists of a porous ceramic tip that is introduced in the soil at a desired depth and is hermetically connected to a tube and a vacuum gage that is graduated in centibars (0–100 cbars).

Tensiometer graph—is a graph of the tensiometer readings. This can help to indicate when to irrigate and the duration of irrigation.

tensiometer, Deaeration—is a process to remove the accumulated air in tensiometer. A manual vacuum pump or a polyethylene tube of 0.125 centimeter (1/8 inch) of diameter can be used for this purpose.

tension, (suction) Soil moisture—suction with which the moisture is retained in the soil. The tension is caused by forces with which the water is attached to soil particles.

tension, Osmatic—retention of water by soil particles in the soil solution.

term, Loan—time for which the loan is approved by the bank to purchase the irrigation system.

terrains.

Tertiary Treatment—the processes which remove pollutants not adequately removed by secondary treatment, particularly nitrogen and phosphorus; accomplished by means of sand filters, micro straining, or other methods (referring to wastewater treatment).

TEST D.P.D—calorimetric method to measure total chlorine and free chlorine available in the water.

Textural class—classification used to convey an idea of the textural makeup of soils and to give an indication of their physical properties. Three broad groups of these classes are recognized—sands, loams, and clays. Within each group specified textural class names have been devised (i.e. loamy sand); Classification name given a soil because of the particular size groups of particles found in the "A" horizon.

Texture—in clayey soils, the clay predominates. In coarse texture, sand predominates.

Texture (soil)—relative proportions of sand, silt and clay in a given soil; Relative proportions of the various soil separates in a soil as described by the classes of soil texture.

TFC—thin film composite.

TH—Total hardness. The sum of calcium and magnesium hardness, expressed as a calcium carbonate equivalent.

Thermal expansion—an decrease in water density of oceans that results from global warming. This leads to an expansion of the ocean volume and hence an increase in sea level.

Thermal pollution—discharge of heated effluents at temperatures that can be detrimental to aquatic life. Usually a concern with cooling water discharges from electrical generating stations and industries using high temperature processes.

Thin film composite membrane—a class of reverse osmosis membranes made with polyamide-based polymer and fabricated with different materials in the separation and support layers.

THM—Trihalomethanes. Toxic chemical substances that consist of a methane molecule and one of the halogen elements fluorine, bromine, chlorine and iodine attached to three positions of the molecule. They usually have carcinogenic properties.

Thornthwaite methodo—Thornthwaite developed an empirical equation based on the monthly average temperature to calculate evapotranspiration.

Thrust block—normally, concrete poured in place at changes in direction of water flow in piping systems (tees, ells) to prevent movement of the pipe.

Tidal excursion—the displacement of a parcel of water (or a float) during one half-tidal cycle.

Tidal flats—marshy or muddy areas that are covered and uncovered by the rise and fall of the tide; the vegetated parts are called tidal marshes.

Tidal marshes—vegetated areas that are covered and uncovered by the rise and fall of the tide.

Tide—the alternative ding and falling of water surfaces caused by the pull of the Moon, and to a lesser degree, by the pull of the sun.

Tide gauge—device at a coastal location or deep-sea location for continuous measurement of the sea level with respect to the adjacent land.

Tilth—physical condition of the soil in relation to plant growth.

Time of concentration—the time it takes for water to flow from the most distant part of the drainage basin to the measuring point.

Time of travel—refers to the rate of movement of water, or waterborne materials, through a defined reach of stream channel for steady or gradually varied flow conditions. Determined by dye tracing methods where dye is injected at some location on a stream and detected at other locations downstream.

Time weighted average—computed by multiplying the number of days in the sampling period by the concentrations of individual constituents for the corresponding period and dividing the sum of the products by the total number of days. A time-weighted average represents the composition of water that would be contained in a vessel or reservoir that had received equal quantities of water from the stream each day of the year.

time, Maximum—highest value of all observations to take a known quantity of water from drippers. This value is used to evaluate uniformity of application of water.

time, Minimum—lowest value of all observations to take a known amount of water from the dripper.

Titration—An analytical technique to determine how much of a substance is present in a water sample by adding another substance and measuring how much of that substance must be added to produce a reaction.

TOC—Total organic carbon.

Tons per acre-foot—indicates the dry mass of dissolved solids in 1 acre-foot of water. It is computed by multiplying the concentration in milligrams per liter by 0.00136.

Tons per day—the quantity of substance in solution or suspension that passes a stream section during a 24-hour day.

Topography—the lay of the land, particularly its slope and drainage patterns; the science of drawing maps and charts or otherwise representing the surface features of a region or site, including its natural and man-made features.

Tortuous path—water flow through channels, which are constricted and marked by repeated twists, bends and winding turns.

Total chlorine—the total concentration of the chlorine in a water, including the combined available chlorine and the free available chlorine.

Total dissolved solids (TDS)—the total weight of the solids that are dissolved in the water, given in ppm per unit volume of water. TDS is determined by filtering a given volume of water (usually through a 0.45 micron filter), evaporating it at a defined temperature (usually 103–105°C) and then weighing the residue.

Total hardness (TH)—the total of the amounts of divalent metallic cations, principally calcium hardness and magnesium hardness, expressed in terms of calcium carbonate equivalent.

Total in bottom material—the total amount of a given constituent in representative sample of bottom material. This term is used only when the analytical procedure assures measurement of at least 95% of the constituent determined. A knowledge of the expected form of the constituent in the sample, as well as the analytical methodology used, is required to judge when the results should be reported as "total in bottom material."

Total Kjeldahl Nitrogen (TKN)—the organic nitrogen, as found in plant cells, plus the ammonia nitrogen.

Total load—(tons) the total quantity of any individual constituent, as measured by dry volume, that is dissolved in a specific amount of water (discharge) during a given time. It is computed by multiplying the total discharge in ft³/s, times the mg/l of the constituent, times the factor 0.0027, times the number of days.

Total matter—the sum of all suspended and dissolved matter in a water sample.

Total phosphorus (TP)—all of the phosphorus forms in the water sample (mg/l as P)

Total pressure head—the sum of all the factors, which increase or decrease the available water pressure.

Total solids—the mineral, cells, etc. left in wastewater after evaporation of the water fraction at 103°C. Usually measured in mg/L; All the solids in wastewater or sewage water, including suspended solids and filterable solids. It is usually determined by evaporation. The total weight concerns both dissolved and suspended organic and inorganic matter.

Total solids (TS)—the weight of all organic and inorganic solids, both dissolved and suspended, per unit volume of water.

Total suspended solids—the mineral, cells, etc. in wastewater retained on a standard filter paper after filtration followed by drying at 103°C. Usually measured in mg/L.

Total suspended solids (TSS)—a measurement of the solids that either float on the surface of, or are in suspension in, water or wastewater. A measure of wastewater strength often used in conjunction with biochemical oxygen demand (BOD).

total, Recoverable—the amount of a given constituent that is in solution after a representative water-suspended sediment sample has been digested by a method (usually using a dilute acid solution) that results in dissolution of only readily soluble substances. Complete dissolution of all particulate matter is not achieved by the digestion treatment, and this determination represents something less than the "total" amount (that is, less than 95 percent) of the constituent present in the dissolved and suspended phases of the sample. To achieve comparability of analytical data, equivalent digestion procedures would be required of all laboratories performing such analyzes because different digestion procedures are likely to produce different analytical results.

Toxic—poisonous (to human beings); capable of producing disease or otherwise harmful to human health when taken into the body.

Toxic—poisonous, or otherwise directly harmful to life.

Toxic substance—chemicals such as pesticides (bug killers) and mercury that can make people and animals ill or die.

Toxic water pollutants—compounds that are not naturally found in water at the given concentrations and that cause death, disease, or birth defects in organisms that ingest or absorb them.

Toxicity—the adverse effect which a biologically active substance has, at some concentration, on a living entity.

Toxicity test—a test which determines the potency of a toxic substance by measuring the intensity of a biological response.

Trace elements—elements of compounds that are necessary for the operation of living systems, but that are present in the environment only in minute quantities.

Trace metals—trace metallic elements found in surface waters and sediments; the principal ones in the Potomac Basin: strontium, aluminum, boron, barium, and zinc.

Trace substance (or trace)—a substance which is found during water analysis in a small concentration, high enough to be detected, but to low to be quantified accurately by standard testing methods.

Tradeable water rights—means water access rights; water delivery rights, or irrigation rights. Note: Some components within water access rights are not tradeable.

Trading zone—zones established to simplify administration of a trade by setting out the known supply source or management arrangements and the physical realities of relevant supply systems within the zone.

Transaction costs—includes search, negotiation and enforcement costs including, but not limited to, all government water transfer fees and charges applicable to water trade, conveyance charges and professional service fees (such as accountants, brokers, lawyers).

Trajectory—the trajectory of a sprinkler is the measurement (in degrees) of the angle of the water projecting out from the sprinkler's nozzle. A trajectory of 0° would indicate a flat projection of water out from the nozzle.

Transformation—price of establishing correspondence between elements in one set of data to element in another set of date, such that each element in the first set corresponds to a unique element in the second set.

Translocation—movement of water to other areas than where it was applied.

Transmission lines—pipelines that transport raw water from its source to a water treatment plant.

Transmissivity—the rate at which water of the prevailing kinematic viscosity is transmitted through a unit width of saturated thickness of an aquifer under a unit hydraulic gradient.

Transpiration—process of plant water uptake and use, beginning with absorption through the roots and ending with transpiration at the leaf surfaces. See also *evapotranspiration*; Liquid movement of water from the soil, into the roots, up the plant stems, and finally out of the plant leaves into the air as vapor.

Trap—a device used to prevent a material flowing or carried through a conduit from reversing its direction of flow or movement or from passing a given point; a device to prevent the escape of air from sewers through a plumbing fixture or catch basin.

Trash and debris removal—mechanical removal of debris, snags, and trash deposits from the stream banks to improve the appearance of the stream.

Traveler—a sprinkler unit which propels itself along via the water pressure operating the sprinkler. Various types are manufactured which either operates via a pull wire or follow the layout of the hose supplying the water. Most units have an automatic shut off at the end of the run.

Treatment plant—a structure built to treat wastewater before discharging it into the environment.

treatment, Acid—to prevent partial obstruction of drippers because of salts. The irrigation system can be washed with some acids that are not detrimental to plant growth.

treatment, Hot water—water at 80°C is allowed through the irrigation system to avoid damage by freezing.

tree, Witness—tree used by surveyors to mark the location of a survey corner; the tree is located near the survey corner and is inscribed with survey data. Also known as a bearing tree.

tree, Wolf—large rough tree, generally not good for lumber.

Triangle spacing—the term given to a sprinkler head layout pattern where the sprinklers, when viewed from above, appear as a more or less equilateral triangle with one sprinkler in each corner. Triangular spacing results in the most uniform and efficient water application using sprinklers.

Tributary—a stream or other body of water, which contributes its water to another and larger stream or body of water.

Trichloroethylene (TCE)—a toxic volatile organic compound often found as a solvent.

Trickle irrigation—another name for drip irrigation. Probably a more accurate name since for most drip irrigation systems the water flow is more of a trickle than a drip.

Trickling filter—a filter consisting of an artificial bed of coarse material, such as broken stone, clinkers, slate, slats, brush, or plastic materials, over which wastewater is distributed or applied in drops, films, or spray from troughs, drippers, moving distributors, or fixed nozzles, and through which it trickles to the under drains, giving opportunity for the formation of zoological slims which remove dissolved organic matter from the wastewater and reduce the BOD5.

Tsunami—from the Japanese words tsu ("horbor") and nami ("wave"), a seismic sea wave caused by underwater earthquakes or volcanic eruptions.

Tube settler—device using bundles of tubes to let solids in water settle to the bottom for removal by sludge.

Tube wall (double wall or biwall)—are drip lines that consist of two chambers. The inner chamber maintains a high pressure (50–150 cbs) in a diameter of approximately 1.25 cm. The outer chamber contains orifices in the wall of a tube at given intervals. The water leaves the tube at low pressure through the orifices to wet the soil.

Turbidity—the cloudiness of water. It is determined by the presence of suspended matter such as clay, silt, organic matter, and living organisms. High turbidity may reduce light transmission, and therefore reduce photosynthesis of aquatic plants.

Turbidity currents—the fast movement of a water and sediment mix down steep slopes in the oceans. They are often triggered by earthquakes and are associated with the formation of submarine canyons.

Turbine pump for a deep well—has an impeller with radial vanes. It may be suspended vertically inside the unloading pipe of a well. The impeller can be centrifugal, axial, or of an intermediate type depending on the desired pressure for a given flow rate.

Turbulent flow—flow in which the fluid particles move in an irregular random manner, in which the head loss is approximately proportional to the second power of velocity.

Turbulent flow emitter—emitters with a series of channels that force water to flow faster not allowing particles to

Turgid—state of a plant cell when the cell wall is rigid due to the hydrostatic pressure of liquid in the cell.

Turnover time—the ratio of the mass M of a gaseous compound in the atmosphere and the total rate of removal S of the compound: $T = M/S$. For each removal process separate turnover times can be defined.

Type II water—water prepared by using a still (deionized supply water may be necessary) designed to produce a distilate having a conductivity of less than 1.0 umho.cm at 25°C and a maximum total matter content of 0.1 mg/L.

Typical sewage—Arizona regulatory terminology that means sewage in which the total suspended solids (TSS) content does not exceed 430 mg/L, the five-day biochemical oxygen demand (BOD) does not exceed 380 mg/L, and the content of fats, oils, and greases (FOG) does not exceed 75 mg/L.

Ultrafiltration—a method of cross-flow filtration (similar to reverse osmosis but using lower pressure), which uses a membrane to separate small colloids and large molecules from water and other liquids.

Ultrapure water—highly treated water that is deionized and mineral free with high resistivity and no organics; it is usually used in the semiconductor and pharmaceutical industries. Ultrapure water is not considered biologically pure (potable) or sterile.

Ultraviolet (UV) light—radiation (light) having a wavelength shorter than 3900 angstroms, the wavelengths of visible light, and longer than 100 angstroms, the wavelength of x-rays.

Ultraviolet chamber—the area where the water is irradiated with ultraviolet rays.

Ultraviolet demand—the amount of ultraviolet rays required to inactivate certain microorganisms.

Ultraviolet dosage—the amount of disinfectant ultraviolet rays delivered to the organisms in the water being disinfected. Dosage is a combination of UV intensity times the contact time and is measured in watt-seconds per square centimeter.

Ultraviolet oxidation—a process using extremely short wave-length light that can kill microorganisms (disinfection) or cleave organic molecules (photo oxidation) rendering them polarized or ionized and thus more easily removed from the water.

Ultraviolet radiation—radiation having a wavelength shorter than that of visible light; used in disinfection processes.

Unavailable soil water—portion of water in a soil held so tightly by adhesion and other soil forces that it cannot be absorbed by plants rapidly enough to sustain growth.

Uncertainty—An expression of the degree to which a value (future state) is unknown. Contrary to risk, uncertainty suggests unknown probability of occurrence. Uncertainty can result from lack of information or from disagreement about what is known. It may have many types of sources, from quantifiable errors in the data to ambiguously defined concepts or terminology, or uncertain projections of human behavior. Uncertainty can therefore be represented by quantitative measures or by qualitative statements.

Unconfined aquifer—an aquifer that has a water table; it contains unconfined ground water.

Unconsolidated—not formed into a compact mass; loosely associated.

Under drain—plastic pipes with holes drilled through the top, installed on the bottom of an infiltration BMP, or sand filter, which are used to collect and remove excess runoff.

Undulating—a periodic rise and fall of a surface; having a wavy outline or appearance.

UNFCCC—United Nations Framework Convention on Climate Change. A treaty was signed at the Rio Earth Summit in 1992 in which 150 countries promised stabilization of greenhouse gas concentrations in the atmosphere at a level that would prevent dangerous anthropogenic interference with the climate system.

Uniform flow—a flow in which the feet per second velocity rates and directions are the same from point to point along the conduit.

Uniformity—term describing how evenly water is applied by overlapping sprinklers; Evenness of precipitation over a given area.

uniformity coefficient, Christiansen's—measure of the uniformity of irrigation water application. The average depth of irrigation water infiltrated minus the average absolute deviation from this depth, all divided by the average depth infiltrated.

Uniformity coefficient (irrigation)—characteristic of the areal distribution of water in a field as the result of an irrigation;

Uniformity of water application—to evaluate a drip irrigation system with respect to the rate of application of water by the emitters.

Union—a pipe fitting used to connect two lengths of pipe in such a way that neither has to be rotated.

Unleveled—change in elevation of the land may cause gain or loss in the pressure. A change of elevation of 2.3 feet is equivalent to change in the pressure of 2.3 psi.

Unloading—the release of the contaminant that was captured by a filter medium.

Unregulated water resource—a water resource that is not controlled through the use of infrastructure to store and release water.

Unsaturated zone—the zone between the land surface and the water table containing water held by capillary water, and containing air or gases generally under atmospheric pressure.

Up flow—a pattern of water flow in which a solution (water or regenerant usually) enters at the bottom of the vessel or column and flows out at the top of the vessel or column during any phase of the treatment unit's operating cycle.

Up gradient—in the direction of increasing static head.

Up gradient well—one or more wells which are placed hydraulically upgradient of the site and re capable of yielding ground-water samples that are representative of original conditions and are not affected by the regulated facility.

Up water—ultra pure water creation demands a specialized way of working. A number of techniques are used among others; membrane filtration, ion exchanges, sub micron filters, ultra violette and ozone systems. The produced water is extremely pure and contains none to very low concentrations of salts, organic/pyrogene components, oxygen, suspended solids and bacteria.

Upper most aquifer—the geologic formation, group of formations, or part of a formation that contains the uppermost potentiometric surface capable of yielding a significant amount of ground water to wells or springs and may include fill material that is saturated. There should be very limited interconnections, based upon pumping tests, between the uppermost aquifer and lower aquifers. Consequently, the uppermost aquifer includes all interconnected water-bearing zones capable of significant yield that overlie the confining layer.

Upwelling—the upward movement of cold bottom water in the ocean that occur when winds or currents displace the lighter surface water.

Urban runoff—storm water from city streets and gutters. It usually contains litter, organic, and bacterial wastes. In the past, urban runoff has gone untreated, but recently it has been recognized as a significant pollutant that needs retention and treatment.

use, Beneficial—beneficial use of water supports the production of crops: food, fiber, oil, landscape, turf, ornamentals, or forage; *nonbeneficial:* Water used in plant growth which can not be attributed as beneficial; *reasonable:* In the context of irrigation performance, all beneficial uses are considered to be reasonable uses. Non-beneficial uses are considered to be reasonable if they are justified under the particular conditions at a particular time and place; *unreasonable:* Unreasonable uses are nonbeneficial uses that, furthermore, are not reasonable; that is, they are without economic, practical, or other justification.

use, Consumptive—total amount of water taken up by vegetation for transpiration or building of plant tissue, plus the unavoidable evaporation of soil moisture, snow, and intercepted precipitation associated with vegetal growth;

UV—ultra Violet. Radiation that has a wavelength shorter than visible light. It is often used to kill bacteria and destroy ozone.

Vacuum distillation—distillation that occurs at a pressure somewhat below atmospheric pressure. Lowering the pressure also lowers the boiling point of water, thus conserving energy by requiring less heat to bring about distillation.

Vacuum filter—a filter consisting of a cylindrical drum mounted on a horizontal -axis, covered with a filter cloth, and revolving with partial submergence in the liquid to be treated. A vacuum is maintained under the cloth for the larger part of a revolution to draw the liquid through the filter cloth. Solids accumulate on the exterior of the drum as "cake" which is scraped off continuously.

Validation—determination upon testing that a representative sample of a water treatment product/model has meet the requirements of a specific standard.

Valve—a device used to control the flow of water valve is like a faucet. Valves respond to commands from the controller. When valves receive a signal to open, water flows to the sprinklers when they receive another signal to close, the flow of water stops. Valves used in pressurized systems include: *alfalfa:* Outlet valve attached to the top of a short vertical pipe (riser) with an opening equal in diameter to the inside diameter of the riser pipe and an adjustable lid or cover to control water flow. A ring around the outside of the valve frame provides a seat and seal for a portable hydrant. Typically used in border or basin irrigation; *angle*: Valve configured with its outlet oriented 90 degrees from its inlet; *air vent (air relief, air release)*: device that releases air from a pipeline automatically without permitting loss of water; *air vacuum or air relief*: device that releases air from a pipeline automatically without permitting loss of water or admits air automatically if the internal pressure becomes less than atmospheric; *back flow prevention*: Check valve that allows flow in one direction; *ball:* Valve in a pipeline used to start or stop flow by rotating a sealed ball with a transverse hole approximately equal to the diameter of the pipeline. Ball rotation is typically 90 degrees for a single-port control; *butterfly:* Valve in a pipeline to start or stop flow by rotating a disk 90°. The disk is about the same diameter as the pipeline; *check:* Valve used in a pipeline which allows flow in only one direction; *chemigation*: Valve especially designed to be used with the injection of chemicals in an irrigation system; *corporation stop*: Quarter turn valve similar to a ball valve with two exceptions. Internally there is a circular disk rather than a ball, and there is no attached handle; *curb stop*: Physically the same as corporation valve but used at a different location; *drain valve automatic*: Spring loaded valve that automatically opens and drains the line when the pressure drops to near zero; *drain valve or flushing:* Valve on the end of a line to flush out dirt and debris. May be incorporated into an end plug or cap; *float valve-* Valve, actuated by a float, that automatically controls the flow of water; *foot valve:* Check valve used on the bottom of the suction pipe to retain the water in the pump when it is not in operation; *flow control*: Valve with automatically adjusts to provide a predetermined downstream flow; *gate*: Valve in a pipeline used to start or stop water flow. May be operated by hand with or with mechanical assistance or by high or low voltage (solenoid) electric controlled mechanical assistance. Gate valves consist of seated slide or gates operated perpendicular to the flow of water. Head loss through a gate valve is typically less than a globe valve, but more than a ball or butterfly valve; *globe*: Valve in a pipeline used to start or stop water flow. Globe valves stop flow by positioning a disk and gasket over a machined seat about the same diameter as the pipe. Globe valves are limited to smaller sizes because of the high velocities and very high head loss through the valve; *hydraulic*: Irrigation zone valve which uses small flexible tubes and water under pressure to provide the actuation signal from the controller to the valve; *isolation valve*—Any mechanical valve used to isolate a section of a piping system; *master:* Valve used to protect the landscape from flooding in case of a ruptured main or malfunctioning downstream valve. The master valve is installed on the mainline after the backflow preventer (in some systems); *orchard valve:* Outlet valve installed inside a short vertical pipe (riser) with an adjustable cover or lid for flow control. Similar to an alfalfa valve, but with lower flow capacity. Typically used in basin irrigation; *pilot valve:* Small valve used to actuate a larger one; *pressure regulating*: Valve designed to automatically provide a preset downstream pressure in a hydraulic system; *pressure relief*: Spring loaded valve set to open at a pressure slightly above the operating pressure, used to relieve excessive pressure and surges; *pressure sustaining*: Valve designed to provide a minimum preset upstream pressure; *quick coupling*: Permanently installed valve which allows direct access to the irrigation mainline. A quick coupling key is used to open the valve; *remote control*: Valve which is actuated by an automatic controller by electric or hydraulic means.

It is synonymous with *Automatic Control Valve*; *surge valve:* Device in a pipe T fitting to provide flow in alternate directions at timed intervals. Used in surge irrigation; *vacuum relief valve*: Valve used to prevent a vacuum in pipelines and avoid collapsing of thin-wall pipe.

Valve zone—an area where the irrigation is all controlled by a single control valve. Each valve zone must be within only one hydrozone.

valve, Air relief (vacuum breaker)—removes air in the irrigation line. It must be installed at the highest location in the field.

valve, Angle—refers to the water flow pattern into and out of the valve. Often used as control valves. Seldom used as isolation valves. The valve inlet is on the bottom of an angle valve and the outlet is on one side. Angle valves as a group tend to be very reliable and have lower friction losses than "globe" valves, the other common style used for control valves.

valve, Anti-siphon—a control valve with a built-in atmospheric vacuum breaker (backflow preventer). Most commonly used in residential irrigation systems.

valve, atmospheric vacuum breaker (AVB)—A type of backflow preventer.

valve, Automatic control—a valve which is activated by an automatic controller using electric or hydraulic means. Synonymous with Remote Control Valve.

valve, Automatic—valve which can be remotely operated. The remote operation method may be either electrical (the most common) or hydraulic. Automatic valves are commonly used as "control valves" for irrigation systems.

valve, Backflow preventer—a mechanical device which prevents backflow. In irrigation, it is used to protect the potable water supply from potentially contaminated irrigation water. There are several types of backflow preventers. The choice of backflow preventer used depends on the degree of hazard and the particular piping arrangement involved. Virtually all regulatory agencies in the United States require backflow prevention devices to protect the domestic water supply from contamination by the backflow of irrigation water. In areas where it is not required, it is highly recommended. Consult local building codes for laws applicable in your area.

valve, Ball—this type of valve controls the water by means of a rotating ball with a hole through the center of it. When the hole is aligned with the water flow the water flows freely through the valve with almost no friction loss. When the ball is rotated so that the hole is not aligned the flow is completely shut off. Ball valves are used primarily as isolation valves. They tend to be very reliable and trouble-free. Ball valves as a group tend to require more effort to turn on and off than other valves. For larger size pipes butterfly valves are usually used rather than ball valves. If you're looking for a dirty joke, you'll need to keep looking.

valve, Butterfly—this type of valve uses a rotating disk to control the water flow. A true butterfly valve has two half-disks, hinged together in the center. When the disks, or "wings" are folded together the water flows freely past them, when folded out into the water stream the wings block the flow. Most "butterfly valves" are really "rotating disk" valves. They have a single, round disk that rotates on an axle. When fully open the disk is rotated so that it is aligned with the water flow. To close, the disk is rotated at a right angle so that it fully blocks the flow. Butterfly valves are used as both isolation and control valves. Butterfly valves tend to be very reliable and trouble free. They are mostly used on larger pipe sizes, seldomly less than 3″ in size. Ball valves are used on smaller size pipes.

valve, Check—a small device allowing water to flow in one direction only. A check valve has a spring that will hold the valve closed, and will not allow water to flow until a preset pressure is achieved in the system. Check valves that are preinstalled in sprinklers, or field installed in the lateral line pipe, take advantage of this spring.

valve, Diaphragm—a globe or angle pattern valve which uses a diaphragm to control the flow of water through the valve.

valve, Drain—a valve used to empty water from a lateral or main line, usually for winterization purposes.

valve, Electric—automatic valve usually controlled by 24 to 30 volt (AC) current.

valve, Flushing—is at the end of the drip lines to clean the system when taking out the dirty water out of lateral lines.

valve, Foot—it stops back flow of water to the water resource. It is made of bronze, galvanized iron, steel, PVC, polyethylene, and so forth.

valve, Gate—refers to the operating mechanism for the valve, which is a sliding gate which moves up or down to block the flow. Often used as isolation valves. Never used as control valves. Because the gate slides it is very subject to wear, and gate valves wear out fast when used often. Some gate valves use a wedge-shaped gate, which holds up better. They are still not designed for regular use, but for emergency shut-off only. Use a cheap gate valve and you will soon learn to "hate the gate."

valve, Gate (ball)—allows certain amount of water through the pipe.

valve, Globe—a valve configured with its outlet oriented 180° from its inlet. In irrigation, these valves are generally installed so that the inlet and outlet are parallel to the ground.

valve, Isolation—a valve used for isolating all or part of the irrigation system for repairs, maintenance, or winter shut-down (winterization). Common types of isolation valves are the ball valve, butterfly valve, and gate valve.

valve, Master—a valve used to protect the landscape from flooding in case of a ruptured main or malfunctioning downstream valve. The master valve is installed on the mainline after the backflow preventer and the control valves.

valve, Solenoid—controls the amount of water in the submain or lateral by an electrical current.

valve, Volumetric—measures volume of water that is applied to a field.

Vapor—the gaseous phase of substances such as water.

vapor pressure gradient—is a difference between the free air vapor content on the crop and the content of vapor in the evaporative surface.

Vaporize—Conversion of a liquid into vapor.

Var.—variety; a taxonomic category below the subspecies level, used to separate members of a species that have special, but similar, differences from other members of their species group.

Velocity—The speed at which water travels through a pipe. The designer should use caution when designing a system where the water velocity exceeds 5 feet per second (see "water hammer").

Ventilation—is a process by which there is exchange of air and other gases of the soil. The index of ventilation of the soil depends on the size and number of soil pores and the soil moisture.

Venturi—a channel that serves the measurement of water flows.

Vertical separation—the depth of unsaturated, original, undisturbed soil of Soil Types 1B-6 between the bottom of a disposal component and the highest seasonal water table, a restrictive layer, or Soil Type 1A.

Viable—capable of living independently and being reproductive.

Viable water treatment process—a water or wastewater treatment process capable of accomplishing the desired water quality.

Virga—particles of water that fall from clouds but evaporate before reaching the ground.

Virus—a class of ultramicroscopic, filterable, infectious agents, chiefly protein in composition, which are typically inert except when in contact with certain living cells.

Viscosity—the property of a fluid describing its resistance to flow. Units of viscosity are newton-seconds per meter squared or pascal seconds. Viscosity is also known as dynamic viscosity.

Voids ratio—ratio of the volume of voids (pores) to the volume of soil.

Volatile—easily changed from a solid or liquid to a gas at relatively low temperatures.

Voltage—force required to push and pull a stream of electrons through a circuit; Amount of electrical potential required to force one amp of current flow in a circuit against one ohm of resistance.

Volts Alternating Current—VAC: Most electric control valves operate on 24 VAC. That's "alternating current," like household electricity, not "direct current" like batteries! However, most valves can be activated using direct current also (snap three 9 volt batteries together in a chain and touch the valve solenoid wires to the end terminals of the battery chain.

Volume—expressed in gallons per minute (GPM), gallons per hour (GPH) cubic feet per second (ft^3/s), cubic meters per hour m^3/h), liters per minute (l/m), or liters per second (l/s). Volume is used to describe either the amount of water available or the amount of water used.

Volume weighted average price (VWAP)—The average value (dollars per megaliter) of the water traded where each trade is weighted proportionally by the volume of water (in mL) involved in the sale. This provides a more accurate representation of the price (i.e., high volume trades generally attract a 'bulk discount').

Volute—refers to the flow path of water and its associated pump casing as it leaves the impeller of a pump.

Warm front—warm it systems, usually from the temperature regions that press into colder air masses.

Waste—any solid or liquid material, product, or combination of them that is intended to be treated or disposed of or that is intended to be stored and then treated or disposed. This does not include recyclables.

Waste disposal facilities—facilities designated for the disposal of liquid or solid wastes.

Waste stabilization pond—the oxidation of waste in ponds by sedimentation, the removal of settleable solids and the decomposition of this resulting sediment by microorganisms. The sludge is converted to inert residues and soluble organic substances. Decomposition of organic matter is the work of microorganisms, either aerobic or anaerobic. It is desirable to maintain aerobic conditions, since aerobic microorganisms cause the most complete oxidation of organic matter. Also referred to as lagooning or polishing, after previous treatment.

Waste water—a combination of liquid and water-carried pollutants from homes, businesses, industries, or farms; a mixture of water and dissolved or suspended solids.

Waste water infrastructure—the plan or network for the collection, treatment, and disposal of sewage in a community.

Waste water reclamation—processing of wastewater for reuse.

Waste water treatment system—a collection of treatment processes designed and built to reduce the amount of suspended solids, bacteria, oxygen-demanding materials, and chemical constituents in wastewater.

Waste water treatment, Advanced—any treatment of sewage water that includes the removal of nutrients such as phosphorus and nitrogen and a high percentage of suspended solids.

Water (H$_2$O)—a odorless, tasteless liquid necessary for keeping the plants alive. Liquid form that exists between ice (at temperatures below 0 degrees Centigrade or 32°F) and water vapor (at temperatures above 100°C or 212°F). The water molecule consists of a central oxygen atom with two hydrogen atoms attached at a separation angle of about 105 degrees. Water is an excellent solvent because of its chemical properties. Nutrients required for plant growth are dissolved in the soil water, which is then taken up by the plant root system. Water is also used for plant cooling.

Water allocation—the specific volume of water allocated to water access entitlements in a given water year or allocated as specified within a water resource plan.

Water allotment—a method to accurately and fairly estimate a total volume of water that should be allocated to a site.

Water allotment adjustment factor—factor used in the equation to predict water allotment.

Water amendments—in addition to fertilizers, pesticides, micro nutrients, and other water additives to improve the crop yield, water amendments are used for chemical treatment of the water to reduce the obstruction of drippers.

Water areas—areas within a land mass persistently covered by water. Examples are bays, estuaries, streams, reservoirs, and lakes. *Census Water:* Streams, sloughs, estuaries, and canals more than 1/8 of a statute mile in width. Lakes, reservoirs, and ponds more than 40 acres in size. *Non-Census Water:* Streams, sloughs, estuaries, and canals more than 120 feet and less than 1/8 of a mile wide. Lakes, reservoirs, and ponds 1 to 40 acres in size.

Water assignment—the transfer of a water right application or permit from one person to another. This can be done in conjunction with the sale of land.

Water available—amount of water in the soil that is available to the plants. It is a soil moisture content between the field capacity and the point of permanent wilting.

Water balance—a measure of the amount of water entering and the amount of water leaving a system. Also referred to as *Hydrologic Budget*.

Water balance in the soil—sum of all moisture entering into the soil system and water loss, in a given period of time, mm/period.

Water bar—a diagonal ditch or hump in a trail that diverts surface water runoff to minimize soil erosion.

Water borne disease—a disease, caused by bacterium or organism able to live in water, which can be transmitted by water.

Water course—the bed and shore of a river, stream, lake, creek, lagoon, swamp, marsh or other natural body of water, or a canal, ditch, reservoir or other artificial surface feature made by humans, whether it contains or conveys water continuously or intermittently.

Water closet—a flushable toilet.

Water column—word used to refer to a water body in its vertical extent.

Water company—a private or public entity which provides water, in most cases to properties by means of pipe lines. Some water companies sell water in containers of various sizes.

Water conditioning—the treatment or processing of water, by any means, to modify enhance or improve its quality to meet a specific water quality need desire or set of standards. Also called water treatment.

Water conservation—the wise use of water with methods ranging from more efficient practices in farm, home and industry to capturing water for use through water storage or conservation projects.

Water content of the soil—is a depth of water contained in the soil. It is a ratio of the mass of water to the mass of dry, or a ratio of the volume of water to a unit of volume of soil, sample. It can be expressed on the basis of wet or dry weight of a soil sample. It is called soil moisture content.

Water control—management of water (both surface and subsurface) to maintain plant growth, water quality, wildlife habitat, and fire control.

Water course crossing—a permanent or temporary crossing and any associated permanent or temporary structures that are or will be constructed to provide access over or through a water body.

Water cycle—transition and movement of water involving evaporation, transpiration, condensation, precipitation, percolation, runoff, and storage.

Water equivalents—one cubic foot per second (cfs) = 450 gallons per minute, or 7–1/2 gallons per second; 1 cfs for 1 day or 1 cfs-day = about 2 acre feet; 1 acre foot = 326,000 gallons; 1 cubic foot weights 62.4 pounds; 1 cubic foot = 7–1/2 gallons; 1 gallon = 8.33 pounds; 1 ton = 240 gallons.

Water flow velocity—the average velocity of water flowing through a cross-section of a stream.

Water fowl—aquatic or semiaquatic birds of the order Anseriformes (ducks, geese, and swans).

Water hammer—the surging of pressure, which occurs when a control valve is suddenly closed. In extreme conditions, this surging will cause the pipes to vibrate or create a pounding noise. Water hammer is most commonly caused by fast closing valves and/or high velocity water flow.

Water harvesting—the capture and use of runoff from rainfall and other precipitation (e.g., the collection of rainwater in cisterns).

Water holding capacity—total amount of water held in the soil per increment of depth. It is the amount of water held between field capacity and oven dry moisture level.

Water horsepower—the energy or power added to water by the pump [hp, kW].

Water management—the protection and conservation of water and aquatic ecosystems, including their associated riparian area.

Water meter—A device that measures the quantity of water used at a house, business, factory, etc. Cities that have implemented a water meter system and charge people according to the amount of water consumed use less water than those cities that charge a flat rate for water.

Water monitoring—the process of constant control of a body of water by means of sampling and analyzes.

Water pollution—the presence in water of enough harmful or objectionable material to damage water quality.

Water quality—the physical, chemical, and biological characteristics of water.

Water quality inlet—best management practice consisting of a three-stage underground retention system designed to remove heavy particulates and absorbed hydrocarbons. Also, known as an OIL/GRIT SEPARATOR.

Water quality standards—fixed limits of certain chemical, physical, and biological parameters in a water body; water quality standards are established for various uses of water (e.g., drinking).

Water recycling—using water again for the same or another process step, after a small form of purification is applied.

Water related data—any data having significance to users of water data or to those conducting water-resource projects or investigations. Examples are meteorological, water use, oceanographic, agricultural, demographic, etc.

water right, Abandoned—a water right which has not been put to Beneficial Use for generally five or more years, in which the owner of the water right states that the water right will not be used, or takes such actions that would prevent the water from being beneficially used.

Water rights—state administered legal rights to use water supplies derived from common law, court decisions, or statutory enactments.

Water shed—the area of land that catches precipitation and drains into a larger body of water such as a marsh, stream, river, or lake. A watershed is often made up of a number of subwatersheds that contribute to its overall drainage.

Water shed Management—the protection and conservation of water and aquatic ecosystems, including their associated riparian area. Because land use activities on the uplands of a watershed can affect ground and surface water quality and quantity, a broader, more comprehensive approach to planning is often required. A Watershed Management Plan may look at water quantity, water quality, aquatic ecosystems, riparian area, as well as a variety of land use issues as they impact water. Watershed management plans require water and land use managers to work together to ensure healthy watersheds. (Partnerships)

Water shed or drainage Area—an area from which water drains to a single point; in a natural basin, the area contributing flow to a given place or a given point on a stream. A region drained by a stream.

Water softening—the reduction /removal of calcium and magnesium ions, which are the principal cause of hardness in water.

Water solubility—the maximum possible concentration of a chemical compound dissolved in water.

Water source—it supplies amount of water needed for irrigation system. It can be a well, lake, river, reservoir tank, or a natural fall of water, and so forth.

Water storage efficiency—ratio of the amount of water stored in the root zone during irrigation to the amount of water needed to fill the root zone to field capacity.

Water storage in soil—quantity of stored water in a root zone due to rainfall, snow, or irrigation applications. This meets crop water requirements in subsequent periods, partially or completely.

Water storage pond—an impound for liquid wastes designed to accomplish some degree of biochemical treatment.

Water stored—the amount of water that is stored in a crop root zone, for use by the crop as a result of an irrigation. It is expressed as a volume or depth of water.

Water supply system—the collection, treatment, storage, and distribution of water from source to consumer.

Water system—a river and all its branches.

Water table—that surface in a ground water body at which the water pressure is atmospheric. It is defined by the levels at which water stands in wells that penetrate the water body just far enough to hold standing water. In wells, which penetrate to greater depths, the water level will stand above the confining upper bed of the aquifer.

Water table well—a vertical excavation that taps a underground source of water in an unconfined aquifer; the level of water in the well reflects the water table of the aquifer.

water tight—a condition existing in water treatment equipment and materials of such precision of construction and fit as to be impermeable to water unless sufficient pressure occurs to cause rupture.

Water use—whenever water is used by an activity or organism, either in the place it is found or by withdrawing it.

Water use efficiency—ratio of the yield per unit area to the applied irrigation water per unit area.

Water uses—the status of Water uses subject to State water laws.

Water uses (domestic, household)—includes household water, watering of lawns, gardens, etc., for Forest Service administrative sites or other residential areas not served by a municipal water system, and not directly related to recreational use. For private residences, includes watering of lawns and gardens up to 1/2 acre.

Water uses (domestic, recreational)—includes camp and picnic grounds, special use concessionaires, recreational residences, lodges, restaurants, etc., not served by a municipal water system. This code should be used for all water systems developed solely to serve recreational purposes regardless of the number of persons served or the number of service connections.

Water uses (fish and wildlife)—includes, but is not limited to, storage of water that is either retained in the reservoir or released downstream to support these purposes. Also includes instream reservation of natural flow for these purposes.

Water uses (flood control)—water (stored, directed or rerouted) for the sole purpose of flood control or abatement.

Water uses (industrial)—water used in manufacture or production.

Water uses (instream)—in-channel flow for purposes other than those under Recreation, Fish and Wildlife, and Navigation, and when there are multiple instream purposes.

Water uses (irrigation)—application of water for the production of crops or the maintenance of large (more than 1/2 acre) areas of lawns, gardens, etc. This includes Forest Service pastures and golf courses.

Water uses (mining)—any application of water to mining processes (e.g., hydraulic), for drilling and on concentrator tables.

Water uses (municipal)—water provided for human consumption and other purposes through a city, town, community, water district or other formalized water system regularly serving at least 25 individuals at least 60 days out of the year or provides at least 15service connections.

Water uses (navigation)—use of water specifically for the maintenance of navigation for commerce or trade. Excludes water used for recreational boat travel.

Water uses (power)—water used for hydroelectric and hydro mechanical power generation.

Water uses (recreation)—any use other than above including impoundment or instream flow specifically intended for recreational use. This also includes water required for Wild and Scenic Rivers.

Water uses (stock water)—use for domestic livestock grazing on National Forest, other public or private lands.

Water width—the width of the top of the water prism from channel bank to channel bank. The line of measure is perpendicular to the flow direction.

Water window—the amount of time in a day that is available for irrigation to occur at a site.

Water year—in U.S. Geological Survey reports and statistics dealing with surface-water supply, the 12-month period of October 1 through September 30 is known as a water year.

water, Aggressive—water which is soft and acidic and can corrode plumbing, piping, and appliances.

water, Black—that portion of the wastewater stream that originates from toilets. It includes feces, urine, and associated flush waters.

water, Bottled—water that is sold in plastic containers for drinking water and/ or domestic use.

water, Brackish—water polluted or contaminated by organic matter, salts or acids, or a combination thereof. Water containing dissolved solids in the range >1,000 to <15,000 ppm.

water, Capillary—water that is retained by capillary pores of the soil. It is an amount of water that is retained in the soil between the field capacity and the hygroscopic tension.

water, Domestic—potable or drinking water. It can be used as a source of irrigation water, but once water enters an irrigation system it is no longer considered domestic or potable.

water, Gravitational—water that moves into, through, or out of a soil under the influence of gravity forces. Gravitational water drains rapidly from the soil and is not considered to be available for plant use. Generally, this water exists in a soil profile at suctions less than field capacity (i.e. less than approximately 0.1 to 0.33 bars).

water, Grey—that portion of the wastewater stream that originates in sinks, tubs, showers, laundry; i.e., all portions of the wastewater stream excluding toilet wastes.

water, Hygroscopic—Water that is strongly bound to the soil particles at suctions greater than 31 bars. This water cannot be used by plants. The suction limit is, again, approximate.

water, Industrial waste—Wastewater from industrial processes or contaminated with wastewater from industrial processes.

water, Non-available—water retained in the soil and that is not available to the plants.

water, Potable—water that is of known adequate quality to be used for human consumption. Domestic or drinkable water. It can be used as a source of irrigation water if protection is provided to prevent contamination to the domestic supply. Once water enters an irrigation system it is no longer considered potable.

water, Process—water that is used in an industrial process and is not intended for human consumption.

water, Readily available (RAW)—the amount of water in a crop root zone that can be easily extracted by the crop. RAW is the fraction of the difference between FC and PWP that crops can use such that no stress caused by a lack of water occurs. Although crops can theoretically use all the water between FC and PWP, as PWP is approached the water becomes harder to extract and water stress and yield reductions will occur. Therefore, irrigations are scheduled when a fraction of the difference is used. The volume that the crop is "allowed" to use prior to another irrigation is RAW. RAW is expressed as a volume of water per unit of irrigated area, or as a depth of water per unit area in the same units as rainfall.

water, Surface—water on the surface of the ground (lakes, rivers, ponds, floodwater, oceans, etc.); precipitation which does not soak into the ground or return to the atmosphere by evaporation or transpiration.

water, Total available (TAW)—the total water in the crop root zone that is available for crop use. TAW is the difference between field capacity (FC) and the permanent wilting point (PWP) multiplied by the depth of the root zone. Units of expression are generally a volume per unit irrigated area (e.g., acre–inches/acre or inches).

Watering days—the specific days of the week on which watering will take place. For example, every Monday, Wednesday and Friday, or every third day.

Watering point—a waterworks system that provides potable water in bulk to the public.

Waters of the United States—Federal regulatory terminology that means: 1. All waters that are currently used, were used in the past, or may be susceptible to use in interstate or foreign commerce, including all waters that are subject to the ebb and flow of the tide; 2. All interstate waters, including interstate wetlands; 3. All other waters such as intrastate lakes, rivers, streams (including intermittent streams), mudflats, sandflats, wetlands, sloughs, prairie potholes, wet meadows, playa lakes, or natural ponds the use, degradation, or destruction of which would affect or could affect interstate or foreign commerce including any waters (that are or could be used by interstate or foreign travelers for recreational or other purposes; from which fish or shellfish are or could be taken and sold in interstate or foreign commerce; or that are used or could be used for industrial purposes by industries in interstate commerce); 4. All impoundments of waters defined as waters of the United States under this definition; 5. Tributaries of waters identified in subsections (1) through (4); 5. The territorial sea; and 6. Wetlands adjacent to waters (other than waters that are themselves wetlands) identified in subsections (1) through (6).

Watershed—A land area from which water drains to a particular water body.

Weather—short-term changes in temperature, humidity, rainfall, and barometric pressure in the atmosphere. It is usually in reference to local atmosphere. It is usually in reference to local atmospheric changes.

Weathering—decomposition, mechanical and chemical, of rock material used the influence of climatic factors of water, heat, and air.

Wedge theory—the theory of opportunities for limiting or diminishing the greenhouse effect and global warming. Wedges include: decarbonized electricity, decarbonized fuels, fuel displacement by alternative energy sources (e.g., solar energy, nuclear energy), methane management, and natural carbon sinks (e.g., forests).

Weeds—are unwanted plants in the vicinity of a given crop.

Weighted average—generally refers to discharge-weighted average. It is computed by multiplying the discharge for a sampling period by the concentrations of individual constituents for the corresponding period and dividing the sum of the products by the sum of the discharges. A discharge weighted average approximates the composition of water passing a given location during the water year after thorough mixing in the reservoir.

Weir—flow measuring device for open-channel flow. Weirs can be either sharp-crested or broad-crested. Flow opening may be rectangular, triangular, trapezoidal, or specially shaped to make the discharge linear with flow depth. Calibration is based on laboratory ratings.

Well—a deep hole with the purpose to reach underground water supplies; a shaft or pit dug or bored into the earth, generally of a cylindrical form, and often walled with tubing or pipe to prevent the earth from caving in.

Well casing—a solid piece of pipe, typically steel or PVC plastic, used to keep a well open in either unconsolidated materials or unstable rock.

Well cluster—a well cluster consists of two or more wells completed (screened) to different depths on a single borehole or a series of boreholes in close proximity to each other. From these wells, water samples that are representative of the different horizons within one or more aquifers can be collected.

Well depth—the greatest depth below land surface at which water can enter the well will be reported. For screened or perforated wells, the depth to the bottom of the screen or to the lowest perforation will be reported. The open hole or open-end wells, the total depth will be reported.

Well development—the process whereby a well is pumped or surged to remove any fine material that may be blocking the well screen or the aquifer outside the well screen.

Well evacuation—process of removing stagnant water from a well prior to sampling.

Well interference—the result of two or more pumping wells, the drawdown cones of which intercept. At a given location, the total well interference is the sum of the drawdowns due to each individual well.

Well screen—a tubular device with either slots, holes, gauze, or continuous wire wrap; used at the end of a well casing to complete a well. The water enters the well through the well screen.

well, Abandoned—a well which is no longer used or a well removed from service; a well whose use has been permanently discontinued or which is in a state of such disrepair that it cannot be used for its intended purpose. Generally, abandoned wells will be filled with concrete or cement grout to protect groundwater from waste and contamination.

well, Fully penetrating—a well drilled to the bottom of an aquifer, constructs d in such a way that it withdraws water from the entire thickness of the aquifer.

well, Partially penetrating—the well constructed in such a way that it draws water directly from a fractional part of the total thickness of the aquifer. The fractional part may be located at the top or the bottom or anywhere in between in the aquifer.

Wet land—land that is saturated with water or submerged, at least during most of the growing season; wetlands generally include swamps, marshes and bogs.

Wet land migration—regulatory requirement to replace wetland areas destroyed by proposed land disturbances with artificially created wetland areas.

Wet weight (of soil sample)—weight of soil sample and included soil moisture.

Wettability—the relative degree to which a fluid will spread into solid surface in the presence of other immiscible fluids.

Wetted area—surface area wetted at completion of irrigation.

Wetting agent—chemical used to reduce the surface tension of a liquid causing it to make better contact with the desired target.

WHO—World Health Organization.

Whole tree—all components of a tree, except the stump. Also known as a full tree.

Wild fire—unplanned fire requiring suppression action. Can be contrasted with a prescribed fire that burns within prepared lines, enclosing a designated area, under predetermined conditions.

Wild life—a broad term that includes nondomesticated vertebrates, especially mammals, birds, and fish.

Wild life & fish user days—a 12-hour day in which a person participates in a wildlife- or fish-related recreation activity.

wilting percentage, Permanent—is a seasonal wilting shown by the plants due to the deficiency of water to compensate the loss by transpiration. This is called incipient wilting.

Wilting point—the soil moisture content when the soil has dried to the point that evaporation and transpiration cease. This is a property that varies among different soil types.

Winch—steel spool connected to a power source. Used for reeling or unreeling cable. Also known as drum.

Wind—air in natural motion parallel to the surface of the earth. Data parameters commonly measured are velocity (miles per hours) and directions in degrees from true north (clockwise).

Wind fall—tree or trees that have been uprooted or broken off by the wind. Also known as blow down.

Wind row—a long, narrow row of vegetation, debris, and some soil created during site preparation and clearing operations.

Wind shake—crack or separation in a tree's wood structure caused by high winds.

Wind speed—the wind speed at the height of interest (surface, plume height, or upper air).

Wind throw—trees uprooted by excessive wind. Shallow-rooted trees are almost always affected.

Wind velocity—speed of wind at a height of two meters from the soil surface in an environment without barriers.

Wing Deflector—a triangular structure constructed from stone or logs that is used to narrow and deepen streams and protect eroding banks. Double wing deflectors are constructed on opposite bank.

Winterization—the process of removing water from an irrigation system before the onset of freezing temperatures. Necessary to prevent damage to the sprinkler system that can be caused by expansion due to freezing water in the pipes.

Wire—in an automatic sprinkler system, low voltage direct burial wire is used to connect the automatic control valves to the controller. The most frequently used wire for commercial applications is single strand, heavy gauge direct burial copper wire (The larger the gauge number, the thinner the wire). The most frequently used wire for the home sprinkler system is multistrand. Color-coded, multistrand sprinkler wire has several coated wires together in one protective jacket.

Wire gauge—standard unit of measure for wire size. The larger the gauge number, the smaller the wire.

Withdrawals—lands that have been removed or segregated from the operation of some or all of the public land laws through Executive or Congressional action.

Witness tree—tree used by surveyors to mark the location of a survey corner; the tree is located near the survey corner and is inscribed with survey data. Also known as a bearing tree.

Wolf tree—large rough tree, generally not good for lumber.

Wood conversion—transformation of natural timber into any kind of commercial product. Includes all activities from commercial timber (log) delivery to the log yard at the initial commercial processing facility to the final product form offered for commercial sale as a consumer product.

Work—work done by a force on a particle is defined as the product of the magnitude of the force and the distance through which the particle moves. In hydraulic systems, it can be calculated as the product of the pressure and flow rate. *brake hp:* Measure of work input into a pump or other device; *water hp:* Measure of work output from a pump.

Working pressure—the pressure of the irrigation system during operation. Synonymous with Dynamic Pressure.

Working storage—amount of water available in the soil profile for plant use after consideration of MAD. Preferred term is *Allowable Depletion* [AD].

X-ray diffraction—an analytical technique used for mineralogical characterization. A sample is exposed to a filtered and monochromatic beam of X-rays and the reflected energy is measured and used to identify soil colloid-type, degree of interleaving, or in tertrtification, and variations in interplatelet spacings.

X-rays—electromagnetic radiation with a very short wavelength (0.01 to 12 nanometers), shorter than ultraviolet radiation.

Xenobiotic—any biological substance, displaced from its normal habitat; a chemical foreign to a biological system.

Xeriscape—a landscaping method that employs drought-resistant plants in an effort to conserve water.

Year class—part of a species population consisting of all individuals produced during a given year.

Yield—The rate of production of cake from a dewatering device; the amount of product water produced by a water treatment process.

Zeolites—hydrated sodium alumina silicates, either naturally occurring mined products or synthetic product, with ion exchange properties.

Zero discharge—complete recycling of water; discharge of essentially pure water; discharge of a treated effluent containing no substance at a concentration higher than that found normally in the local environment.

Zero discharge water—the principle of "zero discharge" is recycling of all industrial wastewater. This means that wastewater will be treated and used again in the process. Because of the water reuse wastewater will not be released on the sewer system or surface water.

zero soft water—water produced by cation exchange process and measuring less than 1.0 grain per U.S. gallon (17.1 ppm or 17.1 mg/L) as calcium carbonate.

Zeta potential—an electrokinetic measurement which can be used for the control of coagulation processes.

Zone—a zone is the area to be watered by a single control valve. Zones are ideally comprised of similar sprinkler types and plants with similar water requirements. This term is usually used with domestic sprinkler systems.

Zone of aeration—layer of rock or soil (or both) not fully saturated with water but instead containing both air and water in its pores; also called unsaturated zone.

Zone of potential contaminant migration—any subsurface formation or layer which is permeable and would preferentially channel the flow of contaminants away from a regulated facility.

Zone of saturation—the layer in the ground in which all available interstitial voids, cracks, crevices, holes are filled with water. The level of the top of this zone is the water table.

Zoning—a legal form of land-use control and building regulations usually exercised by a municipal authority; usually involves setting aside of distinct land areas for specific purposes, such as commercial, educational or residential development.

Zooplankton—group of animals whose distributions in the water column are governed primarily by currents; sing. -zoo-plankter: include protozoans, entomostracans, and various larvae.

Zwitter ions—act as cations or as anions according to the environment in which they find themselves. In water technology they are usually organic macromolecules.

REFERENCES

1. Agriculture, **1997**. *Agriculture: A Glossary of Terms, Programs, Laws*. Committee for the National Institute for the Environment. Congressional Research Service. Report to Congress. Washington, DC.
2. Alberta Environment, **2002**. *Glossary of Terms Relating to Governing and Governance and the Tools Associated with Each.*
3. Alberta Environment, **2008**. *Alberta's Drinking Water Program: A 'Source to Tap, Multi-barrier' Approach.*
4. Alberta Environment, **2008**. *Watershed Management Planning.*
5. Alberta Environment. *Glossary of Water Management Terminology.*
6. Alberta Environment. *Wetland Restoration Program Guide.*
7. Alberta Water Council, **2007**. *Wetland Consultation Workbook: Talking with Albertans about a new Wetland Policy and*

8. American Society of Agricultural and Biological Engineers, **1985**. *ASAE Standards.*
9. ASAE Engineering Practice EP40: Design and Installation of Micro irrigation Systems. American Society of Agricultural Engineers.
10. ASAE Standard EP405. ASAE, St. Joseph—Michigan. *An Introduction to Cross Connection. Foundation for Cross-Connection Control and Hydraulic Research.* University of Southern California. Los Angeles, CA.
11. ASAE Standard EP405.1. *Design and Installation of Microirrigation Systems.* ASAE Standards 2000. American Society of Agricultural Engineers. St. Joseph, MI.
12. ASAE Standard S376.1, **1985**. *Design, Installation and Performance of Underground Thermoplastic Irrigation Pipelines.*
13. ASAE Standard S526.1, **1998**. *Soil and Water Terminology.* ASAE St. Joseph, MI 49085–9659 USA
14. ASAE Standards, **2000**. American Society of Agricultural Engineers. St. Joseph, MI.
15. ASAE/ANSI Standard S261.7, **1996**. *Design and Installation of Nonreinforced Concrete Irrigation Pipe Systems.*
16. ASAE/ANSI Standard S376.2, January **1998**. *Design, Installation and Performance of Underground, Thermoplastic Irrigation Pipelines.* ASAE Standards 2000. American Society of Agricultural Engineers. St. Joseph, MI.
17. ASCE, **1990**. *ASCE Manual and Reports on Engineering Practice 70.* American Society of Civil Engineering.
18. Bates, R. L.; Jackson, J. A. Eds., 1980. *Glossary of Geology.* 2nd edition. American Geological Institute, Falls Church –VA. 749 p.
19. Black, C. A., **1965**. *Methods of Soil Analysis, Part 1.* Number 9, Agronomy. American Society of Agronomy. Madison, WI.
20. Brady, N.C., **1990**. *The Nature and Properties of Soils.* MacMillan Publishing Company. New York.
21. Buckman, H. O.; Brady, N. C. **1961**. *The Nature and Properties of Soils.* The Macmillan Co., New York.
22. Burt, C. M., **1998**. *Ag-Irrigation Management.* Irrigation Training and Research Center. California Polytechnic State University. San Luis Obispo, CA
23. Burt, C. M.; Clemmons, A. J.; Strelkoff, T. S.; Solomon, K. H.; Bliesner, R. D.; Hardy, L. A.; Howell, T. A.; Eisenhauer, D. E. **1997**. *Irrigation performance measures: Efficiency and uniformity.* Journal of Irrigation and Drainage Engineering. November. Chemigation Symposium by The Irrigation Association. Falls Church, VA, USA.
24. *Certification Program Reference Manual for Step 2 and 3 Examinations,* **1998**. The Irrigation Association. Falls Church, VA, USA, 22042–26638.
25. Contractor, **1999**. *Certified Irrigation Contractor Workbook.* Second Edition. The Irrigation Association. Falls Church, VA, USA, 22042–26638.
26. Derryberry, B., **1994**. *Trouble Shooting Irrigation Control Systems.* Derryberry Irrigation. Scottsdale, AZ.
27. Development, **2001**. *Implementation Plan.*
28. Drip, **1997**, and **1999**, *Drip Irrigation in the Landscape.* The Irrigation Association. Falls Church, VA, USA 22042–6638.
29. Efficiency, **1997**, *High Efficiency Design.* The Irrigation Association. Falls Church, VA, USA 22042–6638.
30. Giambelluca, T.; *Water in the Environment.* Course GEOG-405.
31. *Glossary of Soil Science Terms,* **1997**, Soil Science Society of America.
32. Goyal, Megh R.; Editor, **2012**, *Glossary of Technical Terms.* In: Management of Drip/Trickle or Micro Irrigation. Apple Academic Press Inc.; New Jersey—USA.
33. Head Layout, **1998**, *Advanced Head Layout and Zoning.* The Irrigation Association. Falls Church, VA, USA 22042–6638.
34. Hess, T. M.; (Editor), **1999**, *Glossary of Soil Water and Plant Terminology.* Institute of Water and Environment. Cranfield.
35. ITRC, **2001**, *Pump efficiency Testing.* Irrigation Training and Research Center, California Polytechnic State University. San Luis Obispo, CA 93407.
36. James, L. G.; **1988**, *Principles of Farm Irrigation System Design.* John Wiley and Sons.
37. Jensen, M. E.; Editor. **1980**, *Design and Operation of Farm Irrigation Systems.* American Society of Agricultural Engineers. St. Joseph, MI.

38. Keesen, L.; **1995,** *The Complete Irrigation Workbook.* Franzak & Foster, div. of, G. I. E.; Inc.
39. Landscape, **1996,** *Landscape Irrigation Auditor Training Manual.* The Irrigation Association. Falls Church, VA, USA 22042–6638
40. Landscape, **2000,** *Landscape Irrigation Auditor Training Manual.* The Irrigation Association. Falls Church, VA 22042–6638.
41. LD, **1977,** *Landscape Drainage.* The Irrigation Association. Falls Church, VA, USA 22042–26638.
42. Monroe, B. R.; **1993,** *Precipitation Rates and Sprinkler Irrigation (Instructor's Manual).* Hunter Industries Irrigation Design Education Module. Hunter Industries. San Marcos, CA.
43. Monroe, B. R.; **1995,** *Irrigation Hydraulics.* (Instructor's Manual). Hunter Industries Irrigation Design Education Module. Hunter Industries. San Marcos, CA.
44. National Technical Information Service, **1980,** *Water Resources Thesaurus.*
45. NRCS, **1997,** *Glossary and REFERENCES, Chapter 17. Irrigation Guide, National Engineering Handbook.* USDA, Natural Resource Conservation Service. Washington, DC.
46. Oliphant, Joe C.; **1989,** Modeling Sprinkler Coverage with the SPACE Program. CATI Publication No. 890802. Center for Irrigation Technology. California State University. Fresno, CA.
47. On-Farm Irrigation Committee, **1978,** *Describing Irrigation Efficiency and Uniformity.* Journal of Irrigation and Drainage Division. IR1, March, **1978,** ASCE.
48. *Predicting and Estimating Landscape Water Use,* **2000,** The Irrigation Association. Falls Church, VA.
49. RainBird, **1997,** *Landscape Irrigation Products 1997–1998 Catalog.* Glendora, CA.
50. Resnick, R.; Halliday, D. **1962,** *Physics for Students of Science and Engineering.* Part I.; John Wiley.
51. Rochester, E. W.; **1995,** *Landscape Irrigation Design.* ASAE. St. Joseph, Michigan.
52. Schwab, G. O.; Fangmeier, D. D.; Elliot, W. J.; Frevert, R. K.; **1993,** *Soil and Water Conservation Engineering.* Fourth ed. John Wiley and Sons.
53. Smith, S. W.; **1997,** *Landscape Irrigation.* John Wiley & Sons. New York.
54. Snedecor, G. W.; Cochran, W. G.; **1967,** *Statistical Methods.* Iowa State University Press. Ames, Iowa.
55. Solomon, Kenneth H.; **1988,** *A New Way to View Sprinkler Patterns.* CATI Publication No. 880802. Center for Irrigation Technology. California State University. Fresno, CA.
56. Solomon, Kenneth H.; David F.; Zoldoske, Joe C.; Oliphant. *Laser Optical Measurement of Sprinkler Drop Sizes.* CATI Publication No. 961101. Center for Irrigation Technology. California State University. Fresno, CA.
57. *Sprinkler System Scheduling.* The Irrigation Association. Falls Church, VA, USA 22042–26638
58. SREM Office- Alberta—Canada, **2006,** *Sustainable Resource and Environmental Management Glossary.*
59. Tisdale, S. L.; Nelson, W. L.; **1975,** *Soil Fertility and Fertilizer.* 3rd edition. Macmillan Publishing Co. New York.
60. *Understanding Pumps, Controls and Wells.* The Irrigation Association. Falls Church, VA, USA 22042–6638
61. *Understanding Pumps.* The Irrigation Association. Falls Church, VA, USA 22042–26638
62. US Geological Survey, **1982,** *Water Resources Data for Maryland and Delaware Water Year **1981,*** Washington, D. C.
63. USDA, **1993,** *USDA Soil Conservation Service—National Engineering Handbook,* Part *623,* chapter 2 Irrigation Water Requirements.
64. Vickers, Amy, **2001,** *Handbook of Water Use and Conservation.* Amherst—MA: WaterPlow Press.
65. *Watershed* by North Saskatchewan Watershed Alliance.
66. Webster, **1981,** Webster's New Collegiate Dictionary. G. C.; Merriam Company. Springfield, Massachusetts.
67. Weinberg, S. S.; J. M.; Roberts (Editors), **1988,** *The Handbook of Landscape Architectural Construction.* Volume III: Irrigation. Landscape Architecture Foundation. Washington, DC.
68. Zoldoske, D. F.; Solomon, K. H.; Norum, E. M.; **1994,** CATI Publication No. 941102. Center for Irrigation Technology. California State University. Fresno, CA.
69. http://edis.ifas.ufl.edu Forrest T. Izuno and Dorota Z. Haman, 2002. *Basic irrigation terminology.* Publication # AE66, one of a series of the Agricultural and Biological Engineering Department, Florida Cooperative Extension Service, Institute of Food and Agricultural Sciences, University of Florida. Revised July, **2002.**
70. http://environment.alberta.ca/documents/FinalReport.pdf *Recommendations for Septage Management in Alberta* by Alberta Environment: Septage Management Advisory Committee, **2004.**

71. http://environment.alberta.ca/documents/SSRB_Plan_Phase2.pdf *Approved Water Management Plan for the South Saskatchewan River Basin (Alberta)* by Alberta Environment, **2006**.

72. http://environment.alberta.ca/forecasting/FAQ/definitions.html *Flow Forecasting Frequently Asked Questions—Definitions*/Terminology by Alberta Environment.

73. http://environment.gov.ab.ca/info/library/5713.pdf *Surface Water Quality Guidelines for Use in Alberta* by Alberta Environment, **1999**.

74. http://environment.gov.ab.ca/info/library/6367.pdf *Framework for Water Management Planning* by Alberta Environment, **1999**.

75. http://environment.gov.ab.ca/info/library/6979.pdf *Standards and Guidelines for Municipal Waterworks, Wastewater and Storm Drainage Systems* by Alberta Environment, 2006.

76. http://environment.gov.ab.ca/info/library/7255.pdf *Water-Quality Based Effluent Limits Procedures Manual* by Alberta Environment, **1995.**

77. http://environment.gov.ab.ca/info/library/7712.pdf *Glossary of Terms Related to Shared Environment Management* by Alberta Environment, **2007**.

78. http://environment.gov.ab.ca/info/library/7864.pdf *The Water Management Framework for the Industrial Heartland and Capital Region* by Alberta Environment, **2008**.

79. http://environment.gov.ab.ca/info/posting.asp?assetid=6209&categoryid=5 *Inspection of Small Dams* by Alberta Environment, 1998

80. http://espwaterproducts.com/water-purification-terms.htm

81. http://forestry.about.com/library/glossary/

82. http://forestry.about.com/library/glossary/ Tree glossary

83. http://forestry.about.com/od/environmentalissues/a/forest_ecosystem.htm

84. http://irrigation.punjab.gov.pk/IrrigationGlossary.aspx

85. http://irrigation.wsu.edu/Content/Resources/Irrigation-Glossary.php

86. http://laws.justice.gc.ca/en/ showFullDoc/cs/F-14///en Canada Fisheries Act by Government of Canada, **1985**.

87. http://www.aadnc-aandc.gc.ca/eng/1100100028092/1100100028094

88. http://www.albertawatercouncil.ca/Portals/0/pdfs/CEP_Definitions_Final_Report.pdf *Water Conservation, Efficiency, and Productivity: Principles, Definitions, Performance Measures, and Indicators* by Alberta Water Council, **2007**.

89. http://www.asla.org/nonmembers/publicrelations/glossary.htm

90. http://www.awchome.ca *Strengthening Partnerships: A Shared Governance Framework for Water for Life Collaborative Partnerships* by Alberta Water Council, **2008**.

91. http://www.bom.gov.au/water/awid/initial-a.shtml

92. http://www.brbc.ab.ca/pdfs/Guidebook.pdf *Guidebook to Water Management: Background Information on Organizations, Policies, Legislation, Programs, and Projects in the Bow River Basin* by Bow River Basin Council, **2002**.

93. http://www.calgary.ca/CSPS/Parks/Pages/Planning-and-Operations/Water-Management/Park-irrigation-glossary.aspx

94. http://www.cawater-info.net/bk/glossary/water_management/en/index.htm

95. http://www.cefns.nau.edu/Projects/WDP/resources/Definitions.htm#references *Glossary of Terms for Onsite Wastewater Treatment and Dispersal.*

96. http://www.chrs.ca/Main_e.htm Canadian Heritage River Society.

97. http://www.climatenetwork.org

98. http://www.dripworks.com/product/glossary/

99. http://www.environment.alberta.ca/documents/Provincial_Wetland_Restoration_Compen sation_Guide_Feb_2007.pdf *Wetland Restoration/Compensation Guide* by Alberta NAWMP Partnership and Alberta Environment, **2007**.

100. http://www.environment.gov.ab.ca/stakeholder/search.aspx?region=Provincewide *Getting Involved: Initiatives* by Alberta Environment.

101. http://www.epa.gov/nps/watershed_handbook/pdf/glossary.pdf *Handbook for Developing Watershed Plans to Restore and Protect Our Waters, Glossary* by United States Environmental Protection Agency, **2008**.

102. http://www.ext.colostate.edu/pubs/crops/04717.html

103. http://www.irrigation.org/Resources/Irrigation_Glossary_Pages/A-C.aspx

104. http://www.irrigationtutorials.com/glossary.htm

105. http://www.lenntech.com/dictionary/absolute-rating.htm
106. http://www.lenntech.com/greenhouse-effect/climate-change-glossary.htm#ixzz28FtCimIL
107. http://www.lenntech.com/water-glossary.htm#ixzz28Fsh02He
108. http://www.municipalaffairs.gov.ab.ca/Handbook_index.cfm *Alberta Private Sewage Systems Standard of Practice Handbook* by Alberta Municipal Affairs, 2000.
109. http://www.nalms.org/Resources/Glossary.aspx *Water Words Glossary* by North American Lake Management Society.
110. http://www.nationalwatermarket.gov.au/site-information/water-words.html
111. http://www.nswa.ab.ca/pdfs/muni_guide_glossary.pdf *Municipal Guide: Planning for a Healthy and Sustainable North Saskatchewan.*
112. http://www.nwb.nunavut.ca
113. http://www.qp.gov.ab.ca/documents/Acts/A07.cfm?frm_isbn=0779742621 *Agricultural Operation Practices Act* by Government of Alberta, 2000
114. http://www.qp.gov.ab.ca/documents/Acts/E12.cfm?frm_isbn=0779717287 *Environmental Protection and Enhancement Act* by Government of Alberta, 2000
115. http://www.qp.gov.ab.ca/documents/Acts/W03.cfm?frm_isbn=0779711424 *Water Act* by Government of Alberta, 1999
116. http://www.qp.gov.ab.ca/documents/Regs/2003_276.cfm?frm_isbn=0779750616 *Activities Designation Regulation* by Government of Alberta, 2003
117. http://www.rainbird.com/homeowner/education/glossary.htm
118. http://www.silsoe.cranfield.ac.uk/watermanagement/documents/glossary.pdf.
119. http://www.sun showerservices.com/irrigation-glossary.html
120. http://www.usc.edu/dept/fcchr/intro.html
121. http://www.waterforlife.alberta.ca/ *Glossary of terms related to water and watershed management in Alberta.* First edition. By Partnerships & Strategies Section of Alberta Environment, November **2008**.
122. http://www.waterforlife.gov.ab.ca *Water for Life: Alberta's Strategy for Sustainability* by Government of Alberta, **2003**.
123. http://www.waterforlife.gov.ab.ca/docs/EnablingPartnerships.pdf *Enabling Partnerships: A framework in support of Alberta's Water for Life Strategy for Sustainability* by Government of Alberta, **2005**.
124. http://www.waterforlife.gov.ab.ca/docs/Oilfield_Injection_GUIDELINE.pdf *Water Conservation and Allocation Guideline for the Oilfield Industry* by Government of Alberta, **2006**.
125. http://www.waterforlife.gov.ab.ca/docs/Oilfield_Injection_Policy.pdf *Water Conservation and Allocation Policy for Oilfield Injection* by Government of Alberta, **2006**.
126. http://www.waterforlife.gov.ab.ca/html/background4.html *Valuing water: pricing and other economic* instruments by Government of Alberta.
127. http://www1.agric.gov.ab.ca/$department/deptdocs.nsf/all/wat2451 *Water Quality Terms and Definitions* by Alberta Agriculture and Rural.

APPENDICES

(Modified and reprinted with permission from: Goyal, Megh R., 2012. Appendices. Pages 317–332. In: *Management of Drip/Trickle or Micro Irrigation* edited by Megh R. Goyal. New Jersey, USA: Apple Academic Press Inc.)

APPENDIX A

CONVERSION SI AND NON-SI UNITS

To convert the Column 1 in the Column 2, Multiply by	Column 1 Unit SI	Column 2 Unit Non-SI	To convert the Column 2 in the Column 1 Multiply by

LINEAR

0.621 ------ kilometer, km (10^3 m)	miles, mi ------------------	1.609
1.094 ------ meter, m	yard, yd --------------------	0.914
3.28 ------- meter, m	feet, ft ----------------------	0.304
3.94×10^{-2} ---- millimeter, mm (10^{-3})	inch, in -------------------	25.4

SQUARES

2.47 ------- hectare, he	acre ---------------------	0.405
2.47 ------- square kilometer, km²	acre ---------------------	4.05×10^{-3}
0.386 -------- square kilometer, km²	square mile, mi² ------------	2.590
2.47×10^{-4} ---- square meter, m²	acre ---------------------	4.05×10^{-3}
10.76 -------- square meter, m²	square feet, ft² --------------	9.29×10^{-2}
1.55×10^{-3} ---- mm²	square inch, in² --------------	645

CUBICS

9.73×10^{-3} ---- cubic meter, m³	inch-acre -----------------	102.8
35.3 -------- cubic meter, m³	cubic-feet, ft³ ----------------	2.83×10^{-2}
6.10×10^4 ---- cubic meter, m³	cubic inch, in³ -------------	1.64×10^{-5}
2.84×10^{-2} ---- liter, L (10^{-3} m³)	bushel, bu ------------------	35.24
1.057 -------- liter, L	liquid quarts, qt ------------	0.946
3.53×10^{-2} ---- liter, L	cubic feet, ft³ --------------	28.3
0.265 -------- liter, L	gallon --------------------	3.78
33.78 -------- liter, L	fluid ounce, oz ------------	2.96×10^{-2}
2.11 ------- liter, L	fluid dot, dt --------------	0.473

WEIGHT

2.20×10^{-3} ---- gram, g (10^{-3} kg)	pound, --------------------	454
3.52×10^{-2} ---- gram, g (10^{-3} kg)	ounce, oz ------------------	28.4
2.205 ------ kilogram, kg	pound, lb -----------------	0.454
10^{-2} ------- kilogram, kg	quintal (metric), q ----------	100
1.10×10^{-3} ---- kilogram, kg	ton (2000 lbs), ton ----------	907
1.102 ------ mega gram, mg	ton (US), ton --------------	0.907
1.102 ------ metric ton, t	ton (US), ton --------------	0.907

YIELD AND RATE

0.893 ------- kilogram per hectare	pound per acre ------------	1.12
7.77×10^{-2} --- kilogram per cubic meter	pound per fanega ----------	12.87
1.49×10^{-2} --- kilogram per hectare	pound per acre, 60 lb -----	67.19
1.59×10^{-2} --- kilogram per hectare	pound per acre, 56 lb -----	62.71
1.86×10^{-2} --- kilogram per hectare	pound per acre, 48 lb -----	53.75
0.107 ------- liter per hectare	gallon per acre ---------	9.35
893 ---------- ton per hectare	pound per acre ----------	1.12×10^{-3}
893 ---------- mega gram per hectare	pound per acre ----------	1.12×10^{-3}
0.446------- ton per hectare	ton (2000 lb) per acre -----	2.24
2.24 ---------- meter per second	mile per hour ------------	0.447

SPECIFIC SURFACE

10 --------- square meter per kilogram	square centimeter per gram ------------------	0.1
10^3 ---------- square meter per kilogram	square millimeter per gram ------------------	10^{-3}

PRESSURE

9.90 ---------- megapascal, MPa	atmosphere -----------	0.101
10 --------- megapascal	bar -------------------	0.1
1.0 ---------- megagram per cubic meter	gram per cubic centimeter --------------	1.00
2.09×10^{-2} ---- pascal, Pa	pound per square feet ------	47.9
1.45×10^{-4} ---- pascal, Pa	pound per square inch -----	6.90×10^3

To convert the column 1 in the Column 2, Multiply by	Column 1 Unit SI	Column 2 Unit Non-SI	To convert the column 2 in the column 1 Multiply by

TEMPERATURE

1.00 (K-273)--- Kelvin, K   centigrade, °C -------- 1.00 (C+273)
(1.8 C + 32)--- centigrade, °C   Fahrenheit, °F -------- (F--32)/1.8

ENERGY

9.52×10^{-4} ---- Joule J   BTU ------------------ 1.05×10^3
0.239 -------- Joule, J   calories, cal ------------ 4.19
0.735 -------- Joule, J   feet-pound ------------ 1.36
2.387×10^5 --- Joule per square meter   calories per square centimeter --- 4.19×10^4
10^5 ---------- Newton, N   dynes ----------------- 10^{-5}

WATER REQUIREMENTS

9.73×10^{-3} --- cubic meter   inch acre --------------- 102.8
9.81×10^{-3} --- cubic meter per hour   cubic feet per second ------   101.9
4.40 ---------- cubic meter per hour   galloon (US) per minute ----   0.227
8.11 ---------- hectare-meter   acre-feet --------------- 0.123
97.28 ------- hectare-meter   acre-inch --------------- 1.03×10^{-2}
8.1×10^{-2} ---- hectare centimeter   acre-feet --------------- 12.33

CONCENTRATION

1 ------------ centimol per kilogram   milliequivalents per
                                               100 grams -------------- 1
0.1 --------- gram per kilogram   percents ---------------- 10
1 ------------ milligram per kilogram   parts per million --------- 1

NUTRIENTS FOR PLANTS

2.29 -------- P P_2O_5 --------------------   0.437
1.20 -------- K K_2O --------------------   0.830
1.39 -------- Ca   CaO --------------------   0.715
1.66 ------- Mg   MgO -----------------   0.602

NUTRIENT EQUIVALENTS

| | | Conversion | Equivalent |
Column A	Column B	A to B	B to A
N	NH_3	1.216	0.822
	NO_3	4.429	0.226
	KNO_3	7.221	0.1385
	$Ca(NO_3)_2$	5.861	0.171

Column A	Column B	Conversion A to B	Equivalent B to A
	$(NH_4)_2SO_4$	4.721	0.212
	NH_4NO_3	5.718	0.175
	$(NH_4)_2HPO_4$	4.718	0.212
P	P_2O_5	2.292	0.436
	PO_4	3.066	0.326
	KH_2PO_4	4.394	0.228
	$(NH_4)_2HPO_4$	4.255	0.235
	H_3PO_4	3.164	0.316
K	K_2O	1.205	0.83
	KNO_3	2.586	0.387
	KH_2PO_4	3.481	0.287
	Kcl	1.907	0.524
	K_2SO_4	2.229	0.449
Ca	CaO	1.399	0.715
	$Ca(NO_3)_2$	4.094	0.244
	$CaCl_2 \times 6H_2O$	5.467	0.183
	$CaSO_4 \times 2H_2O$	4.296	0.233
Mg	MgO	1.658	0.603
	$MgSO_4 \times 7H_2O$	1.014	0.0986
S	H_2SO_4	3.059	0.327
	$(NH_4)_2SO_4$	4.124	0.2425

Column A	Column B	Conversion A to B	Equivalent B to A
	K_2SO_4	5.437	0.184
	$MgSO_4 \times 7H_2O$	7.689	0.13
	$CaSO_4 \times 2H_2O$	5.371	0.186

APPENDIX B

PIPE AND CONDUIT FLOW

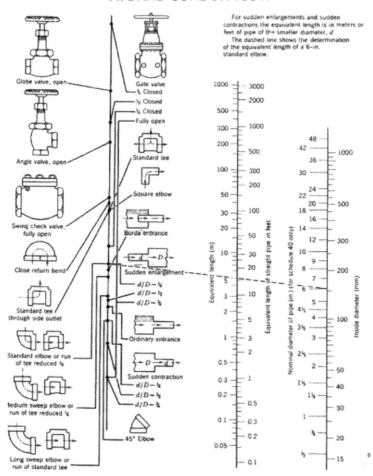

For sudden enlargements and sudden contractions the equivalent length is in meters or feet of pipe of the smaller diameter, d. The dashed line shows the determination of the equivalent length of a 6-in. standard elbow.

APPENDIX C

PERCENTAGE OF DAILY SUNSHINE HOURS: FOR NORTH AND SOUTH HEMISPHERES

Latitude	Jan	Feb	Mar	Apr	May	Jun	Jul	Aug	Sep	Oct	Nov	Dec
					NORTH							
0	8.50	7.66	8.49	8.21	8.50	8.22	8.50	8.49	8.21	8.50	8.22	8.50
5	8.32	7.57	8.47	3.29	8.65	8.41	8.67	8.60	8.23	8.42	8.07	8.30
10	8.13	7.47	8.45	8.37	8.81	8.60	8.86	8.71	8.25	8.34	7.91	8.10
15	7.94	7.36	8.43	8.44	8.98	8.80	9.05	8.83	8.28	8.20	7.75	7.88
20	7.74	7.25	8.41	8.52	9.15	9.00	9.25	8.96	8.30	8.18	7.58	7.66
25	7.53	7.14	8.39	8.61	9.33	9.23	9.45	9.09	8.32	8.09	7.40	7.52
30	7.30	7.03	8.38	8.71	9.53	9.49	9.67	9.22	8.33	7.99	7.19	7.15
32	7.20	6.97	8.37	8.76	9.62	9.59	9.77	9.27	8.34	7.95	7.11	7.05
34	7.10	6.91	8.36	8.80	9.72	9.70	9.88	9.33	8.36	7.90	7.02	6.92
36	6.99	6.85	8.35	8.85	9.82	9.82	9.99	9.40	8.37	7.85	6.92	6.79
38	6.87	6.79	8.34	8.90	9.92	9.95	10.1	9.47	3.38	7.80	6.82	6.66
40	6.76	6.72	8.33	8.95	10.0	10.1	10.2	9.54	8.39	7.75	6.72	7.52
42	6.63	6.65	8.31	9.00	10.1	10.2	10.4	9.62	8.40	7.69	6.62	6.37
44	6.49	6.58	8.30	9.06	10.3	10.4	10.5	9.70	8.41	7.63	6.49	6.21
46	6.34	6.50	8.29	9.12	10.4	10.5	10.6	9.79	8.42	7.57	6.36	6.04
48	6.17	6.41	8.27	9.18	10.5	10.7	10.8	9.89	8.44	7.51	6.23	5.86
50	5.98	6.30	8.24	9.24	10.7	10.9	11.0	10.0	8.35	7.45	6.10	5.64
52	5.77	6.19	8.21	9.29	10.9	11.1	11.2	10.1	8.49	7.39	5.93	5.43
54	5.55	6.08	8.18	9.36	11.0	11.4	11.4	10.3	8.51	7.20	5.74	5.18
56	5.30	5.95	8.15	9.45	11.2	11.7	11.6	10.4	8.53	7.21	5.54	4.89
58	5.01	5.81	8.12	9.55	11.5	12.0	12.0	10.6	8.55	7.10	4.31	4.56
60	4.67	5.65	8.08	9.65	11.7	12.4	12.3	10.7	8.57	6.98	5.04	4.22
					SOUTH							
0	8.50	7.66	8.49	8.21	8.50	8.22	8.50	8.49	8.21	8.50	8.22	8.50
5	8.68	7.76	8.51	8.15	8.34	8.05	8.33	8.38	8.19	8.56	8.37	8.68
10	8.86	7.87	8.53	8.09	8.18	7.86	8.14	8.27	8.17	8.62	8.53	8.88
15	9.05	7.98	8.55	8.02	8.02	7.65	7.95	8.15	8.15	8.68	8.70	9.10
20	9.24	8.09	8.57	7.94	7.85	7.43	7.76	8.03	8.13	8.76	8.87	9.33
25	9.46	8.21	8.60	7.74	7.66	7.20	7.54	7.90	8.11	8.86	9.04	9.58
30	9.70	8.33	8.62	7.73	7.45	6.96	7.31	7.76	8.07	8.97	9.24	9.85
32	9.81	8.39	8.63	7.69	7.36	6.85	7.21	7.70	8.06	9.01	9.33	9.96
34	9.92	8.45	8.64	7.64	7.27	6.74	7.10	7.63	8.05	9.06	9.42	10.1
36	10.0	8.51	8.65	7.59	7.18	6.62	6.99	7.56	8.04	9.11	9.35	10.2
38	10.2	8.57	8.66	7.54	7.08	6.50	6.87	7.49	8.03	9.16	9.61	10.3

40	10.3	8.63	8.67	7.49	6.97	6.37	6.76	7.41	8.02	9.21	9.71	10.5
42	10.4	8.70	8.68	7.44	6.85	6.23	6.64	7.33	8.01	9.26	9.8	10.6
44	10.5	8.78	8.69	7.38	6.73	6.08	6.51	7.25	7.99	9.31	9.94	10.8
46	10.7	8.86	8.90	7.32	6.61	5.92	6.37	7.16	7.96	9.37	10.1	11.0

APPENDIX D

PSYCHOMETRIC CONSTANT (Γ) FOR DIFFERENT ALTITUDES (Z)

$$\gamma = 10^{-3} \left[(C_p \cdot P) \div (\varepsilon.\lambda) \right] = (0.00163) \times [P \div \lambda]$$

γ, psychrometric constant [kPa C^{-1}] c_p, specific heat of moist air = 1.013

[kJ kg^{-10}C^{-1}] P, atmospheric pressure [kPa].

ε, ratio molecular weight of water vapor/dry air = 0.622 λ, latent heat of vaporization [MJ kg^{-1}]

= 2.45 MJ kg^{-1} at 20°C.

Z (m)	γ kPa/°C	z (m)	γ kPa/°C	z (m)	γ kPa/°C	z (m)	γ kPa/°C
0	0.067	1000	0.060	2000	0.053	3000	0.047
100	0.067	1100	0.059	2100	0.052	3100	0.046
200	0.066	1200	0.058	2200	0.052	3200	0.046
300	0.065	1300	0.058	2300	0.051	3300	0.045
400	0.064	1400	0.057	2400	0.051	3400	0.045
500	0.064	1500	0.056	2500	0.050	3500	0.044
600	0.063	1600	0.056	2600	0.049	3600	0.043
700	0.062	1700	0.055	2700	0.049	3700	0.043
800	0.061	1800	0.054	2800	0.048	3800	0.042
900	0.061	1900	0.054	2900	0.047	3900	0.042
1000	0.060	2000	0.053	3000	0.047	4000	0.041

APPENDIX E

SATURATION VAPOR PRESSURE [e_s] FOR DIFFERENT TEMPERATURES (T)

Vapor pressure function = $e_s = [0.6108]*\exp\{[17.27*T]/[T + 237.3]\}$

T °C	e_s kPa	T °C	e_s kPa	T °C	e_s kPa	T °C	e_s kPa
1.0	0.657	13.0	1.498	25.0	3.168	37.0	6.275
1.5	0.681	13.5	1.547	25.5	3.263	37.5	6.448
2.0	0.706	14.0	1.599	26.0	3.361	38.0	6.625
2.5	0.731	14.5	1.651	26.5	3.462	38.5	6.806

3.0	0.758	**15.0**	1.705	**27.0**	3.565	**39.0**	6.991
3.5	0.785	**15.5**	1.761	**27.5**	3.671	**39.5**	7.181
4.0	0.813	**16.0**	1.818	**28.0**	3.780	**40.0**	7.376
4.5	0.842	**16.5**	1.877	**28.5**	3.891	**40.5**	7.574
5.0	0.872	**17.0**	1.938	**29.0**	4.006	**41.0**	7.778
5.5	0.903	**17.5**	2.000	**29.5**	4.123	**41.5**	7.986
6.0	0.935	**18.0**	2.064	**30.0**	4.243	**42.0**	8.199
6.5	0.968	**18.5**	2.130	**30.5**	4.366	**42.5**	8.417
7.0	1.002	**19.0**	2.197	**31.0**	4.493	**43.0**	8.640
7.5	1.037	**19.5**	2.267	**31.5**	4.622	**43.5**	8.867
8.0	1.073	**20.0**	2.338	**32.0**	4.755	**44.0**	9.101
8.5	1.110	**20.5**	2.412	**32.5**	4.891	**44.5**	9.339
9.0	1.148	**21.0**	2.487	**33.0**	5.030	**45.0**	9.582
9.5	1.187	**21.5**	2.564	**33.5**	5.173	**45.5**	9.832
10.0	1.228	**22.0**	2.644	**34.0**	5.319	**46.0**	10.086
10.5	1.270	**22.5**	2.726	**34.5**	5.469	**46.5**	10.347
11.0	1.313	**23.0**	2.809	**35.0**	5.623	**47.0**	10.613
11.5	1.357	**23.5**	2.896	**35.5**	5.780	**47.5**	10.885
12.0	1.403	**24.0**	2.984	**36.0**	5.941	**48.0**	11.163
12.5	1.449	**24.5**	3.075	**36.5**	6.106	**48.5**	11.447

APPENDIX F

SLOPE OF VAPOR PRESSURE CURVE (Δ) FOR DIFFERENT TEMPERATURES (T)

$$\Delta = [4098.\ e^0(T)] \div [T + 237.3]^2$$
$$= 2504\{\exp[(17.27T) \div (T + 237.2)]\} \div [T + 237.3]^2$$

T °C	Δ kPa/°C	T °C	Δ kPa/°C	T °C	Δ kPa/°C	T °C	Δ kPa/°C
1.0	0.047	13.0	0.098	25.0	0.189	37.0	0.342
1.5	0.049	13.5	0.101	25.5	0.194	37.5	0.350
2.0	0.050	14.0	0.104	26.0	0.199	38.0	0.358
2.5	0.052	14.5	0.107	26.5	0.204	38.5	0.367
3.0	0.054	15.0	0.110	27.0	0.209	39.0	0.375
3.5	0.055	15.5	0.113	27.5	0.215	39.5	0.384
4.0	0.057	16.0	0.116	28.0	0.220	40.0	0.393
4.5	0.059	16.5	0.119	28.5	0.226	40.5	0.402
5.0	0.061	17.0	0.123	29.0	0.231	41.0	0.412
5.5	0.063	17.5	0.126	29.5	0.237	41.5	0.421
6.0	0.065	18.0	0.130	30.0	0.243	42.0	0.431

6.5	0.067	18.5	0.133	30.5	0.249	42.5	0.441
7.0	0.069	19.0	0.137	31.0	0.256	43.0	0.451
7.5	0.071	19.5	0.141	31.5	0.262	43.5	0.461
8.0	0.073	20.0	0.145	32.0	0.269	44.0	0.471
8.5	0.075	20.5	0.149	32.5	0.275	44.5	0.482
9.0	0.078	21.0	0.153	33.0	0.282	45.0	0.493
9.5	0.080	21.5	0.157	33.5	0.289	45.5	0.504
10.0	0.082	22.0	0.161	34.0	0.296	46.0	0.515
10.5	0.085	22.5	0.165	34.5	0.303	46.5	0.526
11.0	0.087	23.0	0.170	35.0	0.311	47.0	0.538
11.5	0.090	23.5	0.174	35.5	0.318	47.5	0.550
12.0	0.092	24.0	0.179	36.0	0.326	48.0	0.562
12.5	0.095	24.5	0.184	36.5	0.334	48.5	0.574

APPENDIX G
NUMBER OF THE DAY IN THE YEAR (JULIAN DAY)

Day	Jan	Feb	Mar	Apr	May	Jun	Jul	Aug	Sep	Oct	Nov	Dec
1	1	32	60	91	121	152	182	213	244	274	305	335
2	2	33	61	92	122	153	183	214	245	275	306	336
3	3	34	62	93	123	154	184	215	246	276	307	337
4	4	35	63	94	124	155	185	216	247	277	308	338
5	5	36	64	95	125	156	186	217	248	278	309	339
6	6	37	65	96	126	157	187	218	249	279	310	340
7	7	38	66	97	127	158	188	219	250	280	311	341
8	8	39	67	98	128	159	189	220	251	281	312	342
9	9	40	68	99	129	160	190	221	252	282	313	343
10	10	41	69	100	130	161	191	222	253	283	314	344
11	11	42	70	101	131	162	192	223	254	284	315	345
12	12	43	71	102	132	163	193	224	255	285	316	346
13	13	44	72	103	133	164	194	225	256	286	317	347
14	14	45	73	104	134	165	195	226	257	287	318	348
15	15	46	74	105	135	166	196	227	258	288	319	349
16	16	47	75	106	136	167	197	228	259	289	320	350
17	17	48	76	107	137	168	198	229	260	290	321	351
18	18	49	77	108	138	169	199	230	261	291	322	352
19	19	50	78	109	139	170	200	231	262	292	323	353
20	20	51	79	110	140	171	201	232	263	293	324	354
21	21	52	80	111	141	172	202	233	264	294	325	355

Day	Jan	Feb	Mar	Apr	May	Jun	Jul	Aug	Sep	Oct	Nov	Dec
22	22	53	81	112	142	173	203	234	265	295	326	356
23	23	54	82	113	143	174	204	235	266	296	327	357
24	24	55	83	114	144	175	205	236	267	297	328	358
25	25	56	84	115	145	176	206	237	268	298	329	359
26	26	57	85	116	146	177	207	238	269	299	330	360
27	27	58	86	117	147	178	208	239	270	300	331	361
28	28	59	87	118	148	179	209	240	271	301	332	362
29	29	(60)	88	119	149	180	210	241	272	302	333	363
30	30	—	89	120	150	181	211	242	273	303	334	364
31	31	—	90	—	151	—	212	243	—	304	—	365

APPENDIX H

STEFAN-BOLTZMANN LAW AT DIFFERENT TEMPERATURES (T):

$[\sigma^*(T_K)^4] = [4.903 \times 10^{-9}]$, MJ K^{-4} m^{-2} day^{-1}
Where: $T_K = \{T[°C] + 273.16\}$

T	$\sigma^*(T_K)^4$	T	$\sigma^*(T_K)^4$	T	$\sigma^*(T_K)^4$
			Units		
°C	MJ m^{-2} d^{-1}	°C	MJ m^{-2} d^{-1}	°C	MJ m^{-2} d^{-1}
1.0	27.70	17.0	34.75	33.0	43.08
1.5	27.90	17.5	34.99	33.5	43.36
2.0	28.11	18.0	35.24	34.0	43.64
2.5	28.31	18.5	35.48	34.5	43.93
3.0	28.52	19.0	35.72	35.0	44.21
3.5	28.72	19.5	35.97	35.5	44.50
4.0	28.93	20.0	36.21	36.0	44.79
4.5	29.14	20.5	36.46	36.5	45.08
5.0	29.35	21.0	36.71	37.0	45.37
5.5	29.56	21.5	36.96	37.5	45.67
6.0	29.78	22.0	37.21	38.0	45.96
6.5	29.99	22.5	37.47	38.5	46.26
7.0	30.21	23.0	37.72	39.0	46.56
7.5	30.42	23.5	37.98	39.5	46.85
8.0	30.64	24.0	38.23	40.0	47.15
8.5	30.86	24.5	38.49	40.5	47.46
9.0	31.08	25.0	38.75	41.0	47.76
9.5	31.30	25.5	39.01	41.5	48.06
10.0	31.52	26.0	39.27	42.0	48.37

T	$\sigma*(T_K)^4$	T	$\sigma*(T_K)^4$	T	$\sigma*(T_K)^4$
		Units			
10.5	31.74	26.5	39.53	42.5	48.68
11.0	31.97	27.0	39.80	43.0	48.99
11.5	32.19	27.5	40.06	43.5	49.30
12.0	32.42	28.0	40.33	44.0	49.61
12.5	32.65	28.5	40.60	44.5	49.92
13.0	32.88	29.0	40.87	45.0	50.24
13.5	33.11	29.5	41.14	45.5	50.56
14.0	33.34	30.0	41.41	46.0	50.87
14.5	33.57	30.5	41.69	46.5	51.19
15.0	33.81	31.0	41.96	47.0	51.51
15.5	34.04	31.5	42.24	47.5	51.84
16.0	34.28	32.0	42.52	48.0	52.16
16.5	34,52	32.5	42.80	48.5	52.49

APPENDIX I

THERMODYNAMIC PROPERTIES OF AIR AND WATER

1. Latent Heat of Vaporization (λ)

$$\lambda = [2.501-(2.361 \times 10^{-3})\, T]$$

Where: λ = latent heat of vaporization [MJ kg^{-1}]; and T = air temperature [°C].

The value of the latent heat varies only slightly over normal temperature ranges. A single value may be taken (for ambient temperature = 20°C): $\lambda = 2.45$ MJ kg^{-1}.

2. Atmospheric Pressure (P)

$$P = P_0\, [\{T_{Ko}-\alpha(Z-Z_0)\} \div \{T_{Ko}\}]^{(g/(\alpha.R))}$$

Where: P, atmospheric pressure at elevation z [kPa]

P_0, atmospheric pressure at sea level = 101.3 [kPa]

z, elevation [m]

z_0, elevation at reference level [m]

g, gravitational acceleration = 9.807 [m s^{-2}]

R, specific gas constant == 287 [J $kg^{-1} K^{-1}$]

α, constant lapse rate for moist air = 0.0065 [K m^{-1}]

T_{Ko}, reference temperature [K] at elevation z_0 = 273.16 + T

T, means air temperature for the time period of calculation [°C]

When assuming P_0 = 101.3 [kPa] at z_0 = 0, and T_{Ko} = 293 [K] for T = 20 [°C], above equation reduces to:

$$P = 101.3[(293-0.0065Z)\ (293)]^{5.26}$$

3. Atmospheric Density (ρ)

$$\rho = [1000P] \div [T_{Kv}\ R] = [3.486P] \div [T_{Kv}], \text{ and } T_{Kv} = T_K[1-0.378(e_a)/P]^{-1}$$

Where: ρ, atmospheric density [kg m^{-3}]

R, specific gas constant = 287 [J kg$^{-1\,K-1}$]

T_{Kv}, virtual temperature [K]

T_K, absolute temperature [K]: $T_K = 273.16 + T$ [°C]

e_a, actual vapor pressure [kPa]

T, mean daily temperature for 24-hour calculation time steps.

For average conditions (e_a in the range 1–5 kPa and P between 80–100 kPa), T_{Kv} can be substituted by: $T_{Kv} \approx 1.01\ (T + 273)$

4. Saturation Vapor Pressure function (es)

$$e_s = [0.6108]*\exp\{[17.27*T]/[T + 237.3]\}$$

Where: e_s, saturation vapor pressure function [kPa]

T, air temperature [°C]

5. Slope Vapor Pressure Curve (Δ)

$$\Delta = [4098.\ e°(T)] \div [T + 237.3]^2$$
$$= 2504\{\exp[(17.27T) \div (T + 237.2)]\} \div [T + 237.3]^2$$

Where: Δ, slope vapor pressure curve [kPa C^{-1}]

T, air temperature [°C]

$e^0(T)$, saturation vapor pressure at temperature T [kPa]

In 24-hour calculations, Δ is calculated using mean daily air temperature. In hourly calculations T refers to the hourly mean, T_{hr}.

6. Psychrometric Constant (γ)

$$\gamma = 10^{-3}\ [(C_p.P) \div (\varepsilon.\lambda)] = (0.00163) \times [P \div \lambda]$$

Where: γ, psychrometric constant [kPa C^{-1}]

c_p, specific heat of moist air = 1.013 [kJ kg^{-10}C^{-1}]

P, atmospheric pressure [kPa]: equations 2 or 4

ε, ratio molecular weight of water vapor/dry air = 0.622

λ, latent heat of vaporization [MJ kg^{-1}]

7. Dew Point Temperature (Tdew)

When data is not available, T_{dew} can be computed from e_a by:

$$T_{dew} = [\{116.91 + 237.3 Log_e(e_a)\} \div \{16.78-Log_e(e_a)\}]$$

Where: T_{dew}, dew point temperature [°C]

e_a, actual vapor pressure [kPa]

For the case of measurements with the Assmann psychrometer, T_{dew} can be calculated from:

$$T_{dew} = (112 + 0.9T_{wet})[e_a \div (e^0\ T_{wet})]^{0.125}-[112-0.1T_{wet}]$$

8. Short Wave Radiation on a Clear-Sky Day (Rso)

The calculation of R_{so} is required for computing net long wave radiation and for checking calibration of pyranometers and integrity of R_{so} data. A good approximation for R_{so} for daily and hourly periods is:

$$R_{so} = (0.75 + 2 \times 10^{-5}\, z)R_a$$

Where: z, station elevation [m]

R_a, extraterrestrial radiation [MJ m^{-2} d^{-1}]

Equation is valid for station elevations less than 6000 m having low air turbidity. The equation was developed by linearizing Beer's radiation extinction law as a function of station elevation and assuming that the average angle of the sun above the horizon is about 50°.

For areas of high turbidity caused by pollution or airborne dust or for regions where the sun angle is significantly less than 50° so that the path length of radiation through the atmosphere is increased, an adoption of Beer's law can be employed where P is used to represent atmospheric mass:

$$\mathbf{R_{so} = (R_a)\ exp[(-0.0018P) \div (K_t \sin(\Phi))]}$$

Where: K_t, turbidity coefficient, $0 < K_t \leq 1.0$ where $K_t = 1.0$ for clean air and $K_t = 1.0$ for extremely turbid, dusty or polluted air.

P, atmospheric pressure [kPa]

Φ, angle of the sun above the horizon [rad]

R_a, extraterrestrial radiation [MJ m^{-2} d^{-1}]

For hourly or shorter periods, Φ is calculated as:

$\sin \Phi = \sin \varphi \sin \delta + \cos \varphi \cos \delta \cos \omega$

Where: φ, latitude [rad]

δ, solar declination [rad] (Eq. (24) in Chapter 3)

ω, solar time angle at midpoint of hourly or shorter period [rad]

For 24-hour periods, the mean daily sun angle, weighted according to R_a, can be approximated as:

$$\sin(\Phi_{24}) = \sin[0.85 + 0.3\ \varphi\ \sin\{(2\pi J/365)-1.39\}-0.42\ \varphi^2]$$

Where: Φ_{24}, average Φ during the daylight period, weighted according to R_a [rad]

φ, latitude [rad]

J, day in the year

The Φ_{24} variable is used to represent the average sun angle during daylight hours and has been weighted to represent integrated 24-hour transmission effects on 24-hour R_{so} by the atmosphere. Φ_{24} should be limited to ≥ 0. In some situations, the estimation for R_{so} can be improved by modifying to consider the effects of water vapor on short wave absorption, so that: $R_{so} = (K_B + K_D)\, R_a$ where:

$K_B = 0.98exp[\{(-0.00146P) \div (K_t \sin \Phi)\}-0.091\{w/\sin \Phi\}^{0.25}]$

Where: K_B, the clearness index for direct beam radiation

K_D, the corresponding index for diffuse beam radiation

$K_D = 0.35-0.33\ K_B$ for $K_B \geq 0.15$

$K_D = 0.18 + 0.82\ K_B$ for $K_B < 0.15$

R_a, extraterrestrial radiation [MJ m^{-2} d^{-1}]

K_t, turbidity coefficient, $0 < K_t \leq 1.0$ where $K_t = 1.0$ for clean air and $K_t = 1.0$ for extremely turbid, dusty or polluted air.

P, atmospheric pressure [kPa]

Φ, angle of the sun above the horizon [rad]

W, perceptible water in the atmosphere [mm] $= 0.14\ e_a\ P + 2.1$

e_a, actual vapor pressure [kPa]

P, atmospheric pressure [kPa]

APPENDIX J

PSYCHROMETRIC CHART AT SEA LEVEL

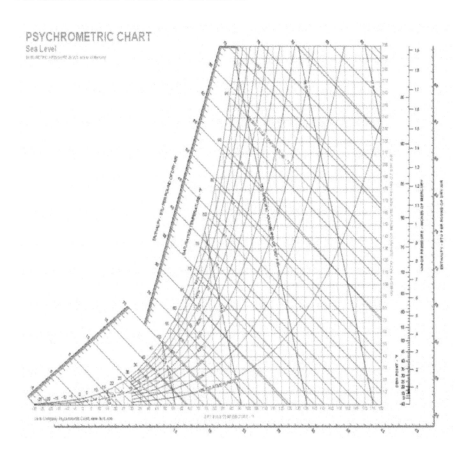

INDEX

A

Active roots zone, 181
Adequate engineering design, 152
Affordable Micro Irrigation Technologies (AMITS), 12, 16
Agricultural commodities, 88
Agricultural techniques, 22
Agriculture Development Officer (ADO), 61
Agro-biological calculations, 125, 137
Agro-climatic conditions, 97
Agroforestry, 100
Air-relief valve, 257
Alfalfa, 162
Algae, 38, 201, 218, 247, 300
Alkaline, 251
Aluminum pipes, 2
Aluminum, 2, 238
American Society of Agricultural Engineers, 40
Ammonium phosphate, 242
Anaerobic decomposition, 181
Antidrain emitters, 178
Antivacuum valves, 129
Apple Academic Press, 199
Aquatic plants, 247
    algae, 247
Arad County, 188
Arecanut farmers, 105
Arid Lands Information Network (ALIN), 82
Autotrophic bacteria, 246
Avocado, 162

B

Basic irrigation systems, 2
    surface/gravity, 2
    subsurface, 2, 88, 125–134, 141, 149
Basin-wide water resources, 32
Batch tank, 276–278

Benefit-cost (BC) ratio, 125
Benefit-cost calculation, 53
Bhil tribals, 59
Bicarbonates, 233, 234, 242, 280
Biloela Research Station, 216
Binding constraint, 54
Biocide, 245, 255, 300
Bi-products (bagasse), 135
Black recycled rubber laterals, 60
Blue–green algae, 247, 248
    anabaena, 247
    aphanizomenon, 247
    gleotrichia, 247
    gomphosphaeria, 247
    microcystis, 247
    nodularia, 247
    nostoc rivulare, 247
Borne fruit, 52
Boron, 233
Broccoli, 162
Bronze, 238
Brown algae, 248, 250
    Ectocarpus, 250
    Fucus, 250
    Sargassum, 250

C

Calcium carbonate, 242, 252, 283
    lime, 283
Calcium hypochlorite, 245, 251, 252, 260
Calcium nitrate, 242, 283
Calcium oxalate, 252
Capital intensive, 67, 97
Carbon black, 38
Carbonates, 218, 233, 246, 252
Catholic Relief Services, 79, 82
Cavity radius, 167, 173
Cellulose, 249, 250
Central and East European Countries (CEEC), 14
Centrifugal pump, 202, 258

Chapin Living Waters Foundation (CLWF), 79
Cheap recycled plastic, 46
Chemical clogging agents, 208
    minerals, 208
    salts, 208
Chemical injection unit, 259
    check valve, 259
    chemigation pump, 259
    chemigation tank, 259
    connecting tubes, 259
    gate valve, 259
    pressure gage, 259
    pressure regulator, 259
Chemigation tank, 238, 257, 260
Chemigation, 242
    Anhydrous ammonia, 242
    Liquid ammonia, 242
    Ammonium sulfate, 242
    Urea, 242
    Ammonium nitrate, 242
    Calcium nitrate, 242
Chimney effect, 166
Chlorine, 245
    microorganisms, 245
    oxidation agent, 245
Chlorophyll, 248–250
    green color, 248
Christian Children Fund, 82
Christiansen coefficient, 168
Chronic freshwater shortages, 25
Clay pots, 25, 198
Climatic hazard, 188
Close-knit groups, 60
Cocopith, 130
    coconut coir waste, 130
Cogeneration of power, 126
Colorado State University, 25, 198
Commercial banks, 92
Commercial farmers, 53, 57
Common perception, 50
Complex decision-making, 152
Computer spreadsheet program, 41
Concrete-roofing tiles, 83
Continuous move types, 2
    center pivots, 2, 11, 12
    linear move, 2, 12

Control equipment, 27, 148
Controlled Environment Agriculture (CEA), 13
Conventional lay-flat hose, 13
Cooperation's Report, 35, 83
Copper sulfate, 247
Copper, 238, 243, 247
Corals, 249
Core booster, 192
Corn, 153, 162
Corrosion resistant, 256–258
Cost-effective role, 83
Cost-effectiveness, 37, 84
Crickets, 218
Crop agronomy, 148
Crop canopy, 12, 216
Crop productivity, 15, 101
    income generation, 15, 36, 74
    employment opportunities, 15, 122
    food security, 15, 78, 100
Crop root zone, 179
Crop water, 42, 91, 148, 177
Cropping cycle, 51
Cumbersome procedure, 92
Custom-built kit, 59
Custom-built systems, 51
Cylinder-piston, 279

D

Dead seedlings, 130
Deciduous tree fruit, 38
Deep percolation, 101, 148, 152, 164, 181, 266
Demonstration effect, 60
Design of irrigation systems, 149, 152, 155
Diaphragm, 165, 241, 258, 266
Diesel engines, 203
    high maintenance cost, 203
    noise during operation, 203
Diesel-powered pumps, 34
Digital manometers, 171
    Woltman, 171
Discharge regulations, 148
Discomforts, 247
    skin rashes, 247
    headaches, 247
    nausea, 247

vomiting, 247
diarrhea, 247
fever, 247
muscular pains, 247
eye, 247
nose, 247
throat irritation, 247
District Agricultural Development Officer, 55
Diversification of crops, 7
Diversification strategy, 55
Divisibility, 13
Dramatic expansion, 56
Dream Drip Kit, 81
Drip tape, 199
   Dew Hose, 199
   Richard Chapin, 199
Drip-irrigated corn, 150
   Zea mays, 150

E

Earwigs, 218
Economic viability, 59, 91
Ecosystems, 22
Elastomer, 169, 170
Electrical motors, 203
   ease of automation, 203
   quiet operation, 203
   least maintenance, 203
   efficiency, 203
Electricity tariff, 54
Emission Uniformity (EU), 40
Emitter discharge equation, 176, 177
Emitter Line Length, 306, 307
Energy line, 175, 176
Enterobacter, 246
Entrepreneurial grape farmers, 91
Erosion, 54, 189, 200, 304
Euglena, 248, 249
   ponds, 249
   pools, 249
European funds, 191, 194
Evaporation, 32, 123, 148, 163, 215, 232, 304
Expandability, 13

F

Farm and Ranch Irrigation Survey, 11, 163
Farm ponds, 54, 247
Farmyard manure, 131
Far-sighted president, 55
Fertigation application scheduling, 144
Fertigation equipment, 88, 105, 108, 124, 129
Fertilizer injector pump, 108
Fiber glass, 258
Field emission uniformity, 41
Field layout, 45, 110
Field oriented research, 93, 106
Field practice process, 152
Field test data, 46
Field-notes, 50
Filamentous algae, 246, 250
   pond scum, 250
   pond moss, 250
   frog spittle, 250
   water net, 250
Filtration systems, 201
   sand separator, 201
   screen filter, 201
   media filter, 201
   manifolds, 201
   backwash controller, 201
Financial assistance, 14
Financial capacity, 33, 54
Financial investment, 93
Floating pumping station, 189–195
Flood irrigation, 53, 54, 58
Flow meters, 165
Flushing pipe, 164, 175, 298
Fodder crops, 15
Foot valve, 242
Foreseeable future, 13
Fresh Produce Exporters Association of Kenya (FPEAK), 81
Friction coefficient, 313
Fruit trees, 38, 92, 106, 115, 153, 162, 209
Fucoxanthin, 250
Full ongoing costs, 36
Full/partial control irrigation technique, 3
Full-scale assessment, 50

G

Gentlemen farmers, 50
Geographic Information Systems (GIS), 151
Geometries, 165
Global scale, 7, 24
Golden–brown algae, 250
    Plankton, 250
    Nanoplankton, 250
Grapes, 53, 97, 108, 114, 199
Grass, 162, 304
Gravity irrigation, 94, 101, 123, 125
Gravity systems, 11
Green revolution, 32
Greenhouse irrigation, 12
Greenhouses, 13, 16, 25
    Plastic houses, 13
    Glasshouses, 13
Gujarat Agricultural University, 114
Gypsum, 280

H

Hand calculator, 41
Hand-dug-open, 34
Hand-held spraying-head, 13
Hard hose products, 214
    Kinking, 214
    Punctures, 214
    Rodent damage, 214
    Herbaceous crops, 162
    Lettuce, 162
    Celery, 162
    Asparagus, 162
    Garlic, 162
Heavy clayey soils, 129
High demographic growth, 14
High-value crops, 214
    Fruits, 214
    Vegetables, 214
    Nuts, 214
    Sugarcane, 214
Home gardeners, 33
Horsepower (HP), 125
Horticultural crops, 12, 35, 38, 199
Horticultural tree crops, 38
Hydraulic structures, 192
    Bridges, 192

Crossing, 192
Weirs, 192
Shock devices, 192
Hydrochloric acid, 218, 256, 300
Hydroelectric power, 141
Hydrogen sulfide, 246, 251, 254
Hydrologic cycle, 111
Hydrotechnique, 189
Hydrus-2D, 176, 180
Hypochloric acid, 251
Hypochlorite, 245, 251, 260, 300
Hypothetical depletion, 154

I

Ideal candidates, 60
Indian agriculture, 122, 199
Inexpensive machinery, 39
Inlet pressure, 36, 39, 42, 45, 171, 173–179, 277
Insecticides and fumigants, 13
Intensification policies, 78
Internal algae buildup, 38
Internal diameters (IDS), 38
International Commission on Irrigation and Drainage (ICID), 2
International Development Enterprises (IDE), 33, 83, 98
Internode length, 130
Iron bacteria, 218, 234
Iron oxide, 218
Iron sulfide, 218
Irrigated agriculture accounts, 109
Irrigation Engineering, 199
    Megh R. Goyal, 199
Irrigation water use efficiency (IWUE), 149
ISI marked branded material, 65
Isolated buyers, 60, 65
Iterative solution, 42

J

Jain irrigation, 50, 65, 68, 166, 199
Jellyfish, 249
Jewish Yishuv, 198

K

Kinematic water viscosity, 172

L

Laboratories (NARL), 80
Land topography, 148, 206
Landscape and park irrigation, 12
Larger *net* area, 55
Lay-flat lateral tubing, 35, 38
Layout flexibility, 37
L-Cm-drip system, 34–42, 45
Lead Farmer Concept, 60
Least-cost option, 64
Linear low density PE (LLDPE), 38
Loamy, 169, 170, 179, 180
Local dam project, 55
Local Initiatives Support Program (LISP), 73
Long labyrinths, 207
Long-term capital investment, 66
Low-end adopters, 54
Low-energy precision application (LEPA), 150
Low-pressure center pivots, 12
Lucrative market, 68
Lysimeters, 148

M

Magnesium salts, 242
Major up-scaling, 54
Manganese oxide, 218
Mango orchard, 199
Marathwada Agricultural University, 114
Marginal farm families, 12
  Marine algae, 249
Market bottlenecks, 55
Marketing elitism, 68
Marketing program, 36, 84
Marketing Supervisor, 69, 70
Mass media advertising campaign, 81
Maturity of stalk population, 130
Mechanized harvesting, 126
Mediterranean, 150, 189
Mennonite Church, 82
Mental model, 70
Methane gas, 181
Metropolitan area, 111
Michigan State University, 25, 198
Micro irrigation system, 201
  gate or ball valve, 201

flushing valve, 201
union, 201
nipple, 201
adapters, 201
reducers, 201
tee, 201
coupling, 201
universal, 201
elbow, 201
double union, 201
cross, 201
one way valve, 201
safety valve, 201
Micro irrigation, 304
  trickle irrigation, 304
  daily irrigation, 304
  daily flow irrigation, 304
Microenterprise manufacturers, 37, 84
Microspray heads, 198, 201, 304
Microtube emitters, 35, 37–39, 42, 45
Microtube system, 52, 62
Micro-tube technology, 52
Microtubes, 35, 40, 64, 199,
Mis-deliver water, 60
Modern irrigation techniques, 3
Moneymaker treadle pump, 80
Mountain Marketshed, 56
Mulberry leaves, 63
Mures River's plain, 188
Mushrooms, 250

N

National Agricultural Research, 80, 82
National Agricultural Statistics Service (NASS), 11
National Committee on Plasticulture Application in Horticulture (NCPAH), 11
National Committees of ICID, 2, 27
Natural Resources Monitoring, Modeling, and Management (NRM[3]) Project, 81
Nepal Federation of Women Group, 61
New technical terms, 238
  chemigation, 238
  chloration, 238
  pestigation, 238
  insectigation, 238
  fungigation, 238

nemagation, 238
herbigation, 238
Nitrogen, 241, 247, 251, 261
Nongovernment development organizations (NGOs), 33
Nonrice irrigated area, 8
Nonrotating type, 2
    micro-sprayers, 2
    static micro jets, 2
Novel technique, 288
Nutriments, 247
    carbon dioxide, 247
    nitrogen, 247
    phosphorus, 247

O

Offset pipe friction, 40
Oilseed crops, 15
On-farm water application systems, 26
Onion, 3, 97, 149, 153, 288
Open canal conveyance, 101
Organic matter, 218, 244, 256, 300
Orifices, 206, 215, 304
Ornamental crops, 13
Osmotic potential, 108, 235

P

Paired-row method, 125, 126, 128
Pan evaporimeter, 130
Paramylon, 249
Pathogen drift, 163
Pay-back period, 95
Peddle low quality products, 65
Pepcee or Pepsi, 37
Pepper, 105, 162
Perennial crops, 164, 217
    lucerne, 217
Perforated pipe, 25, 198
Phosphoric acid, 256, 267, 272
Phosphorus, 243, 247
    organic phosphate, 243
    glycerophosphates, 243
Photosynthetic pigments, 249, 250
    chlorophyll α, 250
    chlorophyll β, 250
Photosynthetic radiation, 232
Phreotophytes, 101

Phylum, 249–251
Physical clogging agents, 208
    sand, 208
    silt, 208
    clay, 208
Physiographic conditions, 88
Piedmont, 189
Pineapple, 112, 113, 162
Pioneering efforts, 50
Pipeline hydraulics, 41
Piping system, 14
Pit hardening, 154, 155
Pit-method of cultivation, 126
Plastic drip tubing, 25
Plastic mulch, 150
Plastic Sea, 13
Plentiful resources, 111
Policy instruments, 93, 105, 116
Pollution, 238, 247
Polyethylene (PE) tubing, 37
Polyethylene pipe, 172, 246, 309
Polyethylene tube, 117, 209, 212
Polyethylene, 162
    polyvinyl chloride, 162
Polysaccharide, 249, 250
Poor sericulturists, 63
Portable small diameter, 34
Postinstallation evaluation, 40
Potassium sulfate, 243, 280
Potassium, 243
    potassium sulfate, 243
    potassium chloride, 243
    potassium nitrate, 243
Potato, 97, 108, 150, 162
Potential adopters, 61, 66, 68
Potential micro irrigation customer, 32
Poverty alleviation, 57, 83
Preengineered design tables, 41
Pressure range, 307
Primary irrigation method, 29
Prime natural resource, 110
Protective clothes, 258
    boots, 258
    goggles, 258
    gloves, 258
Protists, 247
    bacteria, 247

fungi, 247
protozoa, 247
algae, 247
Pseudomonas, 246
Pyrrhophyta, 249

Q

Quality control, 14, 68

R

Ratoon crops, 125
Real losses, 34
Red Sea, 251
Regulated deficit irrigation (RDI), 153
Remote sensing, 148, 151, 155
Research Institution, 93–95, 102, 109, 113
Reynolds numbers, 172
Richards' equation, 176
River linking project, 11
Rock bottom prices, 61, 66, 68
Romania, 172, 188, 194, 315
Root exploration, 155
Root intrusion, 129, 164, 168, 172, 214, 218

S

Saltwater forms, 248
    brown algae, 248
    red algae, 248
Science of survival, 22, 88
Second-generation issues, 55
Self-propelled center pivot, 11
Semiarid zones of countries, 3
    Argentina, 3
    Chile, 3
    Peru, 3, 25
    Ecuador, 3
Semilay-flat tubing, 40
Shady players, 65
Simulation models, 149, 181
Single-budded-chips, 130
Skilled installer, 39
Skilled laborers, 14
Slash-and-burn farming, 59
Sloping fields, 42
Small Irrigation Market Network (SIMInet), 35

Smaller root system, 150
Small-holder market segment, 65
Small-scale systems, 7
    hose-reel systems, 7
Social Welfare Center, 56
Socioeconomic issues, 82
Sodium hypochlorite, 251, 252
    Whitener for clothes, 252
Soft-drink giant, 63
Soil deterioration, 107
Soil pits method, 177
Solar radiation, 232
Solutionizer machine, 279, 280
Sophisticated drip-tape, 34
Sophisticated technology, 25, 97
Soviet Union, 6
Spain, 9, 13, 232
Spray-head Irrigation, 13
    pedal pump, 13
Sprinkler irrigation, 3, 57, 71, 90, 215, 266
Sprinkler method, 2
    hand move, 2, 7, 11, 15
    hose pull, 2
    solid state/permanent types, 2
    stationary big guns, 2
Stainless steel, 238, 258, 299
State-of-the-art, 12, 16, 32
Strategic goal, 69
Strawberry, 113, 214
Street-smart market maneuvering, 68
Stumbling blocks, 14
Submersible pumps, 193
Sub-Saharan Africa (SSA), 14
Subsidy program, 61
Subsidy scheme, 62, 65, 92
Subsurface drip fertigation (SSDF), 134
Sugar cane, 142, 143, 162
Sugarcane Breeding Institute (SBI), 131
Sulfur bacteria, 234, 245
Sulfur, 131, 218, 246, 289, 299
Sulfuric acid, 256, 299
Supply channel, 189
Surface irrigation systems, 34, 88, 149–152
    border irrigation, 151
    contour irrigation, 151
Surface irrigation,

Sustainable sugarcane initiative (SSI)
   methods, 125
Swiss Agency for Development, 35, 83
Swiss Development Corporation funding,
   81

T

Technical guidance, 93, 105, 116
Technological and financial constraints, 6
Technological innovations, 88
Thermal imaging, 155
Tisa Plain, 189
Tomatoes, 30, 150
   Lycopersicon esculentum, 150
Topography, 41, 149, 206, 306
Tortured (zigzag) path, 207
Total harvested area, 7
Triangular mesh, 176
Tri-annual magazine *Baobab, 82*
Trucking vegetables, 61

U

Ultrasonic flow meter, 171
Ultraviolet light, 38
Uniform topography, 41
Unique Selling Proposition (USP), 51
Unobstructed passageways, 45

V

Vacuum breaker, 205, 257
Vacuum cleaner, 193
Variograms, 173
Venturi principle, 241, 277
Victoria Driplines, 81

Vietnam data, 4
Village Development Committees, 55
Virgin film grade plastic, 38
Volatilization, 107

W

Warm-blooded animals, 248
Wastewater, 59, 181, 214
Water engineer, 198
   blass, 198
Water Resources Management (WAREM),
   81
Water scarcity zone, 135
Water storage tanks, 81
Watering regimes, 154
Watermatics, 13, 79, 198, 199
Water-saving irrigation techniques, 4
Water-scarce region, 50
Well water, 201
White roots, 130
Wide spaced perennial crop, 106
Wide-scale adoption, 15
Wireworm larvae, 218
Woody crops, 162
   apple trees, 162
   citrus, 162
   olive trees, 162
World Resource Institute, 100, 111

Y

Yield potential, 123

Z

Zinc, 243